2013 UPDATE TO ESOPTRICS' TRY TO END THE NOTION OF THE CONTINUUM & THE "ABSURD" MATH IT BEGETS

The author's colorful, distinct voice comes through loud and clear in this attempt to marry science and theology. *2013 Update to Esoptrics' Try to End the Notion of the Continuum & the "Absurd"Math It Begets*, by Edward N. Haas, attempts to use mathematics, philosophy, and theology to explain what the universe looked like prior to the big bang. . . . Quirky, unique, and full of personality . . . . Haas's style is baroque and not boring, strung with metaphors and lively turns of phrase. . . . . *2013 Update...* has plenty of character thanks to the author's distinct voice.
——**ANNA CALL:** review for ***Clarion Review*** (In her "what the universe looked like prior to the big bang", Ms. Call is mentioning only one of the at least 2 dozen cosmological issues to which the book gives Esoptrics' highly detailed and mathematically precise explanations and, to several of which [***ex. gr.:*** the ultramicroscopic makeup of space, time, matter, locomotion & change], Science gives no explanation at all. For more parts of this review and my replies to them, see pages XXVIII, 26, 66, 88, 110, & 226.)

Haas' writing provides evidence of substantial compositional skill . . . .
——**KIRKUS INDIE: *Kirkus Reviews*** (For more parts of this review and my replies to them, see pages VIII, XXIX & XXX.).

Dear Edward,

Yes! Yes! Yes! I love your thesis, Edward. Now you are speaking my language. Not only is it my language but also it's the language of nature. I agree wholeheartedly with your philosophical approach.

Yes! I, too, believe that "inward reflection" and concentration can draw out principles for all the basic laws of the universe. Yet knowing the natural steps that need to be taken can be the trick of it all. Your concept is an exciting and huge leap for mankind in the 21st century. I wish you well in your task.

Certainly, it's a system of beauty. Like no other I have ever witnessed!
——**TOM HOLMAN**: His e-mail to me dated Sep. 30, 2001 (I've never been able to establish whether or not this Tom Holman is the same as the well-known Tomlinson [Tom] Holman, developer of the Lucasfilm THX [***i.e.:*** Tom Holman Xtra] Sound System, and, for a while, professor of film sound at the University of Southern California.).

. . . . Matter ends up being an insubstantial substrata [Sic! "substrata" for "Substratum" - ENH], parasitic on the ultimate particles—see later. Whether correct or incorrect, this is ingenious.

. . . .

Our impression is that, from the point of view of speculative cosmology in the Whiteheadian tradition, the system [Esoptrics – ENH], may possibly do as well or better than the original Whitehead in explaining how big and small fit together. . . .

. . . .

By way of a final point: both of the present reviewers are surprised at the seemingly almost total neglect in the responding literature of the phenomenological basis of the schema. We think this is a shame.
——**REV. DRS. DAVID A. BOILEAU & GREGORY J. MOSES**. Departments of Philosophy, Loyola University, New Orleans, LA, USA, & St. Paul's Theological College, Banyo, Queensland, Australia, respectively.

2013 UPDATE TO ESOPTRICS' TRY TO END THE NOTION OF THE CONTINUUM & THE "ABSURD" MATH IT BEGETS

By

Edward N. Haas – Haaswood, La.

AuthorHouse™ LLC
1663 Liberty Drive
Bloomington, IN 47403
www.authorhouse.com
Phone: 1-800-839-8640

Published by AuthorHouse 05/02/2014

ISBN: 978-1-4918-4699-5 (sc)
ISBN: 978-1-4918-4698-8 (e)

Library of Congress Control Number: 2013923328

ABOUT THE BOOK

Because it's the logic of a *mirror*, Esoptrics' name comes from εσοπτρον, (***i.e.:*** esoptron), the Greek word for mirror. It's the logic of a *mirror* because its effort to detail the origin and structure of the Universe starts as an algebraic logic working with multiples of 2 (especially 2^0, 2^1, 2^2, 2^4, 2^8, 2^{16}, 2^{32}, 2^{64}, 2^{128}, & 2^{256}) and with reverse images, just as a mirror doubles and reverses what faces it.

On pages IX, 6, 335, 486, and 493 of his book ***The Fabric Of The Cosmos***, Prof. Brian Greene basically admits scientists "have no idea" what the "ultramicroscopic makeup of space, time and matter" might be. On page 471, he concedes current views on space and time may be "mere allusions" to what's "more profound" and "fundamental" at the base "underlying physical reality". On page 352, he grants that makeup can't be confirmed by direct observation because the Planck scale, 10^{-33} cm., is "nearly a billion billion times" smaller than current technology can probe. On page 352, he adds what's that tiny is "so removed from direct empirical testing", some insist the search for the ultramicroscopic makeup "lies in the realm of philosophy or theology but not physics."

That's Esoptrics in a nutshell. Primarily Philosophy and Theology, it starts as an introspective analysis of the necessary 3-way structure of consciousness (***i.e.:*** encounterer, encountered, & the act joining them); theorizes that structure is the result of a self-mirroring mirroring activity out to encounter itself by mirroring itself; and is then led to a finite vs. an infinite version (***i.e.:*** God) of that 3-part quest. In that quest, it then sees: (1) 6 ways for God to excite the finite version (This produces *observable* 3 dimensionality.), (2) every single mirror's pre-disposition to multiples of 2, and (3) 2^{256} levels of intensity for each of the 6 ways to be excited by God (This yields history's first and only highly detailed and mathematically precise description of a 4th dimension which is neither time nor observable.). Step by step, it then leads to the conclusion that the ultramicroscopic makeup of space and locomotion is at a level we'd call c. $7.34683969 \times 10^{-47}$ cm. (***i.e.:*** 2^{-129} x the hydrogen dia. of 5×10^{-8} cm.). The ultimate units of space (***i.e.:*** u-spaces) found there are more properly called *ultramicrostates of excitation*. Each of them is neither p*hysically* extended nor separate from the others *in space*; each is *logically* separate and extended in a 7-way *logical sequence* inherent to the unique internality given each by God's 7 *a priori* coordinates. Though only *logically* apart, they *collectively* add up to 3 dimensional shapes *in real effect*. Time's least unit = what we'd call c. $7.201789375 \times 10^{-96}$ seconds. Esoptrics alone thus presents a step by step, highly detailed, and mathematically precise description of exactly what is the ultramicroscopic makeup of space, time, and matter—a makeup it then ties into a multitude of Theoretical Physics' favorite topics.

Esoptrics replaces the notion of space with c. 2×10^{231} "u-spaces envelopes" of $(2x1)^3$ or $(2x2)^3$ or $(2x3)^3$ or up to $(2x2^{256})^3$ u-spaces (***i.e.:*** more properly called microstates of excitation). Each envelope is the actuality of a "form" (***i.e.:*** force particle). Each form is at one of the 2^{256} levels of "ontological distance" (***i.e.:*** "OD", "levels of intensity" and "the 4th dimension"). Each form is *logically concentric* with a "carrying generator" (***i.e.:*** matter particle). The result is a "*duo*-combo" and "ultimate occupant of the Universe". Some forms, by rising to an OD above the OD at which created, make from c. 10^{38} to c. 10^{77} duo-combos logically concentric in Esoptrics' "*multi*-combos". Thru its generator, each duo- or multi-combo moves thru the u-spaces envelope of a form who's current OD (***i.e.:*** 4th dimension) level is higher than its own.

On his pages 286 & 318, Prof. Green says the Big Bang's inflation burst "is best thought of" as occurring in a "preexisting universe" instead of as the creation event itself. On page 302, he states we have no convincing answer to whether there was a pre-inflation era's pre-existing Universe or what it was like. On page 337, he contends "we still have no insight into what happened at the very beginning of the universe." Esoptrics gives a highly detailed explanation saying that, in the Universe's first c. 10^{-38} sec., and prior to the <u>Big</u> Bang, God—in the 8 successive <u>mini</u> bangs of the Universe's 1st 8 epochs—created c. 5.25×10^{115} duo-combos in a Universe only c. 2.75×10^{-8} cm. (***i.e.:*** half an atom's dia.) and then, in the next c. 10^{-19} sec. of the 9th epoch, added another 2×10^{231} duo-combos in a Universe suddenly trillions of light years in dia. within c. 6.22×10^{-19} sec. from time zero. Esoptrics alone thus presents a step by step, highly detailed, and mathematically precise account of exactly "what conditions were like" (Greene pg. 322) in the pre-inflation era's pre-existing Universe.

BOOKS IN PRINT BY EDWARD N. HAAS:

The Lovesong Tree, A Fairy Tale Portrait Of God

The Story Of Drawden The Pig

Introspective Cosmology II

Letters And Thoughts On Homosexuality

The Nature And Origins Of Murder Worship, The Ultimate Disease

Pieces Of Moral And Dogmatic Theology

In The Beginning Was The Internet

Two Letters For 1993

Letters Against Murder Worship

A Letter From A Father To His Son In 1994

On Philosophy: One Long And Four Short

Miscellaneous Letters

Substantial Substrata Vs. Insubstantial Substrata

Parasitic Vs. Non-Parasitic Substrata

Letters To A Prison Inmate – Volumes One & Two

A Reply To Pope Benedict XVI And More

Esoptrics: The Logic Of The Mirror

What Happens In Death

The 4 Stages Of Butterflies & Humans

Each is available at any bookstore by special order or on the Internet at web sites such as authorhouse.com or bn.com or amazon.com or others.

Dedicated to all who, wittingly or unwittingly, assisted directly in the development of Esoptrics. Dedicated first to my brother Gordon for his many bits of input including a computer program developed exclusively for Esoptrics' needs. Dedicated secondly to my late brother-in-law, David Kasten, and my brother Brian who, in 1971, conjointly provided me with many of the powers of 2 long before the Internet made such calculations easy. Dedicated thirdly to Po-Han Lin, Mark of markknowsnothing.com, and Marek Kyncl for offline and online math programs which made it easy to work with huge numbers. Dedicated fourthly to Albert Einstein, Michael Atiyah, John Polkinghorne, Stephen Hawking, Leonard Mlodinow, and especially Brian Greene who, by their writings in the field of Theoretical Physics, provided me with concepts and verbalizations which greatly enhanced my ability to perceive and verbalize Esoptrics' concepts by contrasting its with theirs. Dedicated also to David H. Freedman and Abraham Kaplan who granted me an unconditional right to quote from their writings.

Lastly, if this notion of immaterial spirit may have, perhaps, some difficulties in it not easily to be explained, we have therefore no more reason to deny or doubt the existence of such spirits, than we have to deny or doubt the existence of body; because the notion of body is cumbered with some difficulties very hard, and perhaps impossible to be explained or understood by us. For I would fain have instanced anything in our notion of spirit more perplexed, or nearer a contradiction, than the very notion of body includes in it; the divisibility *in infinitum* of any finite extension involving us, whether we grant or deny it, in consequences impossible to be explicated or made in our apprehensions consistent; consequences that carry greater difficulty, and more apparent absurdity than anything can follow from the notion of an immaterial knowing substance.
——**JOHN LOCKE** (1632-1704)**: *Concerning Human Understanding***, Book II: Chap. XXIII: Sec. 31, as found on page 212 left & right of Vol. 35 of ***Great Books Of The Western World*** as published by Encyclopædia Britannica, Inc.; Chicago, 1952 (I dare suggest, Mr. Locke, no one can give you an instance of *any notion whatsoever* more so evidently a contradiction than the maximally absurd math precipitated by the infinite divisibility inseparable from the concept of the continuum. On my page XXVII, see David Hume using the terms "absurdity" and "absurd" in his echo of Mr. Locke's above criticism. In this book's title, I have placed "Absurd" in quotation marks to indicate it's taken unchanged from Hume and, by contraction, from Locke.).

In addition to Haas' unusual ideas, he presents them in a scattershot way, peppering his notes and chapters with personal asides that have little bearing on the work. For instance, a note in Chapter 9 mentions a list of powers of two that his brother provided and trails off into talking about his brother's divorce.
——**KIRKUS INDIE: *Kirkus Review*** (In the footnote on my page 71, I credited a deceased *brother-in-law's* contribution to Esoptrics as justice demands and briefly mentioned his divorce from one of my sisters, in order to clarify his connection to me. What do you suppose it says of a reviewer's ability to forge a meaningful review when he or she can't notice the difference between "brother" and "brother-in-law"? And is there anything *censurable* in that "peppering"?! I don't think so. I'd call it a refreshing break from the tedium of the main text—a kind of entertaining R&R. If, then, the above is a *censure*, I'd call it "scraping the bottom of the barrel" in an attempt to find any pretext, however flimsy, to dismiss what one plainly has not enough mental ability to evaluate properly. Also, this reviewer refers to this as my twenty-*second* book. It's my twenty-*first* one. What are such factual errors, if not powerful evidence this reviewer read my book with as little attention as possible to what's in it?).

CONTENTS

CONTENTS

DESCRIPTION **PAGE NO.**

CONTENTS

DESCRIPTION PAGE NO.

CONTENTS

DESCRIPTION **PAGE NO.**

CHARTS (Chap. & #)

CONTENTS

DESCRIPTION **PAGE NO.**

DRAWINGS

CONTENTS

DESCRIPTION **PAGE NO.**

CONTENTS

DESCRIPTION **PAGE NO.**

23-1: How Gravity Is Exerted At 1Φ Apart . . . . . 191.
23-2: How Gravity Is Exerted At 2Φ Apart . . . . . 192.
23-3: How Gravity Is Exerted At 3Φ Apart . . . . . 193.
23-4: How Gravity Is Exerted At 4Φ Apart . . . . . 194.
25-1: Shape Of The Universe With A Fully Developed Reverse Category #2. . . . . . 204.
25-2: Shape Of The Universe With Fully Developed Reverse Categories #2 & #3. . . . . . 206.
25-3: Shape Of The Universe With Fully Developed Reverse Categories #2, #3 & #4. . . . . . 207.
25-4: Key To The 4 Positive & 4 Negative Quadrants Producing The 8 Octants Of The Universe. . . . . . 208.

FORMULAS

5-1: $P_3 = (2D_C)^3$ = Number of U-Spaces Per form. . . . . . 35.
8-1: $S_T = 2D_C$ = Shifts (Pauses) Per Rotation In The Tertiary Plane . . . . . . 58.
8-2: $S_P = \sqrt{D_N}\updownarrow$ = Shifts Per Rotation In The Primary Plane. . . . . . 58.
8-3: $S_S = \sqrt{D_N}\updownarrow$ = Shifts Per Rotation In The Secondary Plane. . . . . . 59.
8-4: $K_{SR} = D_N$ = Duration Of Each Rotation In The Secondary Plane. . . . . 59.
8-5: $K_{SS} = \sqrt{D_N}$ = Duration Of Each Pause In The Secondary Plane . . . . . . 59.
8-6: $S_{PR2} = D_N$ = Primary Plane Pauses (Shifts) Per Rotation In The Secondary Plane . . . . . . 59.
8-7: $S_C = 2(D_c \times D_N)$ = Primary Plane Pauses (Shifts) Per Rotation In The Tertiary Plane . . . . . . 60.
8-8: $T_{CR} = 2(D_C \times D_N)K$ = Duration Per Rational Form's Complete Cycle. . 60.
8-9: $T_{Ci} = 2[(D_C)^2]K$ = Duration Per Irrational Form's Complete Cycle When In A Multi-combo. . . . . . 61.
9-1: $F_{M1} = 2D_C\Phi$ = Major Form's Diameter Per First Phase Of Each Cycle. 68.
9-2: $F_{M2} = (2D_C + 2D_N)\Phi$ = Major Form's Diameter Per Second Phase Of Each Cycle . . . . . . 69.
9-3: $F_{m1} = 2D_N\Phi$ = Minor Form's Diameter Per First Phase of Each Cycle. 69.
9-4: $F_{m2} = 2D_C\Phi$ = Minor Form's Diameter Per Second Phase Of Each Cycle. . . . . . 69.
11-1: $P_{LW} = 2(Z_G+N)$ = OD Of The Form Providing U-Spaces To A Generator Carrying A Multi-Combo Not A Multi-A-Combo . . . . . . 91.
11-2: $P_{MT} = N(Z_G)$ = OD Of The Form Providing U-Spaces To A Generator Carrying A Multi-A-Combo. . . . . . 92.
11-3: $V = N(C/Z_G)$ = Velocity Of A Multi-Combo's Generator Where C = The Speed Of Light. . . . . . 92.

CONTENTS

DESCRIPTION **PAGE NO.**

CRUCIAL PRINCIPLES

LEGAL NOTICE OF PERMISSION TO QUOTE IN A SCHOLARLY DISSERTATION

As I understand the U. S. Copyright Office's definition of the principle of fair use: "There is no specific number of words, lines, or notes that may safely be taken without permission", from the writings of another where the quoting is for the sake of "criticism, comment, news reporting, teaching, scholarship, and research." Still, since "what is fair use and what is infringement in a particular case will not always be clear or easily defined", I—in an attempt to preclude all such confusion—state as follows:

To every party of one or more persons composing—in voice, print, or electronic media and whether for profit or not—a single scholarly dissertation either upon some scholarly dissertation of mine or upon ideas relevant to ideas expressed by me in some scholarly dissertation of mine whether published or unpublished, I, Edward N. Haas, hereby grant the right to quote—without fee or further permission whether in writing or otherwise and regardless of the manner or the extent to which the quoted words might be crucial to their source—up to a total of five thousand (5,000) of my ***own*** words or translations as found in some scholarly dissertation of mine and to quote them anywhere in the course of the dissertation whether in its body, its footnotes, its endnotes, or as fill between chapters or as chapter headings. This grant also extends to charts, drawings, and figures created by me in the course of my scholarly dissertations and, where no word count is possible in such charts, drawings, and figures, each single chart, drawing, and figure shall, in that case, be counted as one hundred (100) words. To facilitate the word counting process, each complete number no matter how large or small shall be counted as one word, and each formula no matter how many or few its components shall be counted as one word.

This grant is valid provided only that the quoting party meets these following five (5) conditions: (1) What's quoted from me must be quoted in what is truly a scholarly dissertation and is one containing one or more ideas relevant to what is quoted from me (Lewd works posing as scholarly examinations into sexual behavior do not qualify.); (2) the source of my words, charts, etc. and the pertinent copyright notice are fully described; (3) in the grantee's scholarly dissertation, the ratio of: (a) the total number of words written by the grantee, to (b) the total number of quoted words written by myself, is no less than twenty to one (***i.e.:*** 20/1); (4) in the act of availing himself (herself or themselves) of this grant, the grantee does thereby automatically and permanently extend—and forever bind all the grantee's heirs, assigns, publishers, and successors in title to grantee's copy rights to extend—to all other composers of scholarly dissertations the same privilege with regard to the *grantee's* scholarly dissertations as the grantee has accepted from me with regard to *my* scholarly dissertations; and (5) the grantee places near the front of the grantee's scholarly dissertation a statement (written if possible) acknowledging that the grantee accepts these conditions. As long as these five (5) conditions are met, this grant is permanent and irrevocable and remains forever binding upon myself, my heirs, assigns, publishers, and any successors in title to my copyrights.

✡✞✡✞✡✞✡✞✡✞✡✞✡✞✡✞✡✞✡✞✡✞✡

In all I shall say in this book, I submit to what is taught by Our Mother, the Holy Roman Church. If there be anything in it contrary to that, it shall be without my knowledge. Therefore, for the love of Our Lord, I implore the learned men who are to examine this book to look at it very carefully and to amend any faults of that kind which may be herein and the many others which it shall have of other kinds. If there be anything good in it, let that be to the glory and honor of God and in service to His most sacred Mother, our Patroness and Lady, whose habit—though all unworthily—I wear.
——**ST. TERESA OF AVILA** (1515 – 1582 A. D.): ***Book Called Way Of Perfection*** as given in the "Protestation". My re-translation – ENH (My sentiments exactly, dear saint!).

In the interest of clearness, it appeared to be inevitable that I should repeat myself frequently, without paying the slightest attention to the elegance of the presentation.
——**ALBERT EINTEIN:** ***Relativity, The Special And The General Theory***, pg. V of the hardbound copy by Bonanza Books, New York, 1961. ©1961 The Estate Of Albert Einstein (I suspect the frequency with which I repeat myself is quite formidable, to say the least. I pray my reader—assuming there shall ever be such—will find it helpful rather than intolerably exasperating.).

Yet in the whole history of philosophy it would be hard to find a first-rate thinker who resisted better than St. Thomas the temptation to coin his own philosophical vocabulary. He employed the traditional terminology in current use at the time, no matter how difficult it was for that old terminology to express radically new thought. For posterity, this practice of St. Thomas has perhaps been singularly unfortunate.
——**JOSEPH OWENS, C. Ss. R.:** ***St. Thomas Aquinas on the Existence of God, Collected Papers of Joseph Owens, Edited by John R. Catan***, page 70 of the paperback edition by the State University of New York Press, Albany. ©1980 State University of New York Press (Being only a fifth-rate thinker, I have indulged to the hilt rather than resist the temptation to coin my own vocabulary. Here again, then, I must pray my reader—assuming there shall ever be such—will find my verbal profligacy the very opposite of "singularly unfortunate".).[1]

---

[1] As others often do, I insist every chapter or the like begin on an odd numbered page (***ex. gr.:*** on the Internet, use the search string "start chapters on odd page".). If then, a chapter *ends* on an odd page, there's bound to be a blank even numbered one between it and the next chapter. Being opposed to wasting paper, I put such large empty areas to good use by throwing in one or more quotes relevant to some part of the whole and, in most cases, accompanied by a comment or two on such fillers. Hopefully, such asides shall help rather than hinder the reader's struggle to get an accurate grasp of Esoptrics. Maybe they'll even prove to be entertaining "vacations" from the main text. Incidentally, when I was an instructor in the Air Force in the late 50's, blank pages were anathema and, at the very least, had to be marked "intentionally blank".

Preface:

For at least 2,500 years now, one of the most—if not <u>*the*</u> most—hotly, widely, and often contested issues is the notion of the continuum. It's a notion which looks upon time, space, locomotion, and change as continuous, which is to say it implies there is no limit to the smallness of the smallest segment of each. Thereby, it necessarily invokes the notion of infinite divisibility. The latter then demands mathematical conclusions so manifestly self-contradictory, they boggle the mind no less than does the notion of a square circle. No wonder, then, they move many a renowned thinker, such as John Locke and David Hume, to proclaim those conclusions "absurd". See my quotes from them on pages VIII & XXVII.

The existence of the continuum is by no means a *scientific fact*: It cannot be confirmed by *observation*. Human technology is nowhere close to being able to prove by observation that there is no limit to the smallness of the smallest segment of either time or space or locomotion or change. For example, it cannot be observed that an object moving 1 cm. per sec. is also moving 10^{-40} cm. per 10^{-40} seconds. That involves measurements exceedingly far below what tiny levels science can currently probe. Even if science could, would that prove empirically it's also moving $10^{-\text{googleplex}}$ cm. per $10^{-\text{googleplex}}$ seconds? Whether or not there is any kind of change below c. 10^{-18} sec. cannot be confirmed by *observation*.

As widely known and admitted as that is, the vast majority of history's thinkers—whether philosophers or scientists—have most adamantly clung to, and still cling to, the notion of the continuum. Most, if not all, do so on the grounds that it is so powerfully implied by what we do observe, only a mindless fool would deny it. For 2,000 years, Geocentrism's defenders (and until c. the 16th century, the vast majority of thinkers were such defenders) hurled the same charge at those tiny few who dared to suggest Earth is a sphere orbiting the Sun. Going by what one's sense images *imply* is risky business.

Some might add that, so far, no one has ever come even remotely close to devising a cosmological theory obviating the need for *continuous* time, space, locomotion and change. Esoptrics now ends that lack and does it most impressively. How is it able to do so? It has uncovered thru *introspection* (***i.e.:*** Philosophy's methodology), and hands the world, a series of *God-given* coordinates utterly unlike any of the *man-made* ones ever previously known to history. Instead of coordinates which are *as continuous time, space, locomotion, and change supposedly say they must be* (***i.e.:*** per the minds of sense experience's addicts), Esoptrics gives the world *God-given coordinates* dictating how time, space, locomotion and change must be microscopically—namely: not *actually* divided or divisible beyond $7.201789375 \times 10^{-96}$ sec. for time and change & $7.34683969 \times 10^{-47}$ cm. for space and locomotion. Esoptrics thus replaces Science's (***i.e.:*** Empiricism's) *a posteriori* coordinates (***i.e.:*** ones coming into play <u>*after*</u>, and <u>*prompted by*</u>, our sense experiences of time, space, locomotion and change) with Philosophy's (***i.e.:*** introspection's) *a priori* ones (***i.e.:*** ones coming into play *before*, and *producing* our sense experiences of time, space, locomotion and change). Exactly what does that mean? Here's my preliminary elaboration:

Since the start of humanity's quest for intellectual advancement, the vast majority of thinkers have assumed that the coordinates whereby we demarcate objects from one another cannot be, unless, *first of all*, those objects are spatially (say physically, if you prefer) extended, which is to say their internality—no matter how tiny the segment of it—necessarily, and by its very nature (***i.e.:*** inherently), includes the 3 spatial dimensions of physical length, width, and depth as they necessarily occupy different locations in space—something (exactly what is not clear) which itself necessarily, and inherently, includes physical length, width, and depth in its internality no matter how tiny the segment of it. Esoptrics reverses that and insists neither space nor its occupants—*at their tiniest level*—are *really* extended in the above *inherent* manner; they merely *imply* otherwise *to the senses of those ignorant of any other way for particles and areas of space to be outside of one another*. They can do that because, *first of all*, each ultimate has from God—*and only as long as God provides it*—a unique set of 7 *a priori* coordinates giving each ultimate an internality (***i.e.:*** internal content = essence) to which *logical* dimensions, logical location, and a 7-way logical sequence are inherent. So unique is each God-given internality and what's inherent to it, each ultimate *really is* separate and distinct from all the rest without any of them having any trace of physical length, width, and depth and without any one of them being outside of the others in anything having physical length, width, and depth.

As far as *physical* length, width, and depth are concerned, neither the Universe as a whole nor any of what's in it—regardless of what that might be—has any trace of such *physical* dimensions and, therefore, is *physically* infinitesimal (***i.e.:*** *physically* dimensionless); but, because every ultimate thing in the Universe has *logical* dimensions, not a one of them is *logically* infinitesimal (***i.e.:*** *logically* dimensionless). Such they are as long as God wills God's 7 *a priori* coordinates to instill those logical dimensions from outside those ultimates.[1] Those logical dimensions, though, are more than sufficient to make every ultimate *fully* distinct from, *fully* separate from, and *fully* outside of, all the other ultimates in the Universe. Such holds true even as those 7 God-given *a priori* coordinates make the Universe's ultimates sometimes logically *concentric* and sometimes not.[2]

However illogical it may seem to sensation limited minds, *divinely* devised logical separation is so effective, it excludes all need for any other kind. Separation, then, is not the effect of 3-dimensionality, space, spatially extended things, spatial dimensions, or different locations in space. It's the other way around: Those factors are the effects of what's inherent to the unique logically extended internality continuously given to each and every ultimate by God's 7 *a priori* coordinates. They are not effects in the sense of illusions and hallucinations. They are effects just as real as the fiery result of igniting gasoline. For, God's coordinates make the ultimate units of space and matter <u>so</u> *really* outside of one another, they—though *individually* devoid of every trace of spatial extension—*really* do *collectively* add up to 3-dimensional shapes despite producing such "only" as *effects*.

In sum then: Some (scientists in particular) say the coordinates by which <u>*we distinguish for ourselves*</u> between object and object are *a posteriori* with regard to spatial dimensions and time (***i.e.:*** *subsequent* to them and, thus, a result dictated by them) and are the only coordinates we can know. In other words, for them, physical length, width, depth and 3-dimensionality must always and everywhere be inherent to the internality of every one of the Universe's truly real realities, or there's no way to tell them apart and no way for them

---

[1] **<u>NOTE OF JANUARY 23, 2013</u>:** How different that is from all other cosmological systems which at least effectively maintain that every ultimate has physical dimensions coming to it from *within* it! Whether intentionally or not, they thus make every ultimate so inherently self-sufficient, God is wholly excluded from the picture and Atheism made almost inevitable.

[2] **<u>NOTE OF MARCH 19, 2013</u>:** No ultimate can have *logical* dimensions *by its very nature*. Logical dimensions *must* be continuously imparted by an external source so good, all ultimates are drawn to It naturally with some level of intensity just as a magnet naturally draws iron with an intensity proportionate to its magnetic field's strength.

to ***be*** separate from one another. Esoptrics says physical length, width, depth, and 3-dimensionality are never *actually* inherent—and ever only *mistakenly inferred* to be inherent—to the internality of each of the Universe's ultimates. Still, they are separate, but by having coordinates by which they *distinguish themselves* from one another—namely: the logical dimensions and locations incessantly imparted to them from the outside by those 7 God-given coordinates ever *a priori* to spatial dimensions and locations (***i.e.:*** prior to and producing them). More importantly, those 7 divine coordinates can be known by us. All that's needed is the right amount of motivation and effort in the right direction.[1]

Naturally, many will protest: "How can there possibly be coordinates *a priori* to spatial extension? What might they be, and how might one possibly come to know them?"

Esoptrics' answer is extremely detailed and mathematically precise as it uses Geometry to explain *metaphorically* what is purely algebraic in nature. It's so detailed, it explains how these 7 God-given, *a priori* coordinates—and the unique sets of logical dimensions they instill one set per ultimate—relate to such factors as the fundamental nature of fermions vs. bosons (***i.e.:*** carrying generators vs. piggyback forms in Esoptrics' terminology); acceleration by the latter vs. the former; how acceleration by the former absorbs light; the relativity of time for the latter vs. for the former; gravity; leptons vs. quarks; electrons and their shells; hydrogen protons; molecules; the planets; the Sun; the galaxies; the clusters of galaxies; the clusters of such clusters; the super clusters; radiation; photons; the Big Bang and creation prior to it (something no other theory can even so little as begin to do); black holes; inflation; entanglement; the double slit experiments; ultimate time (***i.e.:*** u-time); ultimate space (***i.e.:*** u-spaces); stop and go locomotion; stop and go inertial systems vs. Newton's kind; Esoptrics first law of motion vs. Newton's; and other factors.

In the lengthy exposition of its answer, Esoptrics, if correct, forever lays to rest 2 monumental issues by giving striking evidences of this: (1) The only way to uncover the God-given *a priori* coordinates is to resort to the *philosophical* methodology of attentively

---

[1] **NOTE OF JANUARY 23, 2013:** Is it possible to give even the *remotest* example of how things can be logically separate from one another without the aid of spatial extension? Consider the numbers 1, 2, 3, 4, 5, and on and on for as far as you wish to go. We do not have to picture them as outside of one another in space in order to tell them apart. Why?! Each has a unique internality. It's thus their "essence" which sets them logically apart and logically outside of one another long before one needs to drag in space. Furthermore, if one must always count by 1's, then, there is—among the members of that endless set of numbers—a logical sequence which allows us to say what each number's logical distance is from each and every one of its fellow numbers. For instance, 1 is +1 *count* from 2, +2 *counts* from 3, +99 *counts* from 100, etc.. Oppositely, 100 is –99 *counts* from 1, etc.. Suppose, though, at each step in that sequence, we conceive of a second ancillary sequence which runs I, II, III, IV, and on and on for as far as you wish to go. We can—without *necessarily* having to resort to Geometry's pictures—define the distance from 1 to the IV of 100 as: +99+IV. We can of course—if we need to clarify things for the weak minded—use a two dimensional picture composed of some number of columns each of an equal number of rows. The result would be a square composed of small squares. In each of those small squares, we can then place a notation such as 1/I or 1/II or 1/III, 2/IX, 100/XL, etc. and, thereby, describe its logical location and essence in *counts*. Geometry's pictures then become a *figurative* way of describing what—because counting is purely algebraic in nature—is too difficult for some to make any sense out of it without drawings in space. Next, suppose, ancillary to each of the small squares, we introduced a third numbering system such as A, B, C, D . . . . Z, AA, AB, AC, AD . . . . AZ, BA, BB, BC, etc.. We would then have a large cube composed of small cubes, and, in each of those small cubes, we can then place a notation such as 1/I/A or 2/II/AA, etc. and, again, thereby describe its logical location and essence in counts. Now—though the distance between any two cubes is a number of cubes in *Geometry's realm of drawings*—that distance is far from such in the realm which the drawing does no more than express metaphorically: *Algebra's realm of counting numbers and kinds of counts*.

focusing *inwardly* upon the fundamental nature of an act of consciousness and, from such *introspective* observation, to deduce what is implied regarding the properties (***i.e.:*** ultramicroscopic makeup) of fundamental realities; they can never be uncovered by the *scientific* methodology of focusing *outwardly* upon the sense images and, from such "*extro*spective" observations, to deduce what is implied regarding the properties of fundamental realities (***i.e.:*** only Philosophy can discover what truly are the *most* basic of "first principles".). (2) The existence of God is always and everywhere a mathematically necessary influence without Whom there can be no continuously supplied, *a priori* coordinates and, thus, never any dimensions of any kind and never two or more realities distinct from one another. For, the coordinates making all that possible come only from the ways God continuously influences the ultimate realities. Never, as sensation addicted minds dream, are coordinates, separation and multiplicity due to a kind of time, space, locomotion and change which are—at least after their creation—now forever continuous "all by their lonesome" as they endlessly enjoy a continuousness which never again requires any antecedent, external influence to give them whatever uniqueness, separation, divisibility, and 3-dimensionality they have.

How amazing! Atheists cannot tolerate the notion of an *Infinite* Being Who, to be what He is, requires no kind of antecedent influence whatsoever; but, they are quite comfortable with the notion of *finite* fermions and bosons which require no kind of antecedent influence whatsoever to be infinitely divisible and to have all it takes to build the indescribably vast and variegated Universe we observe. Never was a *deus ex machina* more fantastic than theirs. And how obvious that shall be, when Esoptrics gets done with them!

In an e-mail dated 9/28/12, Gordon, the youngest of my 3 male siblings, commented on the original version of this preface so:

> This is beautiful Edward. You have grown so much in your understanding of what you are seeing. There is no doubt that, if you continue to be able to bridge the scientific tenets as this claims, your theory will have to be looked at. You are, of course, fighting one of the two grand manipulators of truth—ego (the other being fear), and to complicate matters you are ultimately (Esoptrics that is) subordinating physical reality to consciousness.

To that, I, the same day, rebutted so: Subordinating reality to consciousness???!!! I would never agree to that; rather, I would say this: I'm subordinating our ability to uncover the *a priori* coordinates of reality to our ability to understand consciousness. That's a world of difference. Your response implies the *a priori* coordinates of reality are *determined* by consciousness. Esoptrics implies only that the ability to *discover* those *a priori* coordinates is *dependent* upon awareness of how consciousness works.

To that, Gordon, in an e-mail dated 9/29/2012, remarked:

> Perhaps without knowing it, Edward, but this correction in my statement has produced the core element of the "mindset" one must have to start the trek of "seeing" Esoptrics. This should be the opening statement of your theory! You can thank me later!

My e-mail of 9/29/2012 seconded the above so: I agree with that. Indeed, before your response arrived, it was my plan to make my little essay the preface to my upcoming 21st book, and to also use it as the back cover blurb. After your response arrived and precipitated my response, I decided that the preface (though not the rear cover) should also include your response and my reply to it. In short, dear, dear brother, you hit the nail square on the head. It's almost as if you were reading my mind by "action at a distance". LOL.

His 9/29/2012 response was:

I guess we are both "nuts" enough to arrive at the same tree![1]

It is maintained that perhaps the success of the Heisenberg method points to a purely algebraical method of description of nature, that is, to the elimination of continuous functions from physics. Then, however, we must also give up, on principle, the space-time continuum. It is conceivable that human ingenuity will some day find methods which will make it possible to proceed along such a path. At the present time, however, such a program looks like an attempt to breathe in empty space.
——**ALBERT EINSTEIN:** ***Physics And Reality***, originally published in ***The Journal Of The Franklin Institute***, Vol. 221: No. 3: March, 1936, and found on pg. 319 of ***Ideas And Opinions*** as published by Bonanza Books, New York, 1954 (Ah, Prof. Einstein! Would to God you were alive today to see how marvelously successful is Esoptrics' "attempt to breathe in empty space"! Or as I would prefer to put it: Esoptrics' attempt to breathe in the world of "a purely algebraical method of description of nature" rather than in one of a geometrical method. What enables me to do that, Prof.? The world I breathe in is not "empty space" but space as defined by God's *a priori* coordinates—ones of which neither you nor any of your fellow scientists ever had even the slightest inkling. Incidentally, I was born April 13, 1936, and, so, was alive in my mother's womb as you aired your "prophesy".).

[1] **NOTE OF JANUARY 26, 2013:** There is perhaps another profound difference which holds not only between Esoptrics and all other *cosmological* theories, but also between Esoptrics and all other *philosophical* systems as well. It goes like this: At least the vast majority of scientists and philosophers (if not all of them) assume infinitesimals are—and necessarily so—also absolutely *indivisible*. They say that because they can think only in the terms of *physical* dimensions and *physical* division and never in the terms of *logical* dimensions and *logical* division. As a result, in their crippled minds' eyes, division is possible only where one can—at least in one's mind—distinguish between 2 areas each of which has, by its very nature, physical length, width, and depth; but, the physical length, width, and depth of one area are not coterminous with those of the other. Oppositely, Esoptrics' infinitesimals—because of the external influence of God's 7 *a priori* coordinates—have 7 *logical* dimensions which first divide them internally into the 6 simultaneous ways of being related to God (***i.e.:*** in Esoptrics' terms, "in potency to" God) and also into which level of intensity—out of the 2^{256} possible ones—each of those 6 "potencies" is currently exhibiting. These 6 turn out to be states of excitation due to the fact that every ultimate created by God is, by its very nature, always reaching excitedly (This doesn't imply any kind of awareness.) for unity with The Infinitely Attractive in each of the 6 God-given ways to reach and with one of the 2^{256} levels of intensity possible to each of the 6 God-given ways to reach. In short, for most if not all thinkers of every discipline other than Esoptrics, all infinitesimals are also *indivisibles,* because—while they rightly think all infinitesimals are necessarily *physically* so—they wrongly think no infinitesimal can possibly be in any other way extended and thus no other way divisible. Oppositely, none of Esoptrics' infinitesimals are indivisibles, because all are *logically* extended and divisible into the 6 ways of being excited by God, The Infinitely Exciting, plus the 2^{256} levels of intensity possible to each of those 6 states of excitation.

In fact, if general relativity were not taken into account in GPS satellite navigation systems, errors in global positions would accumulate at a rate of about ten kilometers each day! However, the real importance of general relativity is not its application in devices that guide you to new restaurants, but rather that it is a very different model of the universe, which predicts new effects such as gravitational waves and black holes. And so general relativity has transformed physics into geometry.
——**STEPHEN HAWKING and LEONARD MLODINOW:** ***The Grand Design***; as found on page 102 of the paperback published by Bantam Books Trade Paperbacks; New York, 2012 edition (Yes, that's one of the paramount issues: Is it more accurate to transform physics into Geometry or, as Esoptrics does, to transform it into Algebra? Which can describe reality as it really is: Geometry or Algebra? To be sure, Esoptrics uses Geometry's pictures, but only as teaching aids making it easier for readers to grasp thru geometrical metaphors "a purely algebraical method of description of nature" [Einstein] which, otherwise, would be far too abstract for the vast majority.).

Mathematicians would say that the continuous had been replaced by the discrete. Bohr saw that this might be a very general tendency in the new kind of physics that was slowly coming to birth. He applied to atoms similar principles to those that Planck had applied to radiation. A classical physicist would have supposed that electrons encircling a nucleus would do so in orbits whose radii could take any value. Bohr proposed the replacement of this continuous possibility by the discrete requirement that the radii could only take a series of distinct values that one could enumerate (first, second, third, . . .).
——**SIR JOHN POLKINGHORNE:** ***Quantum Theory, A Very Short Introduction***, as given on page 12 of the 2002, paperback edition by Oxford University Press. ©2002 John Polkinghorne (Yes, Sir John, in Esoptrics, "the continuous" "replaced by the discrete" is far more than merely "a very general tendency". For, in Esoptrics—though every ultimate constituent's *existence* and *individuality* remain continuous despite accidental changes at regular intervals—there is nothing continuous about how often they change from one particular act of existence [***i.e.:*** temporarily changeless state of excitation] of such-and-such an intensity to a logically sequential one; instead, yielding to the discrete, there is a definite limit to the number of change rates available to them. So, too, in Esoptrics, do time, space, locomotion, change, and consciousness give up their continuousness to the discrete. And what theory other than Esoptrics, Sir John, gives even so little as the slightest trace of an explanation of how supposedly continuous time, space, locomotion, and change can be discrete and, yet, leave intact the continuous *existence* and *individuality* of each and every one of the Universe's ultimate constituents? [***Cf.:*** Sir Oliver Joseph Lodge's words on my page 246.]).

PREFACE

My e-mail to a friend sent June 15, 2013:

Greetings, JF: In my previous e-mail, I wrote:

> Except for Esoptrics, all theoretical Physics is an attempt to figure out what kinds of coordinates are implied by what outward looking minds have learned about time and space. With Esoptrics, theoretical Physics is an attempt to figure out what kinds of time and space are implied by what an inward looking mind has learned about the 7 coordinates God imparts to creation. Except for Esoptrics, then, all theoretical Physics deals with coordinates which are *a posteriori* (***i.e.:*** after the fact) with regard to time and space; whereas, Esoptrics deals with coordinates which are *a priori* (***i.e.:*** before the fact) with regard to time and space.

In reply, you wrote:

> . . . . "a priori" properly means independent of experience and "a posteriori" properly means based upon experience.[1]

In reply to that, let me present you with this from Wikipedia:

> 13th century philosopher Thomas Aquinas, in the Summa Theologica, writes: "Demonstration can be made in two ways: One is through the cause, and is called 'a priori,' and this is to argue from what is prior absolutely. The other is through the effect, and is called a demonstration 'a posteriori'; this is to argue from what is prior relatively only to us. When an effect is better known to us than its cause, from the effect we proceed to the knowledge of the cause. And from every effect the existence of its proper cause can be demonstrated, so long as its effects are better known to us; because since every effect depends upon its cause, if the effect exists, the cause must preexist. ***Summa Theologica***. I, Q2, A2, Gen. Resp.

---

[1] **NOTE OF AUG. 19, 2013:** JF may be confusing "*a priori* coordinate" with *a priori* proposition". A coordinate is far from being a proposition. A *coordinate* is truly *a priori* if it is prior to, and causes, the internality of what it precedes; a *proposition* is truly *a priori* it is true by definition. Still, the latter is not "independent of experience". After all, even in the case of a proposition as true by definition as is 2+3=5, the only way I know it's true by definition is that, thru numerous sensory experiences, many others have assured me that all educated people agree on defining the terms that way. As some put it: *A priori* propositions are not independent of the experience of learning the language in which they are cast. My copy of the 1996 edition of ***Webster's New Universal Unabridged Dictionary*** says that, in Philosophy, "experience" means: "the totality of the cognitions given by perception". It then defines "perception" as: "the act or faculty of apprehending by means of the senses or the mind". If correct, that equates "experience" with whatever we know no matter how. If so, to what in a human mind can *a priori* possibly refer, if it refers to what's independent of *every* kind of experience? It can only refer to *a priori* coordinates in God's mind, since they're there regardless of whether or not they are known by anyone else. On the other hand, they cannot be *known* by us save thru some kind of experience on our part. Esoptrics says that, initially, that means the kind of suprasensory experience a highly reflexive and introspective human intellect produces. Of course, once some individual does that, he can then communicate that knowledge to others by means of their sensory experience of his words. In doing that, though, he is—shades of Plato!—making them aware of what they could have found on their own in their own minds at any time had they wanted to.

2013 ESOPTRICS UPDATE

For St. Thomas, *a priori* vs. *a posteriori* has to do with cause vs. effect (***i.e.:*** between what, as cause, is absolutely *a priori* vs. what—as an effect of that cause upon us—is *a priori* only in relation to us who start out knowing nothing of the cause save its effect, but then use knowledge of that effect as a means by which to gain knowledge of its *a priori* cause). In St. Thomas' terminology, then, Esoptrics' coordinates are the *a priori* cause of the effect Esoptrics calls "time & space as they really are"; whereas, the coordinates invoked by Science are what Science infers from the *a posteriori* effect which the *a priori* cause, "time and space as they really are", have upon human senses. According to Esoptrics, Science's coordinates are way off base because the scientists are unaware of the fact that the *a posteriori* effect of time and space upon human sense experience is far from being the same as what time and space really are. So great is that discrepancy, says Esoptrics, the only way to know time and space as they really are is to turn away from the scientific method of observing (**i.e.:** accruing sense experience of) the extra-mental world and to turn to the philosophical method of introspecting (***i.e.:*** accruing reflexive experience of) the innermost depths of the nature of consciousness itself.

Wikipedia's article on *a priori* vs. *a posteriori* is very interesting—especially in detail—and basically says that the distinction between *a priori* vs. *a posteriori* is by no means as cut a dry as you seem to suggest it is. It all depends upon what school's terminology you prefer. Still, whatever school's terminology one prefers to use, one should not be so limited in knowledge as to insist that one's chosen terminology is the only terminology there is. Above all, one should not be so *un*limited in hubris as to insist that it's the only terminology there is for *properly informed* people.

Esoptrics is not an attempt to overthrow—and, indeed, does not overthrow—any of the equations fundamental to Relativity or Quantum Theory. It's an attempt to overthrow—and, hopefully, does—every current explanation of *why* those equations are accurate. So contrary to current Science are Esoptrics explanations of the "ultramicroscopic makeup" (Prof. Greene) of time, space, locomotion, and change, it cannot fail to overthrow all those explanations, if, that is, its own are correct.[1] May God grant they are. In the meantime, let no one imagine me unaware that Esoptrics is *Theoretical* Physics and, as such, is mostly a set of *inferences* based on what's *implied* by observed facts. History makes it so clear how often even the most widely, loudly, and long touted inferences prove wrong, I claim no certitude about what's *real*, save this: Esoptrics' starting point is the empirical fact of my persistent observation of the necessarily tripartite structure of my every moment of consciousness. If others see not the same in themselves, so be it. I couldn't care less what they might say either about what they observe in themselves or what I observe in myself. I will be to no extent deterred by those who deny the reality of what's incessantly staring me in my memory's face.

✡✞✡✞✡✞✡✞✡✞✡✞✡✞✡✞✡✞✡✞✡✞✡

[1] **NOTE OF NOV. 8, 2013:** That contrariness is clearest in this: For all but Esoptrics, locality comes first, and things are different from one another because first of all there are different places in space to occupy. For Esoptrics, internality (***i.e.:*** essence) comes first and there are different places in space to occupy because first of all each ultimate has from God an absolutely unique internality. Unique internality begets locality, not v.v.. For all but myself, it's laughably untenable. Such it would be did not Esoptrics' highly detailed, mathematically precise, graphic explanation of the "ultra-microscopic makeup" show exactly how such is possible and possible without any reliance on the notion of the continuum and the absurd math which—as many admit—that notion necessarily begets. Needless to say, it cannot be proven by observation that locality precedes internality even at 10^{-47} cm..

The chief objection against all *abstract* reasonings is derived from the ideas of space and time; ideas, which, in common life and to a careless view, are very clear and intelligible, but when they pass through the scrutiny of the profound sciences (and they are the chief object of these sciences) afford principles, which seem full of absurdity and contradiction. No priestly *dogmas*, invented on purpose to tame and subdue the rebellious reason of mankind, ever shocked common sense more than the doctrine of the infinitive divisibility of extension, with its consequences; as they are pompously displayed by all geometricians and metaphysicians, with a kind of triumph and exultation. A real quantity, infinitely less than any finite quantity, containing quantities infinitely less than itself, and so on *in infinitum*; this is an edifice so bold and prodigious, that it is too weighty for any pretended demonstration to support, because it shocks the clearest and most natural principles of human reason. But what renders the matter more extraordinary, is, that these seemingly absurd opinions are supported by a chain of reasoning, the clearest and most natural; nor is it possible for us to allow the premises without admitting the consequences. Nothing can be more convincing and satisfactory than all the conclusions concerning the properties of circles and triangles; and yet, when these are once received, how can we deny, that the angle of contact between a circle and its tangent is infinitely less than any rectilineal angle, that as you may increase the diameter of the circle *in infinitum*, this angle of contact becomes still less, even *in infinitum*, and that the angle of contact between other curves and their tangents may be infinitely less than those between any circle and its tangent, and so on, *in infinitum*? The demonstration of these principles seems as unexceptionable as that which proves the three angles of a triangle to be equal to two right ones, though the latter opinion be natural and easy, and the former big with contradiction and absurdity. Reason here seems to be thrown into a kind of amazement and suspence, which, without the suggestions of any sceptic, gives her a diffidence of herself, and of the ground on which she treads. She sees a full light, which illuminates certain places; but that light borders upon the most profound darkness. And between these she is so dazzled and confounded, that she scarcely can pronounce with certainty and assurance concerning any one object.

123. The absurdity of these bold determinations of the abstract sciences seems to become, if possible, still more palpable with regard to time than extension. An infinite number of real parts of time, passing in succession, and exhausted time after another, appears so evident a contradiction, that no man, one should think, whose judgement is not corrupted, instead of being improved, by the sciences, would ever be able to admit of it.

——**DAVID HUME** (1711-1776)**:** ***Concerning Human Understanding***, Sect. XII: Part II, as found on page 506 of Vol. 35 of ***Great Books Of The Western World*** as published by Encyclopædia Britannica, Inc.; Chicago, 1952 (You said it, Mr. Hume: If men can cling to the "profound darkness" of the continuum even as they "admit of" it and its math being as "absurd" as you, Mr. Locke, and many others, insist it is, it's because their judgement has been *corrupted* rather than *improved* by the sciences. Or should we perhaps say that, as a result of Original Sin, they have at least instinctively *prostituted* the sciences to the corruption of their judgement for some secret and pathetic purpose inherited from the author of that sin?).

The book also cites *Wikipedia* on a number of occasions, a source considered verboten for most research-based texts.
——**ANNA CALL: *Clarion Review*** (Far less than everyone agrees with that assessment of Wikipedia. See its article ***Reliability Of Wikipedia*** and the authoritative sources it quotes as upholding its accuracy against Britannica. It should be more than enough of a reply—except, of course for *sycophantic* supporters of Britannica. Note, too, there's a financial reason to demean Wikipedia. It's free; Britannica is not.)

The book tends to accompany quotations with an exuberant agreement or reply. One example:

> *"Guess what, Mr. Freedman! I dare suggest to you quite confidently that, with Esoptrics, that clue [to the ultimate laws o f the universe] has now burst upon the world most thunderously."*

This colorful reaction fits well with the personality of the rest of the text, but its location, tacked onto a quote presented independently at the end of a chapter, is structurally odd. There seems to be no reason for its presence except as a means for a reply to the person being quoted.
——**ANNA CALL: *Clarion Review*** (Ms. Call here seems unfamiliar with the widespread practice of starting each chapter on an odd numbered page, and then—to avoid wasting a sheet—of placing one or more fillers onto the blank, even numbered page which must fall between when a chapter is preceded by one *ending* on an odd numbered page [See my footnote on page XVIII.]. As commonplace as that practice is, how could a professional book reviewer be unaware of it? It's also commonplace to throw in a filler or 2 where much of a chapter's last page is blank. Is Ms. Call unaware of that too?)

All well-meaning people should try to contribute as much as possible to improving such mutual understanding. It is in this spirit that I should like to ask my Russian colleagues and any other reader to accept the following answer to their letter. It is the reply of a man who anxiously tries to find a feasible solution without having the illusion that he himself knows "the truth" or "the right path" to follow. If in the following I shall express my views somewhat dogmatically, I do it only for the sake of clarity and simplicity.
——**ALBERT EINSTEIN:** page 140 of the hardbound issue of ***Ideas And Opinions*** by Bonanza Books, New York, 1954. ©1954 Crown Publishers Inc. (Me too! I freely acknowledge, as I always have, that Esoptrics, save for its starting point [***viz.:*** my observation—one totally indifferent to what's true of others or what they might say of me—of ***my*** consciousness' necessary, tripartite structure], is merely a *theory* which my mind cannot possibly carry beyond the theoretical. Subsequent generations may do so; but, as mentally limited as I am, I cannot. Every reader should bear that in mind and, as a result, realize that—when I seem pontificating with dogmatic arrogance—I am, like Einstein, doing it "only for the sake of clarity and simplicity.")

Haas' theory is far more rooted in theology than science, and this bias leads him to attempt to use the tools and concepts of science—hypothesis, empirical observation, objectivity—to try to disprove its effectiveness. Much of Haas' disdain is reserved for empirical observation; in his view, modern physics is limited by its insistence on empiricism, while Haas' views require systems, such as God's *a priori* coordinates, that are unknowable and untestable yet somehow far superior to detailed observations about what we can perceive.
——**KIRKUS INDIE: *Kirkus Review*** (The reviewer seems not to know much about "unknowable". If God's coordinates are "unknowable", how am I able to describe them in detail? They are no more *unknowable* than—and just as *unobservable* as—String Theory's strings which, as Prof. Greene explains on page 352 of ***The Fabric Of The Cosmos***, are, at 10^{-33} cm., "a billion billion times" smaller than current technology can probe. Whether Esoptrics' *a priori* coordinates or String Theory's strings, both are *indirectly* known by those who see them as *inferred* by what's observed. In Esoptrics' case, it means what *I* see inferred by my empirical observation of the necessary, tripartite structure of my consciousness, and this book spells out that lengthy inferential process in mathematically precise detail but with no illusions *all* others will agree the process is logically coherent. Scientists widely disdain my *intra*-mental use of empirical observation; but, I've never disdained their *extra*-mental use of it or the effectiveness of it, *except where knowledge of the ultramicroscopic* is the goal [Since Esoptrics' starting point and ultimate basis is empirical observation of consciousness' tripartite make-up, how crazy it is to say I disdain empirical observation! But, it's exactly what to expect from one blind to the *intra*-mental use of empirical observation.]. Doing that, I merely repeated what, on his page 352, Prof. Greene admits even "Some scientists argue vociferously"—namely: Seeking knowledge of the ultramicroscopic at or below the Planck scale of 10^{-33} cm., one is dealing with what's "so removed from direct empirical testing", it's what "lies in the realm of philosophy or theology, but not physics" [See my pages 54&174.]. It's not a matter of disdain for empiricism or of what's "far superior to detailed observations"; it's a matter of accepting what empirical observation itself makes obvious about itself—namely: that there is no such thing for us as empirical observation below c. 10^{-15} cm. and 10^{-18} sec.. How am I guilty of "bias" in what I state, when even some scientists, including Prof. Greene, publicly admit empirical observation cannot transcend those limits. All Science can do is *indirectly infer* what empirical observation *implies* about what's below those limits. But, if their *extra*-mental, scientific kind of empirical observation can *infer* it, so can Esoptrics' *intra*-mental, philosophical kind, and, by Prof. Greene's admission, "Some scientists argue vociferously" that *only such* a methodology can infer it—*not Physics*. Am I then guilty of "bias" because I dare to repeat what some of empirical observation's own supporters "argue vociferously"? As for testability, time must determine that by measuring whose effectiveness at inferring what's below the limits of 10^{-15} cm. and 10^{-18} sec. is greater: Esoptrics' observation of the *intra*-mental, or Science's observation of the *extra*-mental. No less than the latter, "Haas' views require systems" *indirectly* testable by noting the long-term fruitfulness of the explanations they infer. In the quest for knowledge of the *total chain of causes* behind the effects we observe [especially the unobservable causes], see: Which comes closer to a *total* chain of causes [***i.e.:*** which starts furthest down into the realm of the too tiny to observe] and then eventually accounts for the broader range of observed effects: Esoptrics' theorizing based on intramental empirical observation or Science's based on the extra-mental kind? *That's* what testing is all about. To say I "require systems" "unknowable and untestable" bespeaks a mind ridiculously ignorant of that *indirect* knowing and testing to which even many scientists often appeal (See pg. 175.), despite the admitted fact it often draws them into notoriously erroneous inferences not recognized as mistakes for many years, if not centuries. "Bias", indeed! And for this travesty, I paid $1,000.00???!!!).

At turns faux humble—"I do not pretend to know"—and disdainful of those he refers to as "weak minded," Haas writes with a surety against, and casual dismissal of, those unable to see the truth of his work, even as it flies in the faces of theorists and experimental scientists alike.
——**KIRKUS INDIE: *Kirkus Review*** (The phrase "I do not pretend to know" occurs only once in my book, and that's on page 63 where I wrote: "That same one manner of rotation may be what I call a pattern, or it may be what I call an anti-pattern. I do not pretend to know for sure which way such a rotation should be described." The phrase thus has nothing to do with humility whether "faux" or otherwise; it has solely to do with a single issue (***viz.:*** how to characterize a rotation's orientation) and the *fact* that I am undecided regarding how best to do so. The term "weak minded" occurs only once in my book, and that's in the footnote on page XXI where I referred to those unable to grasp some highly abstract concepts without the aid of Geometry's pictures [On this very issue, see my quote from Sir Michael Atiyah on my page 22.]. No "disdain" for such people was intended in the term. It's simply a statement of the widely admitted fact that some people's minds are considerably less able than others to grasp highly abstract concepts without the aid of visual images. I have never at any time entertained or expressed any kind of negativity toward "those unable to see the truth" of my work. How could I? I have never presented Esoptrics as anything other than a *theory*, and "theory" means what's purely a *speculation* which, as such, must await time and the efforts of others to determine whether or not there is any *truth* to it. Did this reviewer not read, on page XXVI, the preface's closing paragraph [On my pages XXVIII & 229, see my quote from Einstein regarding this very issue and my response to it.]? Any negativity on my part has ever been toward those who—whether thru indifference or gutter level slurs—dismiss Esoptrics as *not worthy of any consideration whatsoever.* On my page II, see the quote from two Philosophy professors who both agreed that the academic world's "almost total neglect" of my effort is "a shame". On page XXXIII, see a third Philosophy teacher echoing basically the same lament. How is Kirkus' review not a malevolent attempt to make me look like a contemptible phony? And if such, why?! Possibly this: Some people instinctively perceive this-or-that theory a threat. Unable, though, to grasp it well enough to refute it, they turn to one of the oldest tactics in the book: *ad hominem* argumentation: If you lack the wherewithal to address an author's *thoughts*, then defame *him*. Make him look bad, and it shall follow for many that no good thoughts can come from someone as shameful as he is. Thus is his theory refuted without a word about it. It's as if one—stymied by Relativity's details—imagined Relativity can be refuted by pointing out that Einstein divorced his first wife. As for "flies in the faces" etc., if Esoptrics does, it does so no more than did Heliocentrism, Relativity, and Big Bang theory when they first appeared. And what about Modified Newtonian Dynamics?! [See my page 110.]).

NOTE ON "ESSENCE: For many (most?) "essence" signifies the one or more out of something's many characteristics they deem *essential* to their efforts to *classify* or *characterize* it meaningfully. I call that select few of something's characteristics its *taxonomic* essence and not its essence *strictly speaking*. In my view, *every one* of any thing's *internal* characteristics is *essential* to its *individuality, distinctiveness* and *real* essence, which is to say to what its internality's uniqueness is, as opposed to what it is *in the way others classify or characterize it*. *Taxonomic* essence is expressed in such phrases as "rational animal", "Theist", "writer", "male", etc. ; but, the *real essence* and essence *stricte dicta* of any given thing is *all* of its internality, and that's particularly true of every intelligent being. For me, then, the strictest, truest, primary, most definite, and unqualified sense of "essence" = real essence = every iota of a given thing's internality. One can then distinguish between *developing* essence (***i.e.:*** every one of the internal characteristics had so far and to which more are still being added), and *finalized* essence (***i.e.:*** the sum total of all the internal characteristic ever had or ever to be had = an essence only God and the rest of infinity's inhabitants observe, because they alone transcend time). In this book, unless stated otherwise, "essence" signifies every iota of any thing's developing internality. To put it another way: It means every one of the characteristics present within some developing thing's extremities.

Naa- just stick your head up the dark place and the lack of oxygen and stench will have you seeing as clearly as Eddy!
——**USER NAME CHRIS:** On the Internet, on April 6, 2014, at Yahoo's home page, there was a news article written by one Karl W. Giberson and entitled ***Creationists Hate This Catholic***. In response to it, I posted a comment in the form of a brief bit about Esoptrics. In reply to my comment, someone with the user name Paul replied asking if he would need a PhD to read my book. The above is how Chris replied to Paul. It's an excellent example of the gutter level manner in which the intellectually superior, extraordinarily unbiased, open-minded paragons of impeccable rationality and morality often react to Esoptrics when their reaction is not one of complete indifference. I suspect *all* we humans *sometimes* find life inadequately fulfilling, unless we can feel we've reduced one or more of our fellow humans to the level of fecal matter. In some, though, it happens far more often than it does in others—particularly, it seems, in the face of my cosmological theorizing. To make matters worse, if, in response to such spittle, I dare to say anything the slightest bit caustic in reaction to such character assassins repeatedly calling me fecal matter, suddenly I am the far worse monster worthy of even more virulent character assassination.

2013 ESOPTRICS UPDATE

NOTE ON "EMPIRICAL OBSERVATION": As Anna Call seems to do, many (most?) limit it to observing the sensations deemed wrought in us by what's *extra*-mental—thus making it *mono*-directional. See, low on my page 6, the quote in which Einstein describes "the empirical procedure" as "exclusively a working-over of the raw material furnished by the senses" and, low on my page 40, his quote stating: "physics treats directly only with sense experience". Einstein & Call are mistaken. Empirical observation does not deal *exclusively* with "the raw material furnished by the senses." It's not *mono*-directional; it's *bi*-directional; and so, Esoptrics can back off in the opposite direction inward away from the sense images and their extra-mental cause and to use *empirical observation* in an *introspective* and intensely *reflexive* effort to peer into the internality of whatever is doing the bit of experiencing and "working-over" the senses' raw material. As for those who deny there's any such use of empirical observation, their denial probably comes from Empiricism's rampant fallacy that all knowledge derives from the senses. As some prefer to put it: Nothing's present in the mind unless first present in the senses. Empirical observation of my consciousness' activity (and, thereby, its makeup) is present in my mind only if some kind of sensation is *also* there (I know not what sensory deprivation experiments might say on this point.); but, the former is not *in* the latter; it's in what I observe myself *doing to* the latter. What's doing the act of being aware of, and "working-over", sensation is not itself just another sensation. It is but absurd to say my many sensations are what my sensations are sensing of themselves. And, if it is of the nature of feelings to feel themselves, then how are there many feelings at once many parts of the same whole, orderly panorama rather than a chaotic mass of sensations each feeling only itself? Isn't that basically the point Immanuel Kant tried to make? By giving me what to *act upon*, my sense images are the *catalyst* facilitating my act of reflecting upon, and having knowledge of, that highly illusive self able to feel, to corral, and to work-over, sensations while simultaneously noting in abstraction from them what it's doing to them. My sense images are thus *the means by which* I observe my consciousness capable self's activity; but, they are not *what* I observe of that activity; they are what I observe of that activity's *results*. That's what *I* most repeatedly observe in *me* ; and so, I am absolutely certain of it (***i.e.:*** certain of it in that strictest, truest, primary, most definite, and unqualified sense of "certain": "observed", "gazed upon", "encountered", "peered into", and "met face to face".). I am, therefore, so unshakably sure of what I observe in me, I couldn't care less if others insist I can't observe what I observe, since they can't. If a blind man denies visual sense images are possible, should I deny I see them? From one standpoint, then, Esoptrics boils down to this: an attempt to uncover what inferences regarding ultimate realities—especially regarding what ultimate realities make up the Universe—can be drawn from an intensely *introspective* and protracted *empirical observation* of that something (***i.e.:*** "the encounterer" in Esoptrics' terminology) *acting upon* rather than merely *present in* my sense images. Still, if some insist only Science's half-baked *mono*-directional use of empirical observation is possible, let them stick to it. It's of no consequence to me; though, unfortunately, it may be lethal to their ability to learn anything about what time, space, matter, locomotion, and change *really* are, which is to say what they are at that ultra-microscopic level beyond which—in Prof. Greene's words on my page 54—"'shrinking smaller'" is as meaningless as asking if the number nine is happy." Pardon me, then, if I make bold to suggest: Esoptrics—in the inferences it draws from a *bi*-directional empirical observation's *reflexive* scrutiny of consciousness' tripartite makeup—is, by very far, the most detailed, mathematically precise, noteworthy theory ever regarding what time, space, matter, locomotion and change *really* are.

I can honestly say that, in all my years of studying philosophical literature, I have never encountered anything quite like Edward Haas' opus ***Introspective Cosmology***. It is a truly exceptional work—so extraordinary that I find it difficult to evaluate and even describe in terms that will do it justice. The best that I can hope to do here is to introduce you to some of its salient features and to encourage you to experience the work for yourself. The scope, but at the same time the depth of the work, is truly amazing. The book is incredibly wide-ranging, and yet the author never simply waves his hand at an issue. The book portrays a broad landscape, yet there are no broad brush strokes. I have never seen such a multifaceted and eclectic work that nevertheless gives a deep and intensive treatment of every subject/issue mentioned. The author deals with topics from many different fields of study, but he manages to deal with all of these topics in the same informed and articulate manner. This is all the more astounding when one learns that the author has no formal education beyond the high-school level.
. . . one might characterize Mr. Haas' Introspective Cosmology as a cross between Aquinas' Summa Theologiae and Newton's Principia Mathematica. Indeed, this text might just as well have been titled The Mathematical Principles of Metaphysics. It is the only work in metaphysics that I have ever encountered which reads like a physics textbook. The author's use of mathematics to express metaphysical principles is one of the things that make this book unique. The author's employment of mathematics is rather extensive, and this can be somewhat bewildering at times. However, the level of complexity of the mathematics employed is not bewildering. The math does not go beyond simple arithmetic, geometry, and algebra; thus, the book should remain perfectly accessible to the typical scholar or student of the humanities.
. . . . The extent to which Mr. Haas has been misunderstood or ignored by the professional philosophical community is reprehensible. Introspective Cosmology is the result of forty-two years of effort on the part of it author—and it shows. Whether or not the Haasian system is on the mark, it is a well-thought-out and beautiful system.
——**RALPH AQUILA:** In the 4th quarter, 2004, edition of ***Prima Philosophia***, published by Cuxhaven: Junghans (Cuxhaven is a city of c. 51,000 on the shore of the North Sea and at the mouth of the Elbe river in the state of Lower Saxony, North-West Germany), the above is part of his 2004 review of an earlier book of mine setting forth the principles of Esoptrics and titled: ***Introspective Cosmology II***. He wrote his review when he was an assistant teacher of Philosophy at Tulane University, New Orleans, La, and working on his Ph. D. in Philosophy.

2013 ESOPTRICS UPDATE

NOTE ON ESOPTRICS' 4TH DIMENSION: The meaning of it is difficult to grasp, because of how widely known is Einstein's use of "4th dimension" to signify time. In Esoptrics, time is a 5th dimension and "4th dimension" signifies the result of one of the 7 characteristics to be found in every one of the ultimate units of space, matter, and locomotion found at what sensation dependent (and, thus, Geometry dependent) minds would call c. 10^{-47} centimeters (cm. for short). In Esoptrics, every one of these ultimate units is generally referred to as a "u-space" for economy's sake; but, to be more technically correct, each should be called either an actual or a potential ultra-microscopic state of excitation, because, in the final analysis, no u-space is a unit of *space* in the commonplace sense of "physically extended in 3 dimensions". In other words, no u-space has any trace of the kind of length, width, and depth with which sensation dependent (and, thus, Geometry dependent) minds are familiar. *Collectively*, they have such *in effect*; but, do not have such *individually*. One of this book's major goals is to explain how what's *individually* devoid of three-dimensionality can have such in *collective effect*, and that explanation boils down to a detailed explanation of how 6 of the 7 coordinates imparted to the u-spaces by God affect them. The 7th of those 6 coordinates accounts for the 4th dimension of the u-spaces. What it means is this: Without ceasing to have a "size" of only c. 10^{-47} cm. in dia.—according to the way sensation and Geometry depends minds speak—each u-space has anywhere from 1 to 2^{256} (***i.e.:*** 1.1579×10^{77}) different levels at which it can house anywhere from 1 to 2^{256} ultimate occupants of the Universe each logically concentric with all the others (These ultimates are called "duo-combos" in Esoptrics and, according to Esoptrics, are the same as what Science calls neutrinos.). Even where there are 2^{256} of them confined to the same 10^{-47} cm., they—whether individually or collectively—are to no extent outside of one another in space. They are only *logically* rather than *physically* separate, because each of them has a discrete level of intensity from the standpoint of the 6 other coordinates imparted to them by God. The lowest intensity of each of the 6 is X, then 2X, then 3X, then 4X, etc., until the highest intensity of each of the 6 is 2^{256}X. There are no fractional intensities. The specific number of 4th dimension levels in a given u-space is determined by the level of acceleration achieved by the "leading form" (Say "topmost one", if you prefer.) in a group of duo-combos made logically concentric in what Esoptrics calls a "multi-combo". As for what is meant by acceleration on the part of a "leading form", pages to follow shall explain that in great and mathematically precise detail, and it shall have to do with exactly what God did in creating its ultimate occupants in the Universe's first c. 10^{-18} seconds. For now, what is to be firmly kept in one's mind is: No matter how many 4th dimension levels a particular u-space has, every one of them *remains* within the confines of that particular u-space's c. 10^{-47} cm., and that's how some u-spaces can be *astronomically* more massive than others, despite the fact all u-spaces always everywhere are identical in "size".

Chapter One:

INTRODUCTORY SYNOPSIS:

"Esoptrics" derives from "*esoptron*", the ancient Greek word for our "mirror" (***cf.:*** the Greek of ***I Cor.*** 13:12 & ***James*** 1:23). Esoptrics is "The Logic Of The Mirror" because, as with a mirror, Esoptrics deals with reverse images and multiples of 2. I say that because, if I observe myself facing some particular mirror (I did not say 2 mirrors facing each other: *the* mirror ≠ *two* mirrors.), I observe this: If I hold a single object in front of that one particular mirror, I now have 2 instances of that single object: One is in my hand, the other is in the mirror, and those 2 are mirrored (***i.e.:*** reversed) images of one another. If I could take that second object out of the mirror and hold it and the original in front of the mirror, there would now be 2 sets each of 2 mutually mirrored objects and each set would be the mirrored image of the other. If I could then take that second set out of the mirror, etc.—thus multiplying by 2 indefinitely. A *dynamic* mirror would do such on its own.

"Esoptrics, The Logic Of The Mirror" is the Philosophy and Theology-based Cosmology I've been honing since 1957. It's *Cosmology* in that it seeks to surpass Physics' ability to describe the fundamental natures of time, space, matter, locomotion, gravity, and the like. It's *Philosophy-based* in that it focuses first on what the mind can learn by "mere reflection" (Einstein's phrase) upon its own operations. It's *Theology-based* in that it explains how God continuously produces: (1) the fundamental traits of the Universe's ultimate constituents, and: (2) the 7 kinds of *a priori* (prior to time & space) coordinates producing the 7 kinds of logical sequence responsible for each ultimate's unique logical dimensions.

By: (1) artfully deducing what the absolutely necessary, *a priori*, tri-partite structure of my every act of consciousness (***i.e.:*** encounterer, encountered, and the act joining them) implies about the infinite and finite versions of Being as the act of a dynamic mirror mirroring itself, and (2) then noting the math inherent to a dynamic mirror—Esoptrics is history's first and only detailed and mathematically precise explanation of:

(1) how the Universe and its ultimates are only *logically* rather than *spatially* extended (***i.e.:*** how infinitesimals are logically divisible, divided, separate & contiguous);

(2) <u>*ultimate*</u> time (***i.e.:*** u-time) vs. <u>*individual*</u> time (***i.e.:*** i-time): which makes <u>**u-time**</u>: (a) the measure of the frequency of internal change for whatever changes internally at the universally and forever fastest rate of internal change possible for all always and everywhere; (b) a *sequence* of ultimate units called alphakronons (each K for short) and defined by sensation dependent minds as roughly 7.35×10^{-96} sec. each; (c) *intermittent* since there is no u-time as long as the fastest changing is not changing internally; (d) *indivisible* since, as the measure of *change*, it measures what occurs intermittently in a *durationless instant*; but which makes <u>**i-time**</u>: (a) the measure of the frequency of internal change on the part of whatever changes internally at <u>*less*</u> than u-time's fastest rate; and (b) definable in the terms of the ratio of the number of successive u-times per one i-time.

(3) how—for any given one of the Universe's ultimate occupants—there is no *effective* moment of u-time or i-time *for it* save in the durationless instant of changing from per-

forming one temporarily changeless state of excitation to performing a different one;[1]

(4) *ultimate* spaces (***i.e.:*** u-spaces = microstates of excitation): the *physically* infinitesimal and indivisible, smallest units of space sensation dependent minds call c. 10^{-47} cm. (***i.e.:*** 1 alphatopon = $\Phi = 2^{-129}$ x the hydrogen atom's dia. with the electron in its outermost orbit), and which *collectively* add up to *physically* discontinuous 3-dimensional areas since no single u-space is *spatially* extended, and adjacent ones—per the 7 orders of logical sequence set by God's 7 *a priori* coordinates—are only *logically* rather than *spatially* adjacent;[2]

(5) what it says are the two great classes of the Universe's ultimate realities:

(a) *carrying generators* (Compare with Science's fermions.): what perform *de*tectable *micro*states of excitation (detectable because they generate units of force in the sense of tension) and which, though not spatially extended, are described by sensation dependent minds as roughly 10^{-47} cm. in dia.; and

(b) *piggyback forms* (Compare with Science's bosons.): what perform *un*detectable *macro*states of excitation and which, though not spatially extended, are described by sensation dependent minds as varying integrally from 1 to 2^{256} x 10^{-46} cm. = from 10^{-46} to 10^{31} cm. in dia. max (***i.e.:*** from 2^{-128} to 2^{128} x the dia. of the hydrogen atom with the electron in its outermost orbit);[3]

(6) how every generator generates 1 to 3 kinds of units of tension which our sensation dependent minds experience as what we see or hear or touch or taste or smell;[4]

---

[1] Here's an example of what that means: A particular ultimate (Call it Alphie.) changes from performing one temporarily changeless state of excitation to another at the rate of one change per K (***i.e.:*** every one of Alphie's states of excitation remains changeless for no more than what we would describe as roughly 10^{-96} seconds). A different particular ultimate (Call it Gea.) changes from performing one temporarily changeless state of excitation to another at the rate of one change per 10^{96} successive units of u-time (***i.e.:*** every one of Gea's states of excitation remains changeless for what we would describe as .05 seconds). For Gea, 10^{96} successive K are experienced as going by in the twinkling of an eye. For Alphie, 10^{96} successive K are experienced as going by in what we would describe as 10^{96} sec. (***i.e.:*** roughly 10^{88} of our Earth years). Esoptrics' explanation of the nature of consciousness spells out in mathematical detail how that difference is possible.

[2] Ultimately, space—whether the tiny "empty" area of a needle's eye or the whole cosmos—is a vast set of physically infinitesimal, physically indivisible microstates of excitation (***i.e.:*** there are no fractional ones) only logically outside of one another in a definite sequence, just as 1, 2, 3, etc. are concepts separate in our minds and in a definite sequence solely due to each one's unique internality.

[3] $2^{256} = 1.15792x10^{77}$; $2^{128} = 3.4x10^{38}$; and 10^{31} cm. = 17,984,400,000,000 light years = 10^{26} miles.

[4] For Aristotle and St. Thomas Aquinas, et al, material substances (***ex. gr.:*** this man, this horse, this rock) are a combination of matter and form. Such a theory is called Hylemorphism (***a/k/a:*** Hylomorphism). For Esoptrics, material substances are a combination of matter, form, *and the generators of matter*. Such a theory can be called Hyle*gene*morphism. Hylemorphism views every material substance as a combination of *two*, but Esoptrics as a combination of *three* principles. For Esoptrics, matter, *stricte dicta*, = the units of tension generated by the generators in the patterns dictated by the forms. For Esoptrics, the "stuff" of which both matter and our sense images are made = the units of tension generated by the generators. That's why, when a tree falls, the sound of it all is still there whether or not anyone hears it, because the generators associated with the tree generate the units of tension which are the very "stuff" of which sound is made regardless of whether or not it is present in an ear. For St. Thomas, the difference between the form of a material substance and the form of an immaterial substance (***i.e.:*** angel) is that the former is *"signed with matter"*. For Esoptrics, the difference is that the former *is*, and the latter *not*, *logically concentric with a generator of matter,* which is to say the former is, and the latter not, hypostatically united to a carrying generator, and that is why the former *can*, and the latter *not* engage in locomotion *stricte dicta*.

(7) how each generator is forever logically concentric with a particular piggyback form in a *duo-combo* which, due to that union, goes wherever its carrying generator goes as the latter regularly makes what sensation dependent minds wrongly dub instantaneous quantum "leaps" of 10^{-47} cm. as it goes from one u-space (***i.e.:*** from performing one micro-state of excitation) to one logically sequential to the prior one according to God's 7 kinds of logical sequence;[1]

(8) how the piggyback forms are what make instant action at a distance possible;

(9) how 6 of the 7 kinds of logical sequence determine which 26 micro states of the same OD level are logically adjacent to the one a generator is currently performing;

(10) how the 7th kind determines which 2 macrostates (***i.e.:*** one up vs. one down in intensity) are logically adjacent to the one a piggyback form is currently performing;

(11) how the piggyback forms, by changing internally at different rates, produce the kind of i-time we experience as universally the same, indefinitely divisible flow;

(12) how, for every form, i-time is relative to its rate of change from one temporarily changeless state of excitation to another at a rate slower than the highest possible rate (***i.e.:*** that u-time sensation dependent minds describe as 1 change per c. 10^{-96} sec.);

(13) how u-spaces, with the forms, collectively produce, as an effect, the kind of spatial extendedness and locomotion we experience as continuous and indefinitely divisible;

(14) why there are electrons, why they are points but ingest and emit neutrinos as they rise and fall in energy level, and why they orbit, instead of occupy, the atomic nuclei;

(15) how God created all the Universe's ultimate constituents in its first 10^{-18} sec., and thus gave the Big Bang its bang;

(16) and how the carrying generators, together with the piggyback forms, produce dark matter, electromagnetic radiation, electrons, hydrogen nuclei, hydrogen atoms, black holes, planets, stars, galaxies, clusters of galaxies, clusters of clusters, and all in a Universe expanding outward to what sensation dependent minds would call roughly 18T + 18T + 9T + 4.5T + 1.124T + .0703T + .000274T + .00 000 000 42T light years, where T=1 trillion.[2]

The 6th thru the 8th members in the sequence just described constitute what Esoptrics calls an alpha area of the Universe. There are many alpha areas each expanding to 70.275 billion lt. yrs., as their centers travel outward in many directions from the center of one of the many "beta" areas each expanding to 1.124T lt. yrs. as their centers travel outward in many directions from the center of one of the many "gamma" areas each expanding to 4.5T lt. yrs. etc.. Of the Universe, we detect nothing more than that particular one of its many alpha areas which we inhabit, and Esoptrics—in its explanation of dark matter—gives a rather detailed explanation of why that is so.

Esoptrics is history's first and only theory to explain how Newton's first law of motion does not apply to the Universe's *ultimates*. Newton says every body continues in its state of rest, or uniform motion in a straight line, unless forced to change its state by forces exerted upon it. Esoptrics says every *ultimate*, at the level Science calls c. 10^{-47} cm., continues at its current *rate* of change (***i.e.:*** NK per change) from temporarily changeless state of

---

[1] In Esoptrics, there are forms not logically concentric with a carrying generator. Each is a species of angel, and, because there are 2^{256} species of them, there are 2^{256} species of angels and, probably, one angel per species as St. Thomas says. Not being logically concentric with a carrying generator, all angelic forms forever remain concentric with one another at the Universe's center. For all of that, they remain distinct from one another because each is at a discrete level of what Esoptrics calls "ontological distance", "OD" & "the 4th dimension". No "theory of everything" is truly such, unless it explains how angels fit into the Universe, and only Esoptrics explains that or even *tries* to explain it.

[2] In the above sequence, the progression is: 18T + $(18T/2^0)$ + $(18T/2^1)$ + $(18T/2^2)$ + $(18T/2^4)$ + $(18T/2^8)$ + $(18T/2^{16})$ + $(18T/2^{32})$. The sequence among the exponents is thus: 0, 1, 2, 4, 8, 16, and 32.

excitation to another and in its current pattern (as set by God's 7 kinds of logical sequences), unless forced to change that rate and/or pattern by forces exerted upon it. Newton vs. Esoptrics = tendency to stay with the current state vs. with the current rate.

For Esoptrics, that Newton's law applies to what we observe, with our senses and/or scientific instruments, is merely an effect produced by the fact that what happens at 10^{-47} cm. is much smaller than anything Science can observe. Because it cannot be proven by observation that Newton's law holds even at the level of 10^{-47} cm., it cannot be proven by observation that Esoptrics' principle is incorrect. More importantly, as long as Science cannot extend its observations to segments of space and motion as small as 10^{-47} cm., Newton's first law of motion remains merely an *assumption* which—as were the convictions that Earth is flat and at the center of the orbits of the Sun, the Moon, and the stars—is no more than *suggested* by what we can observe. It's strictly an *inference* and—like anything smaller than 10^{-15} cm. or 10^{-18} sec.—is far from being an *empirical fact*.

For Einstein, *space curves* near matter. For Esoptrics, matter itself is the result of *space thickening* as the 4th dimension of one or more *u-spaces increases in number of levels*, because the number of ultimates simultaneously in the same one u-space (***i.e.:*** the number of duo-combos logically concentric in the same multi-combo) increases as the square root of the multi-combo's mass and force of gravity. The latter two thus vary as the square of the number of duo-combos concentric with one another in the same multi-combo at the same u-space's center, and the force of gravity diminishes as the inverse of the square of the distance from the u-space's center in number of logically sequential u-spaces.

Every u-space can "thicken" due to the 4th dimension Esoptrics calls "ontological distance" and "OD" for short. OD means: Though every generator's every microstate has the same, one, minimum *level of intensity*, their piggyback forms do not. A form's macrostate of excitation may have one of 2^{256} *levels of intensity* describable as 1E, 2E, 3E . . . . . 2^{256}E, where E= the least intensity. Thus, up to 2^{256} duo-combos may be concentric in the same multi-combo and u-space; and yet, they're distinct from one another since each form is at a different OD (***i.e.:*** each one's macrostate has a discrete intensity) thus putting it and its generator at a unique level of the same u-space's 4th dimension (***i.e.:*** each generator is performing the same microstate but at a different level of that u-space's 4th dimension.).

A multi-combo that "thick" would be the Universe's most massive occupant. Its mass would be c. 2^{58} (***i.e.:*** 2.9×10^{17}) times our Sun's (***i.e.:*** $[2^{256}]^2$ vs. $[2^{227}]^2$ if the Sun's main multi-combo = c. 2^{227} duo-combos) and c. 10^7 times the biggest known black hole.

Perhaps, in the final analysis, Esoptrics' main achievement lies in what it says regarding consciousness. No scientific theory known to me offers—or even *tries* to offer—even so little as the slightest *trace* of an explanation of why, on their own, our minds are: (1) limited to observing a realm in which everything is at least many times larger than the hydrogen atom, and (2) cannot come close to observing what occupies the sub-atomic realms. Esoptrics, though, in its mathematically detailed explanation of how consciousness works, gives far more than a mere *trace* of such an explanation.

> . . . . metaphysics is of no use in furthering output of experimental science. Discoveries and inventions in the land of phenomena? It can boast of none. Its heuristic value, as they say, is absolutely nil in that area (pgs. 3-4).
> . . . . we note that philosophy does indeed provide a deductive science of corporeal being, but that it is incapable of providing a deductive science of the phenomena of nature (pg. 38).
> . . . . The ancients thought their philosophy of nature was a science of the phenomena of nature. That was their misfortune (pg. 175).
> ——**JACQUES MARITAIN:** ***The Degrees Of Knowledge*** on the indicated

pages in the hardbound edition by Charles Scribner's Sons, New York, 1959.

There are countries where no professor of any science could hold his job for a month if he started teaching that he does not know what is true about the very science he is supposed to teach, but where a man finds it hard to be appointed as a professor of philosophy if he professes to believe in the truth of the philosophy he teaches. The only dogmatic tenet still held as valid in such philosophical circles is that, if a philosopher feels reasonably sure of being right, then it is a sure thing that he is wrong, because it is of the very essence of philosophical knowledge merely to express "a certain attitude, purpose and temper of conjoined intellect and will."
——**ETIENNE GILSON:** ***Being And Some Philosophers***, as given on pages viii and ix of the hardbound copy published by Garden City Press Co-operative, Toronto, Canada, 1949.

But some have maintained that Metaphysics, considered as a quest for real explanations, is a fool's errand. Kant, writing in 18th Century Germany, labeled it a "transcendental illusion." Since Kant, many thinkers have taken it for granted that Metaphysics can at best achieve a subjectively convenient construction—a schematization—of the pattern of reality. To ask it to provide an objective grasp of the way things really are is to demand the impossible.
——**AVERY DULLES S. J., JAMES M. DEMSKE, S. J., & ROBERT J. O'CONNELL, S. J.:** ***Introductory Metaphysics*** as given on page 9 of the hardbound edition by Sheed & Ward, New York, 1955.

Is philosophy totally useless, then, as compared with science? Yes, it is, if we confine ourselves to the use of knowledge or understanding for the sake of producing things. Philosophy bakes no cakes and builds no bridges.
——**MORTIMER J. ADLER:** ***Aristotle For Everybody*** as given on page 66 of the hardbound edition published by Macmillan Publishing Co., Inc., New York. ©1978 Mortimer J. Adler.

The synthetic propositions that make up the content of the sciences can be known only on the basis of experience; science depends ultimately, and in a fundamental way, on observation. This is the *empiricism* of analytic philosophy, central to the attack on traditional metaphysics, carried on, for instance, by Reichenbach. In his interpretation, metaphysicians like Kant took for granted the existence of synthetic *a priori* propositions—that is, propositions which can be known independently of experience, but which nevertheless are not mere tautologies. The task of the metaphysician was only to discover which they are, and to explain how it is that we can come to know them. It is this conception, Reichenbach argued, that gave philosophy a deservedly bad name among scientists, for the philosopher pretended that his armchair speculations, if sufficiently profound, could answer questions about the nature of things, while the scientist, asking much more modest questions, was constrained to go into his laboratory for experiment and observation.
——**ABRAHAM KAPLAN:** ***The New World Of Philosophy*** as given on page 81 of the Vintage Book published by Alfred A. Knopf, Inc. and Random House, Inc., New York. ©1961 Abraham Kaplan.

During philosophy's childhood it was rather generally believed that it is possible to find everything which can be known by means of mere reflection. It was an illusion which anyone can easily understand if, for a moment, he dismisses what he has learned from later philosophy and from natural science; [Every *first principle* is not the same as "everything" – ENH.]
——**ALBERT EINSTEIN:** ***Ideas And Opinions*** on pages 19&20 of the hardbound edition by Bonanza Books, New York, 1954. ©1954 Crown Publishers.

The Aristotelian tradition also held that one could work out all the laws that govern the universe by pure thought: it was not necessary to check by observation. So no one until Galileo bothered to see whether bodies of different weight did in fact fall at different speeds.
——**STEPHEN HAWKING:** ***A Brief History Of Time*** page 15 of the paperback edition of the updated and expanded tenth anniversary edition as published by Bantam Books, New York, 1998. ©1988 & 1996 Stephen Hawking.

In the eighteenth century, philosophers considered the whole of human knowledge, including science, to be their field and discussed questions such as: did the universe have a beginning? However, in the nineteenth and twentieth centuries, science became too technical and mathematical for the philosophers, or anyone else except a few specialists. Philosophers reduced the scope of their inquiries so much that Wittgenstein, the most famous philosopher of this century, said, "The Sole remaining task for philosophy is the analysis of language." What a comedown from the great tradition of philosophy from Aristotle to Kant.
——**STEPHEN HAWKING:** ***op. cit.*** pgs. 190 & 191.

In the above 8 quotes: For Einstein, Philosophy is "mere reflection" and "pure speculation". For Hawking, it's "pure thought". For Kaplan, it's "armchair speculation". That is certainly what Esoptrics is as I go about trying to uncover the most fundamental principles (***i.e.:*** ultramicroscopic makeup) of time, space, matter, locomotion, gravity, etc., by focusing almost exclusively upon the innermost depths of my own mind. Has that been a "misfortune" (Maritain), a "fool's errand" (Dulles), an "illusion" (Einstein), "totally useless" (Adler), a "sure thing that" I am "wrong" (Gilson), or a pursuit of "the impossible" (Dulles)? By no means! On the contrary, I dare to assert that what Esoptrics' introspective method has uncovered is impressive enough to explode most totally all the many attempts (such as the above) to laugh Philosophy off the stage of Cosmology. After all Esoptrics has accomplished, let's see who laughs last: Science or Philosophy?

Gradually the conviction gained recognition that all knowledge about things is exclusively a working-over of the raw material furnished by the senses. In this general (and intentionally somewhat vaguely stated) form this sentence is probably today commonly accepted. But this conviction does not rest on the supposition that anyone has actually proved the impossibility of gaining knowledge of reality by means of pure speculation, but rather upon the fact that the empirical (in the above-mentioned sense) procedure alone has shown its capacity to be the source of knowledge.
——**ALBERT EINSTEIN:** ***Ideas And Opinions*** on pages 19&20 of the hardbound edition by Bonanza Books, New York, 1954. ©1954 Crown Publishing.

Well, my dear, highly and rightly revered sir, let me assure you that, with the coming of Esoptrics, it is no longer a *fact* that "the empirical procedure alone has shown its capacity to be the source of knowledge" regarding the Universe's ultramicroscopic level.

> What a great practical joke nature will have played on us if all the thinking that has gone in to discover the ultimate laws of the universe turns out to reveal that one of the biggest clues was woven all along into the very fabric of thought itself.
> ——**DAVID H. FREEDMAN:** ***Quantum Consciousness*** as found on page 98 of the June, 1994, edition of ***Discover*** magazine.

Guess what, Mr. Freedman! I dare suggest to you quite confidently that, with Esoptrics, that clue has now burst upon the world with a deafening roar.

> Continuing our study of knowledge in general, we discover another principle which is too often overlooked even in Thomistic textbooks. "The more perfect knowledge is, the more it comes from within," and vice versa, "The more knowledge comes from within, the more perfect it is."
> ——**J. F. DONCEEL, S. J.:** ***Philosophical Psychology***, pg. 169 of the hardbound edition by Sheed & Ward; New York, 1965. ©1961 Sheed & Ward.

How true it is, Fr. Donceel! For, with what insights it has developed by focusing within, if Esoptrics tells us anything, it most loudly trumpets: "The more knowledge comes from within, the more perfect it is" at least with regard to first principles, which is to say with regard to the Universe's ultramicroscopic makeup.

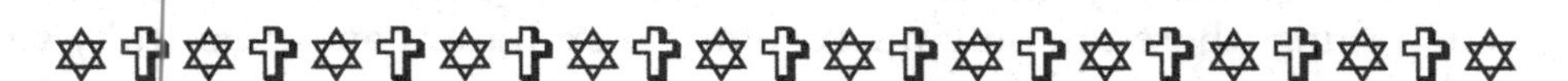

REMEMBER: In Esoptrics:
$7.34683969 \times 10^{-47}$ cm. = 1 u-space = 1 alphatopon = 1Φ = our
Sensation dependent minds' way to describe *physically* infinitesimal
But *logically* divided microstates (a/k/a ultramicroscopic states) of excitation
Each being, or able to be, performed by a carrying generator;
$7.201789375 \times 10^{-96}$ sec. = for us, 1 ultimate unit of u-time =
1 alphakronon = 1K for short & 1 second = $1.388543802 \times 10^{95}$K &
Measures internal change's fastest rate but i-time all slower rates;
1 centimeter = $1.361129468 \times 10^{46}$Φ;
Speed of light = $1\Phi/2^{128}$K = 29,979,245,800 cm./sec.;
Ontological distance = OD = the Universe's 4th dimension (time = the 5th) =
Which discrete level of intensity, out of 2^{256}, is currently being utilized
By a given piggyback form in each of the 6 states of excitation
Composing its current *Macro*-state of excitation;
The Universe's every ultimate *constituent* is either a generator or a form;
Its every ultimate *occupant* is: one form forever logically concentric with
One generator in an indestructible duo-combo in which
One of each of those 2 kinds of ultimate *constituents* are eternally joined.

Let no one suppose, however, that the mighty work of Newton can really be superseded by this or any other theory. His great and lucid ideas will retain their unique significance for all time as the foundation of our whole modern conceptual structure in the sphere of natural philosophy.
——**ALBERT EINSTEIN:** ***What Is The Theory Of Relativity*** as published Nov. 28, 1919, in ***The London Times*** and found on page 232 of the hardbound copy of ***Ideas And Opinions*** published by Bonanza Books, New York, 1954 (I, of course, would add that Newton's first law of motion also serves as "the foundation of our whole modern conceptual structure in the sphere of" *atheistic* Science. For, let us suppose it is true what Newton effectively says—namely: Until a force external to them [but not external to the Universe] causes a change, then: (1) whatever is at rest remains uninterruptedly so, and (2) whatever is in motion is uninterruptedly so. As long as that remains true, then the Universe is totally self-sufficient and no more needs God than it needs a tooth fairy. Esoptrics, however, puts forth two lines of thought utterly lethal to Newton's first law. **ONE:** All rectilinear locomotion = pauses interspersed with durationless instants of change as generators—without passing thru any smaller segments of *space* [For, at c. 10^{-47} cm., there is no such thing; there are only microstates of excitation astronomically vast in number and thus making the ultramicroscopic makeup of space hetero- rather than homogeneous.]—periodically execute a series of what our sensation dependent minds would erroneously describe as instantaneous quantum leaps of 1Φ per X number of successive K—the number of which varies inversely as the velocity of the generator's locomotion. **TWO:** there cannot possibly be anything real in this Universe save what is ultimately either currently an excited state of excitation or what can be transformed into such. But, unless there is, from *outside* the Universe, a *divine* influence setting up and continuously maintaining the 7 kinds of logical sequence, there cannot possibly be either: (1) even so little as a *single* state of excitation, let alone *myriads of myriads* of them logically outside of one another, or: (2) that logically sequential arrangement of "u-spaces" [***i.e.:*** physically but not logically infinitesimal or logically indivisible microstates of excitation] which alone makes possible the kind of locomotion we observe. Always remember: In Esoptrics, "u-space" is short for "microstate of excitation" and "ultramicroscopic makeup of space." No u-space is *really* a unit of *space.*).

Chapter Two:

THE INTROSPECTIVE PATH TO THE DISCOVERY OF GOD'S *A PRIORI* COORDINATES:

What methodology does Esoptrics utilize in order to arrive at its conclusions? That brings us to the sense in which Esoptrics is a *Philosophy* based Cosmology. Science based cosmologies derive their principles by means of an extroverted examination of what's generally regarded as *extra*-mental realities. Esoptrics derives its principles by means of an *in*troverted examination of what's as *intra*-mental as intra-mental can be. Thus, Esoptrics is not even so much concerned with sense images as it is with what's the most basic structure (say "nature" or "design", if you prefer) of the act of consciousness itself. How so?!

In the strictest sense of "immediate", I immediately experience (***i.e.:*** encounter, come face to face with, and gaze upon) nothing but my own experiencing self and its states. But, whether recalling a past state or pondering a briefly present one, it is obvious to me that I am face to face with something whose inescapable, absolutely necessary, *a priori* structure (vs. proposition) is that of a triousious *object*. That's a whole undivided within itself (***i.e.:*** my body parts aren't in different rooms) but divided from all others (***i.e.:*** my body's not permanently stuck to my furniture); and yet, its extremities enclose 3 *subjects* inextricably bonded together. Rather than being a *part* of their same, one *object*, each of the 3 must be classed a *subject* and *co-tenant* within their same one whole, much as Siamese triplets (were such possible) inextricably joined would be one whole object but still three subjects so separate in identity, that to call them *parts* of the whole would be misleading.[1]

First, there are the direct encountereds (***i.e.:*** the sense images) and the encounterer (***i.e.:*** what's directly aware of the sense images but unable to be directly aware of itself). The latter is so different from the former, the two can only be described as two kinds of "subjects" so different from one another they are radically different categories of reality in no way a mere characteristic or part of the other. Neither can be predicated of the other; and yet, neither can be split from the other in fact, however much they can be separated in thought, fancy, imagination, etc.. Above all, like Siamese twins, neither can rightly be called a mere *part* either of the other or of the whole produced by their union.

Next, there is what Esoptrics calls "the prime abstractable". The term refers to the all-pervasive continuousness which seems to underlie all motion thru time and space on the part of consciousness and the things of which I'm aware. Following Kant, Esoptrics decides to try assuming that such continuousness is a characteristic limited to the Encounterer himself (***i.e.:*** the "spectator" in Kant's terminology), but which he projects onto virtually everything else he's directly encountering. Esoptrics then refers to the prime

---

[1] In Esoptrics, "subject" and "substance" signify what, *like* accidents, can have characteristics but, *un*like accidents, is *not* merely the characteristic or part of another. As Aristotle would say it: that of which everything else is predicated *while it is not itself predicated of anything*.

abstractable as the "reflex encountered" and deems it to be what the Encounterer knows of himself *on the rebound* as the result of projecting his continuity onto the direct encountereds (***i.e.:*** the sense images).

Next, once it's suggested the prime abstractable, too, is some kind of reality, then—*on the face of it!*—it's so different from the encounterer and the direct encountereds, it must be a *third* category of reality radically different from the other two. It cannot be predicated of the sense images and cannot be predicated of the Encounterer save as some kind of "after effect" of its act of encountering. Above all, as with one of 3 Siamese triplets, neither is it rightly called a mere *part* of the whole produced by its union with the other two.

This bit—bit of three radically different kinds of reality inextricably bonded together within the confines of the same one whole—is the absolutely unavoidable structure of my every act of consciousness. After all, there can be no such thing as an act of consciousness, save where there are: (1) that which is being aware, (2) that of which the former is being aware, and (3) the act of consciousness joining the two. But, there is no such thing as awareness or observation or confirmation by observation save by means of an act of consciousness; and so, I can never observe or be aware of anything save that which is one of the three subjects (***i.e.:*** co-tenants) present within the confines of my acts of awareness.

Therefore, if there is such a thing as that which is *not* one of the three subjects within the confines of the same, one object, then I cannot confirm the existence of it *by observation*. For, the only way I can *observe* it is to make it one of the three subjects within the confines of the same one object. On the face of it, that would make such a sought-after exception the very opposite of what I was trying, by observation, to prove it to be. In short, it is absolutely impossible for me to prove *anything* by *observation* save the universality of the triousious structure I observe in my every act of consciousness. That being the case, no conclusion is reasonable or in conformity with what I incessantly observe save this:

THE PRIME CONCLUSION:
THE NECESSARY, TRIOUSIOUS STRUCTURE (*i.e.:* 3 subjects in 1 object)
OF EVERY ACT OF CONSCIOUSNESS
IS THE NECESSARY STRUCTURE OF EVERY TRULY BASIC REALITY.

By "basic reality", I mean one which cannot in *fact* (vs. in *fancy, imagination, speculation, theory*, or the like) be broken up either into smaller pieces (***ex. gr.:*** as a boulder can be smashed into pebbles) or into more basic realities (***ex. gr.:*** as a molecule can be reduced to atoms, or an atom reduced to leptons and quarks). "Basic reality" thus means an *ultimate* reality. That doesn't mean—as some like Spinoza mistook Aristotle to say—a reality so self-sufficient, it needs no kind of assistance from, or connection with, anything else. It merely means what, unlike every accident, doesn't need to be *intimately bonded* to another to continue existing. *Intimate bonding* is the only dependency "basic reality" excludes.

How, then, is such a tri-partite structure the inescapable structure of every basic reality? Long ago, I pondered the prime abstractable and asked myself: "How can that which is not activity be conscious of activity? How does the encounterer encounter a property of himself which—if the encounterer is something other than activity—must be radically unlike himself? How can any encounterer encounter an act of encountering unless he himself can be described as an act of encountering?" Thus was I driven to conclude this: That which actually moves is itself motion or, as I *prefer* to express it, actuality is what acts.

"But," my mind complained, "how can motion be what moves? To have motion, you must have something moving." I answered myself: "The only way you can grasp the answer to that question, is to encounter *real* motion *itself*. Come face to face with the internal characteristics of *real* motion, and you will see how it is possible that motion is, indeed, a real thing capable of being that which is moving."[1]

From that realization, it followed that the prime abstractable is by no means merely a projection of a characteristic of a kind of activity *associated with* the encounterer. Utterly to the contrary, it's a projection of the very encounterableness of the encounterer.[2] In other words, instead of being a projection of something the encounterer is merely ***doing***, it is a projection of something which the encounterer is also ***being***. In the encounterer's case, to ***do*** a kind of motion and activity characterized by smoothness and continuousness is to ***be*** a kind of "act" and "actuality" characterized by smoothness and continuousness.

How does that projection take place? Is it a thing or an illusion? What is an illusion? Is it a nothingness? My mind once suggested: "You're merely noting the difference between an older versus a newer memory of the same encountered." I replied: "That sounds like being aware of the hole in a doughnut without the doughnut being present."

Long ago, an almost unbelievable answer came to me: The encounterer is an activity best described as the activity of a mirror so *dynamic*, its very nature is to mirror itself. By simply *being* himself, the encounterer thus engages in the act of trying to mirror something, and the thing which he tries to mirror is himself. He's trying to mirror himself in an attempt to come face to face with himself. But, he does not mirror himself with infinite efficiency. As a result, he gets an inside-out picture of himself and, thereby, comes face to face with an encounterableness (***i.e.:*** a set of internal characteristics) *radically* different from the one he set out to encounter. In other words, imperfectly mirroring the activity which he is, he begets a distorted, reversed, negated image of himself—provided that by "negated" you mean "made negative" rather than "annihilated".

But, what is *having* and *exhibiting* this reversed encounterableness? Clearly, it cannot be that the same one, unencountered encounterer—while remaining the same, one encounterable—has two radically different encounterable*nesses* (***i.e.:*** two radically different sets of internal characteristics). The answer is inescapable: It must be that, in mirroring himself with less than infinite efficiency, the encounterer somehow produces a second encounterable with three indispensable features: (1) Its encounterableness is radically dif-

---

[1]In Book One of his ***Summa Contra Gentiles***, Chap. 13, Par. 10, St. Thomas Aquinas contrasts how Plato and Aristotle use the word "motion". Plato, he says, uses it to refer to *any* given operation—even mental acts. For Aristotle, says St. Thomas, "motion" refers only to the kind of operation which "belongs only to divisible bodies" (***Vd.:*** pg. 88 of Vol. 1 of the University of Notre Dame Press edition of 1975). For Aristotle, then, "motion" denotes those changes thru time and space our sense images detect, and that's mainly locomotion. Needless to say, *such* "motion" cannot be that which moves. For "motion" to signify what moves, it must—somewhat as it does for Plato—signify an *operation* and, at that, one of a very unusual kind. In Book III of his ***De Anima***, Chap. 7 (431a 5), Aristotle says *motion* is different from *activity*, since, "in the unqualified sense", activity belongs to "what has been perfected" (***Vd.:*** pg. 663 right of Vol. 19 of the ***Great Books Of The Western World***). Unlike myself, most people will *insist* "motion" be used the way Aristotle uses it. As a result, some will insist that, in the above, I am, like Bergson and Heraclitus, claiming that change is the only real thing. Well, I am *not* using it Aristotle's way. I'm using it, it seems, the way Aristotle uses "activity". If, then, it confuses you to hear me speak of motion moving, then change it to read "activity is what acts" or "act is what acts" or "actuality is what acts". However difficult it is for sensation dependent minds to do so, one must cease thinking of ultimate realities as little lumps of inert, billiard-ball rigidity (***i.e.:*** "little globules" in Locke's terminology) and learn to think of them as so inherently dynamic, every iota of their internality is some kind of "whirling dervish" whose whirl, nevertheless, undergoes no kind of internal change, save when, occasionally, the dervish suddenly stops whirling with one discrete level of intensity and starts to whirl with a different one (***Vd.*** Ftn. #1, my pg. 214).

[2] "Encounterableness" is my way of referring to some subject's internal characteristics (***i.e.:*** internality). Looking in a mirror, you see a copy of your face's encounterableness, essence, and internality.

ferent from that of the encounterer's actuality itself, and (2) it is annihilated, unless (a) the encounterer continues to produce it by continuing to mirror himself imperfectly, and (b) it shares at least some of its extremities with those of the encounterer's actuality.

Let the prefix "non" mean "made negative", and we can say this: The prime abstractable is a non-reality, a non-thing, and a gossamer, almost unreal, mirrored image of the encounterer. Still, this non-thing is not another reality *completely separate from*, and *independent of*, the encounterer. Somehow it's another "side" of the encounterer as the latter serves as the *positive* reality projecting this *negative* one. In short, the encounterer is—simultaneously, and within the confines of a same one *whole* sharing none of *its* extremities with (***i.e.:*** not intimately bonded to) another *whole*—two radically different realities. Thus, though this object and whole itself is to no extent either bonded to another or internally divided in fact, it is two radically divergent kinds of reality inextricably bonded together and each having an internality so different from that of the other, we *must* make these contrasts: If one be called a reality, then the other must be called an *anti*-reality; if one be called a kind of "Being", then the other must be called a kind of "*Anti*-Being"; and, if one be called substantial, then the other must be insubstantial. One is parasitic and the other not even as the latter incessantly projects its parasitic companion and co-tenant.

So greatly do the encounterer and his anti-self differ, we can refer to them as two different *subjects* inextricably coupled within the confines of the same, one whole which is, itself, to no extent inextricably affixed to another whole. It's by mirroring himself imperfectly that the encounterer produces a "*polyousious* object (***i.e.:*** a *multi-subject* whole)" so profoundly sealing two subjects together, one of them (***viz.:*** the anti-self) *inheres* in the other, somewhat after the manner in which a fetus perishes if it ceases to be intimately bonded to its mother.[1]

If the encounterer is the *mirroring* activity of a *dynamic* mirror, all of existence's fundamental laws may reduce to the way a mirror operates. The great key, then, may be some kind of "Logic Of The Mirror". Using the Greek word for "mirror" I call that logic "Esoptrics". That logic most immediately suggests to me these two postulates:

FIRST POSTULATE

EVERY FINITE ENCOUNTER IS—SIMULTANEOUSLY AND WITHIN THE SAME ONE WHOLE'S CONFINES —AN ENCOUNTERER AND HIS NEGATIVE ANTI-SELF PROJECTED BY HIS ATTEMPT TO ENCOUNTER HIMSELF BY MIRRORING HIMSELF.

SECOND POSTULATE

---

[1] In Esoptrics, "to inhere" implies a *parasitic* substratum—a kind of "stuff", subject and substance which, to avoid annihilation, must ever remain *intimately bonded to* (***i.e.:*** share at least some of its extremities with) a non-parasitic substratum. A non-parasitic substratum is a kind of "stuff", subject and substance which does not ***need*** to be intimately bonded to (vs. in any way dependent upon) another in order to avoid annihilation. A parasitic substance is still a substance, because it is not merely a characteristic of the non-parasitic substance to which it must ever remain bonded. Esoptrics' distinction between parasitic and non-parasitic substances never occurred to Aristotle. For him, besides being not predicable of another, every substance is not *present* in another, and by "present in another", he means—as with characteristics (***i.e.:*** accidents)—incapable of existence apart from presence in that other (***Categories***, Chap. 2; 1a:22-23). That second part of his definition of a substance rules out the possibility of a *substance* and *subject* incapable of existence *apart from intimate bonding to another*. I do not thereby say—as some mistook Aristotle to say—"*apart from any and every kind of dependency upon" another*. "Not bonded to" ≠ "in no way dependent on".

IN THE REALM OF FINITE NON-INHERING REALITIES, MIRRORED IMAGES ARE NOT EQUALLY REAL AND, THEREFORE, ARE RADICALLY DIFFERENT FROM ONE ANOTHER FROM THE STANDPOINT OF THEIR ENCOUNTERABLENESS AND KIND OF BEING.[1]

Let's give a figurative explanation of what I mean by an *infinite* encounter, and maybe the picture shall become more intelligible. An infinite encounter is one involving an infinite encounterer. An infinite encounterer is one who engages in the act of encountering with infinite intensity. His act of mirroring himself—in an attempt to encounter himself—involves the *full* development of all his power (Say "energy", if you prefer.) and is, consequently, a *pure act*. If you do not mind metaphors cast in self-contradictory (paradoxical?) language, we can express the result like this: In a manner of speaking, he emits a beam of light traveling at infinite velocity; and so, the beam of light covers an infinite distance in an infinitely small period of time. That means it *circles* infinity *instantly*, comes back to itself unchanged (***i.e.:*** is still "pointed in the same direction"), and encounters itself in the very act of being emitted (***i.e.:*** emission & reception are simultaneous). Such can be expressed in the image below where "E" represents the infinite encounterer emitting the beam, and "I" represents the opposite "ends" of infinity. The arrow to the right of E signifies the beam being emitted, and the arrow to the left of E signifies the beam being received.

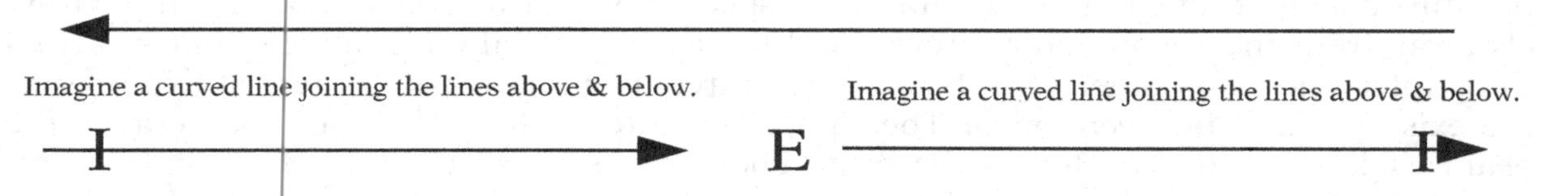
Imagine a curved line joining the lines above & below.
Imagine a curved line joining the lines above & below.
I
E
I

DRAWING #2-1: THE INFINITE ENCOUNTER.

Thus, The Infinite Encounterer encounters a positive, non-reversed, non-negated "picture" of Himself. He thus encounters the internality of His act of encountering exactly as that internality is. By His fundamental activity as a dynamic mirror, He causes Himself to observe no internal characteristics but those of His own fundamental activity itself. Oppositely, the finite encounterer—by his most fundamental activity (***i.e.:*** his "primary act", act of Being" and "act of existence")—causes himself to have *observed* internal characteristics radically different from the unobserved internal characteristics of his fundamental activity itself. As a result, the observed internality of the *finite* encounterer is purely an ethereal mask continually projected by his face; whereas, the observed internality of the *in*finite encounterer is forever His infinitely real face Itself. We can describe the latter result as:

ENCOUNTERING IS ENCOUNTERING ENCOUNTERING.

That means encountering is *both* the encounterer, *and* the encountered, *and* the act uniting the two. You might say we have one act but three activities, or one act but three functions. Thus, at one and the same instant, and within and throughout the confines of

[1] I formulated the above principle slightly before the news broke that Tsung Dao Lee and Chen Ning Yang had won the 1957 Nobel Prize for overturning the law of parity (Of which, I knew nothing.) by proving that, for basic particles, mirrored images are not equally real. It's ever sounded so similar to my principle, it's ever left me wondering if perhaps a rank amateur's *armchair introspection* has *deduced* what 2 scientists could only *induce* using a methodology the opposite of "*mere reflection*".

the same, one act, the *pure* act of *infinite* encountering produces, ***is***, and stands within, three frames of reference as the act of encountering encounters its act of encountering.

The remarks I have made about the First Encounter allow us to formulate what I call the third and fourth postulates of Esoptrics. I state them thusly:

THIRD POSTULATE

IN THE REALM OF THE INFINITE ENCOUNTER, EVERY ENCOUNTERER IS—AT ONE AND THE SAME INSTANT, AND WITHIN THE CONFINES OF THE SAME, ONE, WHOLE—BOTH HIMSELF, HIMSELF MIRRORED (THOUGH NOT NEGATED), AND THE ACT UNITING THE-TWO.

FOURTH POSTULATE

IN THE REALM OF THE INFINITE ENCOUNTER, MIRRORED IMAGES ARE EQUALLY REAL AND, SO, IDENTICAL FROM THE STANDPOINT OF THEIR ENCOUNTERABLENESS AND KIND OF BEING.

Contrast the preceding with the *finite* encounterer. Failing to develop all his power, he emits a beam of light at less than infinite speed. For that reason, the beam simply travels away from the encounterer forever, and the act is eternally incomplete, unless something intervenes to boomerang it back to its source. For that reason, we must hypothesize the existence and intervention of The First Encounterer. It is He Who boomerangs the beam back to its source. But, as the beam boomerangs back, it is turned around and reversed. That is to say, when it is reflected back by The First Mirror (***i.e.:*** God), what comes back is not equally real with regard to what went out. That can be expressed in the *two-stage* figure below (Note that the *infinite* encounter involves only *one* stage. That's very important.). The first stage, of course, signifies the *emission* of the light beam, and the second stage signifies the *subsequent reception* of the light beam.

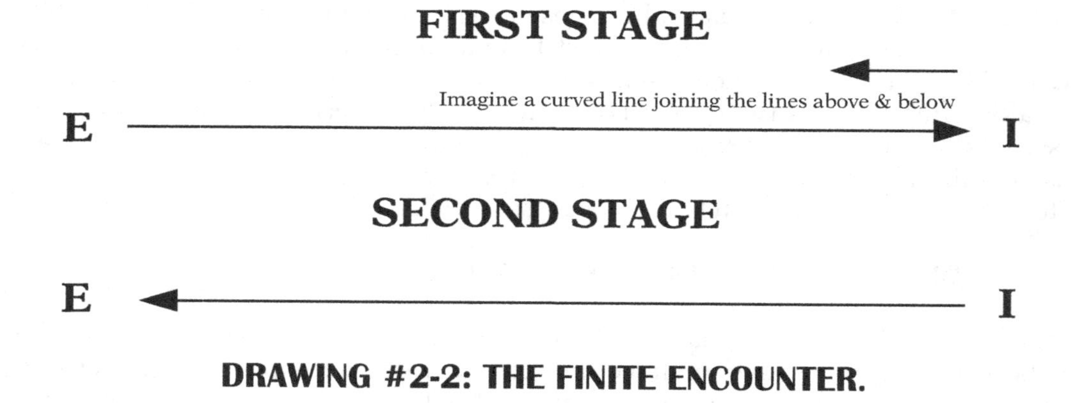

DRAWING #2-2: THE FINITE ENCOUNTER.

Returning to its source, the beam is negative (reversed), and the encounterer encounters a mere ghost of himself. That's why real motion appears to us as nothing more than a gossamer, empty smoothness underlying the encountered transitions (***i.e.:*** *secondary* acts) of the direct encountereds. That's why the *reflexively* encountered *in*substantial internality of motion is radically different from the *substantial* internality it has in the encounterer's actuality (*i.e.:* in his primary act). As St. Thomas Aquinas might put it: In The Infinite E, God, essence and existence are the same; whereas, in the finite E they are not, because The Former does, but the latter does not observe in his self an internality identical to the internality of his mirroring activity (*i.e.:* his actuality = his primary act).

With all the above thoughts racing through my perhaps crazy head, I next said to myself: "It seems to me that the encounterer can be described as a thesis seeking an anti-thesis. In this anti-thesis, he hopes to encounter himself and to achieve a synthesis in which he will then—in the ecstasy of *total* self-discovery—enjoy quiescence from his chief yearning: to come face to face with that *actual* version of himself most properly called either 'an act of existence' or 'an act of encountering' or 'a primary act'." Therefore, every encounterer—whether God, an angel or a human soul—is the tri-functional mirroring activity of a dynamic mirror comprised of a positive and a negative element seeking neutrality as the third element. In God and every occupant of *infinity*, that neutrality is achieved. In humans and every occupant of the *finite* world, that neutrality is *not* achieved.

As early as 1957, and certainly no later than 1961 (Writings of mine from that year confirm it.), I turned to Algebra's symbolic logic to produce the following expression of The Infinite Encounter:

$$(+A) \otimes (-A) \longrightarrow 0$$

DRAWING #2-3: ALGEBRAIC EXPRESSION OF THE INFINITE ENCOUNTER & INFINITE VERSION OF ESOPTRICS' FIRST PRINCIPLE.

When I first produced this drawing, I looked upon the (+A), the (-A) and the zero as the three elements of the encounter. It wasn't until 1994 that I realized the circled X is the third element, and the zero merely expresses what (+A) and (-A) achieved by means of the circled X. It then became the circled X, and not the zero, which represents the act of encountering whereby (+A) comes face to face with Its actual self in (-A). More precisely, it is thru ⊗ that Encountering comes face to face with Itself as three distinct functions of the same, one instance of the same, one set of internal characteristics.

In finite encounters, the all-important neutrality is not always achieved; and so, the fundamental algebraic expression of the finite encounter must be:

$$(+a) \otimes (-a) \longrightarrow ?$$

DRAWING #2-4: ALGEBRAIC EXPRESSION OF THE FINITE ENCOUNTER & FINITE VERSION OF ESOPTRICS' FIRST PRINCIPLE.

Drawing #2-3 is the infinite and #2-4 the finite version of what I call the principle of: (1) the *a priori* necessary structure of every act of consciousness (*i.e.:* before every experience, it dictates every experience's structure), and (2) a mirroring act of the first order. I also call them "the infinite and finite versions of the principle of principles" and "the first and supreme principle of Esoptrics and of Being-as-Being itself". That's because it's the ultimate, supreme definition of all reality and Being, and holds in itself all the diversification possible to reality and Being. That makes it an *a priori* and *synthetic* principle for all who see how it applies to itself. In its infinite version, it defines the *logical* essence of God and, as such, is the ultimate ground for whatever is intrinsically possible.

If there is to be any multiplicity in reality and Being, it can only come about by applying the first principle to itself. That is to say, we must modify the first principle by itself. But, **to be** is to be involved in a mirroring activity, and the first principle is the definition of that mirroring activity (***i.e.:*** the definition of the most basic kind of Being-as-Being itself). Therefore, to multiply it, we must mirror it. Obviously, mirroring it will produce a configuration of *two* such principles. One will be "inside" the mirror, and the other will be "outside" the mirror in the sense that they shall be reverse images of one another. Such will give us the principle of a mirroring act of the second order. It can be called the secondary form of seeking synthesis or the secondary form of seeking neutrality. It can also be called the second kind of Being-as-Being itself. We will generally refer to it, though, as: "the second principle of Esoptrics". Using Algebra's symbolic logic, it can be stated so:

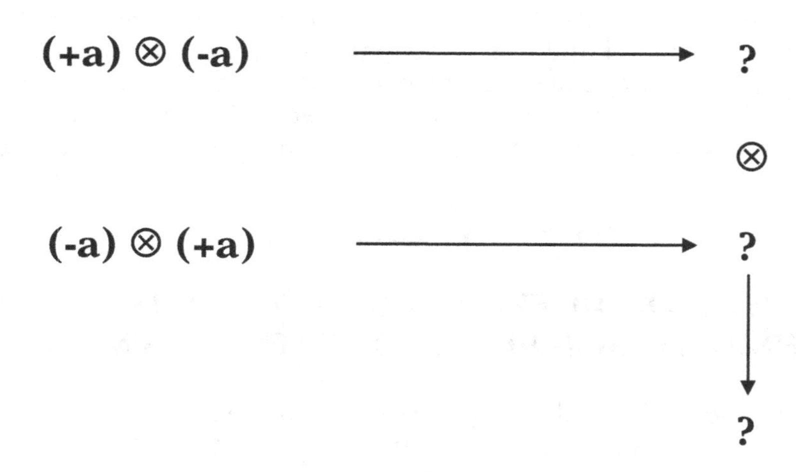

DRAWING #2-5: ALGEBRAIC EXPRESSION OF ESOPTRICS' SECOND PRINCIPLE.

As said, drawings #2-4 and #2-5 give us an *algebraic* expression of Esoptrics' first and second principles. Those two principles can also be given a *geometrical* expression. In that case, though, we must bear in mind what needs to be said repeatedly—namely: Every geometrical expression of Esoptrics' principles is merely a kind of allegory and metaphor presenting us with a purely *figurative* expression of those principles, and the reason for resorting to such pictures is to make it easy for sensation dependent minds to grasp what would otherwise be too abstract for them. The use of Geometry's pictures is far from being an attempt to show what ultimates "look like". On the next page, then, drawings #2-6 and #2-7 present two geometrical attempts to make drawings #4 and #5 more easily intelligible.

We said the necessary, triousious structure of every act of consciousness is the necessary structure of every basic reality and basic kind of Being. Assuming that true, we can formulate what I call the fifth postulate of Esoptrics:

THE FIFTH POSTULATE

EVERY BASIC REALITY IS A NON-INHERING WHOLE AND A TRIOUSIOUS OBJECT ENCASING THREE SUBJECTS—TWO OF WHICH ARE DIRECT OPPOSITES, THE THIRD OF WHICH IS A BRIDGE BE-

TWEEN THE OTHER TWO. [1]

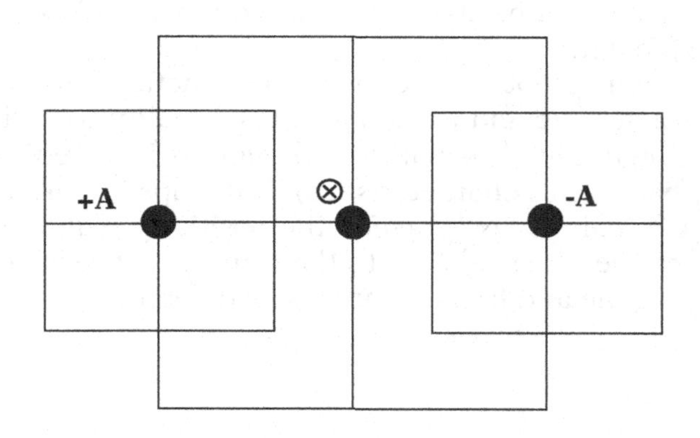

DRAWING #2-6: GEOMETRICAL EXPRESSION OF ESOPTRICS' FIRST PRINCIPLE.

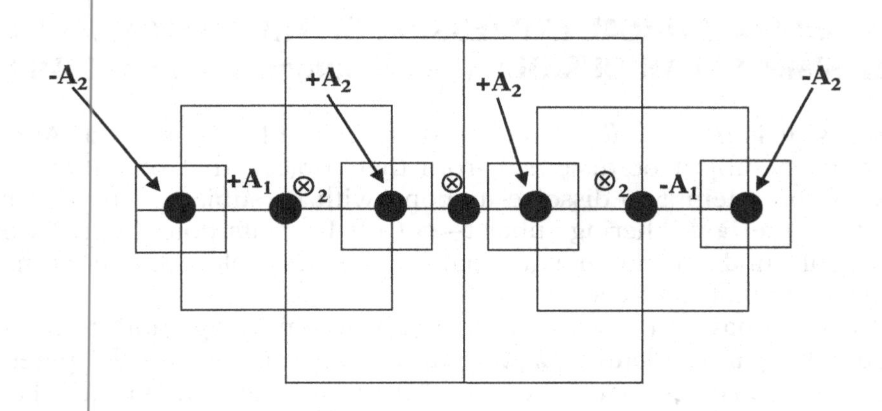

DRAWING #2-7: GEOMETRICAL EXPRESSION OF ESOPTRICS' SECOND PRINCIPLE.

Can one imagine something which would fill that bill? Yes! Turn to the idea of *potentiality* versus *actuality*. By a subject's "actuality," of course, I mean the sum total of all its *current* internal characteristics (***i.e.*** all it is being inside itself); by its "potentiality," I mean the sum total of all its *future* internal characteristics (***i.e.*** all it can become inside itself). That distinction now allows us to turn to the concept *figuratively* depicted below in drawing #2-8.

The small, solid black dot at the center represents one of the co-tenants in a polyousious object. The much larger, transparent circle surrounding the solid black dot repre-

[1] Recall ftn. #1, pg. 12: A non-inhering whole is one which does not *need* to be *intimately bonded to* another to avoid annihilation. It does not *need* to *share it extremities* with another's. That, Mr. Spinoza, does not mean it does *not need to be in any way dependent on* another.

sents another one of the co-tenants in that polyousious object. The black dot represents the *actuality* of a non-inhering subject. What surrounds it represents the *potentiality* of that non-inhering subject and is, itself, a subject—an *inhering* one which is annihilated if it ceases to be intimately bonded to the dot.

This potentiality, then, is not a mere mental abstraction. However transparent, it has at least one internal characteristic and is a *subject*. It's a reality so real it surrounds the non-inhering subject whose potentiality it is without, though, being predicable of it (***i.e.:*** without being merely one of the black dot's characteristics). If it helps, say it's the dot's potentiality in the sense of the negatively real *means by which* the positively real dot shall realize whatever potentiality it has. In a manner of speaking, it's the transparent "gasoline" a driver can use to turn his car into an automobile and himself into a speed demon.

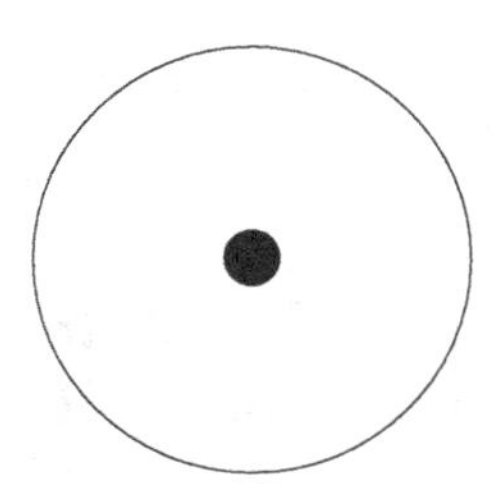

DRAWING #2-8: GEOMETRICAL EXPRESSION OF THE PRELIMINARY CONCEPT OF THE SIMPLEST TRIOUSIOUS (i.e.: 3 subjects in one) OBJECT.

Nevertheless, it is *not equally* as *real* as is the non-inhering subject. After all, if the potentiality of a given subject be disjoined from that subject, that potentiality is nothing, which is to say, if you attempt to dissolve its bond with its subject, you will annihilate it. Should you do that, the *non*-inhering subject—cut off from its potentiality—would *not* be *annihilated*, but could undergo no change whatsoever and would become eternally and totally immobile as in eternally damned.

So far, then, we have an object comprising a non-inhering subject and an inhering one. Each is truly a separate identity (***i.e.:*** separate subject), because the potentiality of a subject—though it *shall become* that subject—is *not yet* that subject. On the other hand, the potentiality of a *given* subject cannot be separated from *that given* subject; otherwise, it is not the potentiality of anything and, therefore, is not anything at all.

We, though, are supposed to be devising a polyousious object having ***three*** co-tenants, which is to say a *tri*ousious object. Where is the *third* subject? It is the unit of tension generated by the drag-inducing differential between the actuality (***i.e.:*** the dot) and the potentiality (***i.e.:*** the remainder) of the object. For now, there's no need to say more than that about the third co-tenant.

We now have a very unusual concept: one whose purely *logical* content can be *pictured*. Still, the result could hardly be called a mathematically precise theory of the universe. Clearly, our concept and "picture" of a triousious object must expand dramatically.

Drawing #2-9 on the next page—if looked at closely—returns us to an important question: What shall we say of a non-inhering subject with *unlimited* potentiality? If its actuality is less than its potentiality, what would be the ratio of the whole to its actuality? Wouldn't it be 10^{infinite} to $10^{-\text{infinite}}$ (Note the minus sign)? Since that strikes me as nonsense, the ratio must be 1 to 1, which is to say it must be that all the potentiality has been converted into actuality, and the whole object is actuality and actuality alone.

In case what I just said above is not clear, consider this example: A group sets out to raise \$1,000,000.00. As they begin, the ratio of money raised to money wanted is obviously: 0/\$1,000,000.00. When they finish raising all the money, the ratio of money raised to money wanted will be \$1,000,000.00/\$1,000,000.00 = 1/1. On the other hand, the ratio of money raised to money *left* to be raised is \$1,000,000.00/0.

Is the infinite, non-inhering subject a triousious object? How *could* it be? And yet it *must* be, *if* the postulates and principles of Esoptrics are correct. Clinging to that insight, I cannot escape the conclusion: It must be the infinite, non-inhering subject is The First *Triousious Object* and, therefore, somehow ***is***, and occupies, three frames of reference each of which—though equal to, and equally related to, the other two—is somehow distinct from the other two. That can be graphically illustrated so:

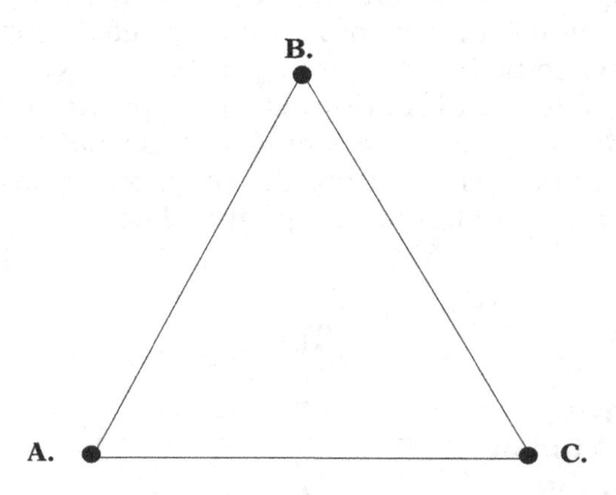

DRAWING #2-9: PRELIMINARY GEOMETRICAL EXPRESSION OF THE FIRST TRIOUSIOUS (*i.e.:* 3 subjects in one) OBJECT.

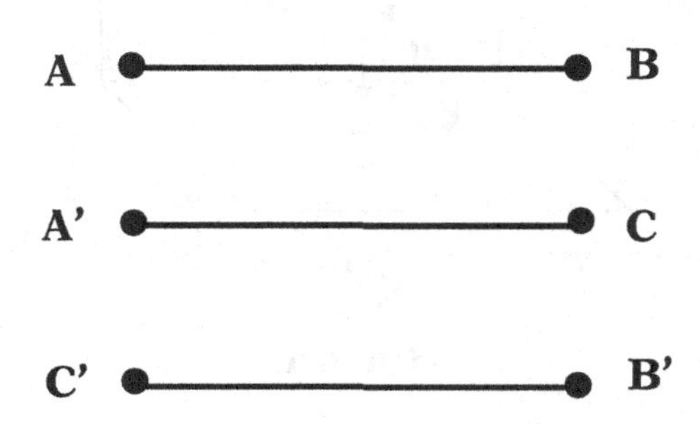

DRAWING #2-10: ADVANCED GEOMETRICAL EXPRESSION OF THE FIRST TRIOUSIOUS (*i.e.:* 3 subjects in one) OBJECT.

In drawing #2-9 above, the points A, B, and C represent the three frames of reference encompassed by The First Triousious Object. I have drawn them as corners of a triangle in order to express the fact each is equally related to the other two. In drawing #2-10, above, I have split the three lines of the triangle apart from one another in order to illustrate that fact even more forcefully. How so?! As you can perhaps readily notice, each of the three points occurs at the ends of two of the three lines.

Dwelling on drawing #2-10, I am inexorably drawn to this: I can look at any one of the three points *two* different ways. For example, A can be considered either as A related to B or as A related to C. Can it be? Is ***"A related to B"*** somehow distinct from ***"A related to C"***? In that case, we can *double* the number of reference points and say: A = A related to B; A' (***i.e.*** A *prime*) = A related to C; B = B related to A; B' (***i.e.*** B *prime*) = B related to C; C = C related to A; and C' (***i.e.*** C *prime*) = C related to B. As you can see in drawing #2-10 on the prior page, it's merely a case of considering the three arms of the triangle in isolation from one another.

The instant I isolate the three arms that way, another insight thrusts itself upon me: The *Infinite* Triousious Object can present every *finite* triousious object (hereafter called a form) with six ways to be related to It. We can express that by saying that the potentiality of every finite form encompasses six potencies (Say "encompasses 6 ways to be in a state of excitation", if you prefer.) which divide into three *sets* each comprised of two potencies (Say "comprised of two ways to be in a state of excitation", if you prefer.). In each set, each of the two potencies (***i.e.:*** states of excitation) is the opposite of the other. Thus, A and B are mutual opposites within the first set; A' and C are mutual opposites within the second set; and C' and B' are mutual opposites within the third set. In keeping with what we've already said, each of these three sets should be equally related to the other two.

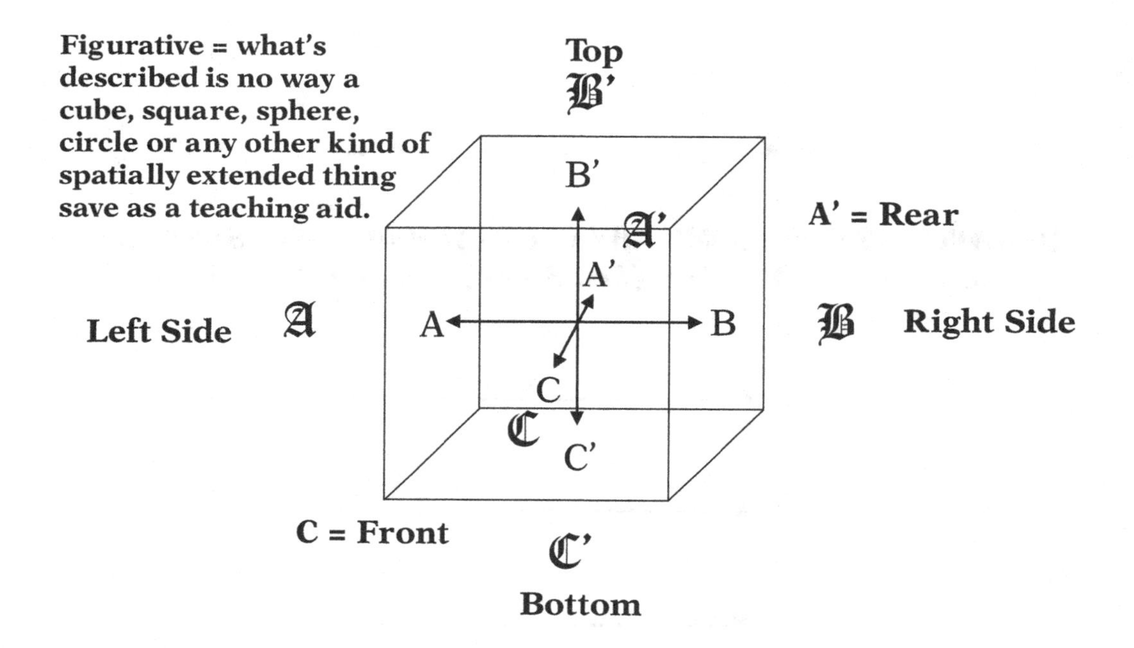

DRAWING #2-11: SIX POSSIBLE WAYS FOR A FINITE FORM'S OVERALL ACT OF EXCITATION TO BE IN A STATE OF EXCITATION RELATIVE TO THE INFINITE TRIOUSIOUS OBJECT

It should take little effort to see that the previous paragraph's concepts can be expressed pictorially by drawing three axes each of which is *perpendicular* to the other two. Making them so, we express the fact that each relates equally to the other two. That's exactly what I've done above in drawing #2-11. Drawing #2-11 can be described as the most primitive *metaphorical* description of a *finite* triousious object (***i.e.:*** form) made possible by the *Infinite* Triousious Object as the former relates to The Latter in the six ways made possible by the fact that The Latter, by Its very nature, is three *frames* of reference each of which is two *points* of reference. We can further describe the result as a finite form *in potency* to (***i.e.:*** seeking its states of excitation from) The Infinite One six ways, and each of those six is a state of excitation. More accurately, every form's *overall* state of excitation, relative to The Infinite One, is a composite of 6 simultaneous states of excitation.

In drawing #2-11 (prior page), the letters **𝔄, 𝔄', 𝔅, 𝔅', ℭ,** and **ℭ'** represent what Esoptrics calls a set of "heavenly benchmarks". The letters A, A', B, B', C, and C' represent what Esoptrics calls a set of "heavenly sextets". For the moment, suffice it to say every set of the latter directly influences some particular form; every set of the former *directly* affects only one or more sextets and, thereby, indirectly affects one or more forms. The former may admit of more than one layer; the latter admit of only one. The former are mostly fixed; whereas, the latter, so to speak, often play "ring around the rosie" with regard to the former and, thereby, cause the forms to rotate. The latter are never an immortal soul; the former may be one, if God so wills it.

As you examine drawing #2-11 (prior page), be sure to bear in mind a crucial qualification it's impossible to repeat too many times: Drawing #2-11 is merely a pictorial, *figurative* expression of the purely *logical* principles constituting every basic reality and basic kind of Being, and it describes them by using Geometry to express *only* how they *relate* to one another. Drawing #2-11 is not an attempt to give you an encounter of the encounterableness (***i.e.:*** observ<u>*able*</u> internality) of a basic reality. In other words, drawing #2-11 is not a "photocopy" of what you would "see" were you to come face to face with the internality of an ultimate. From God's point of view, the points A, B, C, A', B', C', and 0, are <u>*not*</u> outside of one another ***<u>in space</u>***; they are outside of one another ***<u>in logical principle</u>*** and in logical principle ***<u>alone</u>***. Drawing #2-11 depicts the points as outside of one another in space because that's the simplest and most intelligible way for me to go about communicating the logical principles to minds as slow as my own.

If drawing #2-11 applies to God, then 0 (***i.e.:*** the center of the cube) is coterminous with the rest of the cube, and the outer perimeter of the cube represents the "limits" of *infinite* potentiality converted into *infinite* actuality. If drawing #2-11 applies to a form, then 0 may, but rarely is, coterminous with the rest of the cube, and the outer perimeter of the cube represents the limits of whatever finite potentiality God has made available to God's creatures. If Drawing #2-11 applies to a creature, then it applies to a kind of created triousious object which shall hereafter be referred to as a form (***i.e.:*** one of the Universe's ultimate *macro*-constituents). Before it's over, the number of these forms (***i.e.:*** these *macro*-ultimates) and the differences between them shall become astronomical—namely: a total of c. 10^{231} divided into 2^{256} (***i.e.:*** 1.1579209×10^{77}) kinds.

Such, then, are the thoughts which led to Esoptrics and its hypothesis that The Infinite is an extra-cosmic influence which—by means of the 6 points of reference intrinsic to It—imparts a six-fold potency to every finite triousious object (***i.e.:*** form) created by It. As we progress, the number of heavenly sextets, too, shall increase astronomically and be made distinct from one another as a result of each being forever associated with a forever-after unique form. Eventually, we'll explain how each of these "heavenly sextets" shifts roles and, thereby, causes the form associated with it to rotate programmatically in definite patterns with definite ratios between the way it rotates around each of its 3 axes.

Understanding, and making sense of, the world that we see is a very important part of our evolution. Therefore spatial intuition or spatial perception is an enormously powerful tool, and that is why geometry is actually such a powerful part of mathematics not only for things that are obviously geometrical, but even for things that are not. We try to put them into geometrical form because that enables us to use our intuition. (pg.5)

. . . . .

This choice between geometry and algebra has led to hybrids that confuse the two, and the division between algebra and geometry is not as straightforward and naive as I just said. For example, algebraists frequently will use diagrams. What is a diagram except a concession to geometrical intuition? (pg. 7)

——**SIR MICHAEL ATIYAH: *Fields Lectures*,** as given at the World Mathematical Year 2000 Symposium at Toronto, Canada, June 7th thru the 9th. (Yes, Sir Michael, even algebra addicts like myself must often resort to Geometry's diagrams in order to make it easier to grasp Esoptrics' principles. As I do so, though, I beg all to be ever aware I'm using Geometry's pictures "even for things that are not" geometrical.)

Hugh of St. Victor, in the hope of affiliating Mysticism and Scholasticism, collected and arranged systematically the scattered thoughts of St. Bernard favorable to his purpose. With him, the underlying principle of religious science was that one's knowledge of truth is exactly adequate to his interior dispositions. (*Tantum de veritate quisque protest*[1] *videre quantum ipse est*.). The means of arriving at perfect science is contemplation, which was lost through original sin, yet can be recovered by supernatural aids. This fixing of the mental vision on things eternal is what is understood by contemplation in the strict sense. When, on the other hand, the faculties are engaged in the consideration of the objects that meet one in the visible world, the mental operation is called rational meditation.

——**REV. DR. JOHN ALZOG** (1808-1878)**: *Universal Church History***, Vol. 2, pgs. 763-764, as translated by F. J. Pabisch and Thos. S. Byrne, and published in hardbound copy by Robert Clarke & Co., Cincinnati, 1876 (Esoptrics, my dear Fr. Alzog, would seem to confirm what Hugh of St. Victor contended. For, Esoptrics truly is contemplation fixing "the mental vision on things eternal"—which things can only "meet one" in the internal world of one's own mind.).

[1] "Protest" is probably a typo for *potest*. My free translation of the above would be: One's ability to see truth is the same as one's own truthfulness. I myself would say: One's ability to see truth is the same as the intensity of one's devotion to the *whole* of Catholic Doctrine.

Chapter Three:

THE MATH INHERENT TO A MIRROR:

Imagine that you're looking at a mirror. Hold your right hand in front of it. For the sake of comfort, face the palm inward toward the mirror. One will now have two hands—one *in* the mirror and one *outside* of it. Furthermore, the two hands will be reverse images of one another. For, the hand outside the mirror is your right hand, and its thumb is to your left; but, the hand inside the mirror is your other self's left hand, and its thumb is on the right side of your other self's hand.

Suppose we could reach into the mirror, could take hold of the hand in the mirror, and could bring it out to our side of the mirror. We would then have two real hands which are reverse images of one another.

Now hold those two hands in front of the mirror. We will now have two sets each containing two mutually reversed hands, and the two sets will be reverse images of one another. Remove the set in the mirror; add it to the set outside of the mirror; and that will give us a set composed of four hands.

Hold that set of four hands in front of the mirror, and we will now have two sets of four hands each. If we keep that up, we're progressing by the powers of 2, and each new set in the mirror has an internal order the opposite of what's found among the prior sets. We cannot actually perform such steps; a *dynamic* mirror effectively does (***cf.:*** #2-5 pg. 16).

Consider now this symbol: "1^{1m}". It means: "one to the first mirror". That is to say, it is the number 1 followed by an exponent telling us to raise the number 1 to a particular power of 1, and the power to which we must raise the number 1 is described as: "1 mirrored once". Now then, any one object mirrored once produces two such objects. Thus, $1^{1m} = 2$, which is the same as 2^1.

Next, consider this symbol: "1^{2m}". It means: "1 to the second mirror". It tells us we must raise the number 1 to a power described as: "1 mirrored twice". Now then, any object mirrored once produces two such objects, and if you then mirror those two, you will have four such objects. Thus, $1^{2m} = 2^2 = 4$. Continuing such a line of reasoning, we can make these equations:

$$1^{3m} = 2^3 = 8$$
$$1^{4m} = 2^4 = 16$$
$$1^{5m} = 2^5 = 32$$
etc.

To make things a bit more complicated, consider this symbol: "1^{m1}". Note that—in the exponent—the relationship between the letter and the number is reversed. What does such a symbol mean? It means "1 to the first mirror of 1". Thus, the exponent tells us to raise the number 1 to the first mirror of 1, in order to get the exponent of the number we're examining. Thus, $1^{m1} = 1^{2m} = 4$. Suppose, though, the symbol reads thusly: 1^{2m1}. In that case, we must take the second mirror of one in order to establish the exponent of the number 1. Thus, $1^{2m1} = 1^{4m} = 2^4 = 16$. Perhaps the reader can now grasp the following symbols producing successive squares of the powers of 2.

1^{1m1}	$= 1^{m1}$	$= 1^{2m}$	$= 2^{2}$	= 4.
1^{2m1}	$= 1^{m2}$	$= 1^{4m}$	$= 2^{4}$	= 16.
1^{3m1}	$= 1^{m4}$	$= 1^{8m}$	$= 2^{8}$	= 256.
1^{4m1}	$= 1^{m8}$	$= 1^{16m}$	$= 2^{16}$	= 65,536.
1^{5m1}	$= 1^{m16}$	$= 1^{32m}$	$= 2^{32}$	= 4,294,967,296.
1^{6m1}	$= 1^{m32}$	$= 1^{64m}$	$= 2^{64}$	= 18,446,744, 073,709,551,616.
1^{7m1}	$= 1^{m64}$	$= 1^{128m}$	$= 2^{128}$	$= 3.4028236692093846346\text{x}10^{38}$.
1^{8m1}	$= 1^{m128}$	$= 1^{256m}$	$= 2^{256}$	$= 1.1579208923731619542\text{x}10^{77}$.

We have now established two of the progressions with which the logic of the mirror deals. There are two other very important ones. Perhaps I should say that I myself am currently aware of only two more of them. Let's proceed to establish those remaining two.

Take an object and set it to the side. Take another object exactly like it, and mirror it. Take the image from the mirror; couple it to the one in your hand; and set the two next to the first single object. That now gives you three objects (***i.e.:*** $1 + 1^{1m}$). Take another object exactly like the original object, and mirror it. Take the image from the mirror; couple it to the one in your hand; and set the two next to the other three. That now gives you five objects (***i.e.:*** $1 + 1^{1m} + 1^{1m}$). Continuing that process gives you another important progression of Esoptrics.

We now have the following three progressions. I should not have to point out, that—in each progression—the lower lines are each another way of writing what is written in the lines above it.

I. THE GREAT PROGRESSION:

1	$+ 1^{1m}$	$+ 1^{m1}$	$+ 1^{2m1}$	$+ 1^{3m1}$	$+ 1^{4m1}$	$+ 1^{5m1}$	etc. = ↙
1	$+ 2^{1}$	$+ 1^{2m}$	$+ 1^{m2}$	$+ 1^{m4}$	$+ 1^{m8}$	$+ 1^{m16}$	etc. = ↙
1	$+ 2^{1}$	$+ 2^{2}$	$+ 1^{4m}$	$+ 1^{8m}$	$+ 1^{16m}$	$+ 1^{32m}$	etc. = ↙
1	$+ 2^{1}$	$+ 2^{2}$	$+ 2^{4}$	$+ 2^{8}$	$+ 2^{16}$	$+ 2^{32}$	etc.

II. THE SECONDARY PROGRESSION:

1	$+ 1^{1m}$	$+ 1^{2m}$	$+ 1^{3m}$	$+ 1^{4m}$	$+ 1^{5m}$	$+ 1^{6m}$	etc. = ↙
1	$+ 2^{1}$	$+ 2^{2}$	$+ 2^{3}$	$+ 2^{4}$	$+ 2^{5}$	$+ 2^{6}$	etc.

III. THE BASIC PROGRESSION:

1	$+ 1^{1m}$	$+ 1^{1m}$	$+ 1^{1m}$	$+ 1^{1m}$	$+ 1^{1m}$	$+ 1^{1m}$	etc. = ↙
1	+ 2	+ 2	+ 2	+ 2	+ 2	+ 2	etc.

Between the great and secondary progressions, there is obviously another one which should read as follows:

IV. THE INTERMEDIATE PROGRESSION:

1	$+ 1^{m1}$	$+ 1^{m2}$	$+ 1^{m3}$	$+ 1^{m4}$	$+ 1^{m5}$	$+ 1^{m6}$	etc. = ↙
1	$+ 2^2$	$+ 2^4$	$+ 2^6$	$+ 2^8$	$+ 2^{10}$	$+ 2^{12}$	etc.

Such, then, is the math *inherent* to a mirror. I say *inherent* to it, because every mirror, by its very nature, doubles and reverses whatever we place in front of it, unless, of course, we place another mirror in front of it. Esoptrics, though, is based on the mathematical principles I observe when I observe *myself* interacting with ***a*** mirror, which is to say when I observe the relationship between *myself* and some *one* particular mirror. It is, therefore, by no means based upon what I observe when I observe two mirrors interacting with one another.

Since, then, Esoptrics is the logic of a *mirror*, and the math inherent to a mirror involves multiples of 2, Esoptrics, as a mathematical theory, is *necessarily* one dealing with multiples of 2. I emphasize *"necessarily"*, in order to stress the *fact* there is nothing even remotely arbitrary about Esoptrics' fixation upon multiples of 2. Esoptrics, as should be maximally obvious, cannot be *Esoptrics* (***i.e.:*** the logic of the *mirror*) unless, as every mirror invariably does, its sticks to the multiples of 2.[1]

As Esoptrics continues with its fixation upon the multiples of 2, the above progressions shall come into play over and over again with amazing results. One should, therefore, be maximally familiar with those progressions, in order to be aware of when they are coming into play.

✡✞✡✞✡✞✡✞✡✞✡✞✡✞✡✞✡✞✡✞✡✞✡

All things whatsoever—whether those which *are* or those which *can* be—are reflected in God as in a kind of mirror.
——**ST. THOMAS AQUINAS: *Summa Theologica - Pars Prima*,** Question 12: Article 8, Objection 2 (My translation – ENH).

We may as well state, without further ado, that the ultimate ground or foundation of all possibility is found in the Supreme Being, in the *nature of God* (pg. 70).
, , , ,
God's Essence is the ultimate ground of intrinsic possibility.
——**FR. CELESTINE N. BITTLE, O. M. CAP.: *The Domain Of Being—Ontology*,** as given on pages 70&77 of the paperback edition by Bruce Publishing Co., Milwaukee, 1942. ©1939 Bruce Publishing Company.

---

[1] No matter how many times I say that, many of those few—who condescend to give Esoptrics a glance—immediately dismissed it pontificating: "Are you blind to the arbitrariness of your fixation upon multiples of 2? Why not use multiples of 3 or some other number? For, as is utterly manifest, you have not the slightest reason whatsoever to stick with 2 rather than some other integer or even a fraction." One can't help but *suspect* that Esoptrics is—if only instinctively—immediately perceived by some as so great a threat, there is no limit to the diligence with which they will scrape the bottom of the barrel in search of some reason—however flimsy a pretext—to dismiss Esoptrics and its author as "crap"—to use the term hurled at me by one University's Philosophy department.

The choice of the number two as the basis for understanding all reality is never well explained.
——**ANNA CALL:** ***Clarion Review*** (Again???!!! By definition, "mirror" *means* what doubles and reverses what faces it. Thus, like 2+2=4, it's an absolutely certain, true by definition tautology that 2 is the basis for understanding *the mathematical logic of a mirror*. Whether or not 2 is the basis for understanding *all reality* is a different issue. Either all the Universe's realities are governed by the mathematical logic of a mirror, or they are not. If they ***are***, then 2 is the basis for understanding all reality; if they are ***not***, then 2 is not that basis. The question, then, is: Does Esoptrics govern all the Universe's realities? That's not for me to say. All I can say is: Here are the explanations inferred, ***if*** one starts with the *assumption* that Esoptrics ***does*** govern all the Universe's realities. If Esoptrics then yields a more thorough explanation of the Universe's origins and structure than Science does, it'll be powerful evidence Esoptrics started with *a valid hypothesis*. *Does* Esoptrics do the better job? Again, that's not for me to say; history must: In time, Esoptrics must become well known to cosmologists sufficient in number and competence to render a decision few, if any, will challenge. **It's not**: 2 is the basis for understanding all reality; rather, **it's**: Let's *assume* it is, and see if it yields more explanations than Science does. If it does, *that's* when the choice shall be well explained. In demanding the choice be well explained *in the course of this book*, Ms. Call merely exposes her failure to read well.).

Fundamental ideas play the most essential role in forming a physical theory. Books on physics are full of complicated mathematical formulae. But thought and ideas, not formulae, are the beginning of every physical theory. The ideas must later take the mathematical form of a quantitative theory, to make possible the comparison with experiment.
——**ALBERT EINSTEIN & LEOPOLD INFELD:** ***The Evolution Of Physics***, pg. 277 in the sixth paperback edition by Simon and Schuster, New York, 1967. ©1938 by Albert Einstein & Leopold Infeld.

I'd like to emphasize that this idea that time and space should be finite "without boundary" is just a *proposal*: it cannot be deduced from some other principle. Like any other scientific theory, it may initially be put forward for aesthetic or metaphysical reasons, but the real test is whether it makes predictions that agree with observation.
——**STEPHEN HAWKING:** ***A Brief History Of Time***, pgs. 141-142 of the paperback tenth anniversary edition by Bantam Books, New York, 1998. ©1988 & 1996 by Stephen Hawking (Exactly my point! A physical theory must start as "fundamental ideas" "initially put forward for aesthetic or metaphysical reasons" and only *later* yield "the mathematical form of a quantitative theory" which "makes predictions that agree with observation." But, "later" implies the start may come from *one party* putting forth fundamental, metaphysical ideas so interesting as to motivate a *different party* to produce that "mathematical form" able "to make possible the comparison with experiment." Excuse me if I dare suggest Esoptrics is a *stunningly* beautiful system putting forth fundamental, metaphysical ideas so impressive, every thinker *truly* interested in a thorough knowledge of the Universe will readily choose to pursue what Esoptrics must *later* become.).

Chapter Four:

ONTOLOGICAL DISTANCE & THE NUMBER OF INTENSITIES POSSIBLE TO EACH OF THE 6 COMPONENT STATES OF EXCITATION IN EACH OF THE UNIVERSE'S FIRST 9 EPOCHS:

Turn again to drawing #2-11 on page 20. The Infinite Triousious Object is giving a finite triousious object (***i.e.:*** a form) 6 ways to be in a state of excitation, which is to say 6 ways to be in potency to (***i.e.:*** seeking its states of excitation from) infinity. Thus, the form's every particular *overall* state of excitation (***i.e.:*** every primary act) is a *physically* indivisible whole *logically* divisible into 6 component states of excitation so:

(1&2) the A state vs. its opposite the B state;
(3&4) the A' state vs. its opposite the C state;
And
(5&6) the B' state vs. its opposite the C' state.

For all forms, always, everywhere, and in every one of their overall states of excitation (***i.e.:*** in every one of their primary acts), the intensity of each of the 6 states of excitation—simultaneously comprising that overall state—is the same; but, the number of intensities possible to each of them is greater than 1. How much greater than one?!

The Infinite can create forms in which each of the 6 states of excitation—comprising a particular overall state (***i.e.:*** a particular primary act)—has the same absolutely minimal level of intensity. Esoptrics refers to that level as an "ontological distance of one" or "OD1" for short. The term "OD1" means each of the 6 states of excitation—comprising some form's particular overall state of excitation—is one step above zero intensity. In other words: In each of its primary acts, every such form has a six-way *actuality* (***i.e.:*** overall state of excitation) one step above zero in each of its 6 *component* states of excitation.

Next, The Infinite can create forms in which each of the 6 states of excitation comprising an overall state has a level of intensity twice that of the forms at OD1. In each of their primary acts, then, every such form has a six-way actuality which is 2xOD1. Esoptrics expresses that by saying every such form is at OD2 and the 4th dimension's 2nd level.

Next, The Infinite can create forms in which each of the 6 states of excitation comprising an overall state has a level of intensity thrice that of the forms at OD1. In each of their primary acts, then, every such form has a six-way actuality 3xOD1. Esoptrics expresses that by saying every such form is at OD3 and the 4th dimension's 3rd level.

Precisely how many levels of intensity are there? To answer, we must first recall the great progression described on page 24. It's the one which lists successive squares of the powers of 2: 2^0, 2^1, 2^2, 2^4, 2^8, 2^{16}, 2^{32}, 2^{64}, 2^{128}, 2^{256}, and so forth. For Esoptrics, that progression defines: (1) the limits of an epoch of the Universe, and (2) the number of intensities possible—throughout a given epoch—to each of the 6 states of excitation comprising some overall state of excitation (***i.e.:*** primary act) on the part of a form.

For Esoptrics, the progression, 2^0, 2^1, establishes the limits of what it calls the Universe's 1st epoch and says that the "ontological depth" of that epoch was 2—meaning: Throughout that 1st epoch, there were only 2 OD's and, thus, only 2 levels of intensity available to the 6 states of excitation and only 2 levels to the 4th dimension.

For Esoptrics, the progression, 2^0, 2^1, 2^2, establishes the limits of what it calls the Universe's 2nd epoch and says that the "ontological depth" of that epoch was 4—meaning: Throughout that 2nd epoch, there were only 4 OD's (***viz.:*** 1, 2, 3, & 4) and thus only 4 levels of intensity available to the 6 states of excitation and only 4 levels to the 4th dimension.

For Esoptrics, the progression, 2^0, 2^1, 2^2, 2^4, establishes the limits of the Universe's 3rd epoch and says that the "ontological depth" of that epoch was 16—meaning: Throughout that 3rd epoch, there were only 16 OD's (***viz.:*** 1, 2, 3, 4, 5, 6, 7 . . . . 16) and thus only 16 levels of intensity available to the 6 states of excitation and 16 4th dimension levels.

If you're following the progression, we can jump forward and say that, for Esoptrics, the progression, 2^0, 2^1, 2^2, 2^4, 2^8, 2^{16}, 2^{32}, 2^{64}, 2^{128}, 2^{256}, establishes the limits of what it calls the Universe's 9th epoch and says that the "ontological depth" of that epoch is 2^{256}—meaning: Throughout this, the Universe's 9th and current epoch, 2^{256} (***i.e.:*** 1.1579×10^{77}) is the number of OD's present, the number of intensities possible to each of the 6 states of excitation, and the number of the 4th dimension's levels.

The Universe didn't take long by our standards to go from the 1st to the 9th epoch's start (***i.e.:*** roughly 10^{-38} sec.). That's because the 1st epoch, with its ontological depth of 2, lasted only 2 alphakronons (***i.e.:*** K for short). The 2nd epoch, with its ontological depth of 4, lasted only an additional 16K, which is to say 2 times the cube of the square root of its ontological depth of 4. The Universe's 3rd epoch, with its ontological depth of 16, then lasted only an additional 128K, which is to say 2 times the cube of the square root of its ontological depth of 16. Thus, since in Esoptrics $\sqrt{2} = 1$ and $1^3 = 1$, we have:

EPOCH #	**OD'S IN = ONTOLOGICAL DEPTH**	**DURATION IN ALPHAKRONONS (K) = $7.2\text{x}10^{-96}$ SEC. X $2[(\sqrt{\text{Ont. Depth}})^3]$K**
1	1 thru 2^1	$2[(\sqrt{2})^3] = 2$K
2	1 thru 2^2	$2[(\sqrt{2^2})^3] = 2^4 = 16$K
3	1 thru 2^4	$2[(\sqrt{2^4})^3] = 2^7 = 128$K
4	1 thru 2^8	$2[(\sqrt{2^8})^3] = 2^{13} = 8{,}192$K
5	1 thru 2^{16}	$2[(\sqrt{2^{16}})^3] = 2^{25} = 33{,}554{,}432$K
6	1 thru 2^{32}	$2[(\sqrt{2^{32}})^3] = 2^{49} = 5.629499\text{x}10^{14}$K
7	1 thru 2^{64}	$2[(\sqrt{2^{64}})^3] = 2^{97} = 1.5845633\text{x}10^{29}$K
8	1 thru 2^{128}	$2[(\sqrt{2^{128}})^3] = 2^{193} = 1.2554203\text{x}10^{58}$K
9	1 thru 2^{256}	$2[(\sqrt{2^{256}})^3] = 2^{385} = 7.788\text{x}10^{115}\text{K} = 1.798\text{x}10^{13}$ years

CHART #4-1: DURATIONS & ONTOLOGICAL DEPTHS OF THE 9 EPOCHS

Line 2=#1x2^3; #3=#2x2^3; #4=#3x2^6; #5=#4x2^{12}; #6=#5x2^{24}; #7=#6x2^{48}; #8=#7x2^{96}; and #9=#8x2^{192}. I take it you notice the exponents' progression = 3, 3, 6, 12, 24, 48, 96 & 192.

According to Esoptrics, this, the Universe's 9th epoch, shall endure for roughly 18 trillion Earth years, and The Infinite will then initiate the Universe's 10th epoch. In it: (1) 2^{512} ontological distances shall be available (***i.e.:*** the ontological depth and number of 4th dimension levels shall be $2^{512} = [2^{256}]^2$), and (2) 2^{512} intensities shall be available to the 6 states of excitation comprising each overall state of excitation (***i.e.:*** every primary act) on the part of a form. That 10th epoch shall endure for roughly the square of 18 trillion Earth years before the 11th epoch begins, etc..

In that 10th epoch, the contents of the 9th epoch shall be effectively lost as a kind of "bubble" too tiny to be of any significance to the 10th epoch's inhabitants, just as the contents of the 1st thru the 8th epochs are 8 "bubbles" too tiny to be of any significance to us.[1]

If you're wondering why Esoptrics says the Universe is currently in its 9th epoch, you need only look at what Esoptrics says about the duration of each epoch. Since the first 8 epochs lasted for only what our sensation dependent minds would describe as 10^{-38} sec., it must be that the Universe is currently beyond the 8th epoch. Conversely, since the duration of the 9th epoch is what our sensation dependent minds would describe as 18 trillion Earth years, it's highly unlikely the Universe has gone beyond its 9th epoch. There's a second reason for saying 9th epoch; but, I'll not go into it here, save to say this: It has to do with 2^{128} & 2^{-128} times the hydrogen diameter vs. 2^{256} & 2^{-256} times that diameter. Well, there's also the interesting fact that, in Catholic Doctrine, there are 9 choirs of angels.

So then, in this the Universe's 9th epoch, there can be at least 2^{256} different forms (***i.e.:*** macro-ultimates) each distinct and separate from all the others by virtue of the facts that—in each of its overall states of excitation (***i.e.:*** primary acts)—each form has: (1) a discreet OD, and (2) 6 component states of excitation each with the same discrete intensity characteristic of the form's discrete OD.[2]

Since each form is receiving its discreet OD from a set of heavenly sextets associated with it and it alone, there are also at least 2^{256} heavenly sextets each distinct and separate from all the others by virtue of being associated with the one particular form to which it's imparting a discrete OD. In other words, The Infinite creates an astronomical number of heavenly sextets as well as an astronomical number of forms.

For Esoptrics, the OD's of 2^0, 2^1, 2^2, 2^4, 2^8, 2^{16}, 2^{32}, 2^{64}, 2^{128}, and 2^{256}, are *categorical* ones, and every form *native* to one of those OD's (A concept I'll explain later.) are categorical forms. The OD's of 2^3, 2^5, 2^6, 2^7, 2^9, 2^{10}, 2^{11}, 2^{12}, 2^{13}, 2^{14}, etc., are *generic* ones, and every form native to one of them is a generic form. In other words, every OD which is a power of 2 is either a categorical one or a generic one. Every OD which is not a power of 2 is a *specific* one, and every form native to one of them is a specific form. It's a very import distinction. For, in it lies Esoptrics' explanation of why there are electrons, atomic nuclei, solar systems and galaxies, the latter of which are expanding away from one another.

✡✞✡✞✡✞✡✞✡✞✡✞✡✞✡✞✡✞✡✞✡✞✡

[1] In the 10th epoch, the 9th will still exist and have inhabitants viewing their universe as many trillions of light years in diameter, though, to the 10th epoch's inhabitants, that 9th epoch shall appear no larger than what an atom is for us. What I'm saying here shall, of course, be meaningless, until one masters what Esoptrics says about the fundamental nature of acts of consciousness.

[2] The above distinction between forms becomes more complicated as we progress and distinguish between "native" vs. "current" OD—a distinction to be spelled out in due time. For now, to avoid sensory overload, be content with the above over-simplification, and draw from it the one principle which shall not change one iota as we progress—namely: that the current OD of any given form is the same as the current level of intensity for each of the 6 component states of excitation comprising the given form's overall state of excitation, which, as said repeatedly, is also called its primary act.

Lack of attention to externals lowers one's ability to achieve a *thorough* grasp of what we all already accept as fact. That's why we see this: Some live their whole lives keenly focusing upon what the external world presents to us. They beget theories founded on principles ever able to grow in breadth and harmony. Others, having made few external observations, quickly limit themselves to abstruse argumentation. They beget only hasty dogmatism. The former is the *scientific* method of inquiry; the latter is the *dialectical* one. In what follows, the vast difference between those two methods of inquiry will be depicted as we treat the subject now at hand in contrasting ways.
——**ARISTOTLE:** ***On Generation And Corruption***, Book I: Chap. 2; 316a:5-3. My translation – ENH. See pg. 411 vol. 8 of ***Great Books Of The Western World*** (The above strikes me as basically the same complaint [***Cf.:*** Bacon's quote from Heraclitus on my pg. 106] Science has been hurling at Aristotle and Philosophy for roughly the last 400 years; and so, to Aristotle and Science both, I say: If what you say is true, how do you explain what I—in pursuit of the ultramicroscopic makeup of time, space, etc.—have achieved as a result of "having made few external observations" as I focused mainly on "abstruse argumentation" with my own mind? Is it possible Hugh of St. Victor was closer than you to the truth, at least with regard to knowledge of the ultramicroscopic? If you've forgotten what the saint said, review Rev. Dr. John Alzog's words on page 22. Note what else Aristotle's above words suggest: The modern scientific age was not the result of a new *attitude* and *methodology* never before known. For, clearly, Aristotle—over 300 years BC [and Heraclitus over 480 years BC]—advocated the scientific method of external observation above Philosophy's opposite method. I dare suggest, then, that the modern scientific age is mainly set apart from the ancient one by a rise in incentive evoked by technological advances in: (1) telescopes, microscopes, and other instruments which made it much easier than ever before to observe far more widely and minutely; and (2) publishing methods which made it thousands of times faster and cheaper than ever before to record and communicate one's findings. The *great* change was not in the birth of a new attitude or methodology, but in an increased incentive to use what was always there but now much easier to use due to technological advances. A shovel gives far less incentive to dig than does a track hoe.).

The subject of quotation being introduced, Mr. Wilkes censured it as pedantry. JOHNSON. "No, Sir, it is a good thing; there is a community of mind in it. Classical quotation is the *parole* of literary men all over the world."
——**JAMES BOSWELL** (1740-1795)**:** ***Life Of Samuel Johnson***, *April 23, 1781*. As found on pages 474 bottom right-hand & 475 top left of Vol. 44 of ***Great Books Of The Western World*** as published by Encyclopædia Britannica, Inc.; Chicago, 1952.

Nothing gives an author so much pleasure as to find his works respectfully quoted by other learned authors.
——**BENJAMIN FRANKLIN** (1706-1790)**:** ***Pennsylvania Almanac***, pg. 1667: #7, ***The Home Book Of Quotations***, Sixth Edition by Dodd, Mead & Co., New York, 1949 (My sentiments exactly, Messrs. Boswell, Johnson, and Franklin. Hopefully, those I quote will feel the same plus honor the principle of fair use.).

Chapter Five:

FORMS VS. GENERATORS & HOW TOGETHER THEY PRODUCE U-SPACES IN A SIX-WAY LOGICAL SEQUENCE DICTATED BY THE ONTOLOGICAL QUANTUM EQUAL TO THE INVERSE OF THE FORM'S CURRENT OD:

In every form's every primary act (***i.e.:*** every *overall* state of excitation) each of the 6 *component* states of excitation persists unchanged in intensity for a definite number of alphakronons (***i.e.:*** K for short and, for sensation dependent minds, 7.2×10^{-96} sec.). In other words: In any given overall state of excitation on the part of a given form, each of the 6 states of excitation comprising that primary act has, throughout that act, an intensity which is equal to that of the other 5 and which changes not throughout that particular primary act. Throughout this the 9th epoch of the Universe, that intensity may be E, 2E, 3E, 4E, or 5E etc. all the way to 2^{256} E, where E = the least intensity possible.

All of that is the result of the fact that every form, in every one of its primary acts, is in potency to God 6 ways and, thereby, receives from God the 6 states of excitation which comprise each of its primary acts.[1] It's been said, though, that every such form is a *macro*-ultimate. Obviously, that implies a *micro*-ultimate. What, then, is a *micro*-ultimate?

Every micro-ultimate is, in Esoptrics' terminology, a generator. More exactly, each is a *carrying* generator. That's because every generator is forever logically concentric with a form which is forever its one and only *piggyback* form. As a result, wherever a given carrying generator *ever* goes, there the center of its piggyback form also goes.

Hereafter, every such combination of one carrying generator logically concentric forever with its piggyback form shall be called a "duo-centered combo" or simply a "duo-combo". In the final analysis, every one of the Universe's *ultimate* <u>*occupants*</u> is a duo-combo composed of the 2 *ultimate* <u>*constituents*</u>: piggyback forms and carrying generators.

In describing a duo-combo, I'm excluding the angels, each of whom is a form not logically concentric with a carrying generator. Each of them is not a *piggyback* form. For convenience's sake, though, let's not refer again to the angels. That shall allow us to say that, hereafter, the term "form" shall be equivalent to "piggyback form".

Every piggyback form is in potency to God and, thereby, receives from God, The

---

[1] In St. Thomas Aquinas' eyes (as I perhaps misunderstand it), one of the chief problems confronting Philosophy was: to explain how forms, like matter, could be a combination of potency and act and still be different from matter. To solve it, says Etienne Gilson (***cf.: Being And Some Philosophers***, pgs. 174-175), St. Thomas concluded forms are a kind of act in potency to existence. In Esoptrics, that issue turns into one of forms vs. the *generators of matter* rather than forms vs. *matter*, and the problem is easily dispelled by saying that, whereas the generators are in potency to the forms *as forms*, the forms are in potency to God *as The Supreme Act Of Existence*.

Supremely Exciting, the six states of excitation which comprise its every primary act (***i.e.:*** every *overall* state of excitation). Oppositely, every carrying generator is in potency to (***i.e.:*** seeks its states of excitation from) some, one, particular form other than its own piggyback form. It is, however, every carrying generator's own piggyback form which tells it to which other form it is to be in potency. If the generator's piggyback form is native[1] to, and currently at, a *specific* OD (***vd.:*** pg. 29 last par.), that generator shall be in potency to a form which is at the next highest specific OD. If the generator's piggyback form is native to a *generic* OD, that generator shall be in potency to a form which is currently at the next highest generic OD. If the generator's piggyback form is native to a *categorical* OD, that generator shall be in potency to a form which is currently at the next highest categorical OD.

At least, that's the rule as long as there's been no acceleration. That means this: Under some circumstances, a given form native to either a generic or categorical OD is driven to a current OD above its native one. If so, it "swallows" (***i.e.:*** makes logically concentric to itself) one duo-combo per OD thru which it passes. Example: Suppose a form native to OD4 climbs to a current OD of 2^{16} (***i.e.:*** 65,536). Rising, it makes logically concentric to itself 65,532 duo-combos. One of those captured duo-combos shall currently be at OD4; another currently at OD5; another at OD6; another at OD7; etc.. The result is thus what Esoptrics calls a *multi*-concentric combo or multi-combo for short. In this multi-combo, with a leading form at OD2^{16}, its 4th dimension has—within the confines of the same one c. 10^{-47} cm. (Thus the multi-combo's mass but not its "size" increases.)—65,533 levels each housing one duo-combo. If its lead form <u>de</u>celerates, it loses one 4th dimension level and expels one duo-combo per OD thru which it descends. That's a brief preview of how Esoptrics explains electro-magnetic absorption vs. radiation. More in a later chapter!

In potency to God, every form does more than merely receive the 6 states of excitation comprising each of its primary acts. Additionally, it *activates* each of those 6 *equally*. Thus, in each of its primary acts, each of its 6 component states of excitation is, so to speak, "vibrating" with the same intensity. That's far from true of a carrying generator.

From the form to which it's in potency, every carrying generator receives 6 *potential* states of excitation. In each of *its* primary acts, though, no carrying generator can ever simultaneously activate *both* one of those 6 potential states *and its opposite*. For example, for all generators, always, and everywhere, there can never be a primary act (***i.e.:*** *overall* state of excitation) in which it activates both the A and B potencies. By the same token, there can never be a primary act in which it activates both the A' and C potencies, nor a primary act in which it activates both the C' and B' potencies. In each of their primary acts, then, the generators may activate only one of the 6 potencies (***i.e.:*** only one of the 6 component states of excitation) or only 2 of them or only 3 of them. The un-activated ones remain present as a kind of "dead zone" putting a "drag" on the activated ones.

Drawing #5-1, next page, graphically but figuratively expresses the above so: The upper black square—while <u>ever</u> concentric with its larger square and <u>ever</u> symmetrically arranged with regard to its larger square's center—changes solely by periodically expanding outward toward, or contracting inward away from, its larger square's perimeter, and that it does in instantaneous quantum shifts (I didn't say "leaps".) to be explained later. The lower black square—while <u>never</u> concentric with its larger square and <u>never</u> symmetrically arranged with regard to its larger square's center—changes solely by periodically executing a durationless shift from one of the larger square's component squares to an adjacent one. On page 34, drawings #5-2 & 3 are an attempt to give the reader a three-dimensionally graphic but figurative expression of what drawing #5-1 seeks to convey.

---

[1] Chapter 7 spells out native OD's meaning. For now, say this: A form's native OD is the OD at which God created it and is mainly indicated by how it "spins" around its primary and secondary axes.

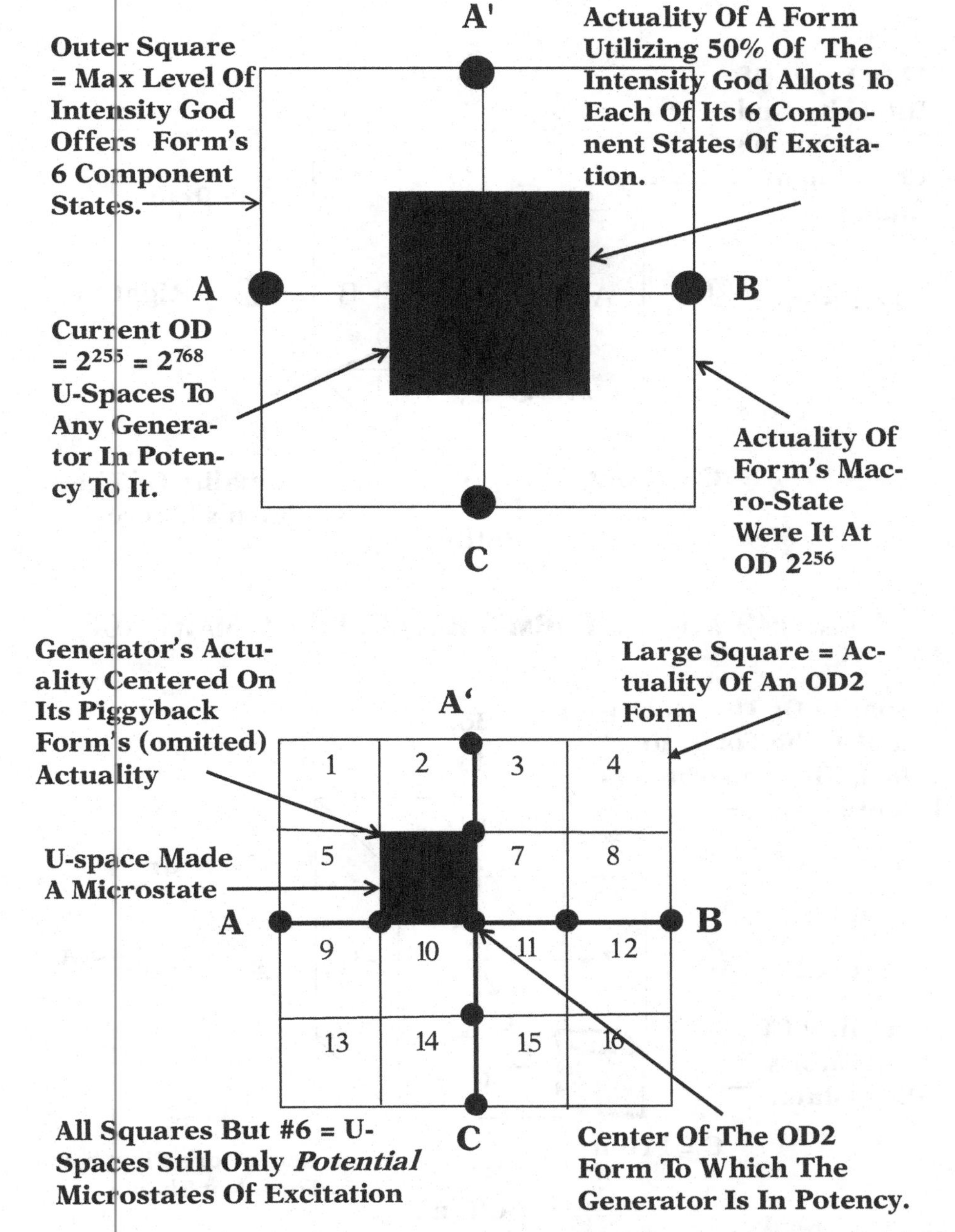

DRAWING #5-1: A FORM'S BALANCED PRIMARY ACT (UPPER BLACK SQ.) VS. A GENERATOR'S UNBALANCED PRIMARY ACT (LOWER BLACK SQ.).

As a result of what's just been said, one of the main differences between the forms and the generators is this: Throughout this the Universe's 9th epoch, the number of discrete kinds of primary acts possible to the forms is exactly 2^{256} (***i.e.:*** 1.15792×10^{77}); but, the number of discrete kinds of primary acts possible to generators activating 3 of their 6 potencies (***i.e.:*** # of 3-way primary acts) is 2^{771} (***i.e.:*** $[2 \times 2^{256}]^3 = 2^{771} = 1.242 \times 10^{232}$) if in potency to a form at $OD2^{256}$, slightly less if in potency to a form at $OD2^{256}$-1, slightly less if in potency to a form at $OD2^{256}$-2, etc.. Why is that?

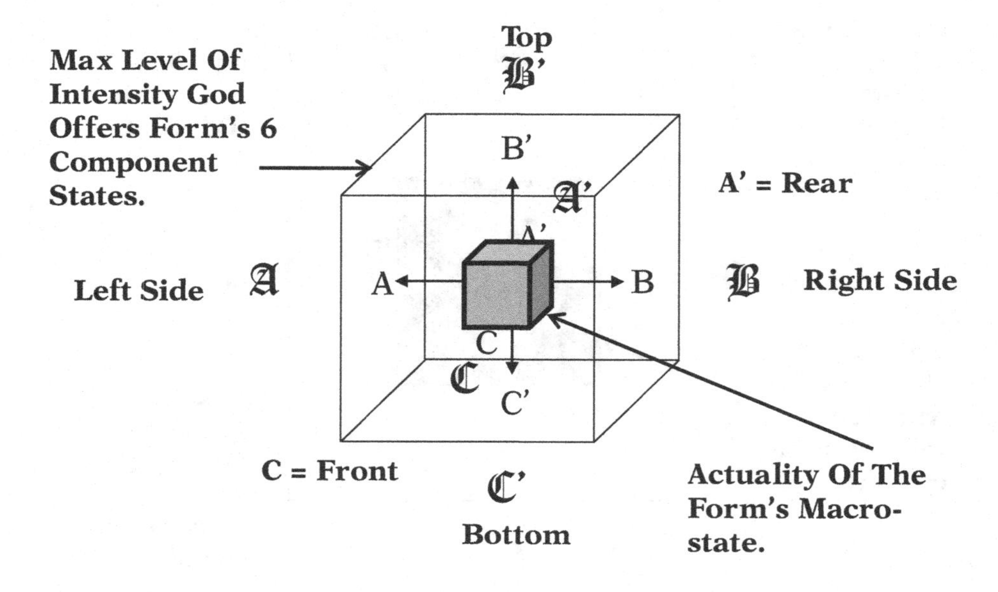

DRAWING #5-2: A FORM'S BALANCED PRIMARY ACT.

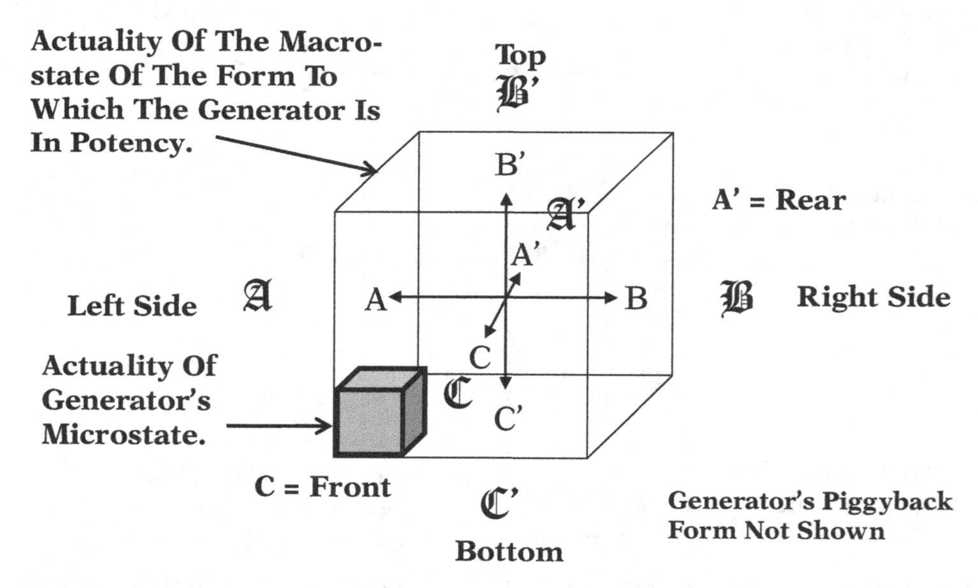

DRAWING #5-3: A GENERATOR'S UNBALANCED PRIMARY ACT.

That brings us to one of Esoptrics' most important formulas. It tells us how many different three-way primary acts a form makes possible to the generators in potency to it. Let P_3 = the number of three-way primary acts (***i.e.:*** overall microstates of excitation in which 3 of the 6 component states are activated) possible to a given generator. Let D_c = the current OD of the form to which the given generator is in potency. In that case, we have:

FORMULA 5-1: $P_3 = (2D_C)^3$

Drawing #5-4 below gives us a graphic but figurative expression of what that means. It pertains to a generator in potency to (***i.e.:*** getting its u-spaces from) a form currently at OD4. Applying the above formula, the drawing should present us with 512 different three-way primary acts available to the generator. That, it cannot do for 2 reasons: (1) that's too many acts to draw, and (2) even if I could draw 512 acts, I certainly wouldn't have the foggiest notion of how to draw a cube composed of 512 cubes. I must, then, implore my reader to be a bit patient and to try with me to invoke a hefty bit of imagination.

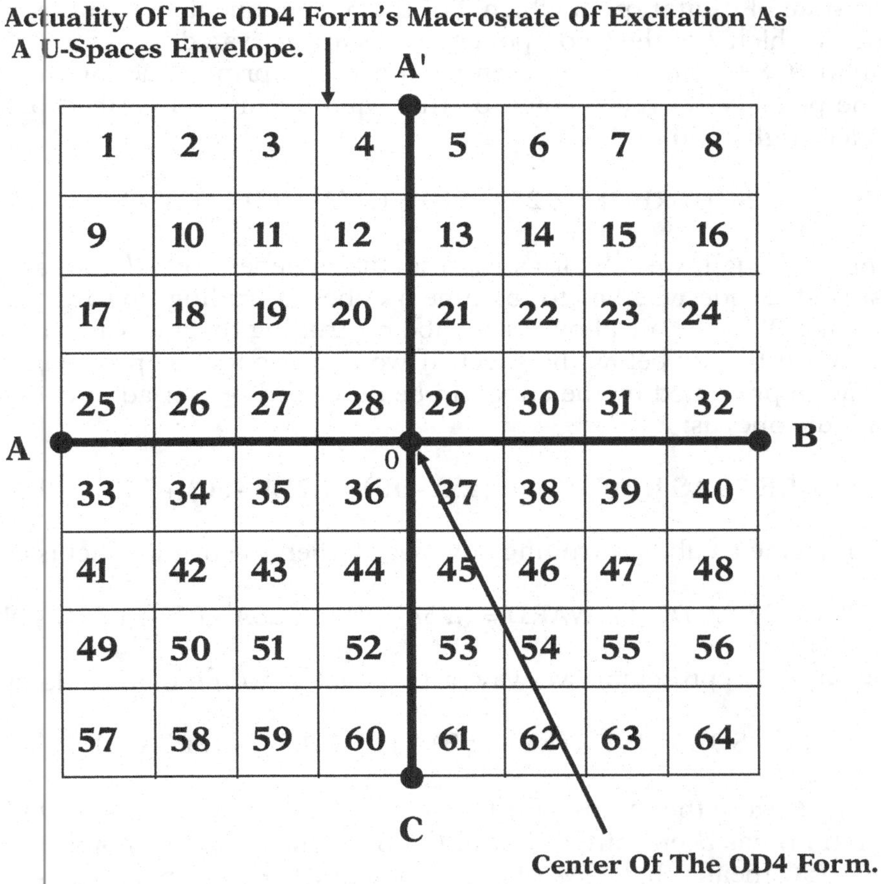

Actuality Of The OD4 Form's Macrostate Of Excitation As A U-Spaces Envelope.

A'

1	2	3	4	5	6	7	8
9	10	11	12	13	14	15	16
17	18	19	20	21	22	23	24
25	26	27	28	29	30	31	32
33	34	35	36	37	38	39	40
41	42	43	44	45	46	47	48
49	50	51	52	53	54	55	56
57	58	59	60	61	62	63	64

A 0 **B**

C

Center Of The OD4 Form.

DRAWING #5-4: U-SPACES (*i.e.*: microstates of excitation) POSSIBLE TO A GENERATOR IN POTENCY TO A FORM CURRENTLY AT OD4

The larger square represents the overall macrostate of excitation (***i.e.:*** actuality) a form is currently performing at OD4. Its 64 smaller squares represent 64 of the 512 possible 3-way microstates of excitation the form offers to any and every generator in potency to it. Every form is thus what Esoptrics calls a "u-spaces envelope", and there is no space save the vast number of such "u-spaces envelopes" each of which is a form's actuality. Every iota of space is thus very much a reality in a way Science does not even begin to understand.

Each of the 64 smaller squares also represents a u-space our sensation dependent minds would describe as c. 10^{-47} cm. in diameter. That makes every u-space—for any generator in potency to the above form—a *potential* microstate of excitation. Until some generator converts a particular u-space into an *actual* microstate (***i.e.:*** *transforms* it rather than merely *occupies* it), that u-space remains what our sensation dependent minds would describe as an area of empty space c. 10^{-47} cm. in diameter. In reality, every such "empty space" is the actuality of a form—an actuality to no extent *spatially* extended. How, then, are the 64 squares (512 cubes) 64 u-spaces outside of one another? Simply put: Immaterial realities are outside of one another merely because each is a unique state of excitation (***i.e.:*** each has a unique set of internal characteristics). How is each state unique?

Focus on square #28. Leave it a square, and use it to explain the difference between a 3-way vs. a 2-way primary act by a generator, which is to say the difference between an overall microstate of excitation in which 2 of the 6 component states of excitation are activated vs. one in which 3 of the 6 component states are activated.[1]

If square #28 be taken as representing a *two*-way primary act on the part of a generator, then the primary act represented by that square can be described by a triad of couplets (***i.e.:*** a left, right, and middle one) as:

SQUARE 28 = (.25A –0B) + (.25A' –0C) + (0B' –0C').

Suppose, though, we take it as representing a generator's *three*-way primary act. If so, we must first decide whether to imagine a cube protruding upward from the front of the sheet toward B' in the air above or a cube protruding downward from the back of the sheet toward C' in the air below the sheet. If we imagine a cube protruding upward, then the primary act represented by the cube can be described by a triad of couplets (***i.e.:*** a left, right, and middle one) as:

CUBE 28 AS UPWARD = (.25A –0B) + (.25A' –0C) + (.25B' –0C').

If we imagine a cube protruding downward, then the primary act is describable as:

CUBE 28 AS DOWNWARD = (.25A –0B) + (.25A' –0C) + (.25C' –0B').

Let square #1 represent a two-way primary act. It would then be describable as:

SQUARE 1 = (1A –.75B) + (1A' –.75C) + (0B' –0C').

The –.75B (Note the minus sign.) means that 75% of the A potency (***i.e.:*** potential A state of excitation) has been shifted over to the B potency (***i.e.:*** the potential B state of excitation). The –.75C means that 75% of the A' potency (***i.e.:*** the potential A' state of excitation) has been shifted over to the C potency (***i.e.:*** the potential C state of excitation). As a result, the B potency is a logically extended, dormant state of excitation putting a drag upon its opposite the A potency; and the C potency is a logically extended, dormant state of excitation putting a drag upon its opposite, the A' potency. There is thus a two-way drag inducing differential causing the generator to generate a two-way unit of tension which sensation dependent minds can experience as some kind of sensation. On the next page, drawing #5-5 gives a graphic but figurative explanation of how that shifting of one potency to its opposite takes place.

What if square #1 is a cube rising into the air above the sheet. In that case, the 3-

---

[1] The number of *two*-way primary acts possible is: $P_2 = 3[(2D_c)^2]$, and the number of *one*-way primary acts possible is: $P_1 = 3(2D_c)$.

way primary act thus represented would be describable as:

CUBE 1 AS UPWARD = (1A –.75B) + (1A' –.75C) + (.25B' –0C').

Here, the –0C' means none of the B' potency has been shifted over to the C' potency thus leaving the C' potency a logically extended, dormant state of excitation but—because it's strictly on its own side of the center point—exerting no dragging effect upon the B' potency. Even as a cube, then, #1 produces only a two-way unit of tension which sensation dependent minds can experience as some kind of sensation.

OD4 Form's A'C Axis

A'

A'

Generator's A'C Axis

1	2	3	4	5	6	7	8
9	10	11	12	13	14	15	16
17	18	19	20	21	22	23	24
25	26	27	28	29	30	31	32
33	34	35	36	37	38	39	40
41	42	43	44	45	46	47	48
49	50	51	52	53	54	55	56
57	58	59	60	61	62	63	

OD4 Form's AB Axis

A

B

75% Of C To A' &
75% Of B To A

A

B

Generator's Microstate Of Excitation

Generator's AB Axis

C

C

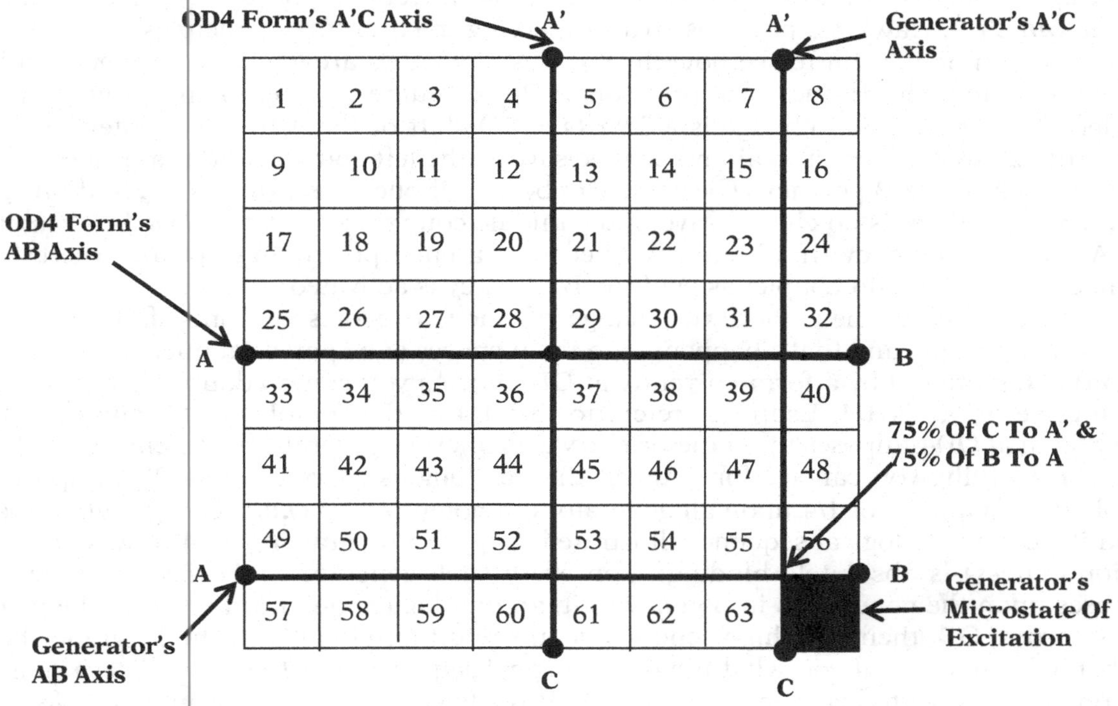

DRAWING #5-5: A GENERATOR IN A MICROSTATE OF EXCITATION DESCRIBABLE AS (B –.75A) + (C –.75A') + (0B' –0C') & SHOWING HOW 75% Of A IS SHIFTED TO B & 75% OF C IS SHIFTED TO A'.

Focus on square #19 in drawing #5-4, page 35, and imagine it's a cube extending upward. Next, imagine there is a cube resting on top if it—a cube we shall call 19+. It represents a three-way microstate of excitation describable as:

CUBE 19+ AS UPWARD = (.5A –.25B) + (.5A' –.25C) + (.5B' –.25C').

We now have the B potency placing a drag on the A potency, the C potency placing a drag on the A' potency, and the C' potency putting a drag on the B' potency (***i.e.:*** B, C, and C' are each now on 2 sides of the center point). The result is a three-way microstate of excitation generating a three-way unit of tension sensation dependent minds can experience as some kind of sensation (tactile?). I decline to describe a one-way microstate.

This bit of one dormant potency putting a drag upon its active opposite never occurs in the *forms*. For, as said, every form, in its every primary act (***i.e.:*** overall macrostate of excitation) equally activates each of its 6 component states (***i.e.:*** its every primary act is always symmetrical at the center point of its 6 potencies) and, thereby, never allows an active potency to be logically adjacent to a dormant opposite crossing the center point. As a result, forms, unlike generators, perform only balanced states of excitation and, thus, never generate units of tension. Balanced vs. unbalanced states of excitation is also why forms never have mass whereas generators—when made concentric by means of their forms—do (***i.e:*** it's the generators which have mass, but the forms which enable them to do so, and it's the forms which transmit that mass' force of gravity). Having no mass, forms can rarely be detected by Science and can only be deduced by those given to the "fool's errand" (Dulles) of trying to "work out all the laws that govern the universe by pure thought" (Hawking). No, not ***all***, Prof. Hawking, just those truly deserving the title "first principles"!

Do you now begin to see how the squares and cubes are logically arranged with regard to one another? If one goes from square 28 to square 27, there is no change save in the left couplet, and that change is a 25% (***i.e.:*** ¼) shift of the A to the B potency. If one goes from 27 to 26, there is again no change save in the left couplet, and that is a second ¼ shift of the A to the B potency. The same can be said if one goes from 26 to 25. If one goes from 25 to 17, there is no change save in the middle couplet, and that change is a ¼ shift of the A' to the C potency. If 17 suddenly becomes a cube protruding upward, there is no change save in the right couplet as ¼ of the B' potency is activated.

In each case, then, there's a change of one-*fourth* of some one of the potencies and/or its opposite, and that's happening to the u-spaces (***i.e.:*** potential microstates of excitation) being offered by a form currently at OD *four*. As you can no doubt see for yourself, the inverse of 4/1 is 1/4. Esoptrics refers to this 1/4 as the "ontological quantum" which every form at OD4 imposes upon the primary acts of every generator in potency to it.

Hopefully, you can see for yourself that, as long as every form at OD4 imposes an ontological quantum of 1/4 upon all generators in potency to it (***i.e.:*** getting their u-spaces from it), the 4 way logical sequence illustrated on pg. 35 by drawing #5-4 (6 way in imagination's *cubes*) is absolutely binding upon every such generator. Likewise, if every form imposes upon the generators in potency to it, an ontological quantum equal to the inverse of its current OD, then even high school students and those in at least the 5th grade should probably be able to tell you what the 6 way logical sequence must be for whatever current OD one might care to bring into the picture. If really good at it, they should be able to tell you which 26 "cubes" are logically adjacent to any cube not at the outer perimeter.

Are you still somewhat in the dark regarding why I refer to logical sequence as a 6 way one? Surely you can see for yourself, that logical sequence takes place with reference to the 6 potencies: A, B, C, A', B', and C', and the only way to accurately describe which u-space is where in logical sequence is to describe it in the terms of three sets of couplets stating which positive or negative percentage of which of those 6 potencies is in play.

Perhaps more importantly, the exact internality of each of those three sets of couplets is determined by the ontological quantum pertinent to the form supplying the u-spaces to the generators in potency to it. How so?! Look again at drawing #5-5 on the prior page and ask: Why is it true that, as a generator shifts from square #28 to square #27, it is not skipping an intermediate square? In other words, Esoptrics is saying the transition is:

FROM: (.25A –0B) + (.25A' –0C) + (0B' –0C') AT SQUARE 28
TO: (.5A –.25B) + (.25A' –0C) + (0B' –0C') AT SQUARE 27.

Why is the shift not:

FROM: (.25A –0B) + (.25A' –0C) + (0B' –0C') AT SQUARE 28
TO: (.375A –.25B) + (.25A' –0C) + (0B' –0C') AT SQUARE 27.5?

It's not such due to the ontological quantum demanding that—going from one u-space provided by a form at OD4 to one of the logically adjacent ones it provides—the change per couplet member must be a multiple of ¼. Fail to see that, and maybe you're thinking of the "squares" as places adjacent to one another in *space*. That, they most certainly are ***not***. They are states of excitation logically adjacent in *logical sequence*. Place in *space* has nothing to do with it because, in Esoptrics, there is no such thing as a place in *space* nor any *spatially* extended thing save as the *collective effect* of the u-spaces and the generators. In Esoptrics, logical proximity is *algebraic* in nature and *not geometrical*; and so, logical proximity is as *Algebra calculates* and not as *Geometry depicts*. No matter what pictures Geometry gives to minds chained to what sensations suggest, Algebra calculates logical proximity in the terms of what is determined by the ontological quantum pertinent to the form providing the u-spaces. Thus, the generator shifting from square #28 to square #27, in no way skips or "leaps" over anything between #28 and #27 because, between them, there is no such thing as space and, because of the ontological quantum, no possibility of an intermediate state of excitation. Is that really all that difficult to grasp?

To close this chapter, let's slightly complicate the issue of logical sequence. In drawing #5-5 on page 37, imagine a generator called Gene is performing the state of excitation represented by square (cube in imagination) #28. Logically adjacent to #28 are the squares (cubes in imagination) #19, 20, 21, 27, 29, 35, 36, 37, plus the 9 "cubes" logically (not spatially) overlying 19, 20, 21, 27, 28, 29, 35, 36, 37, plus the 9 logically (not spatially) underlying those 9. That's a total of 8 + 9 + 9 = 26 from what drawing #5-5 indicates. From another standpoint, the "cubes" logically adjacent to #28 = 26 + 27 +27 = 80. How so?!

The 26 first described are the ones logically adjacent to #28 from the standpoint of the OD4 form providing Gene with Gene's u-spaces. But, there is a case where a form currently at OD5 and another currently at OD3 are each logically concentric with the OD4 form because all 3 are components of a multi-combo. Each of those other 2 provides 27 "cubes" logically adjacent to #28 but only *conditionally* so. What does that mean?

Imagine Gene is persistently shifting from state of excitation to logically adjacent state of excitation at a fixed rate of, say, 1Φ per 2^{158}K (***i.e.:*** roughly .7MPH). As long as that's the case, Gene can use only the u-spaces made available by whatever form is currently providing it with u-spaces. Let's imagine the form doing that is the one at OD4. Suddenly, a collision (the result of Gene and another generator trying to activate the same u-space simultaneously) causes Gene to <u>ac</u>celerate slightly and, as a result, to change from state of excitation to logically adjacent state of excitation at the new rate of 1Φ per 2^{158} -1K. Instantly, Gene shall cease getting his u-spaces from the OD4 form and start getting them from the next <u>*highest*</u> form, which is to say the one at OD5. Oppositely, did the collision cause Gene to <u>de</u>celerate slightly to the rate of 1Φ per 2^{158} +1K, then Gene would start getting his u-spaces from the next <u>*lowest*</u> form, which is to say the one at OD3. Thus, as the result of a collision, what u-spaces had been logically adjacent to #28—but only on the condition of a change in Gene's rate of shifting from one state of excitation to another—are now logically adjacent to #28 because such a change has in fact occurred.

The shift from cube #28 provided by the OD4 form in drawing #5-5 to a logically adjacent one provided by an OD5 form would be describable so:

FROM: (.25A –0B) + (.25A' –0C) + (,25B' –0C') AT CUBE #28 OF THE OD4 FORM
TO: (.2A –0B) + (.2A' –0C) + (.2B' –0C') AT CUBE #45 OF THE OD5 FORM.

2013 ESOPTRICS UPDATE

It's a change Geometry cannot depict, since the center of every level of a multi-combo's 4th dimension is wholly limited to the same unit 10^{-47} cm. small. Thus, in a new drawing, the OD5 form's #50 square adjoins the right side of drawing #5-5's square #32; its #49 matches #31; its #48 matches #31; #47 #30; #46 #29; and its #45 square is in *exactly* the same location as drawing #5-5's #28. For, a drawing of the OD5 form would have 10 rows by 10 columns instead of #5-5's 8 rows by 8 columns; but, that affects only the outer perimeter of drawing #5-5 by adding 36 more squares (***viz.:*** 10 above +10 below + 1 to each side of each of #5-5's 8 rows). It doesn't change the location or size of any of the original squares. Still, though Geometry's pictures put the OD5 form's #45 and the OD4 form's #28 in the same location, Algebra's calculations say they are not so, because *location* has nothing to do with what makes #45 and #28 outside of one another. With both inside the same 10^{-47} cm., all that matters is that the triad of couplets defining the internality of #45's state of excitation is not the same as the triad defining the internality of #28's state. *That's* what makes #45 & #28 2 *separate* levels of the 4th dimension. It is thus their *God-given internality* which separates them—not the mass of humanity's long worshipped *Deus ex machina* kind of space. Kant was right about what a mere mental crutch such space is. Would he had seen it's true only of minds crippled by too little knowledge of God!

In their Theoretical Physics, scientists try to deduce *what kind of coordinates* are demanded if attention is focused outwardly upon time, space, and locomotion as presented to one's senses by the *extra*-mental world. In its Theoretical Physics, Esoptrics tries to deduce *what kind of time, space, and locomotion* are demanded if attention is focused inwardly upon the God-given coordinates presented by the *intra*-mental world to an intellect seeking to find—in a total rise above sensation—the *real* starting point. As one scientist admits it:

> The supreme task of the physicist is to arrive at those universal elementary laws from which the cosmos can be built up by pure deduction. There is no logical path to these laws; only intuition, resting on sympathetic understanding of experience, can reach them. . . [Skip!] . . Nobody who has really gone deeply into the matter will deny that in practice the world of phenomena uniquely determines the theoretical system, in spite of the fact that there is no logical bridge between phenomena and their theoretical principles; [Pg. 226]
> . . . . . In contrast to psychology, physics treats directly only with sense experiences and of the "understanding" of their connection. [Pg. 290]
> ——**ALBERT EINSTEIN: *Ideas And Opinions*,** on the cited pages in the hardbound book by Bonanza Books, New York, 1954. ©1954 Crown Publishers, Inc..

Yes, Science "treats directly only with sense experiences" in its efforts "to arrive at those universal elementary laws from which the cosmos can be built up by pure deduction"; and so, "the world of phenomena uniquely determines" <u>its</u> "theoretical system". Esoptrics treats directly only with introspection's contemplative understanding of God's 7 *a priori* coordinates; and so, that's what uniquely determines <u>its</u> theoretical system. Only secondarily does it use sensation to *help* it understand those coordinates' connection to how the cosmos is built up by God. It's mono- vs. bi-directional empirical observation, and I, while Science denies even the possibility of it, have chosen the latter. Let history determine if that makes me a "nobody" who has <u>not</u> "really gone deeply into the matter".

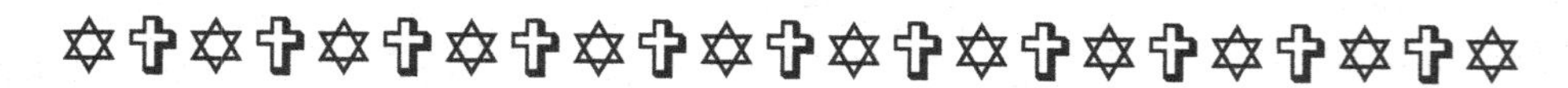

Chapter Six:

THE DUO-COMBO'S RELEVANCE TO NEWTON, EINSTEIN, SUPERSYMMETRY, FERMIONS VS. BOSONS & THE ABSORPTION OF LIGHT:

> The notion "material point" is fundamental for mechanics. If now we seek to develop the mechanics of a bodily object which itself can *not* be treated as a material point—and strictly speaking every object "perceptible to our senses" is of this category—then the question arises: How shall we imagine the object to be built up out of material points, and what forces must we imagine as acting between them? (pg. 301)
>
> . . . .
>
> It is an unsatisfactory feature of classical mechanics that in its fundamental laws the same mass constant appears in two different roles, namely as "inertial mass" in the law of motion, and as "gravitational mass" in the law of gravitation. (pg. 308)
>
> ——**ALBERT EINSTEIN**: ***Physics And Reality***, originally given in ***The Journal Of the Franklin Institute***, Vol. 221, No. 3, March, 1936, and found on the pages cited in the hardbound copy of ***Ideas And Opinions*** published by Bonanza Books, New York, 1954. ©1954 by Crown Publishers, Inc..

Before I take up what I mainly want to say in response to the above, let's talk about what Einstein might mean by "material point". The term "point" is often defined as *infinitesimal* point, meaning something to no extent spatially extended and, therefore, devoid of physical length, width, depth, size, shape, and the like. That's the way I use it whenever I refer to the Universe's ultimates. On the Internet, there is this web site:

youstupidrelativist.com/01Math/01Point/03Infinitesimal.Html

It's a web site at which someone apparently named Bill Gaede—after attributing the infinitesimal definition of "point" to page 6 of the book ***Dr. Math Introduces Geometry*** by Jessica Wolk-Stanley (He seems unaware she merely illustrated it.) of the Drexel University Math Forum—rants against it as he also rails against Newton, Einstein, Hawking, General Relativity, Quantum Mechanics, String Theory, and what he calls "the crazy world of Einstein's Idiots". Wikipedia calls him "the quintessential cold war industrial spy". The articles at the web site are copyrighted by Nila Gaede who is presumably Bill's wife.

According to Mr. Gaede, "point" defined as infinitesimal means points have "0-D" (***i.e.:*** zero dimensions). Esoptrics denies that and says "infinitesimal point" means each u-space (***i.e.:*** microstate of excitation) has zero *spatial* dimensions (because not really units of space) but, nevertheless, has *logical* ones making *groups* of them *collectively* spatial and, thus, an *effect* truly 3 dimensional. In sensing groups of u-spaces and generators as 3 dimensionally shaped, our senses do not lie. For, though the groups' shapes are only *collectively* and *effectively* 3 dimensional, they *are collectively* and *really* shaped as sensed. It's just that our senses do not, and cannot, make manifest the ultra-microscopic *why* (***i.e.:*** divinely imparted *logical* extension) individual u-spaces and generators can produce, *collectively* and *in effect*, what *truly is* the opposite of what they are *individually* and *fundamentally*.

I am far from sure how Einstein is using the term "material point"; but, I seriously doubt he's using it in the sense of *infinitesimal* point. Since he contrasts "material point" with what no "object perceptible to our senses is", I strongly suspect he means that, for Newton, every one of the material objects, large enough for us to observe, is—as is now widely contended—a collection of microscopic *ultimate* particles spatially separated from one another but not themselves composed of parts so separated or to any extent separable save perhaps in thought. In other words, instead of meaning *infinitesimal* point, he means *irreducible* point and, thus, ultimate constituent.

If that is correct, then I think I can recast the above quote from Prof. Einstein so:

> It is an unsatisfactory feature of classical mechanics that in its fundamental laws the same *ultimate irreducible constituent* appears in two different roles, namely as "inertial mass" in the law of motion, and as "gravitational mass" in the law of gravitation.

Whether "ultimate constituent" be taken to mean *infinitesimal* point or not, that "unsatisfactory feature" cannot be found in Esoptrics. For, in Esoptrics, the Universe's every ultimate occupant is a "duo-combo" in which a carrying generator and a piggyback form—the 2 kinds of ultimate constituents—are logically concentric but radically different from one another. As a result of how different they are, the same one ultimate occupant can appear in two different roles *but not under the same "side" of its duality*. As a generator, every duo-combo is an impermeable particle and the '"inertial mass" in the law of motion'; and yet, as a form, every duo-combo is a permeable wave-like reality (due to how forms rotate) and is what exerts the generator's '"gravitational mass" in the law of gravitation.' Indeed, as the form, every duo-combo is the source of instantaneous action at distances as great as 18 trillion light years (in "Geometry-speak"), and is what accounts for all kinds of radiation, electrical charges, and magnetic lines of force. More on that later!

There is still something else which Esoptrics accomplishes with its concept of duo-combos. To illustrate what, let's turn to another very famous scientist who is still alive, though many marvel that he still is. In conjunction with a co-author, he writes:

> One of the important implications of supersymmetry is that force particles [These, they call bosons on pg. 104 – ENH] and matter particles [These, they call fermions on page 104 – ENH], and hence force and matter, are really just two facets of the same thing. Practically speaking, that means that each matter particle, such as a quark, ought to have a partner particle that is a force particle, and each force particle, such as the photon, ought to have a partner particle that is a matter particle.
>
> ——**STEPHEN HAWKING & LEONARD MLODINOW: *The Grand Design***, as found on page 114 of the paperback edition published by Bantam Books Trade Paperbacks, New York, 2012. ©2010 by its 2 authors.

Does that not agree with Esoptrics' contention that—though the Universe's every *ultimate constituent* is either a carrying generator or a piggyback form—every *ultimate occupant* is a *duo*-combo in which a carrying generator is forever logically concentric with (***i.e.:*** partnered with) its own piggyback form? The result is—in the above quote's terminology—"the same thing" which, as the carrying generator, is a unit of mass (***i.e.:*** "matter particle" and "fermion" in the above), even as, in its partner (***i.e.:*** its piggyback form), it is—in the above quote's terminology—the kind of "force particle" and "boson" which, by accelerating, "forces" many duo-combos to become logically concentric in a *multi*-combo. What's unclear here is the *exact* sense in which supersymmetry says the two partners are "the same thing". For, it seems to imply they are still spatially separate rather than logically concentric as Esoptrics says. In other words, in supersymmetry, does "the same thing" mean the same one *whole* or not? In Esoptrics, though, "the same thing" most certainly does mean the same one *whole*.

Here's yet another point of relevance between modern Theoretical Physics and Esoptrics' concept of the duo-combo:

> All known particles in the universe belong to one of two groups, fermions or bosons. Fermions are particles with half-integer spin (such as spin ½), and they make up ordinary matter. Their ground state energies are negative.
>
> Bosons are particles with integer spin (such as 0, 1, 2), and these give rise to forces between the fermions, such as the gravitational force and light. Their ground state energies are positive.
>
> ——**STEPHEN HAWKING:** ***The Universe In A Nutshell***, pg. 50 of the hardbound book published by Bantam Books, New York, 2001. ©2001 by Stephen Hawking.

Generators are *half* states of excitation in that, in every one of their overall *micro*states of excitation (***i.e.:*** primary acts), they never activate both a given one of the 6 component states of excitation and its opposite and, so, at best, activate 3 of the 6 component states. Forms are *whole* states of excitation, in that, in every one of their overall *macro*states of excitation (***i.e.:*** primary acts), they always activate equally all 6 of the 6 component states of excitation. Forms are what give rise to the forces between the generators, such as the gravitational force and light—all of which shall, I hope, become clearer as we progress. Even now, though, perhaps you notice to some extent the similarity between Science's fermions vs. bosons and Esoptrics' carrying generators vs. the piggyback forms logically concentric in every duo-combo.

Finally, there is this point of relevance between today's Theoretical Physics and the result of Esoptrics' concept of the duo- vs. the multi-combo. Regarding Newton's system, we are told:

> The most unsatisfactory side of this system (apart from the difficulties involved in the concept of "absolute space" which have been raised once more quite recently) lay in its description of light, which Newton also conceived, in accordance with his system, as composed of material points. Even at that time the question, What in that case becomes of the material points of which light is composed, when the light is absorbed?, was already a burning one. Moreover, it is unsatisfactory in any case to introduce into the discussion material points of quite a different sort, which had to be postulated for the purpose of representing ponderable matter and light respectively. Later

on, electrical corpuscles were added to these, making a third kind, again with completely different characteristics.
——**ALBERT EINSTEIN: *Maxwell's Influence On The Evolution Of The Idea Of Physical Reality***, first published, 1931 in ***James Clerk Maxwell: A Commemoration Volume***, Cambridge University Press, and found on page 267 of ***Ideas And Opinions*** in the hardbound book published by Bonanza Books; New York, 1954. ©1954 by Crown Publishers, Inc..

What is meant by "ponderable matter"? Does it mean the sensations we see, hear, smell, taste, and feel? Or perhaps it means what's three-dimensional and—if not at least difficult either to penetrate or to split apart—is impossible to penetrate or fragment. In either case, in Esoptrics, "material point" has nothing to do with the generators and piggyback forms which alone are the Universe's ultimate constituents. For, each of them is an immaterial, and only *physically* infinitesimal point; and so, in Esoptrics, "ponderable matter", "material point", and "the stuff of our sensations" must all mean the same kind of reality—namely: the units of tension which the carrying generators generate. Generators and forms cannot possibly be "material points", since—though totally without *spatial* extension and the *physical* dimensions of length, width, and depth—they're not *wholly* without extension. As repeatedly said, they, *individually*, have the *logical* extension and *logical* dimensions uninterruptedly imparted to them by God's 7 *a priori* coordinates.

In Esoptrics, every reality—if not a unit of tension generated by a generator—is either a single duo-combo or two or more such in some way conjoined. What does that mean beyond duo-combos logically concentric in a multi-combo? Esoptrics suggests as follows.

If in no way conjoined with another as it moves thru the Universe, every such single duo-combo is either a neutrino or an anti-neutrino. As we'll see later, these are of more than one kind depending upon the *native* OD of its piggyback form. Naturally, since each is a duo-combo, neutrinos and anti-neutrinos can be absorbed the same way all duo-combos are—namely: by being made logically concentric to a multi-combo and, thereby, housed at one of that multi-combo's 4th dimension levels until that level is dumped by deceleration.

Every photon is two or more duo-combos moving as a *chain* (Whether or not the number is ever greater than two, I know not.). That means this: The links are by no means logically concentric with one another, and the *first* link in the chain serves as a kind of leader for the other links. How so?!

The first link in the chain is a duo-combo whose generator is moving thru the Universe at light speed by using the u-spaces provided to it by one of the forms logically concentric in some multi-combo and serving as a u-spaces envelope. Usually, that means it's using the u-spaces provided by a form whose current OD is at least 2^{194} (***i.e.:*** one having a radius of $2^{194}\Phi$ = c. 22.9 million miles in "Geometry speak"). It persists in using that form's u-spaces until it finds a juncture at which its latest u-space is logically contiguous to one provided by a form currently at the next highest OD (***i.e.:*** at the next highest level of the 4th dimension) and thus able to provide it with a wider field of travel. That form's u-spaces it then persists in using until it finds a juncture at which its latest u-space is logically contiguous to one provided by a form currently at the next highest OD and thus able to provide it with an even wider field of travel. That form's u-spaces it persists in using until it finds a juncture at which its latest u-space is logically contiguous to one in a form currently at the next highest OD, etc., and that's a process which persists until either it winds up using the u-spaces provided by the one form native to OD2^{256} or it is absorbed by, and made logically concentric to, a multi-combo in whose ontological depth there are OD's (***i.e.:*** levels of the 4th dimension) at which no forms are currently present and, therefore, vacancies to be filled by the absorbed photon's small handful of duo-combos.

The second link in the photon's chain is also moving at light speed. It, though, is using the u-spaces provided to it by the first link's piggyback form, and, throughout the "life" of the photon, that's its only source of u-spaces. For that reason, throughout the life of the photon, the second link is, at light speed, shuttling back and forth from the center to the outermost perimeter of the u-spaces provided by the first link's piggyback form. Depending upon the native OD of the first link's piggyback form, that field of travel may be quite small or quite large. Also, that field of travel is rotating at one of an astronomical number of rates. As a result, as the second link shuttles back and forth, it produces a wave as it moves outward for the first half of the first link's rotation and moves inward for the second half of the first link's rotation. Naturally, the length and frequency of the wave will vary as the native OD of the first link. When a photon is absorbed, the *linked* duo-combo composing it become logically concentric with one another and the rest of the duo-combos in the multi-combo.

As for the "electrical corpuscles" mentioned by Prof. Einstein, these are electrons moving at light speed thus making electricity a flow of electrons. In Esoptrics, every electron at light speed is a multi-combo in which 2^{128} (***i.e.:*** 3.4 x 10^{38}) duo-combos are logically concentric with one another (or chained together sometimes?), and each and every one of those duo-combos is one whose piggyback form is native to OD1. For, no categorical form—*other than one native to OD1*—is ever logically concentric within a multi-combo save within one in which it is the leading form, and none of the categorical forms native to OD1 can ever be made logically concentric to any other form save one which, as with itself, is native to OD1, thereby greatly limiting the wave characteristics if the electron travels as a chain of 2^{128} duo-combos. It's very important to remember those two principles.

In Esoptrics, then, there is no need to introduce even so little as *two* let alone *three* kinds of points. There is only the need to distinguish between the various ways in which the duo-combos—the *one* and *only* kind of points—can relate to one another either as: (1) logically concentric (***i.e.:*** absorbed at different levels of a multi-combo's 4th dimension), or (2) as members of a chain, or (3) as single neutrinos and anti-neutrinos. In Esoptrics, there is not the slightest difficulty in accounting for either: (1) how light is absorbed into a multi-combo's 4th dimension, or (2) the difference between particles and waves.[1]

✡✞✡✞✡✞✡✞✡✞✡✞✡✞✡✞✡✞✡✞✡✞✡

[1] Each and every duo-combo is forever the same duo-combo it has been since God created it. For, all the changes each undergoes in the course of its existence is merely an *accidental* rather than a *substantial* one. Yes, there are those who insist somewhat so: "Nothing exists continuously its same self, unless it's continuously changing, since only what changes continuously remains what it is—namely: the continuously changing. Oppositely, whatever changes only intermittently changes from what it was to something else—namely: from the changeless to the changed. Thus, it's no longer the same self." That's merely a way of repeating the ancient doctrine that change is the only reality. I, though, counter that an intermittently changing entity remains itself no less than a continuously changing one does, and the former, unlike the latter, does so without the absurdity of infinite divisibility. It is but a gratuitous and absurd assumption to claim the continuously changing entity remains itself, but the intermittently changing does not. More importantly, the issue is not continuous vs. intermittent change, but accidental vs. substantial change—a distinction which, apparently, even some supposedly intelligent people cannot fathom. But, let them say what they want all they want. *Saying* it doesn't make it true, and I defy them to prove how it is self-contradictory or in any way untenable to say that what changes intermittently can remain itself as long as its changes are only accidental and never substantial.

The parts of pure space are inseparable one from the other; so that the continuity cannot be separated, neither really nor mentally. For I demand of any one to remove any part of it from another, with which it is continued, even so much as in thought. To divide and separate actually is, as I think, by removing the parts one from another, to make two superficies, where before there was a continuity: and to divide mentally is, to make in the mind two superficies, where before there was a continuity, and consider them as removed one from the other; which can only be done in things considered by the mind as capable of being separated; and by separation, of acquiring new distinct superficies, which they then have not, but are capable of. But neither of these ways of separation, whether real or mental, is, as I think, compatible to pure space.

. . . . .

Thus the determined idea of simple space distinguished it plainly and sufficiently from body; since its parts are inseparable, immovable, and without resistance to the motion of body.

If any one ask me *what* this space I speak of *is*, I will tell him when he tells me what his extension is. For to say, as is usually done, that extension is to have *partes extra partes*, is to say only, that extension is extension. For what am I the better informed in the nature of extension, when I am told that extension is to have parts that are extended, exterior to parts that are extended, i.e. extension consists of extended parts? As if one, asking what a fibre was, I should answer him,—that it was a thing made up of several fibres. Would he thereby be enabled to understand what a fibre was better than he did before? Or rather, would he not have reason to think that my design was to make sport with him, rather than seriously to instruct him?

——**JOHN LOCKE: *Concerning Human Understanding*,** Book II: Chap. XIII: Sections 13, 14 & 15 as found on page 151 of Vol. 35 of ***Great Books Of The Western World*** as published by Encyclopædia Britannica, Inc.; Chicago, 1952 (Well, Mr. Locke, were you alive, you would see that Esoptrics—in many drawings each simple enough—shows you: (1) exactly how one *can* mentally separate the ultimate parts of "pure space" into microstates of excitation each logically separated from all the others by virtue of its unique internality; (2) exactly what each of these "u-spaces" *is* [***viz.:*** if not activated = the actuality of some form serving as a u-spaces envelope; if activated = the actuality of some generator]; and (3) how extension can be rightly defined as macro- and micro-states of excitation which—though all contained within the spatially <u>un</u>extended confines of a *physically* infinitesimal and *physically* indivisible point—are *logically* outside of one another in a 7 "dimensional" logical sequence as a result of the unique internality given each by God's 7 *a priori* coordinates. I can't help but wonder though: Were you alive and saw that, would you be "the better informed in the nature of extension", or—like almost everyone who today hears anything regarding Esoptrics—dismiss Esoptrics with indifference at best and gutter level slurs at worst?).

Chapter Seven:

HOW FORMS ENABLE GENERATORS TO ENGAGE IN SEEMINGLY CONTINUOUS RECTILINEAR LOCOMOTION:

Generators engage in rectilinear locomotion only. Curvilinear motion is unique to the u-spaces and occurs when the forms, by rotating, cause the u-spaces to circle the center of the form which, as a u-spaces envelope, is providing them. We'll spell that out later. For now, let's concentrate on the rectilinear locomotion of the generators.

Drawing #7-1, next page, gives us a graphic but figurative expression of how rectilinear locomotion occurs. It does so by figuratively depicting the logical sequence governing all rectilinear locomotion. That means this: Suppose, as in drawing #7-1, a generator is performing the primary act square #27 represents, but must switch to the primary act #63 represents. Because of the logical sequence illustrated by drawing #7-1, the generator cannot merely suddenly stop performing #27's primary act and start performing #63's primary act (***i.e.:*** the ontological quantum principle forbids a "quantum leap" from #27 to #63 skipping all the squares in between). By no means! After performing #27's primary act without any kind of change for a particular number of K, it must, in a durationless instant, have suddenly stopped performing #27's and—if following a "straight" rather than a "zigzag" route—have started performing #36's primary act—something it shall then do without change for a particular number of K. After doing that, it must then, in a durationless instant, have suddenly stopped performing #36's and have started doing #45's primary act. Again, that's something it shall then do without any kind of change for a particular number of K, etc., as—using a stop & go routine—it "leapfrogs" thru the sequence 27, 36, 45, 54, 63. In a manner of speaking, every generator engages in rectilinear locomotion by first *ingesting* a particular u-space which it then *digests* (***i.e.:*** activates and thus turns into a changeless microstate of excitation for a definite number of K), and then, in a durationless instant, *excretes* as it digests another u-space logically adjacent to the prior one according to drawings such as #7-1—drawings which graphically but figuratively set forth what 6 logically intertwined sequences are demanded by 6 of God's 7 *a priori* coordinates per OD. In a manner of speaking, locomotion is God's "stop action animation" presentation.

For decades, I doubted it possible for generators to take "diagonal" paths such as 27, 36, 45, 54, 63. That was because I strongly suspected generators can make only those shifts which involve a change in only 1 of the 3 sets of potency/anti-potency couplets. For example, the change from 28 to 27 would involve:

(.25A –0B) + (.25A' –0C) + (0B' –0C')

followed by:

(.5A –.25B) + (.25A' –0C) + (0B' –0C'),

In the above, I have printed the left couplet in bold print to indicate that such a shift involves no change save in the left couplet. But, the shift from 28 to 19 would involve:

(.25A –0B) + (.25A' –0C) + (0B' –0C')

followed by:

(.5A –.25B) + (.5A' –.25C) + (0B' –0C').

In the above, I have printed both the left and middle couplets in bold print to indicate that such a shift involves simultaneous changes in both the left and middle couplets, and, for decades, I seriously doubted that's allowed. I said that because it seemed to violate what I said about the ontological quantum in the prior chapter. I am now convinced it does not. It merely requires one to say that—in any given shift from microstate to microstate—the ontological quantum may apply to only one of the couplets in the defining triad or it may simultaneously apply to two or it may simultaneously apply to all 3 of the couplets used to define a particular microstate.

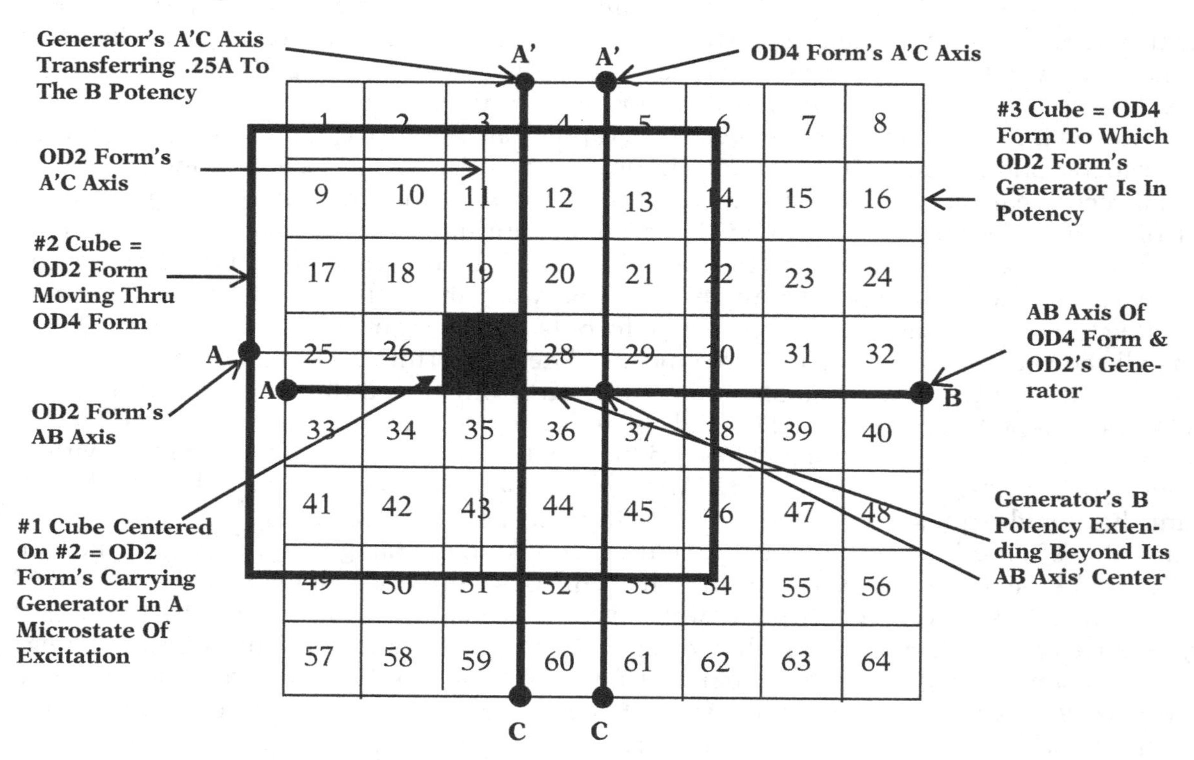

DRAWING #7-1: A GENERATOR CARRYING ITS OD2 PIGGYBACK FORM THRU THE U-SPACES EACH C. 10^{-47} CM. & MADE AVAILABLE BY A FORM AT OD4.

Still, I strongly suspect there are circumstances in which generators take zigzag routes such as, in the above, 27, 35, 36, 44, 45, 53, 54, 62, 63. For now, though, I have no idea what might produce such a pattern which, for sensation dependent minds, is a series of 90^{0} turns, in this case, to the "right", "left", "right", "left", etc..

As long as they truly are *carrying* generators—whether for a free duo-combo or a multi-combo—no two such *carrying* generators can simultaneously "occupy" the same u-space (***i.e.:*** perform what truly is the same one microstate of excitation). If they try, the result is a logical collision—one of the main causes of acceleration (more on that later). Oppositely, up to 2^{256} forms and their generators can occupy the same one u-space though not with any two at the same OD (something which Science can never *in*duce but which Philosophy can *de*duce). Yes, every one of any multi-combo's concentric captives is a duo-combo of form and generator; but, only one of the latter is doing any *carrying*—namely: the one carrying the whole multi-combo because it's the lead form's carrying generator. Compare that with Science's bit about fermions' *in*ability vs. bosons' *ability* to be simultaneously in the same state. Isn't it amazing how much Esoptrics coincides with Physics?[1]

Yes, Esoptrics' stop and go locomotion violates Newton's law which says that every body continues in its state of rest, or uniform motion in a straight line, unless forced to change its state by forces exerted upon it. Remember, though, what I said in chapter one: Esoptrics counters Newton with: Every *ultimate* (***i.e.:*** 1 generator and 1 form logically concentric to one another) continues at its current *rate* of change (a rate stated in units of u-time between each change) from temporarily changeless state of excitation to another and in its current pattern (as set by God's 7 kinds of logical sequences), unless forced to change that rate and/or pattern by forces exerted upon it. I then added: It cannot be proven by observation that Newton's law holds even at the level of 10^{-47} cm.; and so, it cannot be proven by observation that Esoptrics' principle is incorrect.[2]

Have you forgotten what 10^{-47} has to do with it? I have repeatedly said that every state of excitation on the part of a generator—and thus every "piece" of locomotion—takes place at a level our sensation dependent minds would describe as c. 10^{-47} cm., which is to say 2^{-129} times the dia. of the hydrogen atom with its electron in its outermost orbit. In other words, in Esoptrics, every instance of rectilinear locomotion is—at the level of the smallest segment possible to it—a series of "quantum leaps" each c. 10^{-47} cm., ***if*** one de-

---

[1] For all of that, though, Esoptrics has, for 56 years, been almost universally dismissed by the academics as utter nonsense from the mind of a hopelessly dimwitted idiot. Oh and how indignant they are at me and ready to brand me a "muckraker", if I dare to say anything caustic about them!

[2] In other words, for Newton—as for the vast majority of all who ever were, or currently are, native to Earth—locomotion is *absolutely continuous,* which, of course, means such locomotion is not possible save where—throughout its course—there *actually* is no limit to the smallness of the smallest segment of it. As were the theories of a flat Earth and an Earth centered Universe, Newton's first law is *pure conjecture* based upon what is *suggested* by sense imagery to minds too enslaved by sensation to see straight, but which cannot possibly be an empirical fact proven by observation. For, the simple, *supremely* empirical fact is that, neither with our naked senses nor with any of our instruments, can we come close to observing what occurs at 10^{-47} cm., let alone below that. Furthermore, since the notion of absolutely continuous locomotion necessarily entails the mind boggling bit of infinite divisibility (and saying "*potentially* so" does not escape the dilemma of "*actually* no limit to the smallest segment's smallness"), it remains to be seen how anybody could ever have clung to the idea of *continuous* locomotion. It's nonsense so blatantly untenable, I was a mere 6 years old when I clearly saw that, in our world, infinite this-that-and-the other—whether space, locomotion, divisibility, numbers, or what have you—are nothing but brazenly self-contradictory gibberish as impossible as impossible can ever be. And why deny ultimate realities can be of so *dynamic* a nature as to make instantaneous "quantum leaps" (Pardon the Geometry-speak.) on their own at a set rate as long as they're not forced by external influence to change that rate? But, even if one rejects that, the notion of infinite divisibility is still far more untenable than is the Aristotelian-Thomistic notion that no change of any kind is possible save where The First Unmoved Mover initiates it. Yes, I am well aware that even many a Catholic intellectual insisted we must, at all costs, preserve appearances (***i.e.:*** must accept our sense images as "gospel"), lest God be implicated in a lie. I don't care who or how many of them said it. Even if they be canonized saints, I say to them all: "Phooey! Only minds far too carnal for their own good could be duped into swallowing a line as silly and self-contradictory as that pathos most manifestly is to even a 6 year-old's modicum of critical thinking."

scribes the shifts the Geometry-addicted way our sensation dependent minds demand. Esoptrics describes each of those shifts (and u-spaces) as 1 alphatopon and 1Φ for short. Doing that, though, Esoptrics is not *really* talking in *Geometry's* terms. How so?!

Here, I must repeat again what I've probably already repeated *ad nauseam*. My drawings are not photocopies of the Universe's ultimate constituents. They are merely an attempt to use Geometry's pictures as a way of facilitating the communication of highly abstract principles which, strictly speaking, are wholly and entirely *algebraic* in nature and to no extent *geometrical*.

To explain what I'm saying, I can easily imagine one of Geometry's addicts saying: "In drawing #7-2 below, compare the length of the dashed line to the 4 thick black lines running from the center of the overall square to the points A, A', B, and C. It's longer than they are. Figure #7-2 thus tells us that, if a generator were able to move diagonally thru squares 28, 19, 10, and 1, then each of its shifts would cover a distance greater than what would be covered by its shifts thru squares 28, 27, 26, and 25. Your theory is thus self-destructive, unless you draw the microstates of excitation as circles as in figure #7-3 on the next page."

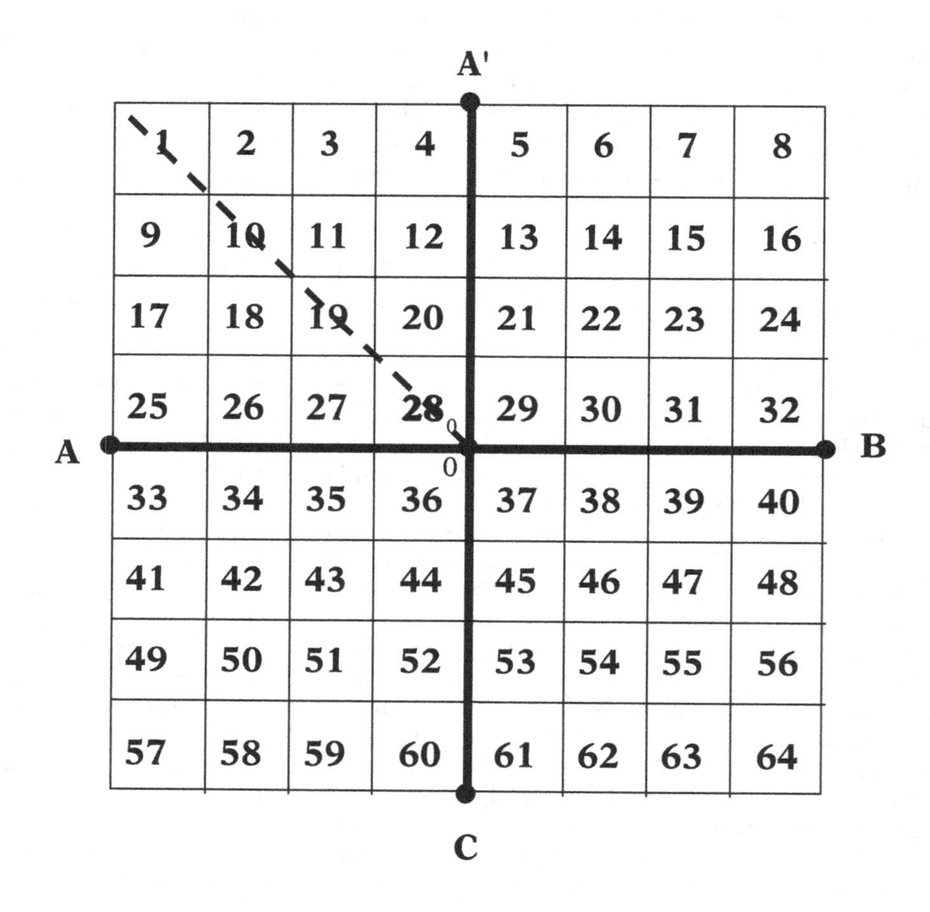

DRAWING #7-2: A DIAGONAL LOCOMOTION

Those who say that indicate they've totally failed to grasp what I have repeatedly emphasized even to the point of exasperation—namely: (1) every one of my drawings is a kind of allegory and metaphor using Geometry's pictures as a means by which to communicate principles which are *algebraical* and by no means *geometrical*, and (2) every temporarily changeless, single state of excitation—whether that of a piggyback form or that of a carrying generator—is a *physically infinitesimal* point in the sense that it is wholly and entirely devoid of every trace of *spatial* though *not logical* extension. No individual state of excitation—macro- or micro—has any parts, areas, segments, or what have you, outside of one another *in space*. Individually, each has only 6 potencies each of which is only *logically* distinct from, and outside of, the other 5. For the forms, it's so because of how they relate to God; for the generators, it's so because of how they relate to the forms.

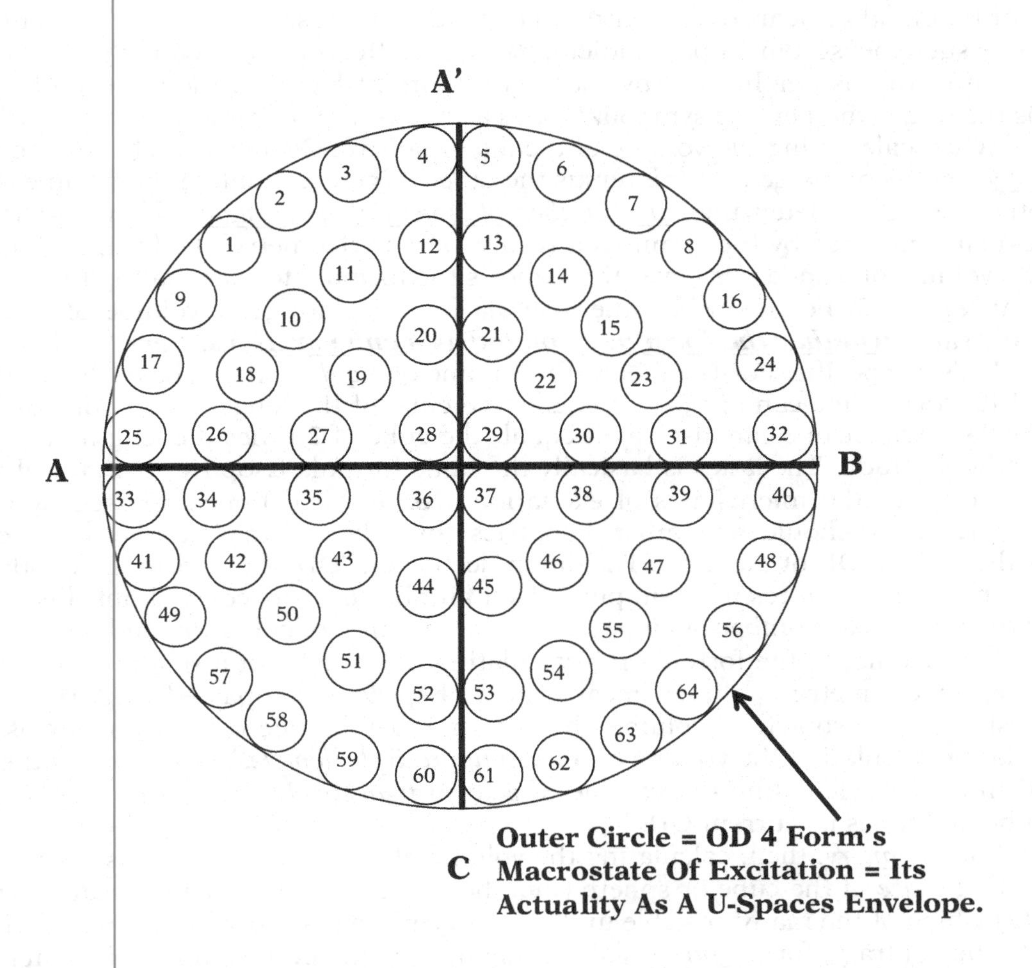

DRAWING #7-3: MICROSTATES OF EXCITATION (U-SPACES) PROVIDED BY AN OD4 FORM'S MACROSTATE & EACH SHOWN METAPHORICALLY AS A CIRCLE BUT COLLECTIVELY A SPHERE IN REAL EFFECT.

Individual macro- and microstates of excitation, then, are not *really* either squares or cubes or circles or spheres (What they are *collectively* and *in effect* is another issue.). Still, each can wisely be *analogized* as either or as both depending upon what lesson one is trying to communicate regarding the basic characteristics, properties and behavior patterns of forms and generators.[1]

Symbolizing the states of excitation as squares (and cubes in imagination) makes it easy to perceive: (1) which state of excitation (***i.e.:*** u-space) is where relative to all the others in logical sequence, (2) how to calculate the algebraic definition of each state, and (3) which shifts involve a one-way vs. a two-way vs. a three-way change in the description of the state which immediately preceded the shift vs. the description of the state which immediately followed it. Symbolizing the states of excitation as circles (spheres in imagination) helps to remind one that, individually, states of excitation are not actually squares or cubes with one length for figure #7-1's heavy black lines and another for its dashed line. In so doing, it helps one to remember that individual states of excitation cannot *really* be measured in the terms of inches, feet, yards, meters, centimeters, or any other measurement of that kind so dear to our sensation dependent minds. They certainly can't be said to have one such measurement perpendicularly and a different one diagonally.

All of that is emphasized by the very important fact that the volume of a form remains the same whether one symbolizes it as a cube or a sphere. How is that?

When calculating the volume of a cube or sphere, Geometry seeks to determine *the area of space* encompassed by one versus the other. When calculating the volume of a form, Esoptrics seeks to determine *the number of potential microstates of excitation* (***i.e.:*** u-spaces) encompassed by the form. As a result: (1) For Geometry, the formula for calculating the volume of a sphere is never the same as the formula for calculating the volume of a cube; ***whereas*** (2) For Esoptrics, the formula for calculating the volume of a form is always the same ***whether the form be depicted as a cube or as a sphere***.

Let's be specific. For Geometry, the volume of a cube = the cube of the length of any one of its edges. One can also say it equals the cube of the length of any one of its 3 perpendicular axes, or one can also say it equals the cube of 2 times the length of any of its 6 perpendicular radii. The latter is basically the same formula Esoptrics uses to calculate the number of potential microstates of excitation (***i.e.:*** u-spaces) encompassed by any given form. For, as you should remember, Esoptrics' formula says that number = the cube of 2 times the current OD of the form. The difference is that "OD of the form" has nothing to do with "length", and the result is simply a fixed number of u-spaces and not this number of cubic *inches* vs. that number of cubic *feet* vs. that number of cubic *cm*. and the like.

For Geometry, the formula for calculating the volume of a sphere is: $4/3\ \pi r^3$. That's because, for Geometry, "volume" means "area of space encompassed". Utterly to the contrary, since, for Esoptrics, "volume" means "number of microstates encompassed", the formula for calculating the volume of a form *figuratively depicted as a sphere* is the same as the formula for calculating the volume of a form *figuratively depicted as a cube*—namely: the cube of 2 times its current OD.

For *Geometry*, then, calculating the volume of a cube or sphere has *everything* to do with: (1) the *size* of the cube or sphere (***i.e.:*** the extent to which each is *spatially extended*), and: (2) which of the many possible units of measurement are used to express that size. Utterly to the contrary, for *Esoptrics*, calculating the volume of a form has absolutely *nothing*

---

[1] U-spaces—and thus generators—are logically outside of one another in such a way (***i.e.:*** a *six*-way logical sequence), that, *collectively*, they produce a real effect which can be observed, though *why* they are observably outside of one another cannot be observed. Conversely, the levels of the 4th dimension (***i.e.:*** the 2^{256} levels of ontological distance) are logically outside of one another in such a way (***i.e.:*** a *one*-way logical sequence), that, even *collectively*, they produce no effect which can be observed. No matter how many levels of the 4th dimension are present in a given multi-combo, the "diameter" of that 4th dimension never exceeds c. 10^{-47} cm., and that remains so even though each level of that 4th dimension has associated with it a u-spaces envelope whose u-spaces *collectively*, *effectively*, *observably* and *really* give that envelope a 3-dimensional dia. unique to that level.

whatsoever to do with its *size* or extent to which it is *spatially extended*. How could it, since no *individual* form or generator is to any extent spatially extended? More importantly, calculating the volume, Esoptrics makes no use of any unit of measurement devised by humans but, rather, invokes the *reality* of *current ontological distance* (***i.e.:*** level of the 4th dimension)—a factor which *Science* can never <u>*in*</u>*duce*, but which *Philosophy* can <u>*de*</u>*duce*. The hydrogen atom is only *verbally* $5x10^{-8}$ cm. in dia., but *really* is $2^{129}\Phi$ (***i.e.:*** u-spaces).

Compare how *Geometry* calculates the *spatial* volume of a cube with how *Esoptrics* calculates the *action* volume of a form (***i.e.:*** the number of primary acts it—as a u-spaces envelope—offers the generators), and it should be rather obvious: When trying to explain to others what volume is with regard to forms, it's best to use the imagery of a cube.

For <u>*Geometry*</u>, "length" of a line means some number of inches or feet or yards or miles or millimeters or centimeters or meters or kilometers or the like; and so, for Geometry, whereas the length is the same for *every* line drawn *from the center of a circle or sphere* to its outermost limit, such is *not* so for *every* line drawn *from the center of a square or cube* to its outermost limit. For, though each of the 6 lines drawn perpendicular to some cube's side are equal to the other 5 in length, those not so drawn have one of an immense number of lengths greater than those of the 6 perpendicular lengths.

Utterly to the contrary, since, in <u>*Esoptrics*</u>, "length" means number of logically sequential microstates of excitation (***i.e.:*** u-spaces), it makes no difference whether you depict the forms, generators, or u-spaces as squares or cubes or circles or spheres. The result is the same: Every line drawn from the center to the perimeter of some given form as a u-spaces envelope has the same identical length—namely: in number of logically sequential microstates of excitation = the current OD of the form. As for every line drawn from the center to the perimeter of some given generator or u-space, every such line has the same, one, purely *imaginary* length of ½Φ. In *reality*, there is no such line and no such length whatsoever, because every generator and u-space is *physically infinitesimal* and, therefore, devoid of every trace of *physical* length, width, and depth.

But, why use the term ".5Φ"? For Geometry, the dia. of the hydrogen atom with its electron in its outermost orbit is *verbally* $5x10^{-8}$ cm.; but, for Esoptrics, it's *really* 2^{129} microstates of excitation. The "dia." of every microstate of excitation (***i.e.:*** u-space) is then 2^{-129} times that $5x10^{-8}$ cm.. If, then, you habitually *verbalize* the hydrogen diameter as $5x10^{-8}$ cm., then it follows you'll habitually *verbalize* the dia. of every Φ (***i.e.:*** every u-space) as the *fractional* 7.35×10^{-47} cm.. In doing so, though, one is—as is all too often the case with Science and Geometry—solidifying the habit of not stating what is *really* there but, rather, what is the most utilitarian way to *verbalize*. In short, ".5Φ" is merely an appeal to utility for the sake of the Geometry addicts' habit of speaking imprecisely—as, of course, every mind crippled by too little knowledge of God *must* speak.

In Science and Geometry, a straight line in a flat plane is one which *appears* to be so. In Esoptrics, a straight line in a flat plane is any series of shifts in which each shift involves changes solely in the same one of the 3 couplets used to define a u-space. In other words, in the terms of the triad of couplets on pages 36, 37, 38, 39, & 40, each of the shifts in a series of shifts *in a straight line* involves changes solely in the left-hand couplet or solely in the middle couplet or solely in the right-hand couplet. Alas, how difficult it is for Science's Geometry addicts to follow Esoptrics' algebraic logic!

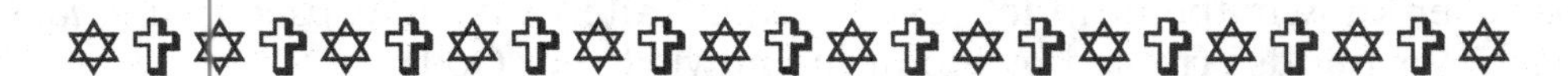

Another possibility is that space and time do not abruptly cease to have meaning on extremely small scales, but instead gradually morph into other, more fundamental concepts. Shrinking smaller than the Planck scale would be off limits not because you run into a fundamental grid, but because the concepts of space and time segue into notions for which "shrinking smaller" is as meaningless as asking whether the number nine is happy. That is, we can envision that as familiar, macroscopic space and time gradually transform into their unfamiliar ultramicroscopic counterparts, many of their usual properties—such as length and duration—become irrelevant or meaningless. Just as you can sensibly study the temperature and viscosity of liquid water—concepts that apply to the macroscopic properties of a fluid—but when you get down to the scale of individual H_2O molecules, these concepts cease to be meaningful, so, perhaps, although you can divide regions of space and durations of time in half and in half again on everyday scales, as you pass the Planck scale they undergo a transformation that renders such division meaningless.
——**BRIAN GREENE:** ***The Fabric Of The Cosmos***, page 272 of the 2005 paperback edition published by Vintage Books, a division of Random House, Inc., New York. ©2004 by Brian Greene (What an incisive rendering, Prof. Greene, of exactly what Esoptrics has been saying in vain for over 50 years! At the level of the alphatopon, length; and, at the level of the alphakronon, duration; and in both cases, division—are "as meaningless as asking whether the number nine is happy". Far more than that, though, Esoptrics also gives the world the only highly detailed and mathematically precise description of those "more fundamental concepts" into which time and space "gradually morph". And how is it able to do that? It's because—instead of being *Science* speaking of how *extra-mental things* affect *the human senses*—it's *Philosophy* and *Theology* speaking of an *intra-mental insight* into how *God*'s 7 *a priori* coordinates affect *the Universe's ultimate constituents*. ***Psalms*** 119:97&99. On his page 333, Prof. Greene gives 10^{-33} cm. as the Planck scale for length and 10^{-43} sec. for time. On page 345, 10^{-33} cm. is string length, and, on page 352, that's a "billion billion" [10^{18}] times too small for Science to observe—meaning: Science observes nothing smaller than c. 10^{-15} cm.. At 10^{-47} cm., Esoptrics' tackles what's nearly 100 trillion times [10^{14}] smaller still. In its article ***Planck Time***, Wikipedia says the smallest time interval detected as of May 2010 = $1.2x10^{-17}$sec. = 12 attoseconds. To confirm use the search string "smallest time interval".).

Motion can neither be, nor be conceived, without space.
——**JOHN LOCKE:** ***Concerning Human Understanding***, Book II: Chap. XIII, Sec. 11, page 151 left of Vol. 35 of ***Great Books Of The Western World*** by Encyclopædia Britannica, Inc.; Chicago, 1952 (Well, Mr. Locke, I dare say that, in the whole of history, no statement was ever more thoroughly refuted than this statement of yours is refuted by Esoptrics. For, certain it is that Esoptrics—in describing every form's actuality as an envelope with a definite number of u-spaces each more properly described as an actual or potential, discrete microstate of excitation—has given an ideal illustration of *exactly how* motion can both be, and be conceived, without the slightest appeal to that only type of space virtually the entire human race has ever known: the infinitely extended, infinitely divisible, *pseudo-divine*, mental crutch [for minds crippled by too little knowledge of God] type.)

Chapter Eight:

HOW HEAVENLY SEXTETS & BENCHMARKS YIELD CURVILINEAR MOTION & NATIVE VS. CURRENT OD:

So far, Esoptrics has described no kind of locomotion save that on the part of the generators, and theirs is strictly rectilinear locomotion. How, then, does Esoptrics account for curvilinear motion? Let's work toward that complicated answer.

Esoptrics has repeatedly said that every form is in potency to God Who—by means of the 6 points of reference intrinsic to The Infinite—imparts a six-fold potency to every form and, thereby, imparts to each form 6 logical dimensions. More exactly, God creates, so to speak, a finite version of God in a series of what Esoptrics labels: "heavenly sextets". It is then thru these heavenly sextets that God gives each form its 6 potencies and logical dimensions. Because each of these heavenly sextets forever applies to only one particular form, each is logically distinct and separate from all the others by virtue of that unique association, and that thus permits the number of heavenly sextets to be astronomical.

Furthermore, as Esoptrics has already said, forms are first and most readily logically distinct and separate from one another by virtue of having a discrete OD (***i.e.:*** a discrete 4th dimension level). They can, however, also be logically distinct and separate from one another by virtue of having a carrying generator performing a unique microstate of excitation. As a result, there can be a multitude of forms each currently having the same discrete OD but remaining logically distinct and separate from one another by virtue of the fact that each has a carrying generator in a unique microstate of excitation. Esoptrics shall say more on that when it's time to describe the Universe's first 10^{-18} seconds.[1]

Focus for now, then, on the huge number of heavenly sextets each having the 6 reference points of A vs. B, A' vs. C, and B' vs. C'. If each of these is a created, finite reflection of God, must each—like its infinite, original version in God—be forever oriented in whatever is its current orientation? In other words, in every heavenly sextet, is member A forever A only with regard to 𝕬 in God; member B forever B only with regard to 𝕭 in God; member A' forever A' only with regard to 𝕬' in God; member C forever C only with regard to 𝕮 in God; member B' forever B' only with regard to 𝕭' in God; and member C' forever C' only with regard to 𝕮' in God? If not, can each of the 6 members of any given heavenly sextet be itself with regard to each of the 6 frames of reference in God? If so, then any given heavenly sextet can "spin" with regard to God and, thereby, impart a spin to the form with which it is forever associated. That's because every form's A arm is forever aligned with the A member of its associated sextet and, thus must go wherever A goes; every form's B arm is

[1] If you need an example of what that means, turn back to figure #7-1 on page 48. In imagination, delete nothing, and add this: Blacken square 64, and say it represents a generator performing the primary act (***i.e.:*** microstate of excitation) characteristic of square #64. Centered on that generator, imagine a form which, like cube #2, is a form native to, and currently at, OD2. The result is two identical forms separate and distinct because of "where" their carrying generators are.

forever aligned with the B member of its associated sextet and, thus, must go wherever B goes; and the same can be said of the other 4 arms and their counterparts in the associated sextet. Exactly what, then, is meant by "spin with regard to God"?

In drawing #8-1 below, I have "bitten the bullet", as they say, and with help from one of the very few philosophers to show interest in my thinking (He forbids me to name him.), I have struggled to draw what I hope can be viewed as a cube. In the drawing, the plain letters A, A', B, B', C, and C' represent the 6 members of a heavenly *sextet*; the ornate letters 𝕬, 𝕭, 𝕮, 𝕬', 𝕭', 𝕮' represent what Esoptrics calls the 6 members of a heavenly *benchmark*. For now, "heavenly benchmark" shall mean the 6 frames of reference in God which are eternally unchanging (There are created sets ignored for now.) and, thereby, allow the heavenly sextets to execute *"patterns and anti-patterns of divine rotation"* whose orientation is always and everywhere infallibly guaranteed by God, The Eternally Unchanging.

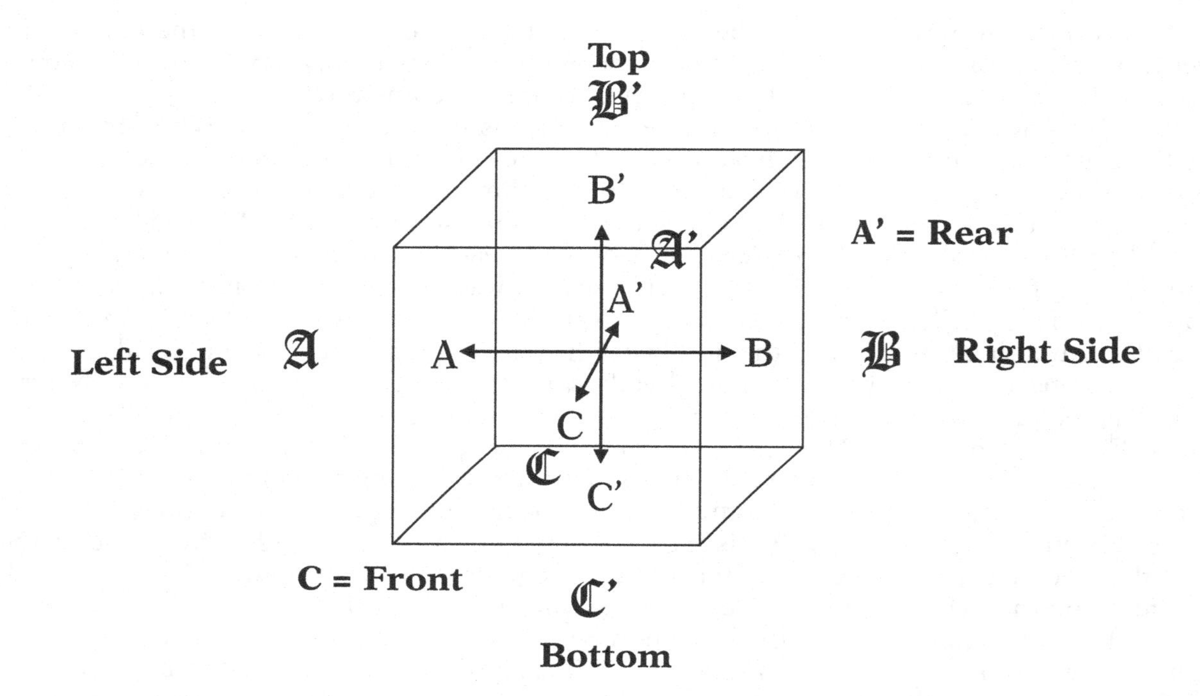

DRAWING #8-1: NON ROTATING HEAVENLY BENCHMARKS VS. ROTATING HEAVENLY SEXTETS

What is meant by "patterns and anti-patterns of divine rotation"? The answer is complicated enough to leave one's head spinning; but, I'll try to make it as painless as my fifth-rate mind can devise.

In the heavenly benchmarks above, we see an eternally fixed arrangement—namely: In "*Geometry* speak", 𝕬 & 𝕭 are always at opposite ends of the same axis; 𝕬' & 𝕮 are always at opposite ends of the same axis; 𝕭' & 𝕮' are always at opposite ends of the same axis; and each of the axes 𝕬𝕭, 𝕬'𝕮, & 𝕭'𝕮' are always perpendicular (***i.e.:*** equally related) to the other two. That says this: If A shifts to 𝕬', then A' must shift to 𝕭, B must shift to 𝕮, and C must shift to 𝕬. In Geometry speak, that's a clockwise (***i.e.:*** right-hand) 90^0 spin around the B'C' axis. To some, that shall immediately and rightly imply a *counter-clockwise* (***i.e.:*** left-hand) 90° spin around that same axis. To some, that shall immediately and rightly imply right- and left-hand spins around the other two axes, AB and A'C.

As I said, it is *Geometry* speak which, for economy's sake, speaks of clockwise (a/k/a right-hand) vs. counter-clockwise (a/k/a left-hand) spins. If we imagine B' is in the air above this sheet as it lies flat on a table top and that C' is in the air underneath it, then Esoptrics' "Algebra speak" must give us these two charts:

PLANE:	WHAT SUPPLANTS WHAT	AROUND
Primary:	A' → B'; B' → C; C → C'; C' → A'	AB
Secondary:	A → B'; B' → B; B → C'; C' → A	A'C
Tertiary:	A → A'; A' → B; B → C; C → A	B'C'

CHART #8-1: *PATTERNS* OF DIVINE ROTATION IN THE 3 PLANES

PLANE:	WHAT SUPPLANTS WHAT	AROUND
Primary:	A' → C'; C' → C; C → B'; B' → A'	AB
Secondary:	A → C'; C' → B; B → B'; B' → A	A'C
Tertiary:	A → C; C → B; B → A'; A' → A	B'C'

CHART #8-2: *ANTI-PATTERNS* OF DIVINE ROTATION IN THE 3 PLANES

Notice how I name the AB axis the axis of the *primary* plane; the A'C axis the axis of the *secondary* plane; and the B'C' axis (perpendicular to this sheet's front & rear) the axis of the *tertiary* plane. For now, it's a purely arbitrary choice. There may be a more accurate approach to it; but, I currently have not the slightest inkling of what that more accurate approach might be; and so, I have no choice but to stick with the labels I've invoked.

For Geometry-speak's need to describe things in the terms of space, the shifts described above are ones of 90° each. For Esoptrics' Algebra-speak, though, that shift is something purely logical in nature (***ex. gr.:*** 50% of the relationship of A to B) and, thus, can take place without any kind of change in *space*. Hmm?! What does that suggest to you? Is there something Geometry would describe as a 45° shift which, however, is also purely logical? How so?! Is not 50% of the relationship A to A' purely logical, just as 50% of counting 1 to 10 = 5? Or shall we say we cannot count 1 to 10, unless we see each of the numbers 1, 2, 3, etc., outside of one another in space? If, then, we're on the right track, what about 10% of that A to A' relationship or .05%, 75%, 90%, 3.1%, etc.? Exactly what percentages are possible?

Where, now, is that taking us? It's taking us to the effects of a form's native (***i.e.:*** created at) OD vs. its current one. What are those effects?

In a perhaps futile attempt to answer as simply as possible, let's start with complete rotations in the tertiary plane around the B'C' axis, and let's examine what that shall involve in the case of a heavenly sextet associated with a form which is currently at, and native to, OD16. Given such a form, how many shifts shall be involved in a complete rotation in the tertiary plane—the plane our senses experience (a very important detail)?

Let S_T = the total number of shifts (and thus pauses) in a complete rotation in the tertiary plane, and let D_C equal the *current* OD of the form with which a given sextet is associated. Esoptrics then says that, for all rotating forms always and everywhere:

FORMULA #8-1: $S_T = 2D_C$

Since, in the current example, we're dealing with a heavenly sextet associated with a form currently at OD16 (Its native OD is here irrelevant.), the above formula tells us a complete rotation in the tertiary plane shall involve 32 shifts. For *Geometry*, that means each of the sextet's members A, A', B, & C shall, in a durationless instant, shift 11.25^{o} around the B'C' axis, pause for a definite number of K, shift another 11.25^{o} around B'C' in a durationless instant, pause for that K, shift another 11.25^{o} around B'C' in a durationless instant, pause for that K, etc., and shall do that 32 times and, thereby, return to their original orientation as shown in drawing #8-1 on page 56, because $11.25^{o} \times 32 = 360^{o}$.

I dare say that's simple enough; but, then comes the more complicated 2/3. Esoptrics says that for each shift in the tertiary plane, there is—with 4 exceptions—a complete rotation in the secondary plane around A'C, and, for each shift in the secondary plane, there is—with 4 exceptions—a complete rotation in the primary plane around AB. How many shifts are we now hauling into the picture?

Let S_P = the number of shifts in a complete rotation in the primary plane, and let D_N equal the *native* OD of the form with which a given sextet is associated. Esoptrics then says that, for all rotating forms always and everywhere:

FORMULA #8-2: $S_P = \sqrt{D_N}$ ↕

The symbol ↕ means: "rounded up or down to the nearest whole number". As is probably known to all with even so little as a high school education, most numbers have no *integral* square root, which is to say one which is a whole number. For example, though $\sqrt{100}$ is the whole number 10, $\sqrt{10} = 3.1622$ and on and on. The above formula says round it down to 3, since that's less of a change than rounding it up to 4. By contrast, $\sqrt{13} = 3.6055$ etc.; and so, the above formula tells us to round it up to 4.

Since, in the current example, we're dealing with a heavenly sextet associated with a form native to OD16, the above formula tells us a complete rotation in the primary plane around the axis AB shall involve: $\sqrt{16} = 4$ shifts (& pauses). For Geometry, that means each of the sextet's members A', B', C, and C' shall, in a durationless instant, shift 90° around the AB axis, pause 1K, shift a second 90° around B'C' in a durationless instant, pause 1K, shift a third 90° around AB in a durationless instant, pause 1K, shift a fourth 90° around B'C' in a durationless instant, and, thereby, complete one rotation in the primary plane.

Why did I, in each instant, say "pause 1K"? It's because one of Esoptrics' oldest and most sacrosanct principles is:

THE FIRST LAW OF MOTION IN THE PRIMARY PLANE

ALWAYS AND EVERYWHERE, FOR EVERY FORM AND SEXTET MEMBER ROTATING IN THE PRIMARY PLANE, EVERY PAUSE IN THE PRIMARY PLANE LASTS FOR ONE ALPHAKRONON (K) THUS

INSURING FOR ALL ALWAYS AND EVERYWHERE THAT THE FATEST RATE OF CHANGE POSSIBLE IN ANY AND EVERY EPOCH OF THE UNIVERSE IS THAT RATE WHICH OCCURS IN THE PRIMARY PLANE—NAMELY: 1 CHANGE PER K.

So much for the primary and tertiary planes! What shall we say of the secondary plane? Let S_S = the number of shifts (& pauses) in a complete rotation in the secondary plane around the axis A'C, and let D_N = the *native* OD of the form with which a given sextet is associated. Esoptrics then says that, for all rotating forms always and everywhere:

FORMULA #8-3: $S_S = \sqrt{D_N}$ ↕

Yes, the *number of shifts per rotation* is the same for both the primary and secondary planes; the *duration* of each *pause between shifts* is not. In the primary plane, it's 1K per pause; in the secondary plane, the K per pause = $\sqrt{D_N}$. Why?! Remember: For every *pause* in the *secondary* plane, there's a *complete rotation* in the *primary* plane. That being so, we are faced with 2 formulas. The first says: Let K_{SR} = the duration in K of each *complete rotation* in the *secondary* plane, and let D_N again = the *native* OD of the form with which a given sextet is associated. Esoptrics then says that, for all rotating forms always and everywhere:

FORMULA #8-4: $K_{SR} = D_N$

The second formula says: Let K_{SS} = the duration in K of each *pause* in the secondary plane, and let D_N again = native OD. Then, for all rotating forms always and everywhere:

FORMULA #8-5: $K_{SS} = \sqrt{D_N}$

Let S_{PR2} = the total number of *shifts* (& pauses) in the *primary* plane in the course of a *complete rotation* in the *secondary* plane, and again let D_N = the native OD of the form to which the sextet is associated. Then, for all rotating forms always and everywhere:

FORMULA #8-6: $S_{PR2} = D_N$

Obviously, formulas 8-4 and 8-6 give the same result: D_N. It's because the duration of each primary plane pause is K; and so, since D_N *pause**s*** in the *primary* plane = $\sqrt{D_N}$ *primary* plane *rotation**s*** as well as one *secondary* plane *rotatio**n***, the latter's duration = D_NK.

By now, you probably perceive "native OD" dictates the rhythmical way a form is made to "spin" around its primary and secondary axes by the heavenly sextet with which it is linked. The rhythmical way a form spins around its primary and secondary axes is thus one of the traits which bespeak the OD at which God created it. For, whenever a form spins rhythmically (I'll explain later what the other trait is and why most forms often do not spin at all.), it spins as demanded by its native (***i. e.:*** created by God at) OD.

Exactly how, though, does God produce this business of native OD and the rhythmical way it requires the heavenly sextets and their associated forms to spin? I'll explain that when, in chapter 10, I describe the Universe's first 10^{-18} seconds.

What is the sum total of shifts in the primary plane in the course of a complete rotation in the *tertiary* plane? In other words, what is that total in the course of a complete *cycle* on the part of a heavenly sextet and the form to which it's linked? Let S_C = that total number of primary plain shifts/pauses per cycle (***i.e.:*** per complete rotation in the *tertiary* plane); let D_C = the *current* OD of the associated form; and let D_N = the *native* OD of that form. In that case, Esoptrics' answer is that, for all rotating forms always and everywhere:

FORMULA #8-7: $S_C = 2(D_C \times D_N)$

For example, turn again to the sextet associated with a form native to, and currently at, OD16. The result is:

FORMULA #8-7_{16}: $S_C = 2(16 x 16) = 2(256) = 512$

What of it?! If that's what you ask, you missed a crucial point. Because the number of *shifts* equals the number of *pauses*—and the duration of each primary plane pause is 1K—formula 8-7 also tell us that the *duration* of one cycle on the part of that form and its associated sextet is 512K. For the Geometry addict's love of seconds, that means 512 x roughly 10^{-96} sec. per cycle. In short, as a whirling dervish, it's a speedy little rascal.

Therefore, let T_{CR} = the duration of a rotating form's complete cycle, and let D_C and D_N = as above. In that case, for all *rotating* forms always and everywhere:

FORMULA #8-8: $T_{CR} = 2(D_C \times D_N)K$

As you can perhaps readily see, if a rotating form is currently at its native OD, then $D_C = D_N$; and so, in #8-8, $T_C = 2[(D_N)^2]K$ in that case. In either case, one half of that result would be the duration of each of every cycle's 2 phases—a division I'll explain shortly.

So far, then, we've said a *rotating* form's *native* (created at) OD is indicated by the way it rotates in its primary and secondary planes, and its *current* OD by the way it rotates in its tertiary plane. If that's clear enough, let's pause to forge 4 important definitions.

In Esoptrics' terminology: If an OD admits of an integral square root (***ex. gr.:*** OD's 1, 4, 9, 16, 25, 36 etc.), every form native to such an OD is a *rational* form; if an OD does not admit of an integral square root (***ex. gr.:*** 2, 3, 5, 6, 7, 8, 10 etc.), every form native to such an OD is an *irrational* form. The importance of that distinction is: Rational forms *always* rotate in each of the 3 planes; irrational forms rotate in the *upper* planes *only as long as* not included in a multi-combo—one in which many duo-combos are captives made logically concentric, 1 per 4th dimension level, by the multi-combo's leading form.

As you can perhaps readily calculate, the vast majority of forms are native to an OD which does not admit of an integral square root; and so, the vast majority of forms are irrational. To be specific about it, of the 2^{256} OD's available in this the Universe's 9th epoch, only 2^{128} admit of an integral square root. For, there is no OD higher than 2^{256}, and the square of $2^{128}+1$ would be greater than 2^{256}. In sum, then, the number of rational kinds of forms = 2^{128}, and the number of irrational kinds of forms = 2^{256} minus 2^{128}—making the number of irrational kinds roughly 10^{77} greater than the number of rational kinds. Of the 2^{128} rational forms, 2^{64} occur at or *below* OD2^{128}, and 2^{128} minus 2^{64} *above* OD2^{128}.

If a form is native to one of the OD's from OD1 thru OD2^{128}, it is, in Esoptrics' terminology, a *major* form. If native to one of the OD's from OD$2^{128}+1$ to OD$2^{256}-1$ (The one form native to OD2^{256} is in a class by itself.), every such form is, in Esoptrics' terminology, a *minor* form. At least, that's the terminology we'll use for now.

Note that formula #8-8 above applies to *rotating* forms. Since all *rational* forms always rotate in each plane, 8-8 always applies to them. Since *irrational* forms never rotate in the upper planes when in a multi-combo, 8-8 often does not apply to them. Still—as I'm about to explain shortly—they do *pulse*, which is to say they still change from one macrostate of excitation to another at a predetermined rate. What is that rate?

Let T_{Ci} = the duration of a cycle (***i.e.:*** duration of 2 successive macrostates of excitation) on the part of an irrational form not rotating in the upper planes (***i.e.:*** captive in a multi-combo), and let D_{Ci} = the current OD of that captive irrational form. In that case, for all captive irrational forms always and everywhere:

FORMULA #8-9: $T_{Ci} = 2[(D_C)^2]K$

Yes, that's basically the same as formula #8-8 where 8-8 applies to a rational form currently at its native OD. It's not an insignificant detail. Here again, of course, one-half of the above result would give the duration of each of every cycle's 2 phases.

For every *major* form—rational or irrational—its native and current OD are also indicated by its 2 "diameters" in Φ (***i.e.:*** the number of u-spaces along each 1 of its 3 axes). That's because, like the human heart, *all* forms pulsate: All shift from one macrostate of excitation to another at regular intervals. Each cycle in the tertiary plane thus has 2 phases, each of which is a macrostate of excitation with—as a u-spaces envelope—a diameter different from that of its partner. I'll elaborate on that in the next chapter.

For now, turn to chart #8-3 on the next page for an illustration of how what we've been saying applies to: (1) forms whose native and current OD is one of the first 32 OD's, and (2) to the associated heavenly sextets. The chart assumes, of course, even the irrational forms are rotating 3 ways—something the *categorical* forms native to OD's 2^0 and 2^1 always do whether or not either is taken as irrational. Doesn't $\sqrt{1}$ = the whole number 1?

Column #1 states the OD to which a particular form is native and at which it currently resides. Column #2 states how many changes (and, thus, pauses) occur in the tertiary plane in the course of a complete rotation in that plane. Column #3 states how many changes (and, thus, pauses) occur in the secondary plane in the course of a complete rotation in the secondary plane. Column #4 states how many changes (and, thus, pauses) occur in the primary plane in the course of one complete rotation in the primary plane. Columns #5 and #6 state how many changes (and, thus, pauses) occur in the primary plane in the course of *one complete rotation in the secondary plane*. That's the same as saying they state the number of changes (and, thus, pauses) which occur in the primary plane in the course of each one of the *pauses* in the *tertiary* plane. Column #6 simply states that total. Column #5 breaks that total down into how many changes occur in the primary plane in the course of each pause in the course of a complete rotation in the secondary plane.

Consider, for example, the form which is native to, and currently at, OD10. Column #2 states that, in the course of one complete rotation in the tertiary plane, it shall pause 20 times and, thus, change 20 times in that tertiary plane.

Column #3 tells us that, for each of the 20 tertiary plane pauses, there's a complete rotation in the secondary plane for a total of 20 secondary plane rotations per tertiary plane rotation, and each of those 20 rotations shall involve 3 changes and, thus, 3 pauses.

Column #4 gives the number of pauses and changes per complete rotation in the primary plane.

Column #5 then tells us that, for 2 of the 3 pauses per secondary plane rotation, there is, in the primary plane, a rotation involving 3 pauses; but, in the course of the 3rd of those 3 pauses per secondary plane rotation, there is, in the primary plane, a rotation involving 4 pauses

Column #6 then gives the sum of: (2 rotations of 3 changes each) + (1 rotation of 4 changes) = 10 changes in the primary plane (and, thus, 10K) per pause in the tertiary plane. Since that's 10K per 20 pauses in the tertiary plane, the duration of each rotation in the tertiary plane is 200K = $2(10^2)$K. For geometers, that means a duration of $200\text{x}10^{-96}$sec..

#1 D_{NC}	#2 Shifts Per Tertiary Rotation	#3 Shifts Per Secondary Rotation	#4 Shifts Per Primary Rotation	#5 Primary Plane Shifts Per Secondary Plane Shifts	#6 P Shifts Per T Shift
1	2	1	1	1 @ 1	= 1
2	4	1	2	2 @ 1	= 2
3	6	2	1 or 2	2 @ 1 + 1 @ 1	= 3
4	8	2	2	2 @ 2	= 4
5	10	2	2 or 3	2 @ 1 + 3 @ 1	= 5
6	12	2	3	3 @ 2	= 6
7	14	3	2 or 3	2 @ 2 + 3 @ 1	= 7
8	16	3	2 or 3	2 @ 1 + 3 @ 2	= 8
9	18	3	3	3 @ 3	= 9
10	20	3	3 or 4	3 @ 2 + 4 @ 1	= 10
11	22	3	3 or 4	3 @ 1 + 4 @ 2	= 11
12	24	3	4	4 @ 3	= 12
13	26	4	3 or 4	3 @ 3 + 4 @ 1	= 13
14	28	4	3 or 4	3 @ 2 + 4 @ 2	= 14
15	30	4	3 or 4	3 @ 1 + 4 @ 3	= 15
16	32	4	4	4 @ 4	= 16
17	34	4	4 or 5	4 @ 3 + 5 @ 1	= 17
18	36	4	4 or 5	4 @ 2 + 5 @ 2	= 18
19	38	4	4 or 5	4 @ 1 + 5 @ 3	= 19
20	40	4	5	5 @ 4	= 20
21	42	5	4 or 5	4 @ 4 + 5 @ 1	= 21
22	44	5	4 or 5	4 @ 3 + 5 @ 2	= 22
23	46	5	4 or 5	4 @ 2 + 5 @ 3	= 23
24	48	5	4 or 5	4 @ 1 + 5 @ 4	= 24
25	50	5	5	5 @ 5	= 25
26	52	5	5 or 6	5 @ 4 + 6 @ 1	= 26
27	54	5	5 or 6	5 @ 3 + 6 @ 2	= 27
28	56	5	5 or 6	5 @ 2 + 6 @ 3	= 28
28	58	5	5 or 6	5 @ 1 + 6 @ 4	= 29
30	60	5	6	6 @ 5	= 30
31	62	6	5 or 6	5 @ 5 + 6 @ 1	= 31
32	64	6	5 or 6	5 @ 4 + 6 @ 2	= 32

CHART #8-3: 3-WAY STRUCTURE OF THE PATTERNS OF DIVINE ROTATION PERTINENT TO FORMS CURRENTLY AT THEIR NATIVE OD

In chart #8-3, lines 1, 2, and 3, between them, present us with the 4 exceptions I mentioned in par. 4 on page 58. First, look at line 1. While a pause is occurring in the tertiary plane, there's only one change in the secondary plane and, simultaneously only one change in the primary plane. Clearly and unequivocally, there can be no such thing as a complete rotation consisting of only one change. In Geometry speak, that would mean a durationless change of 360° which is no change at all. Here, then, there is—for each pause in the tertiary plane—only *half* a rotation in the secondary plane and, simultaneously, only *half* a rotation in the primary plane. That accounts for 2 of the 4 exceptions.

Looking at line 2 of chart #8-3, we see that, for each pause in the tertiary plane, there is only half a rotation in the secondary plane. That's the 3rd of the 4 exceptions.

Looking at line 3 of chart #8-3, we see that, for one of the 2 pauses in each rotation in the secondary plane, there is only 1 change in the primary plane, and that means half of a complete rotation. That's the 4th of the 4 exceptions.

The most important bit of information given to us by chart #8-3 is this: In the case of forms native to OD's 1, 2, 3, & 4, there cannot be any *anti*-patterns in either the primary or the secondary planes, and, only in the case of forms native to OD's greater than 8 is *every* rotation in the primary and secondary planes either a pattern or an anti-pattern of rotation. How is that so? It's so, because, unless a rotation involves more than 2 changes, there is no distinction between a pattern and an anti-pattern.

Take, for example, forms native to OD2 and their associated sextets. Whether in the primary or the secondary planes, there are only 2 changes per rotation. In the case of a rotation in the primary plane, a *pattern* of rotation involves this:

CHANGE #1: A' & C switch as B' & C' switch.
CHANGE #2: A' & C switch back as B' & C' switch back.

An *anti*-pattern would then involve the same procedure:

CHANGE #1: A' & C switch as B' & C' switch.
CHANGE #2: A' & C switch back as B' & C' switch back.

There is, thus, no distinction between pattern and anti-pattern with regard to forms native to OD's 1, 2, 3, and 4 and with regard to their associated sextets. For forms native to OD's 5 & 6 and their associated sextets, there is no pattern/anti-pattern distinction between any of their rotations in the secondary plane, since, as column #3 tells us, each of their rotations in the secondary plane involves only 2 changes (and, thus, 2 pauses). For forms native to OD 5 and their associated sextets, only 1 out of every 2 rotations in the primary plane involves more than 2 changes.

Above all, what that means is this: In the rotations of the three *categorical* forms native to OD1, OD2, and OD4 there is never any pattern/anti-pattern distinction in either their *primary* or their *secondary* planes, and, as long as they remain at OD1, there is—in the rotations of the *categorical* form native to OD1—no pattern/anti-pattern distinction even in the *tertiary* plane.

To me, the latter quickly implies that, when forms native to OD1 accelerate beyond OD1, they all, without exception, rotate in the same one manner in the tertiary plane. That same one manner of rotation may be what I call a pattern, or it may be what I call an anti-pattern. I do not pretend to know for sure which way such a rotation should be described.

What's so important about this distinction between forms which, in the primary and secondary planes, have a definite kind of rotation and those few which do not? It shall allow us to forge a distinction which almost exactly matches modern Science's distinction between leptons and quarks. For, on the one hand, we now have 3 categorical forms (***i.e.:*** those native to OD1, OD2, & OD4) which—when accelerated to become the leading form of a multi-combo—shall resemble the 3 leptons called electrons, muons, and taus and, on the other, we have 6 categorical forms (***i.e.:*** those native to $OD2^4$, $OD2^8$, $OD2^{16}$, $OD2^{32}$, $OD2^{64}$, and $OD2^{128}$) which—when accelerated to become the leading form of a kind of multi-combo—shall resemble the 6 quarks. More on this shall follow later!

I've long been and still am firmly convinced forms native to OD's 1 thru 2^{128} can be *accelerated* to a maximum of 2^{128} times their native OD. Thus, forms native to OD2^{128} can be accelerated as far as OD2^{256}, but, in reality, are never accelerated beyond OD2^{256}–1. Forms native to OD1 can be accelerated as far as OD^{128}. Any attempt to accelerate them beyond that causes them to radiate away at light speed, which, in Esoptrics = 1Φ per 2^{128}K.

Over the years, and even recently, I've vacillated over whether or not: (1) major forms can be <u>de</u>celerated *below* their native OD, and (2) minor forms can be <u>ac</u>celerated *above* their native OD. For the moment, I'm going with the notions: (1) specific major forms can be decelerated to as low as OD1; (2) minor forms can be decelerated but not below OD2^{128}, and (3) minor forms can be accelerated but, as with all forms, are never accelerated to or above OD2^{256}. The one form native to OD2^{256} and the many native to OD1 cannot be made concentric with any but their own kind, and, for Esoptrics, that, in the end, is why there are electrons and why they often *orbit* but never *inhabit* the atomic nucleus.

The chief differences here between acceleration on the part of major forms vs. acceleration on the part of minor forms—are: (1) when the major forms accelerate, each—at least if it's rational and particularly if generic or categorical—produces what Esoptrics calls a multi-combo; but acceleration on the part of the minor forms never produces a multi-combo; (2) the major forms—at least if rational and particularly if generic or categorical—can serve as the lead form of a multi-combo; whereas, the minors can serve only as captives made logically concentric to the lead form by the lead form which then places the captive at some OD (***i.e.:*** discrete level of the multi-combo's 4th dimension) at, above, or below the captive minor's native OD; and (3) the rules governing the 2 different diameters of pulsating major forms is the direct opposite of the rules governing the 2 different diameters of pulsating minor forms. The next chapter shall spell that out in detail.

If the current OD of the lead form in a multi-combo is at, or less than, 2^{129}–1, Esoptrics calls the multi-combo an atomic or sub-atomic multi-combo. If the current OD of the lead form is at, or less than, 2^{161}-1 (***i.e.:*** [2^{129}-1] x [2^{32}]), Esoptrics calls the multi-combo a molecular one. If the current OD of the lead form is at, or greater than, 2^{192}, Esoptrics calls the multi-combo a cosmic multi-combo, and there are 2 kinds of cosmic multi-combos—those producing galaxies and their clusters vs. those producing the stars, planets, etc., which serve as the heavenly bodies contained in a galaxy. Details shall follow later.

Why are there 3 planes of rotation producing tripartite patterns and anti-patterns of rotation? What's the point of such mind-boggling complexity? It's Esoptrics way of explaining what Science calls "electrical charges" and "magnetic lines of force" and why those "magnetic lines" are at right angles to the forward, tertiary-plane-motion of the duo-combos and various kinds of multi-combos. The world we experience is the world of the tertiary plane. In the world of the secondary plane, curvilinear velocities are astronomically greater than the curvilinear and rectilinear velocities we observe in our world of the tertiary plane (***i.e.:*** on the order of 2^{64} times the speed of light); and so, they register on our world somewhat like the traces a water skier leaves in the water over which he skims. In the world of the primary plane, curvilinear velocities are astronomically greater even than those in the secondary plane (***i.e.:*** on the order of 2^{128} times the speed of light).

As we shall eventually see, these high speed rotations in the primary and secondary planes are what account for the chaos which Quantum Theory detects in the sub-atomic world and wrongly interprets as a source of only *probabilistic* conclusions rather than an indication of how little Science knows about the way forms (***i.e.:*** u-spaces envelopes) rotate in the primary and secondary planes. Perhaps more importantly, we shall see that—ever since God finished creating in this the Universe's 9th epoch—these phenomenally greater velocities in the sub-atomic world are not the result of generators switching from u-space to u-space at more than $1\Phi/2^{128}$K; rather, they are the result of the fact that the u-spaces being used by this-or-that generator are themselves rotating wildly around the center of the form providing those u-spaces, even as *its* carrying generator is using u-spaces being provided by a another form currently at a higher OD and spinning at a higher velocity, even as *its* carrying generator, etc. for several steps. As scientists commonly admit,[1] Relativity's light speed limit applies to the ultimate "matter particles" (***i.e.:*** to the generators in Esoptrics' terminology) but not to "space itself". Since, in Esoptrics' terminology, there is no such thing as "space itself", Esoptrics says Relativity's light speed limit—though it applies to the rate at which the generators change from u-space to logically contiguous u-space—does not apply to any of the u-spaces in any of the u-spaces envelopes presented by the forms.

Though I do not yet even begin to have the remotest idea of how one can take advantage of it, rotations in the secondary and primary planes are what make possible relatively instantaneous travel to all parts of the Universe. In short, it's the key to what science fiction would call relatively instantaneous "inter-galactic space travel". Oh, to be a part of an *interstellarized* world rather than confined to this pathetic one limited to a single planet's skies! As the 1936 movie, ***Things To Come***, expressed it: "*All* the universe or *nothing*!" A cry trumpeted the year I was born. Was it, then, perhaps prophetic?

Closing this chapter, note the important question it answers quite definitively: On pages 33-34, 59, 73, and 460, of his book, ***The Fabric Of The Cosmos***, Prof. Greene tackles a question to which Ernst Mach (1838-1916) gave great impetus—namely: How can "spinning" have any meaning either with regard to the entire Universe or with regard to an object which is the sole occupant of the entire Universe. Esoptrics replies: There is no such thing as the Universe, save where there is at least one form performing a macrostate of excitation and, thereby, in its actuality, providing a u-spaces envelope. Neither can there be any such thing as the sole occupant of that Universe, save where there is at least a second form whose carrying generator is utilizing the u-spaces envelope provided by the first one. In either case, "spinning" means each of those forms is rotating around its 3 axes as the heavenly sextet governing that form rotates relative to the stationary heavenly benchmarks governing that heavenly sextet. In short, whether with regard to the Universe or a sole occupant of it, spinning is relative to God.

It's merely a matter of understanding what space *really* is rather than what sense imagery and utilitarian equations *infer* it is. On pages IX, 6, 335, 486, and 493 of his book, Prof. Greene is bold enough to admit Science (as opposed to Esoptrics) doesn't know what space *really* is. On page 471, he concedes the current views on time and space may be no more than "mere allusions" to what's far more profound and fundamental at the base of "physical reality". Oh, and how my heart throbs when, on page 21, he talks of wanting to experience the Universe at its every possible level and not just those we can sense! Unfortunately, it seems most scientists are not about to share Prof. Greene's stance.

✡✞✡✞✡✞✡✞✡✞✡✞✡✞✡✞✡✞✡✞✡✞✡

[1] For example, see: (1) pg. 129 of ***The Grand Design*** by Profs. Stephen Hawking & Leonard Mlodinow, and (2) pg. 523 of Prof. Brian Greene's ***The Fabric Of The Cosmos***. See also page 416 of the latter where Prof. Greene speaks of space as "swirling". It seems I'm not alone after all in every part of my decades-old, crazy view of space.

The book tends to declare math and physics rules arbitrarily.
——**ANNA CALL:** ***Clarion Review*** (***Webster*** defines "arbitrary" as: "subject to individual will or judgment without restriction; contingent solely upon one's discretion." That can't rightly be said of what "math and physics rules" this book declares. For, every one of them is an inference contingent upon what is implied if one assumes 2 is the basis for understanding all reality. Why could Ms. Call not see that? Apparently, she gained too little grasp of the mathematical logic of a mirror to see it as implying any of the math and physics rules declared in this book. What else should one expect from someone who, on the one hand, could not notice that "2 is the basis for understanding all reality" is not the same as "2 is the basis for understanding the mathematical logic of a mirror"; and, on the other, could not notice that Esoptrics is admittedly a theory which, like all theories, is testing the fruitfulness of an opening *hypothesis*—in this case, the *assumption* that 2, the basis for understanding the mathematical logic of a mirror, is also the basis for understanding all reality?)

The school of Thomism defends the view that 'nothing ever passes from potency, whether receptive or operative [A rock has receptive and living things operative, potency – ENH] to act except under the influence of another being which is already in act'; 'no being, under any conditions, can reduce itself from potency to act,' and this applies, therefore, also to the operative potency.

According to the view of modern Thomism, an *antecedent physical influence, or physical premotion* (*praemotio physica*), is required in order that the faculty of a creature can pass from potentiality to actuality. It is only in virtue of this antecedent physical influence that the faculty is immediately capable of operation. This influence is supplied by God to all creatural activity.

. . . . In other words, the Principle of Change is a universal principle without exception. And because of this, a physical premotion on the part of God is necessary in order that the operative potency or faculty can become active.
——**FR. CELESTINE N. BITTLE, O. F. M. Cap.** (1897-1962): ***The Domain Of Being: Ontology***, pages 101 & 102 of the paperback edition by the Bruce Publishing Company, Milwaukee, 1939. ©1939 by the publisher (What then?! Is no generator or form capable of shifting from one state of excitation to another unless God, directly or indirectly [***ex. gr.:*** moving the forms by moving the heavenly sextets and moving the generators by moving the forms] moves it to do so? Is rate of change constant only as long as God, directly or indirectly, maintains it so? In a scenario in which there is no kind of extension, separation, multiplicity, or locomotion whatsoever—particularly not the self-contradictory, hopelessly impossible *continuous* kind—without God's influence, I'm not sure the question has any importance – ENH).

Chapter Nine:

THE MIRROR THRESHOLD (MT FOR SHORT), THE 4 WAYS FORMS BEHAVE & THE EFFECT ON SPACE'S 4TH DIMESION & ON SOME HEAVENLY BODIES' ORBITS:

For Esoptrics, OD2^{128} is the "mirror threshold" (***i.e.:*** MT for short). That's because OD2^{128} divides the Universe into two realms which are reverse (***i.e.:*** mirrored) images of one another. Every OD <u>below</u> 2^{128} is "inside" the mirror and "down" into the <u>*micro*</u>scopic world out of sensation's usual range; whereas, every OD <u>above</u> 2^{128} is "outside" the mirror and "up" into the <u>*macro*</u>scopic world within sensation's usual range. More to the point, below the MT, the lowest OD (***i.e.:*** OD1) is 2^{-128} (Note the minus sign.) times the MT; whereas, above the MT, the highest OD (***i.e.:*** OD2^{256}) is 2^{128} (Note the absence of the minus sign.) times the MT. That's $1/2^{128}$ x MT vs. $2^{128}/1$ x MT. As you no doubt readily see, $1/2^{128}$ and $2^{128}/1$ are reverse images of one another.

The concept of the MT allows us to divide the forms into 3 classes: (1) those native to OD2^{128}; (2) those native to some OD below 2^{128}; and (3) those native to some OD above 2^{128}. The first class can be called MT forms; the second, MT minus (***i.e.:*** MT– for short) forms; and the third MT plus (***i.e.:*** MT+ for short) forms. As said earlier, the MT and MT– forms are collectively "major" forms, and the MT+ forms "minor" ones.

The most important of the ways forms behave is acceleration on the part of the *rational major* forms, since the effects are the production of the multi-combos, the 4th dimension of the u-spaces, and instantaneous action at a distance—whether on the part of gravity or on the part of entangled particles' influence upon one another. The latter two effects shall be discussed later. Drawing #9-1 on the next page illustrates the former two.

In drawing #9-1, we're seeing what's *metaphorically* describable as an "onion" with successive layers of spheres each centered on the same one u-space. This "onion" resulted when a major form, by accelerating above its native OD, ingested a particular number of duo-combos (a number clarified later) and made each one currently at a discrete one of the OD's between its own current OD and at least its native one, if not lower. The big difference here is that each of this "onion's" spheres is a form the actuality of which is a u-spaces envelope relative to the generators. The number of u-spaces in each given envelope = the cube of twice the current OD (***i.e.:*** D_C) of the given form/envelope. The number of forms so concentric is usually at least 2^{120} (***i.e.:*** 1.329×10^{36}) at the atomic and sub-atomic levels and as much as $2^{256}–1$ (***i.e.:*** 1.579×10^{77}) at the center of the largest galaxies. In the u-space at the *center* of the former, the 4th dimension would be 2^{120} layers "thick" and $2^{256}–1$ layers thick at the latter's *center*. The "depth" of each multi-combo's *own* 4th dimension drops one layer per Φ from its center; but, the envelopes of nearby multi-combos do overlap thus complicating the 4th dimension's local thickness. If, *per impossibile*, one could eye u-spaces, one would see no distance *even collectively* twixt the 4th dimension's layers, since—by God's 7th *a priori* coordinate, OD—they are logically separate without the u-spaces' 3D effect.

As a u-spaces envelope, each form's actuality—due to the *collective* impact of its u-spaces—*effectively* has a radius in Φ's equal to its current OD; and so, as one moves away from the "onion's" center one Φ per step, one passes from the influence of one form's actuality to that of the form at the next highest OD and with a radius 1Φ greater than that of the previous one—assuming no overlap. To utilize what a form of a higher OD offers, one must increase one's level in the 4th dimension. How that's done shall be explained later.

A major insight to gain here is this: By accelerating above its native OD and producing a multi-combo, major forms beget a *column* and *ladder* of concentric space envelopes enabling other duo- or multi-combos' carrying generators to take one of 3 courses: (1) by acceleration on the part of its own piggyback form, to climb up the *ladder* to a 4th dimension level allowing it to travel such-and-such a further distance away from the "onion's" center; (2) by no acceleration, to remain at such-and-such a step on the ladder and thus limited to such-and-such a distance from the "onion's" center; or (3) by using an overlap, to step from one multi-combo's shorter ladder to another multi-combo's taller one.

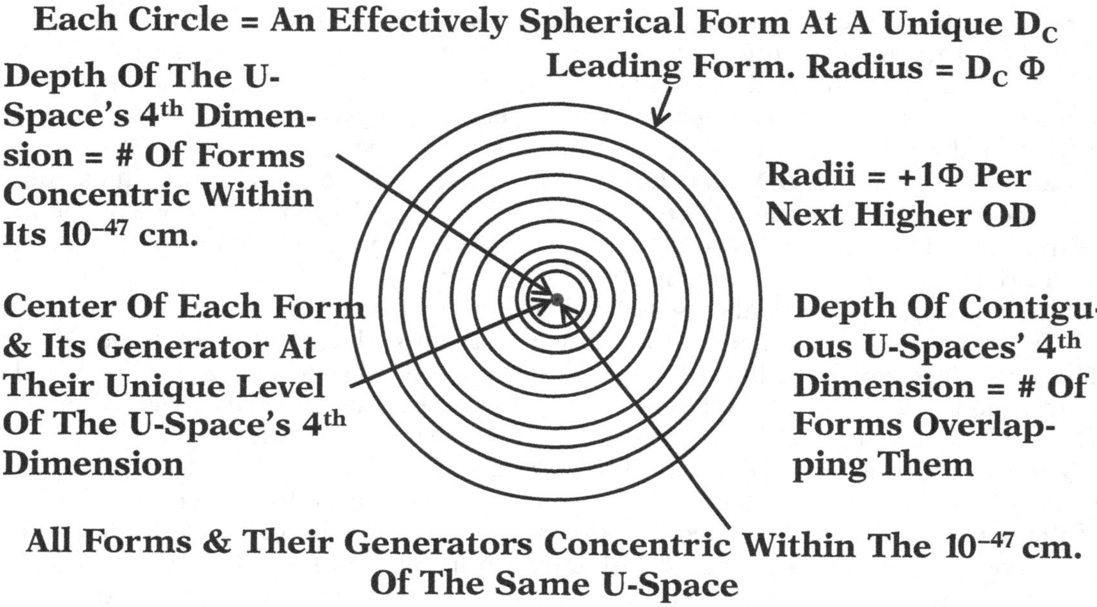

Each Circle = An Effectively Spherical Form At A Unique D_C

Leading Form. Radius = D_C Φ

Depth Of The U-Space's 4th Dimension = # Of Forms Concentric Within Its 10^{-47} cm.

Radii = +1Φ Per Next Higher OD

Center Of Each Form & Its Generator At Their Unique Level Of The U-Space's 4th Dimension

Depth Of Contiguous U-Spaces' 4th Dimension = # Of Forms Overlapping Them

All Forms & Their Generators Concentric Within The 10^{-47} cm. Of The Same U-Space

DRAWING #9-1: A U-SPACE'S 4TH DIMENSION DUE TO A MAJOR FORM'S ACCELERATION ABOVE ITS NATIVE OD.

Grasp the notion that the realm above the MT is the reverse of the realm at and below the MT, and it should rather strongly suggest this to you: The rules governing acceleration and deceleration on the part of the major forms should be the opposite of the rules governing acceleration and deceleration on the part of the minor forms. How so?

Let F_{M1} = some *major* form's dia. throughout the first half of a cycle (***i.e.:*** half of a complete rotation in the tertiary plane), and let D_C = its *current* OD. In that case, for all major forms always and everywhere:

FORMULA #9-1: $F_{M1} = 2D_C\Phi$

For example, if a major form is currently at OD16, then its dia. = 32Φ for the 1st of the 2 phases of a complete rotation in the tertiary plane. To say it another way: Along each of its 3 axes, it's 32 logically successive u-spaces, which is the same as saying that, along each of the *radii* drawn from its center, it's 16 such u-spaces (***i.e.:*** 16Φ).

What, though, of a major form's dia. in the *second* half of a cycle (***i.e.:*** 2nd half of a complete rotation in the tertiary plane)? Let F_{M2} = its dia. throughout the 2nd half of each cycle, and let D_N = its *native* OD. In that case, for all major forms always and everywhere:

FORMULA #9-2: $F_{M2} = (2D_N + 2D_C)\Phi$

Consider an example of what that means: A form is native to, and currently at, $OD2^{128}$. The duration of each cycle is $2(2^{128})^2K = 2^{257}K$ and the duration of each of that cycle's 2 phases is $(2^{128})^2K = 2^{256}K$. For the first $2^{256}K$ of each cycle, its diameter = $2^{129}\Phi$, and for the second $2^{256}K$ of each cycle, its diameter = $2^{129}\Phi + 2^{129}\Phi = 2^{130}\Phi$ which then reverts to $2^{129}\Phi$ for the first half of the next cycle. Suppose, though, it accelerates to $OD2^{256}$ (I ignore the –1 for economy's sake.). Now, the duration of each cycle is $2(2^{256} \times 2^{128})K = 2(2^{384})K = 2^{385}K$. For the first $2^{384}K$, its diameter is $2^{257}\Phi$. For the second $2^{384}K$, its diameter is $2^{257}\Phi + 2^{129}\Phi$. That's a change so minor, it's virtually no change at all.

That latter detail is a highly significant one saying: Whenever a *major* form is highly accelerated above D_N, there is little difference between its dia. in the 1st vs. the 2nd half of each cycle. Indeed, the level of acceleration can be so great, the difference is *astronomically* insignificant. There is virtually no expansion and contraction from phase to phase.

Now contrast that with what's true of the *minor* forms as, for them, their position above MT reverses formulas #9-1 & 9-2. Let F_{m1} (Note the switch from a capital M to a small m.) = some *minor* form's dia. throughout the first half of a cycle, and let D_N = its native OD. In that case, for all minor forms always and everywhere:

FORMULA #9-3: $F_{m1} = 2D_N\Phi$

Formula #9-3 thus reverses the order between #9-1 and #9-2 and replaces D_C in formula #9-1 with D_N. Next, let F_{m2} = some minor form's dia. throughout the second half of a cycle, and D_C = its current OD. In that case, for all minor forms always and everywhere:

FORMULA #9-4: $F_{m2} = 2D_C\Phi$

Formula #9-4 thus reverses formula #9-1 by replacing *native* OD with *current* OD. Consider an example of the above: Suppose a minor form is native to $OD2^{140}$ and has been accelerated to a current OD of 2^{192}. In phase 1 of its cycles, its dia. = $2(2^{140})\Phi = 2^{141}\Phi$. In phase 2, its dia. = $2(2^{192})\Phi = 2^{193}\Phi$. What if its $D_N = 2^{192}$ and $D_C = 2^{140}$. The 2 diameters remain the same. Only their order is reversed, and expansion and contraction remain formidable (***i.e.:*** $2^{52} = 4.5\times10^{15}$). It's the reverse of what's true of major forms.

The crucial importance of the contrast just described is this: Highly accelerated *major* forms produce cycles which, in Geometry-speak, are very close to circular; whereas, *minor* forms, whether highly accelerated or decelerated, produce cycles which, in Geometry-speak, are highly elliptical. You perhaps see where that's going; but, be patient.

Let's digress briefly. If pertinent to a form currently at its native OD, formula #9-1 implies some interesting correlations. First, there's this: While a rotating major form is currently at its native OD, there is a correlation between its *dia. in Φ* in each cycle's first phase and the *number of shifts/pauses per rotation* whether in the primary or the secondary plane.

How so?!

Suppose a rotating form is native to, and currently at, $OD2^{128}$. From what we've just said, its dia. is $2^{129}\Phi$ in each cycle's first phase. The square root of one-half of $2^{129} = \sqrt{2^{128}} = 2^{64}$. Compare that with formulas 8-2 and 8-3 on pages 58 & 59. There, it's the square root of native OD; here, it's the square root of one-half the current dia. in Φ. That dia., of course, is the same number as its current OD doubled.

Next, there's this: While a rotating major form is currently at its native OD, there is a correlation between its *dia. in Φ* in each cycle's first phase and the *duration* of each phase of each cycle. How so?!

Suppose a rotating form is native to, and currently at, $OD2^{128}$. From what we've just said, its dia. is $2^{129}\Phi$ in each cycle's first phase. The square of one-half of $2^{129} = (2^{128})^2 = 2^{256}$, and $2^{256}K$ is the duration of each cycle's each phase. Thus, the duration of each phase in alphakronons = the square of one-half the diameter in each cycle's first phase.

Because of what formula #8-9 told us on page 61, the correlation described in the prior paragraph is even tighter where the major form is an irrational one which, as part of a multi-combo, is not rotating in the upper planes. How so?!

Suppose a form native to $OD2^{127}$, by accelerating to $OD2^{155}$, has become part of a duo-combo and thus not rotating in the upper 2 planes.[1] At $OD2^{155}$, its dia. in each cycle's first phase = $2^{156}\Phi$ (***i.e.:*** c. 3" in dia.) and, in each cycle's second phase, = $2^{156} + 2^{127}\Phi$. The difference between the 2 diameters is thus insignificant. As a result, at $OD2^{155}$, the duration of each phase of each of this irrational form's cycles would = $(2^{155})^2 = 2^{310}K$, which, in turn, = the square of one-half the form's dia. in Φ in each phase of each of its cycles.

The 4 ways forms behave are: acceleration, deceleration, pulsation, and rotation. Now that we've listed and said a few words about each of the 4 ways forms behave, let's turn to a very important result of all 4 of those behaviors working together.

Suppose a first multi-combo's carrying generator is using the u-spaces provided by a second form *native* to $OD2^{128}$. The second, though, is *currently* at $OD2^{252}$ as the lead form of a second multi-combo. As that second form's pulsing moves its outer limit a small amount from phase to phase of each cycle (***i.e.:*** back and forth from a radius of $2^{252}\Phi$ vs. a radius of $2^{252} + 2^{128}\Phi$), the first multi-combo's carrying generator shall routinely cause its multi-combo to perform—relative to the second multi-combo's center—a two phase cycle. In the first phase of that cycle, the first multi-combo shall move inward to a certain minimum distance from the center of the second multi-combo (***i.e.:*** a radius no less than $2^{252}\Phi$), but, in the second phase, move outward to a certain maximum distance from that center (***i.e.:*** a radius no greater than $2^{252} + 2^{128}\Phi$). Such a change in distance from the second multi-combo's center is insignificant, being no greater than one-half the diameter of an atom.

Oppositely, suppose a first multi-combo's carrying generator is using the u-spaces provided by a minor form *native* to $OD2^{147}$ but *currently* at $OD2^{206}$. That's because the *minor* has been captured by a highly accelerated *major* form which—as the lead form of a second multi-combo—has made the minor concentric to itself and has placed it at $OD2^{206}$. As that minor form pulses (***i.e.:*** switches back and forth from one macrostate of excitation to another), its diameter in the first phase of each of its cycles shall be $2^{148}\Phi$, and the diameter in the second phase of each of its cycles shall be $2^{207}\Phi$. That necessarily means a huge

---

[1] As we shall see several chapters from now, by such an acceleration, this *generic* (because native to an OD which is a power of 2) but irrational $OD2^{127}$ form becomes an unusual kind of lead form for an organic molecule. It's an unusual lead form in that, instead of making its "captives" logically *concentric* to itself, it causes a multitude of multi-combos to, so to speak, "dance" around it in close proximity to it thereby producing, at one of the 27 generic levels between $OD2^{128}$ and 2^{155}, a multitude (***i.e.:*** roughly $6^{27} = 10^{21}$) of spiraling double helices.

difference in the number of u-spaces it offers in the first vs. the second phase of each of its cycles—the latter being c. 2^{59} (***i.e.:*** 5.76×10^{17}) times the former in diameter. As a result, the carrying generator of the first multi-combo (***i.e.:*** the one using that minor form's u-spaces) shall routinely cause its multi-combo to perform—relative to that pulsing minor form's center—a two phase cycle. In the second (***i.e.:*** expanded) phase of that cycle, the first multi-combo shall move outward to a certain maximum distance from the center of the pulsing minor form (***i.e.:*** a maximum radius of $2^{207}\Phi$ = c. 187,000,000,000 mi.), but, in the first (***i.e.:*** contracted) phase, move inward to a certain minimum distance from that center (***i.e.:*** a radius no less than $2^{147}\Phi$ = c. .01 cm.). The change in distance, in this case, is exceedingly significant.

It should be somewhat apparent that, in what I just said, I'm taking the four ways forms behave (***viz.:*** acceleration, deceleration, pulsation, and rotation) and using them as Esoptrics' explanation of how and why the planets orbit the Sun, how the Sun orbits the galaxy's center, etc.. It should also be somewhat apparent that Esoptrics' explanation is radically different from those given by Newton and Einstein.[1]

The purpose of the calculations to follow is to give support to my long-standing conclusion that, in Esoptrics, there are two radically different ways for one heavenly body to orbit another. For one, the orbiting body's velocity may be mostly or wholly its own as—thru its carrying generator—it follows a path dictated solely by gravity. For another, its orbital velocity may be mostly that of the u-spaces envelope being used by the carrying generator of the orbiting body's central multi-combo. That's so where the generator is in potency to (***i.e.:*** is receiving its u-spaces from) a form which: (1) is a rational one, (2) is concentric—or near concentric—with the body being orbited, (3) is, as a rational form, rotating around its own center concentric—or near concentric—with the center of the body being orbited, and (4) carries the orbiting body with it by virtue of the fact that the latter's carrying generator is using u-spaces which themselves are doing all the orbiting.

Let's now turn to the concrete examples. As we do so, let me confess up front that even I am far from satisfied with the explanations I'm about to give. I will give them anyhow though, because my hope is that, in spelling them out, either I shall improve my own or someone else's chances of rendering more satisfactory explanations.

Also, for simplicity's sake, I'll be speaking as if there's only one multi-combo per heavenly body and each at its host's center. That *central* multi-combo's carrying generator is then the one carrying the heavenly body thru the Universe. Later, I'll move to the more complicated notion that any given heavenly body may have within it more than one multi-combo no two of which are at, or adjacent to, the body's center. Where that's so, the *off*-center multi-combo's carrying generator may be the one carrying the heavenly body thru the Universe, while the *central* one's carrying generator causes its heavenly body both to rotate around that body's center and to "orbit"—and move in and out toward—the center of the *off*-center multi-combo. It's Esoptrics way of explaining what "barycenters" are.

At the Sun's center is a multi-combo with an ontological depth of c. 2^{227} (***i.e.:*** what might be called a kind of minor "black hole" whose leading form—being somewhere between OD2^{226} and 2^{227}—is a u-spaces envelope with a dia. of between 16,749.175 and 33,498.35 lt.yrs.). Of the roughly 2^{227} duo-combos logically concentric in that multi-combo,

---

[1] My deepest thanks go out to POHANLIN@gmail.com whose little program, Virtual Calc 2000, made it possible for me to give most of the above numbers in full detail. The powers of 2, though, were worked out for me many years ago by my recently deceased and one time brother-in-law (He divorced my sister Janice Haas Kasten many years ago.), Stephen David Israel Kasten. He did so with the help of Brian Haas, the older of my two younger brothers.

there is an MT+ (***i.e.:*** minor) but rational one whose form is native to the OD of:

220,858,380,404,665,916,626,065,753,724,731,835,629,387,025 = 2.21×10^{44} =
The square of 14,861,304 ,801,553,123,270,295 = 1.49×10^{22}.
(2^{147} = 178,405,961,588,244,985,132,285,746,181,186,892,047,843,328 = 1.78×10^{44})

Let's call that form Harry.[1] Accelerated Harry is currently at the OD of:

694,397,960,458,641,148,960,000,000,000,000,000,000,000,000,000,000,000,000,000 =
$6.94397960458641148 96 \times 10^{62}$ (2^{208}= 4.11×10^{62}).

Multiplying his current OD by 2 gives us the number of positions in each of Harry's complete cycles in the tertiary plane. It gives that number as:

1,388,795,920,917,282,297,920,000,000,000,000,000,000,000,000,000,000,000,000,000 =
$1.38879592091728229792 \times 10^{63}$.

To determine the duration of each of those cycles, we take the number of positions per cycle and multiply that number by Harry's native OD. The result is:

306,727,217,806,397,456,665,800,000,000,000,000,000,000,000,000,000,000,000,000,
000,000,000,000,000,000,000,000,000,000,000,000,000,000,000 = $3.067272178064 \text{x} 10^{107}$ K.

To turn that duration into seconds, we divide the above by the number of K/sec. (***i.e.:*** $1.388543802 \text{x} 10^{95}$), and the result is 2,208,984,818,229.000000000119 seconds. Divide that by the number of seconds per Earth year (***i.e.:*** 31,556,925.9747), and result =

70,000.00000000000000000037709 years.

Next, imagine another multi-combo whose carrying generator is in potency to that Sun-centered form called Harry. Call this second multi-combo Hugo. How shall Hugo's generator be affected by Harry's pulsing?

By formula #9-4, page 69, the *radius* of Harry's actuality, during phase two of each cycle, is $6.94397960458641148 96 \text{x} 10^{62}\Phi$, which is to say Harry's *radius* in u-spaces = his OD during that phase. Since 1 cm. = $1.361129468 \text{x} 10^{46}\Phi$, then, in cm., ($6.94397960458 6411489 6 \text{x} 10^{62}$) ÷ ($1.361129468 \text{x} 10^{46}$) = 51,016,304,972,000,000 cm.. Since 1 mile = 160,934.34 cm., 51,016,304,972,000,000 cm. = 317,000,740,625 mi., which is to say that, during phase two of each of his cycles, Harry allows Hugo to be as far as 317,000,740,625 miles from the Sun's center. That conforms closely (***i.e.:*** off .000000234) with the astronomers who tell us Hyakutake's orbit takes it a maximum distance from the Sun of 317,000,000,000 miles.[2]

---

[1] In every multi-combo, no 2 of its component duo-combos is at the same OD; but, I do not see it as necessarily so that each of those duo-combos is at an OD equivalent to its native one. In every multi-combos' efforts to have a duo-combo at each of the OD's in its ontological depth, every multi-combo has the ability, to some extent, to "fill in the gaps" by accelerating (or decelerating?) major or accelerating or decelerating minor forms to whatever OD does "fill in the gap" in its 4th dimension.

[2] According to Wikipedia, the comet Hyakutake now takes 70,000 years to orbit the Sun and, in doing so, goes from a maximum distance of roughly 317,000,000,000 miles to a minimum distance of roughly 21,398,000 miles.

CHAPTER NINE

What happens to Hugo's carrying generator when Harry's actuality suddenly returns to its native OD, and the radius of Harry's actuality thus drops to $2.21\text{x}10^{44}\Phi$= .01625 cm.? Hugo's carrying generator must now spend the next phase of Harry's cycle going from over 300 billion miles from the Sun's center toward a point only .01625 cm. from that center.

The first question that raises is this: How can Hugo's carrying generator engage in such a locomotion during Harry's contracted first phase, if during that phase, Harry's contracted actuality is now offering Hugo's carrying generator no more than an astronomically insignificant fraction of the u-spaces Harry offered when Harry was at $\text{OD}6.9\text{x}10^{62}$? One possible solution goes like this:

One must distinguish between u-spaces which are "in-phase" (Say "ongoing", if you prefer.) and those which are "out-of-phase" (Say "oncoming", if you prefer.). When a form routinely switches back and forth between the expanded and contracted phases of its cycles, the u-spaces made available in the expanded phase don't cease *wholly and entirely* each time the form reverts to its contracted phase. Throughout the course of the contracted phase, each of them remains as a kind of shadowy remnant of what it was and shall routinely become again over and over. It's, so to speak, akin to the "after glow" which remains when a light bulb is turned off. That leaves them of such a nature that they can be used, but only as a means of plummeting toward the fully real u-spaces provided by the contracted actuality of the form. You might say each is a kind of slippery slope inclining its occupant to slide downward toward the form's contracted actuality.

The second question raised by the above it this: Why doesn't Hugo crash into the Sun and remain there as it finally reaches Harry's new actuality only .01625 cm. in radius? Why would it instead—like the comet Hyakutake—go no closer to the Sun's center than 21,398,000 miles on its way to again leaving the Sun 317,000,000,000 miles behind it?

To make the problem even clearer, let's turn once again to the multi-combo at the heart of the Sun and talk about another one of the duo-combos concentric there along with the roughly2^{227} other duo-combos. This particular duo-combo is a minor but rational one *native* to the OD of:

10,907,246,643,114,590,863,624,477,974,731,171,516,134,569 = $1.0907\text{x}10^{43}$ =
The square of 3,302,612,093,951,481,741,837 = Between $\text{OD}2^{143}$ & $\text{OD}2^{144}$.

Let's call it Gene, and say Gene's *current* OD is 2^{197}. Applying formula #8-8 on page 60, we calculate that each of Gene's *cycles* lasts his native OD x 2^{198} (***i.e.:*** 2D_C)=

10,907,246,643,114,590,863,624,477,974,731,171,516,134,569 x
401,734,511,064,747,568,885,490,523,085,290,650,630,550,748,445,698,208,825,344 =
4,381,817,397,234,249,380,939,344,585,975,421,735,165,853,187,858,358,646,903,199,474,
123,826,275,915,230,127,665,956,513,964,921,716,736 =$4.381817397234249381\text{x}10^{102}$ K.[1]

Since 1 sec. = $1.388543802\text{x}10^{95}$K, we divide the above number of K by K/sec. so:

4,381,817,397,234,249,380,939,344,585,975,421,735,165,853,187,858,358,646,903,199,474,
123,826,275,915,230,127,665,956,513,964,921,716,736 ÷ 138,854,380,200,000,000,000,000,

[1] Remember: Whether in the 2nd phase at 2^{197} or in the 1st phase at $\text{OD}1.1\text{x}10^{43}$, the number of positions in the tertiary plane = 2^{197} per phase and 2^{198} per cycle, because 2^{197} is the *current* OD. Because Harry is native to $1.1\text{x}10^{43}$, the number of pauses in Harry's primary plane = $1.1\text{x}10^{43}$ for each of the 2^{198} positions in each of the cycles in the tertiary plane; and so the duration of a complete cycle = $1.1\text{x}10^{43}$ x 2^{198} = 4.38 x 10^{102}K as above stated.

000,000,000,000,000,000,000,000,000,000,000,000,000,000,000,000,000,000,000,000,000,000,000,000,000 = 31,556,925.974699999999995 sec. per 360^0 cycle by Gene.

Since the length of a tropical year is generally given as 31,556,925.9747 sec., Gene's cycle is exactly what it should be to account for the length of an Earth year. Is that close enough?

What is the radius of Gene's actuality in the expanded phase of each of his cycles? As we've repeatedly said it would be his current OD in Φ; therefore, that radius is $2^{197}\Phi$, which is to say $2.00867255532 \times 10^{59}\Phi$. One mile = $2.19052472587 \times 10^{51}\Phi$; and so, if we divide that latter figure into the prior one, the result is 91,698,237.03 miles. Earth's minimum distance from the Sun is generally given as 91,400,000 miles and its maximum as 94,510,000. Subtract Gene's radius from 94,510,000, and Gene's radius is 2,811,978 mi. shorter than it should be. That, though, is a discrepancy of only 2.975%. Is that close enough?

If it is not, it is possible to cut it a lot closer. To do that, we first change Gene's *native* OD to:

10,582,720,889,441,645,717,073,837,217,386,915,948,002,500 =
The square of: 3,253,109,418,608,855,480,050.

Next, we change Gene's *current* OD to:

207,026,975,529,798,659,268,800,000,000,000,000,000,000,000,000,000,000,000,000 =
$2.07026975529798659268 8 \times 10^{59} = 1.030665625321 \times 2^{197}$.

At that current OD, the radius of Gene's actuality in its expanded stage would then be $2.070269755297986592688 \times 10^{59}$ Φ. Divide that by Φ per mile, and:

207,026,975,529,798,659,268,800,000,000,000,000,000,000,000,000,000,000,000,000/
2,190,524,725,870,000,000,000,000,000,000,000,000,000,000,000,000,000 =
94,510,220 miles.

At the current OD above, the number of positions in each one of Gene's orbits in the tertiary plane would be:

414,053,951,059,597,318,537,600,000,000,000,000,000,000,000,000,000,000,000,000 =
$4.14053951059597318537 6 \times 10^{59}$.

Multiply each of those positions by Gene's native OD as given above, and:

414,053,951,059,597,318,537,600,000,000,000,000,000,000,000,000,000,000,000,000 x
10,582,720,889,441,645,717,073,837,217,386,915,948,002,500 =
4,381,817,397,234,249,380,939,217,810,997,514,687,599,083,729,737,082,470,941,144,000,000,000,000,000,000,000,000,000,000,000,000,000,000 = $4.381817397234249381 \text{ X } 10^{102}$K.

As was shown above on page 72, that many K = 31,556,925.974699999999995 sec. per 360^0 cycle by Gene. That may seem impressive; but, it leaves me with a serious problem. How so?!

In due course, I will present my reader with the formulas governing the rectilinear velocity of the carrying generators. Once the reader grasps those formulas, it should become obvious that the above is telling us this: The carrying generator for the multi-combo at Earth's center, is receiving its u-spaces for a form which is 3.0666% above the generic $OD2^{197}$. But, according to Esoptrics' principles regarding rectilinear velocity, that would require Earth's carrying generator to be traveling at 1.5333% of light speed, which is to say 2,851.9 miles per sec. = 10,266,840 miles per hr.. For now, I cannot see how that could possibly be correct. Pardon me, then, if, for now, I stick with the figures which, on page 74 leave me with a 2.975% discrepancy.

Here, now, is the problem with the first set of calculations: What is the radius of Gene's actuality in the contracted phase of each of his cycles? From what's been said, it would, in Φ, = Gene's native OD. Since the latter has been given as between 2^{143} and 2^{144}, the radius of Gene's actuality in each of his second phases would be roughly $2^{144}\Phi$. Dividing that latter by $1.361129468\text{x}10^{46}$ (***i.e.:*** the number of Φ per cm.), we find that radius = .0016384 cm. (or half that if we say 2^{143}). How, then, can such a description of Gene possibly explain how, in the course of each of Gene's cycles, Earth's distance from the Sun varies only from 91,400,000 to 94,510,000 miles?

My current and perhaps unsatisfactory answer first distinguishes between: (1) the *orbital* velocity imparted to a planet by rotation on the part of the form providing its carrying generator with u-spaces, and (2) the inbound/outbound rectilinear velocity on the part of the planet's carrying generator: (a) as that generator—in the first phase of every cycle on the part of the form providing its u-spaces—plunges toward that provider's center in response to that provider's instantaneous contraction at regular intervals, and (b) as that generator—in the second phase of every cycle on the part of the form providing its u-spaces—races away from that provider's center in response to that provider's instantaneous expansion at regular intervals. That latter *rectilinear* velocity is at an angle to the former *curvilinear* one as the latter follows the former around the center of the former's orbit. As you can perhaps readily calculate for yourself, an elliptical orbit is the result of the way the rectilinear and curvilinear motions interlace with one another.

At Earth's center is a multi-combo we'll call Terry. Its carrying generator is one of the parties which actually participate in Earth's orbit of the Sun. Earth's many atoms are not. That's because Terry's generator uses u-spaces provided by a form logically concentric with a multi-combo in the Sun; but, Earth's atoms use the u-spaces provided by the forms logically concentric with Terry. The two motions are independent of one another.

With that distinction made, we continue so: If, as proposed, the duration of Gene's every *cycle* = 31,556,925.9747 sec., then the duration of Gene's every *phase* = 15,778,462.987 sec., which means Terry has that many seconds to go from 94,510,000 miles to .0016384 cm. from the Sun's center before Gene's actuality suddenly expands back to 94,510,000 miles. What velocity would Terry's carrying generator need to have in order to travel 94,510,000 *inbound* miles in 15,778,462.987 sec., if Terry plunges *straight* inward as opposed to using a *zigzag* motion?

Dividing the miles by seconds, the answer = 5.9898 mi./sec. = 21,563.28 mi./hr. average speed. In other words, Terry's velocity may be lesser than that for some number of seconds and greater than that for some number of seconds.

We're told that Earth's farthest distance from the Sun is 94,510,000 miles and its shortest distance 91,400,000. The difference is 3,110,000. What velocity would Terry's carrying generator need to have in order to travel 3,110,000 *inbound* miles in the Sun's direction in 15,778,462.987 sec., if Terry plunges *straight* inward?

Dividing the miles by seconds, the answer = .1971 mi./sec. = 709.56 mi./hr. average speed. In other words, Terry's velocity may be lesser than that for some number of seconds and greater than that for some number of seconds.

Needless to say, if Terry—and, thus, the whole of Earth—plunges toward Gene's center at an average speed of 710 miles per hour, then Earth shall go no closer to the Sun than 91,400,000 miles before Gene's actuality—by suddenly expanding back to $2^{197}\Phi$—presents Terry's carrying generator with an actuality allowing it to return to a distance from the Sun of 94,510,000 miles.

Let's apply that calculation to the relationship between what we called Harry and Hugo. Harry's first phase allowed Hugo to be as far as 317,000,000,000 miles away from the Sun. If, as did the comet Hyakutake, Hugo is going to approach to within 21,398,000 miles of the Sun in the first phase of Harry's cycle, then we subtract 21,398,000 from 317,000,000,000 and calculate Hugo shall travel 316,978,602,000 *inbound* miles toward the Sun in the course of Harry's first phase. Harry's cycle, we said, lasts 70,000 years. Halving that, we calculate Hugo shall travel that 316,978,602,000 miles in 35,000 years. Dividing the latter into the former tells us Hugo shall travel 9,056,531.4857 inbound miles *per year*. Divided by 31,556,925.9747 sec. per yr., that tells us Hugo shall travel inbound at an average speed of .28699 mi./sec. = 1,033.164 mph, if Hugo plummets *straight* inward.

The gist of it all is this: On the issue of whether or not Esoptrics' bit regarding rotating and pulsing forms accounts for the way Earth and Hugo orbit the Sun, the answer depends upon the velocity with which Earth plunges toward the center of Gene and the velocity with which Hugo plunges toward the center of Harry. That, unfortunately, is something I cannot even begin to calculate, because I have not the foggiest idea regarding *how* to go about calculating it. I dare say, though, that I do not *have* to calculate it, because, in the laws of gravity as calculated by Newton and modified by Einstein, that calculation has possibly already been done for me. How, then, does Esoptrics meld with Newton and Einstein on this issue? Let's use drawing #9-2 on the next page to forge my answer.

As even a child can probably appreciate quite easily, I cannot produce a drawing composed of 2^{197} or 2^{144} squares in a line. The best I can do is this: Looking at drawing #9-2, next page, imagine first that the overall square represents Gene with his center at or near the Sun's center. Gene's outer u-spaces are circling that center at roughly 66,616 mph and completing one rotation per $4.381817397\text{x}10^{102}$K, = once per 31,556,925.9747 seconds.

Next, let's turn any particular one of the outermost squares into a solid black one (Don't forget each is logically equidistant from the center.) and imagine it represents the microstate of excitation being performed by Terry's carrying generator. Use square #1. While Terry's generator is there, Gene is at $\text{OD}2^{197}$, and Earth is thus at the outermost perimeter of Gene's actuality in the expanded phase of one of Gene's cycles. Since Gene is rotating around his center, square#1 is doing likewise, and, since Terry's carrying generator is situated at square #1, Earth is being carried around Gene's center and, thereby, being carried around the center of the Sun by Gene. Earth's orbital speed of 66,616 mph is thus Gene's and not Earth's (***i.e.:*** Gene is the *agent* and Earth the *patient* of that motion).[1]

While Earth's generator is resting there for a number of alphakronons dictated by the velocity with which it reached square #1, Gene instantaneously switches to $\text{OD}2^{144}$ leaving Terry's generator "out in the cold". How, now, shall Earth's generator plunge toward Gene's contracted actuality?

On May 9, 2013, at nssdc.gsfc.nasa.gov/planetary/factsheet/earthfact.html on the internet, I came across data from NASA telling me: (1) Earth's mean orbital velocity is 29.78

[1] Over 700 years ago, St. Thomas Aquinas attributed the motion of the heavenly bodies to angels—beings immaterial in nature the same as the forms are. In saying that, he was repeating what others had been saying for hundreds of years before him. Science has long been used to laughing at such an idea. In the above, though, Esoptrics is saying that a thousand year old way of explaining solar orbits of at least some heavenly bodies was much closer to the truth than is modern Science's way of explaining them.

km/sec. (***i.e.:*** 29.78 x 3600 = 107,208 km/hr x .62137 = 66,615.84 mph); (2) its maximum orbital velocity = 30.29 km/sec. = +.51 km/sec. (***i.e.:*** 66,615.84+1,141 mph); and (3) its minimum orbital velocity = 29.29 km/sec. = –.49 km/sec. (***i.e.:*** 66,615.84–1,096 mph).

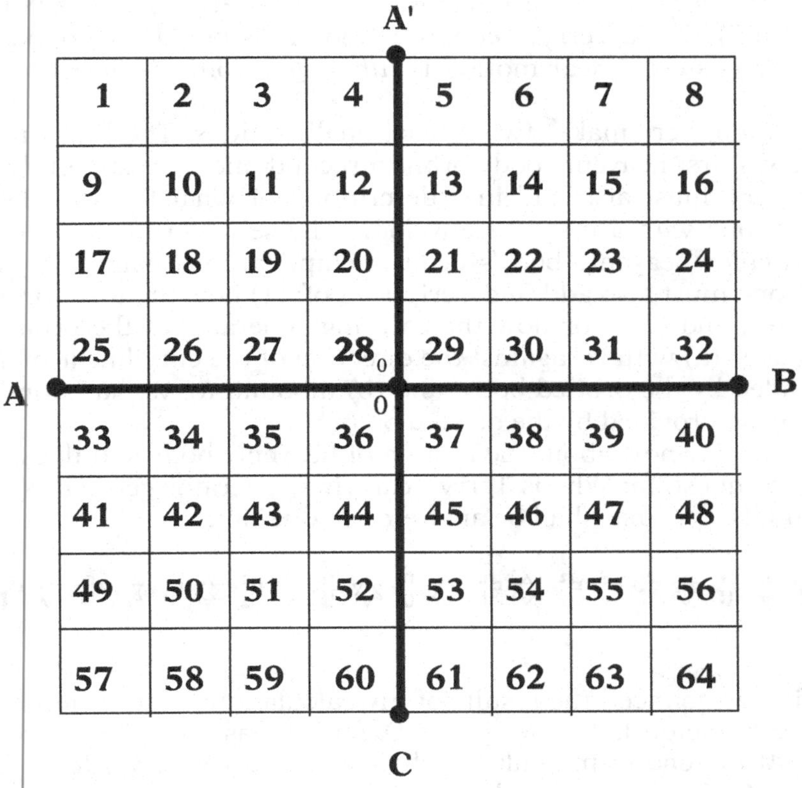

A'

1	2	3	4	5	6	7	8
9	10	11	12	13	14	15	16
17	18	19	20	21	22	23	24
25	26	27	28_0	29	30	31	32
33	34	35	$^{0}36$	37	38	39	40
41	42	43	44	45	46	47	48
49	50	51	52	53	54	55	56
57	58	59	60	61	62	63	64

A B

C

DRAWING #9-2: HOW "GENE" CARRIES EARTH AROUND THE SUN

Esoptrics says the mean orbital velocity of 66,615.84 mph is the velocity of squares 1 thru 8, 9, 17, 25, etc., and the maximum velocity comes about so: As Gene is rotating counter-clockwise at his native OD, Terry's carrying generator—after a definite number of K at square #1—instantaneously shifts to square #9. After a definite number of K at square #9, he instantaneously shift to square #10, then #18, then #19, then #27, then #28, etc.. The shifts from #1 to #9, #10 to #18, and #19 to #27, are rectilinear motions in the same direction as Gene's curvilinear motion. They are thus <u>added</u> to the velocity of Gene's rotation and, thereby, increase Terry's velocity *around* the Sun an additional .51 km/sec. = an additional 1,141 mph. The shifts from #9 to #10, #18 to #19, and #27 to #28, move Terry closer to Gene's center without adding to Terry's forward *rectilinear* velocity.

The velocity of Terry's rectilinear motion relative to Gene (as opposed to that curvilinear velocity which is only apparently Terry's) is thus composed of two velocities (***i.e.:*** one straight ahead and one to the side) intertwined with one another. The ratio of the two is such that, as the velocity reaches a maximum of 1,141 mph, Terry's distance from the center of Gene drops to, but not below, 91,400,000 miles. What is that ratio? I'll not bother to calculate it. I deem it enough to say it's probably already calculated by the laws of gravity as described by Newton & Einstein.

As Terry arrives at 91,400,000 miles from Gene's center (***i.e.:*** some point in the Sun), Gene's first phase gives way to his second one, and the dia. of Gene's actuality instantaneously expands to $2^{197}\Phi$. As you would expect with the logic of a *mirror*, Terry now retraces his steps in reverse order (Shades of Ptolemy's "epicycles"?!). It's now #28 to #27 to #19 to #18 to #10 to #9 to #1. Since Terry's rectilinear motion is now in the opposite direction of Gene's rotation, Terry's rectilinear motion is *subtracted* from the direction of Gene's rotation.

Esoptrics, then, here makes two crucial qualifications. The first runs so: When trying to explain how a first heavenly body orbits a second one, one cannot do so solely in the terms of gravity; one must also add in a description of what form in the second one is providing the first one with a u-spaces envelope. The second runs so: When trying to explain why a particular heavenly body's orbit is elliptical, one cannot do so solely in the terms of gravity; one must also add in descriptions of: (1) how the form provided by the orbited body pulsates, and (2) if or how the carrying generator of the orbiting body's main multi-combo zigzags: (a) with vs. against the direction of the curvilinear motion on the part of the form provided by the orbited body, and (b) inbound to, vs. outbound from, the center of that form being provided by the orbited body.[1]

That's enough for now about the motion of heavenly bodies. It does, of course, leave us with one glaring question: Why is Terry's carrying generator getting its u-spaces from a form currently at $OD2^{197}$? That shall be answered in due time.

✡✞✡✞✡✞✡✞✡✞✡✞✡✞✡✞✡✞✡✞✡✞✡

> When I first announced the results of my calculations at a conference at the Rutherford-Appleton Laboratory near Oxford, I was greeted with general incredulity. At the end of my talk the chairman of the session, John G. Taylor from Kings College, London, claimed it was all nonsense. He even wrote a paper to that effect.
>
> ——**STEPHEN HAWKING:** ***A Brief History Of Time***, as found on pages 115 (bottom) & 116 (top) of the paperback 10th anniversary edition by Bantam Books, New York, 1998. ©1988 & 1996 by Stephen Hawking (Well, Prof. Hawking, when one of your stature provokes such rejection, maybe I should not complain to any extent about the massive rejection I've been experiencing for over half a century.

---

[1] The truth of the matter is that, to this day, May 14, 2013, I continue to vacillate over the question of whether or not the formulas describing the way minor forms pulsate (***i.e.:*** 9-3 & 9-4) should be any different from those describing the way major forms pulsate (***i.e.:*** 9-1 & 9-2). I find my mind repeatedly puzzling: "Why should the elliptical shape of an orbit be due to anything other than gravity? Why should the carrying generators need the additional incentive of a contracted actuality? And isn't it too much of a stretch—and an unnecessary one to boot—to distinguish between ongoing and oncoming u-spaces?" Perhaps there shall come a day when I myself shall be able to answer one way or the other with some confidence. For now, though, I cannot; and so, I can only hope someone else shall accomplish what my fifth-rate mind cannot.

Chapter Ten:

THE UNIVERSE'S FIRST 10^{-18} SECONDS:

In the beginning (***i.e.:*** "time-zero", Greene pg. 519, to 1K), God initiated the Universe's FIRST EPOCH by creating a single form native to OD2. Because there was no form of a higher OD, nothing was presenting that first form's carrying generator with a collection of u-spaces; and so, that first form's center was the Universe's immovable center.

One K later, God created several forms each native to OD1. Because each was created with its carrying generator in potency to the first form, each of these OD1 forms was able to use its carrying generator as a means by which to move around within the parameters (***i.e.:*** 64 u-spaces = the cube of 2x2) offered to them by that first OD2 form.

How many forms did God create native to OD1? For decades, I theorized it would be 6, since forms are in potency to God 6 ways. I thus imagined there would be 6 forms native to OD1. One would set out for the A member of the first form's heavenly sextet; a second for the A' member; a third for the B member; a fourth for C; a fifth for B'; and the sixth for C'. Lately, my mind said to me: "That would be so only if each generator were created in a *one*-way microstate of excitation. They could just as easily be created in a *three*-way microstate. If so, then, instead of starting out at one of the 6 '*arms*' joined at the center of the first form, each would start out at one of the 8 'cubes' joined at that center."[1]

After meditating on that line of thought for the last 5 weeks, I have decided to theorize that God created 8 forms native to OD1. More accurately, that's 8 duo-combos each of which was a generator concentric with a form native to OD1. Each of those 8 duo-combos, by means of its generator, then headed for one of the 8 "corners" of the first form.[2]

With the creation of those 8 forms native to OD1, the Universe's first epoch is now 2K old and has run its course. That leaves the first 8 forms native to OD1 free to move around in the 64 u-spaces provided by the OD2 form serving as the 1st epoch's head.

---

[1] To which 8 do I refer? Look at drawing #5-5 on page 37, and imagine squares 28, 29, 36, & 37 are cubes protruding upward into the air above them. One can then imagine below those 4 cubes another 4 cubes protruding downward into the air below 28, 29, 36 & 37, and we can label them 28-, 29-, 36-, & 37-. Those are the 8 to which I refer. At the center of every form to which a generator is in potency, there is such an 8 fold cluster of u-spaces, and whenever God simultaneously creates 8 forms native to the same OD, each one's generator is created in a different one of those 8 u-spaces. If that's not clear enough, take 8 of the cube shaped building blocks dear to kindergarten tykes, and use them to build a cube. You'll thus produce a large cube 2 tiers deep each of 4 cubes. That's 8 cubes adjacent to the point at the center of the larger cube composed of those 8.

[2] The 8 are *not really* cubes. Each is such only in the imagination of one seeking a teaching aid able to communicate best a principle too abstract to be *really* geometrical. But, the *sphere* they produce *really* is *collectively* a sphere due to how each *really is outside* the other 7, however much only *logically* so. Such is also true of the OD2 form's 64 u-spaces: *Collectively*, they really are a sphere. Our senses lie not about matter's shapes; they just don't manifest the *ultra-microscopic why* behind them.

At 3K into the Universe (***i.e.:*** into a "mini-bang", if you prefer), God initiated the Universe's SECOND EPOCH by creating a single form native to OD4. Again, with no form of a higher OD, nothing was presenting that OD4 form's carrying generator with a collection of u-spaces; and so, that OD4 form's center was now the Universe's immovable center.

Though so nailed in "place", this OD4 form now presents 512 (***i.e.:*** the cube of 2x4) u-spaces to the carrying generator logically concentric with the OD2 form serving as the head of the first epoch; and so, the head of the first epoch is now free to move around within the u-spaces envelope offered by the second epoch's OD4 form and to carry with it the 8 OD1 forms in potency to that head of the first epoch. For 1K, then (but 3K into creation), that first OD2 form's generator is performing its first microstate of excitation.

At 4K into the Universe (2K into epoch #2), God created 8 forms native to OD3 and each with a generator in potency to the OD4 form and each activating one of the 8 u-spaces adjacent to the OD4 form's center. For that to happen, the carrying generator of the original OD2 form (***i.e.:*** the head of the first epoch) must be forced to one of the u-spaces adjacent to whichever one it was using out of the 8 adjacent to the OD4 form's center. But, that generator has been using that u-space for only 1K (***i.e.:*** creation's 3rd K). If it now moves one u-space at 4K into creation, then it thus moves at the fantastic speed of $1\Phi/1K = 2^{128}$ x light speed. Here, we thus glimpse what, in Esoptrics, = the first kind of inflation.

At 5K into the Universe (3K into epoch #2), God created 8 forms native to OD2 and each, thru its generator, in potency to (***i.e.:*** getting its u-spaces from) the one OD4 form and each activating one of the 8 u-spaces adjacent to the center of that one OD4 form. For that to happen, the 8 forms created native to OD3 at 4K must be forced out of the 8 u-spaces adjacent to the OD4 form's center in order to make room for the 8 forms native to OD2. For that to happen, each OD3 form's carrying generator must, in Geometry speak, "quantum leap" to one of the u-spaces logically adjacent to whichever 1 of the 8 it was created using. Here again, generators are moving $1\Phi/1K$—a speed to end when Creation does.

At 6K into the Universe (4K into epoch#2)—as creation thru an $OD2^2$ form yields to creation thru an $OD2^1$ form (Creation is always thru forms native to a power of 2.)—God turns to the 8 categorical OD2 forms and creates 8 OD1 forms in potency to each of those 8 OD2 forms. Thus, God creates 8 sets each of 8 OD1 forms, and each set of 8 is in potency to one of the 8 OD2 forms. They push nothing out of the way, since nothing is in their way.

At 7K into the Universe (5K into epoch #2), God has now ceased from creation (***cf. Genesis*** 2:2) and, in epoch #2, has left behind 1 immobile form native to OD4, 8 native to OD3, 8 native to OD2, 64 native to OD1 plus the remains of epoch #1. Chart #4-1 on page 28 tells us that the duration of epoch #2 is 16K; and so, God will rest from creation for 12K (***i.e.:*** 5K thru 16K of epoch #2 and 7K thru 18K of the Universe).

At 19K into the Universe, God initiated the Universe's THIRD EPOCH by creating a single form native to OD16. Again, because there was no form of a higher OD, nothing was presenting that OD16 form's carrying generator with a collection of u-spaces; and so, this OD16 form's center was now the Universe's immovable center.

Though so nailed in "place", this OD16 form now presents 32,768 (***i.e.:*** the cube of 2x16) u-spaces to the carrying generator logically concentric with the OD4 form serving as the second epoch's head. That head of the second epoch is thus free to move around within the u-spaces envelope offered by the third epoch's OD16 form. Doing that, it carries with it the remains of the first epoch and the second epoch's 8 OD3 forms, 8 OD2 forms, and, by means of the latter, 64 OD1 forms.

At 20 K into the Universe (2K into epoch #3), God created 8 forms native to OD15 and each, by means of its carrying generator, in potency to the OD16 form. Thus, for at least 1K, each is currently at one of the 8 u-spaces clustered around the center of the OD16 form. That fact, of course, forces out of that 8 fold cluster the OD4 form serving as the head of epoch #2. Thus, its carrying generator must move from one u-space to a second one 16K after utilizing the first one. Here, the velocity is 2^{124} x light speed.

At 21 K into the Universe (3K into epoch #3), God created 8 forms native to OD14 and each, by means of its carrying generator, in potency to the OD16 form. Thus, for at least 1K, each is currently at one of the 8 u-spaces clustered around the center of the OD16 form. That fact, of course, forces out of that 8 fold cluster the 8 forms native to OD15. We already know what the high speed effect of that is: more of the first kind of inflation.

To rush a bit, let K_U = number of K into the Universe's creation, and K_3 = number of K into epoch #3. Let "form(s)" include: "each with its generator". We can then say:

At $22K_U$ & $4K_3$, 8 forms each with its generator are created native to OD13 forcing the 8 native to OD14 to move at high speed out of the central cluster of 8 u-spaces and forcing at least some of the 8 native to OD15 to move even further at high speed.

At $23K_U$ & $5K_3$, 8 forms are created native to OD12 forcing the 8 native to OD13 to move at high speed out of the central cluster of 8 u-spaces and forcing at least some of the 8 native to OD14 and some of the 8 native to OD15 to move even further at high speed.

At $24K_U$ & $6K_3$, 8 forms are created native to OD11 forcing the 8 native to OD12 to move at high speed out of the central cluster of 8 u-spaces and forcing the 8 native to OD13, the 8 native to OD14, and the 8 native to OD15 to move even further at high speed.

At $25K_U$ & $7K_3$, 8 forms are created native to OD10 forcing those native to OD's 11, 12, 13, 14, and 15 to continue fleeing at high speed the center of the OD16 form.

At $26K_U$ & $8K_3$, 8 forms are created native to OD9 forcing etc..

At $27K_U$ & $9K_3$, 8 forms are created native to OD8 forcing etc..

At $28K_U$ & $10K_3$, 64 forms are created native to OD7 (***i.e.:*** 8 per each of the 8 OD8 forms), as creation thru the $OD2^4$ form yields to creation thru the 8 $OD2^3$ forms.

At $29K_U$ & $11K_3$, 64 forms are created native to OD6. (***i.e.:*** 8 per each of the 8 OD8 forms) and forcing the OD7 forms into high speed motions away from the center of each of the OD8 forms.

At $30K_U$ & $12K_3$, 64 forms are created native to OD5. (***i.e.:*** 8 per each of the 8 OD8 forms) and forcing the OD7 and OD6 forms into high speed motions away from the center of each of the OD8 forms.

At $31K_U$ & $13K_3$, 64 forms are created native to OD4. (***i.e.:*** 8 per each of the 8 OD8 forms) forcing the OD7, OD6, and OD5 forms into high speed motion away from the center of each of the OD8 forms.

At $32K_U$ & $14K_3$, 512 forms are created native to OD3 (***i.e.:*** 8 per each of the 64 OD4 forms), as creation thru 8 $OD2^3$ forms yields to creation thru 64 $OD2^2$ forms.

At $33K_U$ & $15K_3$, 512 forms are created native to OD2. (***i.e.:*** 8 per each of the 64 OD4 forms) forcing the OD3 forms into high speed motion away from the center of each of the OD4 forms.

At $34K_U$ & $16K_3$, 4,096 forms are created native to OD1 (***i.e.:***.8 per each of the 512 OD2 forms), as creation thru 64 $OD2^2$ forms yields to creation thru 512 $OD2^1$ forms.

Every form native to either a generic or a categorical OD is a "channel of creation". If you remember, every form is a generic one, if it is native to an OD which is a power of 2; every form is a categorical one, if it is native to an OD which is one of the successive squares of the powers of 2. In this the 9th epoch, the latter means the OD's: 2^0, 2^1, 2^2, 2^4, 2^8, 2^{16}, 2^{32}, 2^{64}, 2^{128}, and 2^{256}. Thus, in the above, we see creation "spewing" from the form native to OD16 (***i.e.:*** 2^4) until 8 are created native to OD8 (***i.e.:*** 2^3) and then creation ceases spewing from the form native to OD16 and switches to spewing from the 8 forms native to OD8 (***i.e.:*** 2^3). Creation then spews 8 forms (***i.e.:*** 8 duo-combos) from each of the 8 forms native to OD8 until 64 forms (***i.e.:*** duo-combos) are created native to OD4 (***i.e.:*** 2^2). Creation then ceases spewing from the 8 forms native to OD8 and starts spewing from each of the 64 forms native to OD4. Those 64 each spew 8 forms native to OD3 for a total of 512 native to OD3, then each spews 8 forms native to OD2 for a total of 512 native to OD2. Since that's a power of 2, creation switches from spewing forms thru the 64 forms native to OD4 and commences to spew forms thru the 512 forms native to OD2. Each of those 512 then spews 8 forms native to OD1, and creation then ceases for the 112 successive K remaining to this the first cycle of epoch #3.

The epochs being too complex to continue as above, turn to the following charts:

EPOCH		START EPOCH		STOP EPOCH		STOP CREATION		DURATION
#	DEPTH[1]	K_u	K_E	K_U	K_E	K_U	K_E	K_E
1	2 OD's	1	1	2	2	N/A	N/A	2
2	4 OD's	3	1	18	2^4	7	5	16
3	16 OD's	19	1	152	2^7	40	16	128
4	256 OD's	147	1	8,344	2^{13}	408	256	8,192
5	65,536 OD's	8,339	1	N/A #1	2^{25}	73,875	65,536	33,554,432
6	2^{32} OD's	N/A #2	1	N/A #3	2^{49}	N/A #4	2^{32}	5.6295×10^{14}
7	2^{64} OD's	N/A #5	1	N/A #6	2^{97}	N/A #7	2^{64}	1.58456×10^{29}
8	2^{128} OD's	N/A #8	1	N/A #9	2^{193}	N/A #10	2^{128}	1.25542×10^{58}
9	2^{256} OD's	N/A #11	1	N/A #12	2^{385}	N/A #13	2^{256}	7.788×10^{115}

N/A	N/A = NOT AVAILABLE = NUMBERS TOO LARGE TO FIT THEIR BLOCKS
#1	33,562,770K_u = 2+16+128+8,192 +33,554,432 = sum of the spans of epochs 1 thru 5
#2	33,562,771K_U = N/A#1 +1
#3	562,949,986,984,082K_U = N/A #1+ the span, $2^{49}K_E$, of epoch #6
#4	4,328,530,066K_U = N/A #1 + the span, $2^{32}K_E$, of creation in epoch #6
#5	562,949,986,984,083K_U = N/A #3 +1
#6	562,949,986,984,082K_U + $1.58456325 \times 10^{29} K_U$ = N/A #3 + span, $2^{97}K_E$, of epoch #7
#7	18,447,307,023,695,535,698K_U= N/A #3 + span, $2^{64}K_E$, of creation in epoch #7
#8	158,456,325,028,529,238,137,074,884,755K_U = N/A #6 +1
#9	$1.255420347 \times 10^{58} K_U$ = N/A #6 + the span, $2^{193}K_E$, of epoch #8[2]
#10	$3.4028236692 \times 10^{38} K_u$ = N/A #6 + the span, $2^{128}K_E$, of creation in epoch #8[3]
#11	$1.2554 \times 10^{58} K_U$ = N/A #9 +1 = add 1 to ftn. #3.
#12	$7.78804012392788958424558 08 \times 10^{115} K_U$ = N/A #9 + the span, 2^{385}, of epoch #9[4]
#13	$1.15792089237316195 \times 10^{77} K_U$ = N/A #9 + the span, $2^{256}K_E$, of creation in epoch #9

CHART #10-1: DURATIONS IN ALPHAKRONONS OF THE 9 EPOCHS' CYCLES

---

[1] Ontological depth = # of OD's & 4th dimension levels available in the given epoch's cycles. K_U = # of alphakronons, K, into the Universe's creation. K_E = # of K into the given epoch.

[2] N/A #9 = 12,554,203,470,773,361,527,671,578,846,573,789,157,233,240,127,065,143,910,545K_u.

[3] N/A #10 = 340,282,367,079,394,788,491,903,845,568,843,096,215K_U.

[4] N/A #9 + 2^{385} = 78,804,012,392,788,958,424,558,080,200,287,227,610,159,478,540,930,893,335,909, 141,011,962,216,904,522,092,801,675,106,298,927,064,521,740,321,124,524,183K_U. N/A #9 + 2^{256} = 115,792,089,237,316,195,436,125,188,479,461,269,380,941,563,512,214,353,196,690,824,134,978, 273,550,487K_U.

As you can perhaps see for yourself, chart #10-1 gives us, in alphakronons, K, when the first cycle of each epoch commenced, when it terminated (or shall terminate in the case of epoch #9), and when, in the course of each epoch's first cycle, God ceased adding any additional duo-combos. In the course of one cycle on the part of epoch #9, each of the other 8 epochs shall go thru many additional cycles; but, only in the first cycle of each of those 8 epochs, did God add duo-combos to those epochs. Likewise, only in the first 2^{256} K_E of the 9th epoch did God add any duo-combos to that epoch.

K = 1 ALPHAKRONON = 7.201789375 x 10^{-96} SECONDS	
EPOCH	**START OF EACH EPOCH IN SECONDS FROM UNIVERSE'S BEGINNING**
#1	7.201789375 x 10^{-96} sec.
#2	2.160536812 x 10^{-95} sec. = Kx3 = duration of #1+1K
#3	1.368339981 x 10^{-94} sec. = Kx19 = duration of #1+#2+1K
#4	1.058663038 x 10^{-93} sec. = Kx147 = duration of #1+#2+#3+1K
#5	8.630065158 x 10^{-92} sec. = Kx8,339 = K(147+8,192)
#6	2.417120076 x 10^{-88} sec. = Kx33,562,771 = K(#5+33,554,432)
#7	1.279294971 x 10^{-82} sec. = Kx562,949,986,984,083=K(#6+562,949,953,421,312)
#8	4.544968068 x 10^{-67} sec. = 158,456,325,028,529,238,137,074,884,755=K(#7+2^{97})
#9	5.736556201 x 10^{-38} sec. = K(1.255420347x10^{58})=K(#8+2^{193})
#9A	6.219586696 x 10^{-19} sec. = K(#9+2^{256})=creation's halt until epoch #10 begins.

CHART #10-2: IN SECONDS, WHEN EACH EPOCH STARTED ITS FIRST CYCLE & WHEN GOD CEASED CREATING FORMS IN EPOCH #9

In chart #10-2 above, I've converted the alphakronons into seconds and, thereby, have sought to give my reader a fairly accurate accounting of how far from the beginning of the Universe each epoch began in time as most humans reckon time. Note that the really *big* bang (***i.e.:*** 9th epoch) began at c. 10^{-37} and ended at c. 10^{-18} sec.. I think you'll find that accords rather well with what theoretical physicists say these days.

In the figures given, of course, I make no claim to *perfect* accuracy. Still, I venture to suggest that, even if one were to carry each number out to hundreds of places, the increase in accuracy over the above would be miniscule. As my youngest brother, Gordon, humorously expressed it to me in a recent e-mail: "in terms of significance there simply isn't any change whatsoever unless you is a QUARK!!"

In chart #10-3, on the next page, I've given my reader the exact number of duo-combos created in each of the Universe's first 4 epochs. As for the piggyback forms in those duo-combos, I've listed exactly how many of them, in each of those 4 epochs, are created native to each ontological distance available in a given one of those 4 epochs. Of equal importance is the fact I've given the native OD of the form thru which each duo-combo was created by God. For example, in epoch #3, each of the 512 duo-combos—with forms created native to OD2—were created by God thru one of the 64 forms created native to OD4, and one can say the same of the 512 duo-combos with forms created native to OD3.

Notice how, in each of the epochs, I've given the total number of duo-combos God created in each of them and have then divided that total by the number of duo-combos, in that same epoch, created native to OD1. I then gave running totals of the number of duo-combos created in the first, the first 2, the first 3, and the first 4 epochs, and then, at each step have divided the running total by the total number of duo-combos created native to OD1 in the last epoch involved in the running total. I've done that so as to indicate that the running total of all epochs amounts to roughly 1.33 times the total number of duo-combos created native to OD1 in the latest epoch. For example, the total number of duo-combos created native to OD1 in epoch #4 = 16,777,216, and 1.33364272 times that gives one the sum total of all the duo-combos created in the course of all 4 epochs combined.

In chart #10-4, on page 85, I've extended that exhaustive analysis to the Universe's fifth epoch. I have not extended the exhaustive analysis to epochs 6, 7, 8, and 9, because it would involve far more time and energy than what I speculate is needed to make my reader sufficiently aware of what Esoptrics is saying in this chapter.

EPOCH	OD'S AVAILABLE IN THE EPOCH	FORMS NATIVE TO EACH OD PER EPOCH
#1	$2 = 2^1$ = 1 OD	1
	1 = 1 OD.	8 & thru the one OD2 form
		TOTAL = 9 = 8x1.125
#2	$4 = 2^2$ = 1 OD	1
	3 = 1 OD	8 per the 1 OD listed & thru the OD4 form
	2 = 1 OD	8 per the 1 OD listed & thru the OD4 form
	1 = 1 OD	8^2 = 64 = 8 thru each of the 8 OD2 forms
		TOTAL = 81 = 64x1.265625
		TOTAL OF 1&2 = 90 = 1.4065 x 64
#3	$16 = 2^4$ = 1 OD	1
	15 thru 8 = 8 OD's	8 per the 8 OD's listed & thru the one OD16 form
	7 thru 4 = 4 OD's	8^2 = 64 per the 4 OD's listed & thru the 8 OD8 forms
	3 thru 2 = 2 OD's	8^3 = 512 per the 2 OD's listed & thru the 64 OD4 forms
	1 = 1 OD	8^4 = 4,096 per the 1 OD listed & thru the 512 OD2 forms
		TOTAL = 5,441 = 4,096X1.328369
		TOTAL OF 1,2&3 = 5,531= 1.35034179687x 4,096
#4	$256 = 2^8$ = 1 OD	1
	255 thru 128 = 128 OD's	8 per the 128 OD's listed & thru the one OD256 form
	127 thru 64 = 64 OD's	8^2 = 64 per the 64 OD's listed & thru the 8 OD128 forms
	32 thru 63 = 32 OD's	8^3 = 512 per the 32 OD's listed & thru the 64 OD64 forms
	31 thru 16 = 16 OD's	8^4 = 4,096 per the 16 OD's listed & thru the 512 OD32 forms
	15 thru 8 = 8 OD's	8^5 = 32,768 per those 8 OD's & thru the 4,096 OD16 forms
	7 thru 4 = 4 OD's	8^6 = 262,144 per those 4 OD's & thru the 32,768 OD8 forms
	3 thru 2 = 2 OD's	8^7 = 2,097,152 per the 2 OD's & thru the 262,144 OD4 forms
	1 = 1 OD	8^8 = 16,777,216 per 1 OD & thru the 2,097,152 OD2 forms
		TOTAL = 22,369,281 = 16,777,216x1.3333130478858
		TOTAL OF 1,2,3&4 = 22,374,812 = 1.33364272117x8^8

CHART #10-3: EXACT NUMBER OF FORMS GOD CREATED IN THE UNIVERSE'S FIRST 4 EPOCHS, THE OD TO WHICH EACH IS NATIVE, & THE NATIVE OD OF THE FORM THRU WHICH GOD CREATED EACH FORM.

In chart #10-4, on the next page, notice that the total number of duo-combos created native to OD1= 281,474,976,710,656 (***i.e.:*** 8^{16}), and the total for that 5th epoch = 375,299,968,598,276. The latter = 1.3333333309247 times the former. The running total for the first 5 of the 9 epochs is then 375,299,990,967,557 duo-combos, and that = 1.3333 3341156414 times the number of duo-combos God created native to OD1 in the course of the fifth epoch. That rather strongly suggests that, when creation ends in epoch #9, the total number of duo-combos created in the course of the 9 epochs should equal roughly

1.3333 times the total number of duo-combos God created native to OD1 in the 9th epoch. What is that number?

In chart #10-5, on page 86, I have given a partial analysis of what duo-combos God created in the course of the Universe's 9th epoch. As should be obvious to any and every reader, I would have to list 256 lines in order to give a *full* account of what duo-combos God created in epoch #9. Surely, every reader will agree with me that such is far too formidable a task. If the 32, 64, and 128 lines, required for a full accounting of the 6th, 7th, and 8th epochs respectively, adequately excuse every refusal to list them, then, how much more mightily do 256 lines excuse such a refusal?

EPOCH	OD'S AVAILABLE IN THE EPOCH	FORMS NATIVE TO EACH OD PER EPOCH
#5	65,536 = 2^{16} = 1 OD	1
	65,535 - 32,768 = 32,768 OD's .	8^1 = 8 per 32,768 OD's & thru the one OD2^{16} form
	32,767 - 16,384 = 16,384 OD's .	8^2 = 64 per 16,384 OD's & thru the 8 OD2^{15} forms
	16,383 – 8,192 = 8,192 OD's . . .	8^3 = 512 per 8,192 OD's & thru the 64 OD2^{14} forms
	8,191 – 4,096 = 4,096 OD's	8^4 = 4,096 per 4,096 OD's & thru the 512 OD2^{13} forms
	4,095 – 2,048 = 2,048 OD's	8^5 = 32,768 per 2,048 OD's & thru the 4,096 OD2^{12} forms
	2,047 – 1,024 = 1,024 OD's	8^6 = 262,144 per 1,024 OD's & thru the 32,768 OD^{11} forms
	1,023 – 512 = 512 OD's	8^7 = 2,097,152 per 512 OD's & thru the 262,144 OD2^{10} forms
	511 – 256 = 256 OD's.	8^8 = 16,777,216 @ 256 OD's & by the 2,097,152 OD2^9 forms
	255 – 128 = 128 OD's.	8^9 = 134,217,728 @ 128 OD's & by 16,777,216 OD2^8 forms
	127 – 64 = 64 OD's	8^{10} = 1,073,741,824 @64OD's&by 134,217,728 OD2^7 forms
	63 – 32 = 32 OD's	8^{11} = 8,589,934,592@32OD's& by 1,073,741,824 OD2^6 forms
	31 – 16 = 16 OD's	8^{12} = 68,719,476,736@16OD's by 8,489,934,592 OD2^5 forms
	15 – 8 = 8 OD's	8^{13} = 549,755,813,888@8OD's by 68,719,476,736 OD2^4 forms
	7 – 4 = 4 OD's	8^{14}= 4,398,046,511,104@4OD's by 549,755,813,888 OD2^3 forms
	2 – 3 = 2 OD's	8^{15} = 35,184,372,088,832@2OD's by 4,398,046,511,104 OD4 forms
	1 = 1 OD	8^{16} = 281,474,976,710,656@1OD by35,184,372,088,832 OD2 forms
		TOTAL = 375,299,968,598,276 = 1.3333333309247x8^{16}.
		TOTAL OF 1,2,3,4&5 = 375,299,990,967,557 = 1.333333x8^{16}

CHART #10-4: EXACT NUMBER OF FORMS GOD CREATED IN THE UNIVERSE'S FIFTH EPOCH, THE OD TO WHICH EACH IS NATIVE, & THE NATIVE OD OF THE FORM THRU WHICH GOD CREATED EACH FORM.

Looking at chart #10-5 on the next page, you can see that, in the 9th epoch, God created 8^{256} (***i.e.:*** = 2^{768} = 1.55251809230071x10^{231} rounded to 1.55252x10^{231}) duo-combos each composed of a carrying generator logically concentric with a form native to OD1. In epoch #1, 8 were created native to OD1; in #2, 8^2 were created native to OD1; in #3, 8^4; in #4, 8^8; in #5, 8^{16}; in #6, 8^{32}; in #7, 8^{64}; in #8, 8^{128}; and in #9, 8^{256}. By now, one should be familiar with that progression of exponents 1, 2, 4, 8, 16, 32, 64, 128 & 256.

If one can rely on the ratio of 1.33333, then the total number of duo-combos created by God, in the course of the 9 epochs and the Universe's first 6.22x10^{-19} seconds, was very close to: 1.33333(1.55252x10^{231}) = 2.07002x10^{231} duo-combos. What does that tell us?

Remember: Each duo-combo consists of a piggyback form and its logically concentric carrying generator, and the "size" of each generator is 1Φ. God, then—in the course of the 9 epochs and first 6.22x10^{-19} sec.—created close to 2.07002x10^{231} generators. Take the cube root of that, and we have: 1.274456x10^{77}Φ. That seems to say that it would take a form 1.274456x10^{77}Φ in dia. to provide each of those generators with its own u-space. But, a form of that diameter would—in geometry speak—have a diameter of roughly 9x10^{30} cm. (***i.e.:*** dividing 1.274456x10^{77}Φ by 1.361129468x10^{46}Φ per cm.). That, in turn, would mean that, in 6.22x10^{-19}, the Universe became a sphere which is almost 18 trillion light years in dia. and, in which, every u-space is occupied by a generator (How's that for inflation?). That's what the figures *seem* to say; but, it's not what they *really* say. How so? Chapter 12 shall answer that in its description of what Esoptrics calls "The Great Acceleration".

EPOCH	SOME OF OD'S IN THE EPOCH	FORMS NATIVE TO EACH OD LISTED
#9	2^{256} = 1.1579x10^{77} = 1 OD	1
	2^{256} -1 thru 2^{255} = 2^{255} OD's	8 per 2^{255} OD's & thru the one OD2^{256} form
	2^{255} -1 thru 2^{254} = 2^{254} OD's	8^2 = 64 per 2^{254} OD's & thru the 8 OD2^{255} forms
	2^{254} -1 thru 2^{253} = 2^{253} OD's	8^3 = 512 per 2^{253} OD's & thru the 64 OD2^{254} forms
	2^{253} -1 thru 2^{252} = 2^{252} OD's	8^4 = 4,096 per 2^{252} OD's & thru the 512 OD2^{253} forms
	2^{252} -1 thru 2^{251} = 2^{251} OD's	8^5 = 32,768 per 2^{251} OD's & thru the 4,096 OD2^{252} forms
	2^{251} -1 thru 2^{250} = 2^{250} OD's	8^6 = 262,144 per 2^{250} OD's & thru the 32,768 OD2^{251} forms
	2^{250} -1 thru 2^{249} = 2^{249} OD's	8^7 = 2,097,152 per 2^{249} OD's & thru the 262,144 OD2^{250} forms
	2^{249} -1 thru 2^{248} = 2^{248} OD's	8^8 = 16,777,216 per 2^{248} OD's & thru the many OD2^{249} forms
	2^{241} -1 thru 2^{240} = 2^{240} OD's	8^{16} = 2.8147x10^{14} per 2^{240} OD's & thru the OD2^{241} forms
	2^{225} -1 thru 2^{224} = 2^{224} OD's	8^{32} = 7.9228x10^{28} per 2^{224} OD's & thru the OD2^{225} forms
	2^{193} -1 thru 2^{192} = 2^{192} OD's	8^{64} = 6.2771x10^{57} per 2^{192} OD's & thru the OD2^{193} forms
	2^{129} -1 thru 2^{128} = 2^{128} OD's	8^{128} = 3.9402x10^{115} per 2^{128} OD's & thru the OD2^{130} forms
	2^{65} -1 thru 2^{64} = 2^{64} OD's	8^{192} = 2.4733x10^{173} per 2^{64} OD's & thru the OD2^{65} forms
	2^{33} -1 thru 2^{32} = 2^{32} OD's	8^{224} = 1.95955x10^{202} per 2^{32} OD's & thru the OD2^{33} forms
	2^{17} -1 thru 2^{16} = 2^{16} OD's	8^{240} = 5.51652x10^{216} per 2^{16} OD's & thru the OD2^{17} forms
	2^9 -1 thru 2^8 = 2^8 OD's	8^{248} = 9.25373x10^{223} per 256 OD's & thru the OD2^9 forms
	31 thru 16 = 16 OD's	8^{252} = 3.79033x10^{227} per 16 OD's & thru the OD2^5 forms
	7 thru 4 = 4 OD's	8^{254} = 2.42581x10^{229} per 4 OD's & thru the many OD8 forms
	3 & 2 = 2 OD's	8^{255} = 1.94065x10^{230} per 2 OD's & thru the many OD4 forms
	1 = 1 OD	8^{256} = 1.55252x10^{231} per 1 OD & thru the many OD2 forms
		TOTAL OF EPOCHS 1-9 = 1.33333333x8^{256} = 2.070024 x 10^{231}

CHART #10-5: PARTIAL ANALYSIS OF WHAT FORMS GOD CREATED IN THE UNIVERSE'S NINTH EPOCH, THE OD TO WHICH THE GIVEN FORMS WERE CREATED NATIVE, & THE NATIVE OD OF THE FORM THRU WHICH GOD CREATED THE GIVEN FORMS.

To set the stage for the great acceleration, let's recall and restate in a slightly different way an important chain of thought already presented. In epoch #1, the ontological depth was *two* to the first power, and the number of forms created native to OD1 was *eight* to the first power. In epoch #2, the ontological depth was *two* to the second power, and the number of forms creative native to OD1 was *eight* to the second power. In epoch #3, the ontological depth was *two* to the fourth power, and the number of forms created native to OD1 was *eight* to the fourth power. In epoch #4, the ontological depth was *two* to the eighth power, and the number of forms created native to OD1 was *eight* to the eighth power. In epoch #5, the ontological depth was *two* to the sixteenth power, and the number of forms created native to OD1 was *eight* to the sixteenth power.

Are you following the progression? If so, you see this: In epoch #8, the ontological depth was *two* to the $\underline{128^{th}}$ power, and the number of forms created native to OD1 was *eight* to the $\underline{128^{th}}$ power, which is to say $3.9402006196394x10^{115}$. Multiplying that by 1.3333 tells us that, when creation stopped in epoch #8, God had created a total of close to $5.253469486x\ 10^{115}$ duo-combos. What would be the diameter of the "area" able to hold that many generators each "occupying" a u-space 1Φ in "size"? An online calculator tells me the cube root of $5.253469486x10^{115} = 3.745260854x10^{38}$. That would mean an "area" with a diameter of $3.745260854x10^{38}$Φ would hold the $5.253469486x10^{115}$ duo-combos created in the course of the Universe's first 8 epochs. What does that mean in the cm. so dear to the geometers?

Esoptrics says that—with $7.34683969x10^{-47}$ cm. per Φ—a single cm. = $1.361129468x\ 10^{46}$Φ. So, I divide $1.361129468x10^{46}$Φ (***i.e.:*** 1 cm.) by $3.745260854x10^{38}$ and discover that the latter = 1/36,342,714.767 cm. which, in turn, = $2.75185x10^{-8}$ centimeters. Since the dia. of the hydrogen atom is generally given as $5x10^{-8}$ cm., that puts Esoptrics close to saying that, at $5.736556201x10^{-38}$ sec. since the Universe's beginning (***i.e.:*** the moment at which epoch #9 commenced according to chart #10-2 on page 83): (1) The Universe expanded $3.74x10^{38}$ times going from $7.35x10^{-47}$ cm. in "dia." to $2.75x10^{-8}$ cm., and (2) God had jammed $5.253469486\ x10^{115}$ duo combos into an area a little over half the diameter of an atom. That exceedingly dense atom sized area would then, within the next roughly 10^{-18} sec., explode into so great a number of additional duo-combos, only a Universe trillions of light years in dia. could hold them did they all remain separate from one another.[1] At least, that's the way it would be, did not "The Great Acceleration" produce a vast number of *multi*-combos each composed of an astronomical number of duo-combos.

> At this point, Einstein's general theory of relativity would have broken down, so it cannot be used to predict in what manner the universe began. One is left with the origin of the universe apparently being beyond the scope of science.
> ——**STEPHEN HAWKING:** ***The Universe In A Nutshell***, pg. 79 hardbound edition by Bantam Books, New York, 2001. ©2001 Stephen Hawking.

Yes, Prof. Hawking, the origin is beyond *Science's* scope, but not beyond *Esoptrics'* exercise in introspection (***i.e.:*** the reflexive half of bi-directional empirical observation) prompted by Catholic Theology's praise of Philosophy.

The stage is now *almost* set for "The Great Acceleration" which shall fuse many duo-combos into a vast number of galaxies (***i.e.:*** by means of multi-g-combos), heavenly bodies (***i.e.:*** by means of multi-h-combos), atomic nuclei (***i.e.:*** by means of multi-A, multi-a-, and multi-b-combos), and electrons (***i.e.:*** by means of multi-e-combos).[2] However, before that stage can be *fully* set, we must first spend 3 chapters on a few other concepts.

✡✞✡✞✡✞✡✞✡✞✡✞✡✞✡✞✡✞✡✞✡✞✡

---

[1] That, I dare suggest, is very close to what today's scientists say about the Big Bang and inflation. I know not what devastatingly complex mathematics they employed to arrive at what they calculate. Probably, it was some kind of calculus so technical only those with an IQ above 160 and a decade of college can follow it. As should be rather obvious, though, Esoptrics uses nothing more complex than the kind of simple math which most teen agers still in high school can follow. Yes, the *quantity* of Esoptrics' math may be a bit daunting; but, the *nature* of it is quite the opposite. Were that not true, my fifth-rate intellect could never have made any progress at all with The Logic Of The Mirror.

[2] Be patient, and chapter 11 shall give a detailed description of the difference between the above 6 different kinds of multi-combos.

2013 ESOPTRICS UPDATE

A common misconception is that the big bang provides a theory of cosmic origins. It doesn't. The big bang is a theory, partly described in the last two chapters, that delineates cosmic evolution from a split second after whatever happened to bring the universe into existence, but *it says nothing at all about time zero itself*. And since, according to the big bang theory, the bang is what is supposed to have happened at the beginning, the big bang leaves out the bang. It tells us nothing about what banged, why it banged, how it banged, or, frankly, whether it ever really banged at all.
——**BRIAN GREENE:** ***The Fabric Of The Cosmos***, as found on page 272 of the 2005 paperback edition published by Vintage Books, a division of Random House, Inc., New York. ©2004 by Brian Greene (Esoptrics, though, tells us so much about what happened beginning from "time zero itself" thru the following c. 10^{-18} sec., that we know in great detail what banged, why it banged and how it banged. Doing so, it rejects Prof. Greene's contention, on page 284 of his book, that the Big Bang's bang was supplied by the repulsive gravity of what's called the "Higgs field". For all of that, if my experiences since 1957 are indicative of the future, physicists will universally give no attention whatsoever to Esoptrics or, at best, give it notice only after I've been in my grave for at least 100 years. Well, there's always the possibility of Divine Intervention—the only factor, I dare suspect, which can bring Esoptrics to the world's attention before I pass on to the next life. And why not?! All too often, fame is a curse which eliminates all possibility of any additional creative thinking and writing. And what can be worse than the loss of the time, energy, and capacity to engage in the ecstasy of creative thinking and writing?).

Other concepts, such as the presence of a black hole inside of the sun, have no support in research or theology.
——**ANNA CALL:** ***Clarion Review*** (I never said there's a black hole inside the Sun. At the bottom of page 71, I said there is a "black hole". I put the term in parentheses to indicate I'm not using it in the usual sense. It's a commonplace tactic. For all of that, it apparently never occurred to Ms. Call that such is what might be happening. In Esoptrics, the term black hole does not have the same meaning as it has for Science. That's because Esoptrics proposes Science doesn't understand what a black hole *really* is and shall eventually learn each is a multi-combo in which at least 2^{192} to as much as $2^{256}-1$ duo-combos are logically concentric to a particular kind of leading form. The number of duo-combos so joined sets the black hole's mass. In response to Ms. Call's remark, I've modified the passage in a way I hope shall make it more difficult for others to repeat her misinterpretation of my statement.)

"*All* the universe or *nothing*! Which shall it be, Passworthy? Which shall it be?"
——**RAYMOND MASSEY** (as Oswald Cabal)**:** closing words from the 1936 movie ***Things To Come*** produced by Alexander Korda and directed by William Cameron Menzies. (This is perhaps the most unforgettable utterance I heard as a child. It's me in a nutshell. At least, so I delight to imagine. An inner voice, though, ever insists it is but the deceitful ego-defense mechanism my irrationally proud ego is using to keep me from seeing the unbearably humiliating truth of how much lower and repulsive my goals really are—even if not evil—than I imagine they are.).

Chapter Eleven:

KINDS OF ZONES, KINDS OF MULTI-COMBOS, & THE RECTILINEAR VELOCITY OF THE LATTER:

In Esoptrics, the Universe is a collection of two kinds of zones: categorical and generic. Prior to the great acceleration, there were 10 categorical zones. Both before and after the great acceleration, there were and are 256 generic zones. The original 10 categorical zones here in this the Universe's 9th epoch were these:

C ZONE	ONTOLOGICAL DISTANCES INCLUDED IN THE CATEGORICAL ZONE
1ST	1
2nd	2 & 3
3rd	4, 5, 6, 7, 8, 9, 10, 11, 12, 13, 14 & 15 = 2^2 thru 2^4 -1
4th	16 thru 255 = 2^4 thru 2^8 -1
5th	256 thru 65,535 = 2^4 thru 2^{16} -1
6th	65,536 thru 4,294,967,295 = 2^{16} thru 2^{32} -1
7th	2^{32} thru 2^{64} -1
8th	2^{64} thru 2^{128} -1
9th	2^{128} thru 2^{256} -1
10th	2^{256}

CHART #11-1: THE 10 CATEGORICAL ZONES

G ZONE	ONTOLOGICAL DISTANCES INCLUDED IN THE GENERIC ZONE
1ST	1 = 2^0
2nd	2 & 3 = 2^1 thru 2^2 -1
3rd	4, 5, 6, & 7 = 2^2 thru 2^3 -1
4th	8, 9, 10, 11, 12, 13, 14 & 15 = 2^3 thru 2^4 -1
5th	16, 17, 18, 19, 20, 21, 22, 23, 24, 25, 26, 27, 28, 29, 30 & 31 = 2^4 thru 2^5 -1
6th	32 thru 63 = 2^5 thru 2^6 -1
7th	64 thru 127 = 2^6 thru 2^7 -1
8th	128 thru 255 = 2^7 thru 2^8 -1
9th	256 thru 511= 2^8 thru 2^9 -1
10th	512 thru 1023= 2^9 thru 2^{10} -1
11th	1,024 thru 2,047= 2^{10} thru 2^{11} -1
12th	2,048 thru 4,095= 2^{11} thru 2^{12} -1

CHART #11-2: THE FIRST 12 GENERIC ZONES

2013 ESOPTRICS UPDATE

Because I deem it ridiculous to present a chart 256 lines long, I have limited chart #11-2 to the first 12 generic zones. That should be enough to allow the reader to grasp the basic arrangement—namely: A generic zone is one beginng at an OD which is a power of 2 and then includes it and the OD's *between* it and the OD which is the next power of 2.

Though not listed in chart #11-2, the 129th generic zone—the one running from OD2^{128} to OD2^{129}–1—deserves special mention. It's the generic zone for the MT and serves as the basis for Esoptrics' explanation of gravity vs. anti-gravity. It's an explanation which says that, except for the OD1 type, categorical forms currently within or below the MT's generic zone attract one another, but repel one another if currently above the MT's generic zone, because what's above MT mirrors and reverses what's below it.

OD1 categorical forms cannot pass beyond OD2^{128}; and so, there is no such thing as them acting one way below the MT and an opposite way above it. Always and everywhere, they act one way and one way only—namely: They repel the other 9 categorical kinds in a way causing them to move toward the outermost limits of the form to which they are in potency. That's why, in Esoptrics, they orbit, rather than dwell in, the atomic nucleus.

It's very important to grasp the parameters of the various zones because of the way the 6 kinds of multi-combos move. How is that? To answer, let's first define each of those 6 kinds. As we set out, bear one very important thought in mind: None of the first 5 kinds of multi-combos can ever hold an OD1 form concentric to themselves. Only an OD1 form can capture and make concentric to itself another OD1 form. It's Esoptrics version of electrons.

FIRST KIND: Every multi-g-combo is a galaxy-class multi-combo. Every multi-g-combo serves as the center of a galaxy or cluster of galaxies or cluster of clusters, etc., and, thereby, controls how those galaxies and clusters shall relate to one another. Every multi-g-combo has an ontological depth of at least 2^{192} +1, which is to say that, in every one of the multi-g-combos, there are at least 2^{192} duo-combos logically concentric with their leading form and each at a discrete OD. In every one of them, the *leading* form is native to the categorical OD2^{128} and currently accelerated to at least OD2^{192}+1. The most massive of the multi-g-combos would have a leading form native to OD2^{128} and currently accelerated to OD2^{256}–1 and would have an ontological depth of 2^{256}– 1. That –1 is because no form native to any of the other categorical OD's can be made concentric to an OD2^{128} form; but, it can make concentric to itself—at OD's 2^1, 2^2, 2^4, 2^8, 2^{16}, 2^{32} 2^{64}, 2^{128} and maybe 2^0—non-categorical (and non-generic?) forms which have been either so decelerated (in the case of 2^1 and 2^0, if there is a form at that OD) or so accelerated as to fill in the gaps left by the 9 categorical forms below OD2^{256}. In case you missed it, there's a gap at OD2^{128} in every multi-g-combo because the form native to it has been accelerated well above that OD.[1]

SECOND KIND: Every multi-h-combo is a heavenly-body-class multi-combo. Every multi-h-combo serves as the center of a heavenly body such as a star, a sun, a planet, and perhaps, a comet, or a meteor. Every multi-h-combo has an ontological depth of at least 2^{192}+1, which is to say that, in every one of them, there are at least 2^{192} duo-combos logically concentric with their leading form and each at a discrete OD (***i.e.:*** at a discrete level of the multi-combo's 4th dimension). In every one of them the *leading* form is a major *rational* one native to a non-categorical OD below 2^{128} and accelerated to a current OD of at least 2^{192}+1. That means those forms at every rational OD from just above 2^{64} thru just below 2^{128} and maybe those just above 2^{32} thru just below 2^{64}. Because their acceleration is limited to 2^{128} times their native OD, I'm still puzzled as to whether or not forms native to OD's below 2^{64} could actually serve as the leading form in a multi-h-combo. The most massive of the multi-h-combos would have an ontological depth of 2^{255}–1 and would have a leading form

[1] Actually, I'm often led to think that the gap at OD2^{128} remains because, the instant any form is currently at OD2^{128}, it is emitted at the speed of light. So I currently speculate.

native either to $OD2^{126}$ or some other rational OD between $OD2^{126}$ and $OD2^{128}$.

THIRD KIND: Every multi-A-combo is the central multi-combo in every atom. It is ***THE*** atomic-nucleus-class multi-combo. Every multi-A-combo has a minimum ontological depth of $2^{128}+1$, which is to say that in every one of them, there are at least 2^{128} duo-combos logically concentric with their leading form and each at a discrete OD. In every one of them the leading form itself is native to $OD2^{128}$ and is at *least* at $OD2^{128}+1$ and, *as long as it remains a multi-A-combo*, is at *most* at $OD2^{129}-1$.

FOURTH KIND: Every multi-a-combo—as with the multi-A-combo—is an atomic-nucleus-class multi-combo in which the leading form is native to a categorical OD whether $OD2^1$, 2^2, 2^4, 2^8, 2^{16}, 2^{32}, or 2^{64}. There are thus 8 kinds of atomic-nucleus-class multi-combos to be found in the atomic nucleus. The minimum ontological depth of a multi-a-combo is, I *suspect*, 2^{64}; the maximum anywhere between $2^{121}+1$ and $2^{128}-1$.

FIFTH KIND: Every multi-b-combo is an atomic-nucleus-class multi-combo in which the leading form is one native to an OD which: (1) is below 2^{128}, (2) is a rational OD, and (3) is other than 2^1, 2^2, 2^4, 2^8, 2^{16}, 2^{32}, and 2^{64}. To date, I have no idea what might be the minimum and maximum ontological depths to be found in the multi-b-combos.

SIXTH KIND: Every multi-e-combo is an electron-class multi-combo in which the leading form and every form concentric in the multi-combo is one native to OD1. The minimum ontological depth of a multi-e-combo is, I suspect, 2^{120}-1; the maximum almost certainly 2^{128}-1 as long as its orbiting its atom's nucleus. If the ontological depth of a multi-e-combo rises to 2^{128}, it shall be at maximum acceleration and shall radiate away from the atom at the speed of light (***i.e.:*** $1\Phi/2^{128}K$) as its generator seeks its u-spaces from whatever form, among those available, is currently at the highest OD on up to $OD2^{256}$. That means it can move thru the Universe at light speed and at successively higher OD's as the u-spaces provided by the forms at those higher OD's become logically contiguous.

Hereafter, the multi-g and multi-h-combos can be referred to as weighty multi-combos and W-combos for short. The multi-A-combos can be referred to as MT-combos or A-combos. The other 3 can be referred to as light multi-combos and L-combos for short.

All six of the above are multi-combos in the sense that each is composed of a definite number of *logically concentric* duo-combos. There is, though, a seventh kind of combo which, to avoid confusion, we shall refer to as an m-*cluster*. These are the molecules produced when anywhere from 2 to an astronomical number of L-combos—instead of being concentric—cluster together in response to the influence of irrational generic duo-combos (***ex. gr.:*** those native to $OD2^{127}$, 2^{125}, 2^{123}, etc.) which have accelerated above MT to some current OD between 2^{128} and 2^{160} (***i.e.:*** $2^{128} \times 2^{32}$), if not 2^{192} (***i.e.:*** $2^{128} \times 2^{64}$).

If all of that is sufficiently clear, we can now begin to explain the importance of the various zones and the distinction between the W, L, and MT-combos. Together, those distinctions allow us to determine the OD of the form to which a multi-combo's carrying generator shall turn (or, at least, *try* to turn) for its u-spaces as long as the multi-combo is not moving at light speed. One rule applies to every W and L-combo's carrying generator and a different one to every MT-combo's (***i.e.:*** multi-A-combo's) carrying generator.

Let P_{LW} = the OD of the form to which an L or W-combo's (***i.e.:*** every combo but the MT-combos) carrying generator turns for its u-spaces; let Z_G = the OD of the base of the generic zone in which the L or W-combo's leading form is currently operating; and let N = the number of steps that form has taken into that generic zone. In that case:

FORMULA #11-1: $P_{LW} = 2(Z_G+N)$

That means this: Suppose a multi-g or multi-h-combo's leading form is currently in the 247th generic zone somewhere between $OD2^{246}$ and $OD2^{247}$. In that case, $Z_G = 2^{246}$, because $OD2^{246}$ is where the 247th generic zone starts.[1] Suppose it's currently at $OD2^{246}+1$. Its carrying generator is then receiving its u-spaces from a form currently at $2^{247}+2$, because $P_{LW} = 2(2^{246}+1) = 2^{247}+2$. If an L-combo's leading form is currently at $OD2^{64}+2^{24}$, then its carrying generator's source of u-spaces = a form currently at: $2(2^{64}+2^{24}) = 2^{65}+2^{25}$. Thus, as long as the L or W-combo continues to be currently within the parameters of a particular generic zone, its carrying generator is receiving its u-spaces from a form within the parameters of the next highest generic zone which, even far above MT, may also be a categorical zone as we'll see later when we discuss the reverse categories.

That, then, is the rule for the L and W-combos not at light speed. As for the rule for the MT-combos (***i.e.:*** multi-A-combos) not at light speed, let P_{MT} = the OD of the form to which an MT-combo's carrying generator turns for its u-spaces; let $Z_G = 2^{128}$ = the OD of the base of the generic zone in which every MT-combo's leading form is always operating; and let N = the number of steps that form has taken into that generic zone. In that case:

FORMULA #11-2: $P_{MT} = N(Z_G)$

Consider an example of its meaning: Suppose an MT-combo's leading form is currently in the 129th generic zone somewhere between $OD2^{128}$ and $OD2^{129}$. Let's choose $OD2^{128}+2$. In that case, its carrying generator receives (at least tries to receive) its u-spaces from a form currently at: $2(2^{128}) = 2^{129}$. If that leading form is currently at $OD2^{128}+3$, its carrying generator then receives (at least tries to receive) its u-spaces from a form currently at: $3(2^{128}) = 2^{129}+2^{128}$. If that leading form is currently at $OD2^{128}+4$, its carrying generator receives (at least tries to receive) its u-spaces from a form currently at: $4(2^{128}) = 2^{130}$. If that leading form is currently at $OD2^{128}+2^{64}$, its carrying generator receives (at least tries to receive) its u-spaces from a form currently at $2^{64}(2^{128}) = 2^{192}$. If that leading form is currently at $OD2^{128}+2^{128} = 2^{129}$, the multi-combo is at light speed and its carrying generator shall—using logically contiguous u-spaces—seek its u-spaces from a series of successively higher OD forms as it tries to work its way toward getting its u-spaces from the one form at 2^{256}. That quest—if there's a lack of logically contiguous u-spaces—can prove to be a task so difficult, combos at light speed may often wind up moving thru the Universe for many K at some OD which is at some distance—small or otherwise—below 2^{256}.[2]

But, why would a multi-combo's leading form be at an OD somewhere within a given generic zone's range of OD's? That brings us to the formula for the velocity of a multi-combo, which is to say the formula which describes the frequency with which the multi-combo's carrying generator is changing from u-space to u-space (***i.e.:*** from one temporarily changeless microstate of excitation to a logically contiguous one.). For all multi-combos whatsoever, that formula is:

FORMULA #11-3: $V = N(C/Z_G)$

In the above formula, V = the rectilinear velocity of a multi-combo's carrying gener-

---

[1] Every generic zone's # = the exponent of the next one's lowest OD. Thus, zone 1's # = the exponent of 2^1 the lowest OD in zone 2. Zone 2's # = the exponent of 2^2 the lowest OD in zone 3, etc..

[2] As we'll eventually see, the above statement lies at the root of Esoptrics' explanation of dark matter & energy. Simply put: No given W-combo or the L-combos (***i.e.:*** atoms & sub-atomic particles) clustered around it can absorb or receive combos—duo- or otherwise—moving thru the Universe at an OD (***i.e.:*** 4th dimension level) higher than the OD of the W-combo's leading form.

ator; C = the speed of light; Z_G = the OD at which a given generic zone commences; and N = the number of steps above Z_G at which the generator's piggyback form is currently located. That may sound formidable; but, it's easily explained.

To do that, let's say that at Earth's center is a multi-h-combo we've called Terry. Let's imagine Terry is currently at $OD2^{196}$ +1. Terry is thus operating in the 197th generic zone; Z_G thus = 2^{196}; and his carrying generator is receiving its u-spaces from a form currently in the 198th generic zone (***i.e.:*** $OD2^{197}$ thru $OD2^{198}$ -1) at $OD2^{197}$ +2 according to formula #11-2 above. To determine the rectilinear velocity of Terry's carrying generator, we first take the OD serving as the starting point for the generic zone in which Terry is operating. That OD, as said, is 2^{196}. That being so, we divide the speed of light—c. 670,616,629 MPH—by 2^{196} (***i.e.:*** $1.00433627766186892x10^{59}$), and the astronomically slow result = $6.677212044x10^{-51}$ mph, and let us call that value C_M (***i.e.:*** C minimal) in the 197th generic zone. If Terry is currently at one step above $OD2^{196}$, the rectilinear velocity of his carrying generator = $1C_M$. If Terry moves to $OD2^{196}$ +2, then the rectilinear velocity of his carrying generator = $2C_M$. At $OD2^{196}$ +3, the velocity of his generator = $3C_M$; at +4, $4C_M$; at +5, $5C_M$; etc.. I take it you catch the drift; and so, foresee that, with Terry at $OD2^{197}$ -1, the rectilinear velocity of his carrying generator is a mere $1C_M$ below the speed of light. At $OD2^{197}$, the velocity of Terry's carrying generator is C; but, if Terry then moves to $OD2^{197}$ +1, the rectilinear velocity of his carrying generator drops to C divided by 2^{197}.[1]

Yes, in Esoptrics, the rectilinear velocity of every generator is the speed of light, if its piggyback form: (1) is either the leading form of a multi-combo or the form of a free duo-combo (***i.e.:*** a duo-combo not one of the logically concentric duo-combos in a multi-combo); and (2) has a current OD which is a power of 2.

As in Relativity, the speed of light cannot be exceeded; but, that's true only of *rectilinear* velocity on the part of generators changing from u-space to contiguous u-space. The reason for that limitation is a factor wholly and entirely unknown to Science—namely: If one moves the generator's piggyback form even so little as 1 OD beyond a generic one, the form crosses from one generic zone to another and, thereby, subjects its carrying generator to a different set of parameters. In other words, it's not that the limit is absolute under all circumstances but, rather, that it cannot be exceeded where trying to do so causes the generator's piggyback form to move from the start of one generic zone to one or more steps beyond that start. In Esoptrics, light speed's effect on time depends upon the generic zone in which light speed is attained. Attained in the observer's generic zone, 1 hour for the observer is 30 minutes for the light speed traveler. Attained 10 generic zones above the observer's, 1 hour for the observer is 2^{-10} hours for the light speed traveler.

[1] Light speed =29,979,245,800± 400 cm./sec. =107,925,284,880,000 cm./hr. = $3.931162553x10^{56}$Φ/ sec. =$1.415218519x10^{60}$Φ/hr.. Divided by 2^{196}, C_M, in the 197th generic zone =$2.984980874x10^{-49}$cm./ sec. =$1.07459311468x10^{-45}$cm./hr. =.004062945Φ/sec. (***i.e.:*** =1Φ/246.127sec.) =14.6266Φ/hr. (***i.e.:*** = 15Φ/1.025523hr.). There can be a fractional Φ in our *calculations* but, in *reality*, never.

As we will see, superstring theory starts off by proposing a new answer to an old question: what are the smallest, indivisible constituents of matter? For many decades, the conventional answer has been that matter is composed of particles—electrons and quarks—that can be modeled as dots that are indivisible and that have no size and no internal structure. Conventional theory claims, and experiments confirm, that these particles combine in various ways to produce protons, neutrons, and the wide variety of atoms and modules making up everything we've ever encountered. Superstring theory tells a different story. It does not deny the key role played by electrons, quarks, and the other particle species revealed by experiment, but it does claim that these particles are not dots. Instead, according to superstring theory every particle is composed of a tiny filament of energy, some hundred billion billion times smaller than a single atomic nucleus (much smaller than we can currently probe), which is shaped like a little string. And just as a violin string can vibrate in different patterns, each of which produces a different musical tone, the filaments of superstring theory can also vibrate in different patterns. These vibrations, though, don't produce different musical notes; remarkably, the theory claims that they produce different particle properties. A tiny string vibrating on one pattern would have the mass and the electric charge of an electron; according to the theory, such a vibrating string would *be* what we have traditionally called an electron. A tiny string vibrating in a different pattern would have the requisite properties to identify it as a quark, a neutrino, or any other kind of particle. All species of particles are unified in superstring theory since each arises from a different vibrational pattern executed by the same underlying entity.
——**BRIAN GREENE:** ***The Fabric Of The Cosmos***, pages 17 & 18 of the 2005 paperback edition by Vintage Books, a division of Random House, Inc., New York. ©2004 Brian Greene (The above is rather close to Esoptrics, but only to what it says about the generators. They are found at a level "much smaller than we can currently probe", and each, as a microstate of excitation, can rather rightly be described as "a tiny filament of energy" in a sense "vibrating". They cannot, however, be described as "shaped like a little string." For, they have no *physical* size or shape, and, contrary to what superstring theory—as described by Prof. Greene—says about them, they are *physically* indivisible and *physically* dots even as they are not *logically* such—a distinction no one ignorant of God's 7 *a priori* coordinates can grasp. Again, contrary to what superstring theory says, no sub-atomic particle is a *single* "string"; rather, each is an astronomical number of such "tiny filaments of energy" logically concentric to one another within the confines of what Geometry-speak would describe as c. 10^{-47} cm. in diameter. That again is something Prof. Greene and kind cannot conceive possible, since they know nothing of OD [***i.e.:*** the 4th dimension] and, so, cannot understand how what's only c. 10^{-47} cm. in dia. could have anywhere from 1 to 2^{256} different strata [***i.e.:*** 4th dimension levels] only *logically* & <u>un</u>observably apart. Contrary to what superstring theory says, what distinguishes one subatomic particle from another is <u>*not*</u> how each one's "tiny filament of energy" is "vibrating". It's: (1) the number—which rises & falls with absorption & radiation—of logically concentric duo-combos making up their internal structure [***i.e.:*** how many 4th dimension levels at which "vibrating" is occurring], (2) the native OD of each one's leading form, and (3) how that leading form rotates and pulsates. For Prof. Greene and kind, of course, that's all meaningless, because they know nothing about forms and can know nothing of them without a detailed knowledge of God's 7 *a priori* coordinates.).

Chapter Twelve:

CHAPTER ELEVEN'S CONCEPTS APPLIED TO THE ROTATIONS OF EARTH & MOON:

Let's now try to apply chapter eleven's concepts to the relationship we described between Earth and that Sun-centered form we called Gene (***i.e.:*** one of the forms logically concentric with the leading form of the Sun's main multi-h-combo). About to do so, let me first confess that what is about to follow is exceedingly imprecise due to the facts that: (1) the number of seconds given out for each of Earth's 360° rotations goes not beyond 9 decimal places and varies somewhat in the course of each year,[1] and (2) I'm not sure just how precise are the measurements I give out for K and Φ and, therefore, cannot say how many digits in my calculations are significant. I leave it up to my readers, then, to decide which digits are significant in each calculation and, therefore, to round off where they think that should be done based on the following.

If I take the "dia." of the Φ to be 2^{-129} times the hydrogen diameter given as .00 000 005 cm., the result $=7.34683969 \times 10^{-47}$ cm.. Divide 1 cm. by that figure, and:

1/.000,000,000,000,000,000,000,000,000,000,000,000,000,000,000,0734683969 = 13,611,294,681,727,293,820,943,573,630,636,821,476,636,847,646=$1.361129468 \times 10^{46}$ Φ/cm. & x 160,934.34 cm./mi. = 2,190,524,726,149,292,091,059,632,199,487,940,644,040,376, 495,543,215 = $2.190524726 \times 10^{51}$ Φ/mi., and, in the full number, the last digit rounded up in each case to the nearest integer, since Esoptrics permits no fractional Φ.

Again, if I take the duration of K to be $7.201789375 \times 10^{-96}$ sec., and divide 1 sec. by that figure, I get:

1/.000,000,000,000,000,000,000,000,000,000,000,000,000,000,000,000,000,000,000,000,000, 000,000,000,000,000,000,000,000,000,000,000,007,201789375 = 138,854,380,200,476,218,453, 695,058,250,714,253,914,153,105,873,080,327,345,729,963,117,673,099,124,757,449,602, 585,746,268,093.268 = $1.388543802 \times 10^{95}$ K/sec., and, in the full number, the last digit rounded down to the nearest integer, since Esoptrics permits no fractional K.[2]

---

[1] That suggests to me that the leading form for Earth's main multi-h-combo varies slightly in current OD in the course of each orbit around the Sun.

[2] K's duration per our second comes from Esoptrics' principle that, for every observer, it varies as the inverse of the square of the observer's current OD divided by the number of cycles his consciousness' form executes per second. On that, more shall follow in a later chapter.

2013 ESOPTRICS UPDATE

In keeping with Esoptrics' principle that light speed is $1\Phi/2^{128}$K, I first divide K per sec. by 2^{128} to determine the number of Φ per sec. so:

138,854,380,200,476,218,453,695, 058,250,714,253,914,153,105,873,080,327,345,
729,963,117,673,099,124,757,449,602,585,746,268,093 ÷
34,028,236,692,093,846,346,337,4607,431,768,211,456 =
408,056,348,781,474,710,918,193,347,338,155,691,972,963,856,906,773,089,816 Φ/sec.

I then divide the number of Φ per sec. by the number of Φ per cm. so:

408,056,348,781,474,710,918,193,347,338,155,691,972,963,856,906,773,089,816
÷ 13,611,294,681,727,293,820,943,573,630,636,821,476,636,847,646 =
29,979,245,789.842215429 cm./sec..

With the speed of light usually given as 29,979,245,800 ± 400 cm./sec., I leave it to my readers to determine for themselves how many of the digits in each of my calculations can be accepted as significant. I dare suspect there shall be no universal agreement.

Also be advised that I will start with an explanation in the terms of one multi-h-combo at Earth's center. That explanation, though, shall give way to one in the terms of two multi-h-combos one of which is at Earth's center and the other c. 2,858 miles from that center. That switch shall be prompted by the need to speak of what is commonly called Earth's barycenter.

We said that Gene—throughout the first phase of each of his cycles—is at OD2^{197}. We also said Gene is a captive form centered in the Sun's multi-h-combo and native to an OD not a categorical one. That, of course, means the captive Gene is at *best* a generic rational form and at *least* a non-generic rational one. More importantly, Gene is in the 198th generic zone, which is to say the one including the OD's 2^{197} thru 2^{198} -1.

If Earth is at the outer reaches of Gene's actuality, then, from what we've been saying above, it must be that, at Earth's center, there is a multi-h-combo whose leading form is: (1) native to a rational OD below 2^{128}, (2) is somewhere in the 197th generic zone, (3) has a current OD somewhere between 2^{196} and 2^{197} -1, (3) has a carrying generator receiving its u-spaces from a form somewhere in the next highest generic zone between 2^{197} and 2^{198} -1, and (4) because its native OD is below MT, it, in its every 2nd phase, increases its radius from c. $2^{196}\Phi$ to only c. 2^{196}+ c. $2^{128}\Phi$. In each cycle, it expands and contracts very little. Its name shall be Terry. Terry is Earth's *leading* form and Gene a *captive* one in the Sun.

What is Terry's current OD? On page 75 in chapter 9, it was calculated that the rectilinear velocity of Terry's multi-combo thru his generator must be 709.56 mph to travel the 3,110,000 miles between Earth's apogee and perigee in a half year's 15,775,462.987 sec.. How far into the 197th generic Zone must Terry be for his generator to have a rectilinear velocity of 709.56 mph? To answer we must first determine the value of C_M for the 197th generic zone. That means dividing the speed of light *in mph* by 2^{196} so:

670,616,879.405601067 ÷ 100,433,627,766,186,892,221,372,630,771,322,662,657,637,687,
111,424,552,206,336 = .000,000,000,000,000,000,000,000,000,000,000,000,000,000,000,000,
006,67721453781218925388714 = 6.677214537812189254 x 10^{-51} mph.

That result we divide into 709.56 mph so:

709.56 ÷ .000,000,000,000,000,000,000,000,000,000,000,000,000,000,000,000,006,677214
53781218925388714 =

106,265,868,197,263,227,128,087,051,398,862,301,238,498,348,361,403,914.8 = 1.062658682 x 10^{53} steps into that 197th generic zone running from 2^{196} to $2^{197}-1$.

Therefore, at 709.56 mph, Terry's current OD is 2^{196} + the above number of steps into the 197th generic zone. That gives us:

100,433,<u>627</u>,766,186,892,221,372,630,771,322,662,657,637,687,111,424,552,206,336
+106,265,868,197,263,227,128,087,051,398,862,301,238,498,348,361,403,915 =
100,433,<u>734</u>,032,055,089,484,599,758,858,374,061,519,938,925,609,772,913,610,251 =
1.00433734032055 x 10^{59}.

To find the number of positions in each of Terry's cycles, double the above result so:

100,433,734,032,055,089,484,599,758,858,374,061,519,938,925,609,772,913,610,251 x 2 =
200,867,468,064,110,178,969,199,517,716,748,123,039,877,851,219,545,827,220,502 =
2.00867468 x 10^{59}.

Terry is native to the rational OD of:

340,282,366,920,938,463,426,481,119,284,349,108,225 =
$(2^{64}-1)^2$ (***i.e.:*** 18,446,744,073,709,551,615^2) = 1st rational OD below 2^{128}.

To calculate the duration of each of his cycles, we multiply his positions per cycle by his native OD so:

340,282,366,920,938,463,426,481,119,284,349,108,225 x
200,867,468,064,110,178,969,199,517,716,748,123,039,877,851,219,545,827,220,502 =
68,351,657,470,271,428,775,652,049,305,058,455,601,989,906,722,599,132,701,945,038,491,
003,936,140,413,437,361,407,674,136,828,950 = a duration of 6.835165747x10^{97} K.

Divide that by the number of K/sec. (***i.e.:*** 1.388543802x10^{95}), and the result is:

68,351,657,470,271,428,775,652,049,305,058,455,601,989,906,722,599,132,701,945,038,491,
003,936,140,413,437,361,407,674,136,828,950 ÷ 138,854,380,200,476,218,453,695,058,
250,714,253,914,153,105,873,080,327,345,729,963,117,673,099,124,757,449,602,585,746,26
8,093 = 492.2542 sec..

The duration of Terry's every cycle in the tertiary plane is thus 492.25 sec.. But, Earth's 360° rotations (as opposed to a 360.9856° high-noon-to-high-noon rotation) take 86,164.098903691 sec.; and so, Earth's actual rotation lasts c. 175 times longer than what this paragraph says it should.

Actually, the above is not necessarily saying Earth should rotate 360° every 492.25 sec.; rather, it's saying the *center* of Earth should do that. Esoptrics is thus saying that Earth's centermost core should rotate faster than its surface does. According to Internet sources, scientists have for at least 15 years been saying Earth's inner core rotates faster than its crust does; but, none of those sources come anywhere close to saying the core rotates 175 times as fast as the crust does. In a report dated Feb. 20, 2011, siencedaily.com says previous reports of one extra degree per year have been changed to one extra degree every million years—a telling example of how risky is reliance on Science's "observations".

Is there, perhaps—at a level Science would describe as 10^{-47} cm.—a core of the core which does rotate unusually fast without, however, being in any way the cause of how the *whole* of the planet rotates? If so, would Science be able to observe what goes on at a level that small? That may well be what Esoptrics is suggesting. How so?!

Again, there are 86,164.098903691 seconds to each of Earth's 360^0 rotations. At $1.3885438020047621845\text{x}10^{95}$ K/sec., that many seconds = $1.19642625487635132\ 97\text{x}10^{100}$ K. Because his ontological depth is somewhere between 2^{196} & $2^{197}-1$, Terry—among the many duo-combos concentric with him in his multi-h-combo—could have a captive duo-combo whose form is *currently* at, and *native* to, the rational OD:

77,344,239,284,521,637,719,212,996,502,657,086,968,088,464,480,164 = $7.7344\text{x}10^{49}$ = the square of 8,794,557,367,174,407,016,872,358 = $8.794557\text{x}10^{24}$.[1]

Let's call it Jerry. For Terry's captive buddy, rational Jerry, the number of positions per cycle in the tertiary plane would thus = $2(7.7\text{x}10^{49})$ =

154,688,478,569,043,275,438,425,993,005,314,173,936,176,928,960,328 = $1.54688\text{x}10^{50}$.

Multiplying each of those positions by captive Jerry's native OD, we find that the duration of each full cycle on the part of the captive form Jerry = $2(D_N \text{ x } D_c)K = 2[(D_N)^2]K$ =

11,964,262,701,002,680,355,610,024,234,476,870,978,994,223,891,496,917,820,768,599,119, 634,232,103,021,514,287,539,509,373,498,933,792 = $1.1964262701\text{x}10^{100}$K.

Divide the above by the number of K per second, and the result is:

11,964,262,701,002,680,355,610,024,234,476,870,978,994,223,891,496,917,820,768,599,119, 634,232,103,021,514,287,539,509,373,498,933,792 ÷138,854,380,200,476,218,453,695,058,250,714,253,914,153,105,873,080,327,345,729,963, 117,673,099,124,757,449,602,585,746,268,093 = 86,164.0999997899 sec. per rotation = the duration of a 360^0 rotation on the part of Earth.

Fascinating, is it not?[2]

Terry's captive buddy, Jerry, then, accounts quite closely for the duration of Earth's rotation; but, he does so, only if all the Earth's component atoms have carrying generators receiving their u-spaces from Jerry. There are, though, 2 reasons why that cannot be so.

First, the captive Jerry's current OD = the radius of his actuality in Φ. Divide that by Φ/mi., and the result is:

77,344,239,284,521,637,719,212,996,502,657,086,968,088,464,480,164 ÷ 2,190,524,726,149,292,091,059,632,199,487,940,644,040,376,495,543,215 = a radius of 0.035308544277646 mi. = dia. of .070617 mi..

---

[1] $2^{166} = 9.3536104\text{x}10^{49}$; $\sqrt{2^{166}} = 2^{83}$, and $2^{83} = 9.6714\text{x}10^{24}$. A radius of $2^{166}\Phi$ = c. .0854 miles.

[2] The above calculations—begun on the feast day of St. Patrick, Sunday March 17, 2013, and completed this day, Monday, March 18, 2013—were made possible by the combined use of: (1) Po-Han Lin's Virtual Calc 2000; (2) the square roots calculator to 10,000 places maintained till lately by "Mark" at markknowsnothing.com/cgi-bin/sqrroots; and (3) Marek Kynci's on line calculator at www.calculator-tab.com/slash. Mr. Kynci's calculator gives less than 30 digits per calculation; but, even still, his assistance was crucial to my ability to use the other 2 calculators; and so, without any one of the 3, Esoptrics could likely not have gone but a pitiful fraction of how far it has gone.

With an actuality that small in radius (I'm using Geometry speak.), the captive Jerry is able to provide u-spaces only to those multi-A-combos (***i.e.:*** atoms) which can be packed together within a "sphere" only .0706 miles in "diameter". That's but a tiny fraction of the 3,959 miles generally given as Earth's equatorial radius.

Secondly, every multi-A-combo's carrying generator seeks to get its u-spaces from a form whose current OD equals 2^{128} times its leading form's current number of steps into the 129th generic zone (***i.e.:*** steps above the MT of $OD2^{128}$). The captive Jerry's current OD is not such a multiple. To illustrate that, consider the following two sets of calculations:

SET ONE: ATOM'S LEADING FORM AT 227,294,290,8**69** STEPS ABOVE MT x 2^{128}:

227,294,290,869 x 340,282,366,920,938,463,463,374,607,431,768,211,456 =
77,344,239,284,519,571,040,894,197,149,905,012,925,667,961,995,264 = 7.7344 x10^{49} =
OD of the form to which a multi-A-combo's generator would be in potency
with its leading form at 227,294,290,869 steps into the 129th generic zone.

77,344,239,284,521,637,719,212,996,502,657,086,968,088,464,480,164 minus
77,344,239,284,519,571,040,894,197,149,905,012,925,667,961,995,264 =
2,066,678,318,799,352,752,074,042,420,502,484,900 = 2.066678319 x 10^{36} steps
below captive Jerry's current OD. Number of steps from Jerry to 2^{166} = 1.619x10^{49}.

SET TWO: ATOM'S LEADING FORM AT 227,294,290,8**70** STEPS ABOVE MT x 2^{128}:

227,294,290,870 x 340,282,366,920,938,463,463,374,607,431,768,211,456 =
77,344,239,284,859,853,407,815,135,613,368,387,533,099,730,206,720 =
OD of the form to which a multi-A-combo's generator would be in potency
with its leading form at 227,294,290,870 steps into the 129th generic zone.

77,344,239,284,859,853,407,815,135,613,368,387,533,099,730,206,720 minus
77,344,239,284,521,637,719,212,996,502,657,086,968,088,464,480,164 =
338,215,688,602,139,110,711,300,565,011,265,726,556 = 3.38215688602 x 10^{38} steps
above the captive Jerry's current OD = still c. 4.8x10^{10} steps below 2^{166}.

In Set One, 227,294,290,8**69** steps into the 129th generic zone are multiplied by 2^{128}, and the result is an OD 2.3x10^{36} steps *below* Jerry's OD. In Set Two, the number of steps into the 129th generic zone is raised one step to 227,294,290,870. Multiplied by 2^{128}, that gives us an OD 3.4x10^{38} steps *above* Jerry's OD but still 48,000,000,000 OD's below 2^{166}. Those facts present us with 2 important lines of thought.

In the first place, it tells us this: As a multi-A-combo's (***i.e.:*** atom's) leading form makes the *single* step from 227,294,290,869 steps above $OD2^{128}$ to 227,294,290,870 steps above $OD2^{128}$, its generator moves (at least tries to) 2^{128} steps up above the OD of the form from which it was previously receiving its u-spaces. As you can readily calculate:

77,344,239,284,859,853,407,815,135,613,368,387,533,099,730,206,720 minus
77,344,239,284,519,571,040,894,197,149,905,012,925,667,961,995,264 =
340,282,366,920,938,463,463,374,607,431,768,211,456 = 2^{128}

How does the generator make that switch from a first form to a second one 2^{128} OD's above the first one? It can't do it in one step, because it must always go from a first u-space to a second, *logically contiguous* one. The generator, then, must execute a logical sequence of 2^{128} steps. How long might that take? The astonishing answer is: one step per K and 2^{128} steps in 2^{128}K. Thus, the *OD-linear* velocity of every multi-A-combo's carrying generator thru the 4th dimension (***i.e.:*** OD) is as much as 2^{128} x light speed—depending on how quickly it can find u-spaces logically contiguous despite being at different levels of the 4th dimension. That in no way violates the light speed limit of 2^{128}K/Φ, because it applies to *recti*linear velocity, and *recti*linear and *OD-linear* velocity are not the same. The former moves the generator 10^{-47} cm. in Geometry-speak; the latter moves it zero cm..

In the second place, it tells us this: No multi-A-combo's carrying generator is ever in potency to the captive Jerry *long-term*. Each would, at best, use his u-spaces for 1K as it transitions from the previous to the latest form demanded when an atom's leading form goes from 227,294,290,869 to 227,294,290,870 steps above OD2^{128}.

What, then, shall we say of Earth's many atoms? How do they share rational Jerry's rate of rotating 360° every 86,164.1 sec.? The answer which readily suggests itself to me is: Every form *between* the captive *rational* Jerry and his captor Terry (***i.e.:*** the leading form of Earth's multi-h-combo), is *ir*rational and, thus, only non-rotating forms. It may be, though, that the rotating rational Jerry causes all the irrational forms between him and Terry to rotate at his pace. If so, it would make no difference which of the forms between Jerry and Terry is providing u-spaces to the generators of Earth's many atoms. For, no matter which of them is doing that, they're all being moved by captive Jerry to rotate at his pace.

If so, the leading forms of every Earth atom must be *at least* 227,294,290,870 steps above OD2^{128} just to be in potency to Jerry with his radius of .0353 mi. per page 98. How many steps and what velocity would put them in potency to a form with Earth's radius of c. 4,000 miles? At OD2^{181}, the radius = 2,798.4 and, at OD2^{182}, = 5,596.8 miles.

If a given multi-A-combo is 227,294,290,870 steps above MT, what is its carrying generator's velocity? To answer, we first calculate the value of C_M in the 129th generic zone. Using formula 11-3 on page 92, we divide the speed of light by 2^{128} so:

670,616,879.405601 MPH ÷340,282,366,920,938,463,463,374,607,431,768,211,456 =
.000,000,000,000,000,000,000,000,000,001,97076588326838804 1 =
$1.97076588326838804 1\text{x}10^{-30}$ MPH.

That, then, is the value of C_M in the 129th generic zone. Since we're dealing with a generator carrying a multi-A-combo's leading form currently 227,294,290,870 steps above OD2^{128}, we multiply $1.97076588326838804\text{x}10^{-30}$ by 227,294,290,870 and find the velocity of its generator in potency to a form $3.382\text{x}10^{38}$ steps above the captive Jerry would be:

0.000,000,000,000,000,000,447943833908277457667083485 67 = $4.479\text{x}10^{-19}$ MPH.

Are any of Earth's atoms slower? Even in visibly motionless objects, do not the component atoms' nuclei move more than a quintillionth of a mile per hour? At 2^{54} steps above MT, atoms would be in potency to an OD2^{182} form and have a speed of c. $3.55\text{x}10^{-14}$ MPH. Earth's atoms, then, easily become in potency to a space envelope of Earth's radius.

I can easily imagine someone objecting: "Earth's circumference at its equatorial surface is 24,901 miles, and every atom on that surface travels that 24,901 miles every 86,164.1 seconds. The velocity of each is thus c. 1,040 MPH." I deny that and say, instead, that 1,040 MPH is the *curvilinear* velocity *of the u-spaces being used by those atoms*. Remember what I said earlier about the velocity of the Earth's atoms as Earth circles the Sun: Their velocity is the rate at which they shift from u-space to u-space and has nothing to do with the curvilinear velocity of those u-spaces and the form providing them.

Let's digress for a bit to consider a line of thought which, at first puzzled me greatly:

If a multi-A-combo is 2^{68} steps into the 129th generic zone, its carrying generator would be in potency to a form at $2^{128} \times 2^{68} = 2^{196}$, and the combo's velocity would thus be 2^{68} times the value of C_M in the 129th generic zone. Take the value of C_M in MPH and that gives us this:

.000,000,000,000,000,000,000,000,000,001,970765883268388041 x
295,147,905,179,352,825,856 =
0.000,000,000,581667422045601713292985427305988096 MPH = $5.81667422 \times 10^{-10}$ MPH.

Doesn't that mean that, at a miniscule velocity, every atom should be in potency to Terry, Earth's leading form, and, therefore, produce a planet which, like Terry is rotating 360° every 492.25 sec.? By no means! For one thing, Terry is not exactly at $OD2^{196}$, or the entire planet would be racing thru the Universe at light speed. Terry, therefore, must be at least a few steps into the 197th generic zone. In fact, we have already calculated that, since in his orbit of the Sun, Terry's rectilinear velocity is 709.56 mph, Terry is $1.062658682 \times 10^{53}$ steps into the 197th generic zone and, therefore, has a current OD of:

100,433,734,032,055,089,484,599,758,858,374,061,519,938,925,609,772,913,610,251.

Divide that by 2^{128}, and the result is 295,148,217,466,672,203,627.762. Terry's current OD, then, is not a multiple of 2^{128}, and it's not likely it would ever be such.

One might be tempted to say: "Well, the generator shall try to go beyond Terry but become stalled in potency to Terry because no form higher than Terry is available in Terry." By no means! No generator can *long* be in potency to a form to which its own piggyback form does not to any extent make it in potency; and so, if a carrying generator cannot find the form to which its piggyback form *currently* makes it in potency, then it shall fall back upon whatever available form is one to which its piggyback form *previously* made it in potency. For example, if a generator's piggyback form is at $2^{68}+1$ steps and Terry is only a few steps into the 197th generic zone, then the generator must fall back upon the form its piggyback form called for when it was at 2^{68} steps into the 129th generic zone. If that's not possible, then the generator must fall back upon the form its piggyback form called for when it was at $2^{68}-1$ steps above $OD2^{128}$, then $2^{68}-2$, then -3 etc..

For a dramatic example of how that works, consider the velocity its takes to escape Earth's gravity. That's generally given as 25,053.674 MPH. Divide that by C_M in MPH for the 129th generic zone, and the result is:

25053.674/.000,000,000,000,000,000,000,000,000,001,970765148567959 =
12,712,663,413,093,668,328,136,444,814,996,268.109 et = $1.27126634131 \times 10^{34}$

That tells us this: To move its generator to 25,054 MPH, a multi-A-combo's leading form would have to be 1.27×10^{34} steps into the 129th generic zone. Multiply that by 2^{128}, and that would put its carrying generator in potency to a form at $OD4.325787314 \times 10^{72}$. That turns out to be between $OD2^{241}$ and 2^{242}. Though the multi-g-combo at the center of the Milky Way galaxy might provide such a form, neither the Sun nor the Earth can. Thus, despite the fact the multi-A-combo's piggyback form is calling for one that high in OD, its carrying generator shall be forced to fall back upon a form in Terry at $OD2^{196}$—a multiple of 2^{128} (Remember Terry is at an OD above that.), until the multi-A-combo, by racing outward into "space" (Using Terry's $OD2^{196}$ form, it can go $2^{196}\Phi$ = 91,698,020.5 miles from Earth's center point.), it can arrive at a juncture which makes its current u-space logically contiguous to a u-space provided by a form which—because it's one of the forms logically concentric with a multi-h-combo having an ontological depth much greater than Terry's—

can now allow it to transition up that "taller ladder" to a form whose current OD shall then allow it to travel many light years away from Earth's center point.

Exactly how such junctures take place is not, for now, within my ability to explain, except for two which I deem too simple and obvious to miss: Midway between Earth and the Sun, there is a point logically equidistant from the center of each. At that point, the u-spaces provided by the OD2^{196} form captive in Terry shall be logically contiguous to the u-spaces provided by the OD2^{196} and 2^{196} +1 forms logically concentric in the Sun's center. If the ontological depth of the Sun's multi-h-combo is c. 2^{227}, that would allow the multi-A-combo—after its OD linear velocity takes it upward thru the forms concentric with the Sun's leading form—to travel as far as 17,000 light years away from the Sun's center. Next, mid way between the Sun and the multi-g-combo at the center of the Milky Way, there is a point equidistant from the center of each. I dare say my reader can guess the rest of it.

At that, it pops into my head: "That's true of the mid-point twixt any 2 heavenly bodies." A pop-up retorts: "But, the orientations are reverse. What's A potency for one is B for the other." A pop-up counters: "Maybe, under such a circumstance, every multi-A-combo's carrying generator flips 180°. More specifically, the heavenly sextet governing that form flips A to B and vice versa as A' flips to C and vice versa." I deem it a concept too simple to need elaboration.

Now that I've made that as clear as mud, let me muddy it even further. Astronomers tell me that, as Earth rotates around its center, it and the Moon rotate around what the astronomers call the "barycenter" (***a/k/a*** "barycentre"). It's located c. 4600 km. (***i.e.:*** c. 2,858.3 mi.) from the center of the Earth. That's roughly ¾ of the length of the Earth's equatorial radius usually given as c. 6,371 km. (***i.e.:*** c. 3,959 mi.). I've yet to find anything on the Internet telling me whether or not the barycenter moves or, if so, how much it moves.

According to articles on the Internet, the barycenter is merely the *imaginary* point at which two unequal weights wind up balancing one another. If the man on one end of a seesaw's plank weighs 200 pounds and the man on the opposite end weighs 100, then, to keep the 200 pound man's end from falling permanently to the ground, we must place the 100 pound man much further from the plank's supporting structure than the 200 pound man is. That we do by moving the supporting pipe (***i.e.:*** the fulcrum) much closer to the 200 pounder than to the 100 pounder. The point at which the pipe underlies the plank is then the barycenter which allows the plank to remain level until one of the two men uses his legs to push himself upward and his companion downward.

If I understand Esoptrics correctly, no barycenter is ever merely an *imaginary* point. Every barycenter has to be a multi-h-combo which—pulled away from the center of its host by the force of another body's gravity—is at a point whose distance from its host's center is determined by the gravitational force of some heavenly body which is somehow "entangled" with it.[1]

I dare say some will object: "Sometimes the barycenter is out in space between two

[1] How might Earth and the Moon be "entangled"? The multi-h-combo at the Moon's center was once included in Earth, but was thrown off for some reason c 4.5 billion years ago. That answer has been around for decades. Recently, it's become very popular along with the notion that, prior to the separation, Earth rotated much faster—maybe once every hour or two. That's still much slower than the figure of 492.25 sec. I gave out earlier. I speculate the multi-h-combo at the Moon's center is a truncated one and, thus, without a rational form in it. How, then, does it rotate? Some say it rotates at a velocity *reduced* by Earth's influence. I suspect it's the other way around: Earth's influence imposes upon the Moon's irrational forms a rotation they would otherwise not have, and that's why the rate of rotation equals the orbital rate. Saying the Moon has no rational forms accords with Science's claim the Moon has no magnetic field. For, in Esoptrics, magnetic fields are the result of rational forms rotating at colossal speeds in the primary and secondary planes.

heavenly bodies. Where that's the case, there's no body of any kind for Science to detect." I reply: What's there is only 10^{-47} cm. in dia. in Geometry-speak. It is, therefore, too small for Science to detect any part of its internality, which is to say too small for Science to detect any of the duo-combos logically concentric there. All Science can currently detect is the effect upon the curvilinear motion of the heavenly bodies affected by that barycenter (***i.e.:*** that "dark matter"? or "mini black hole"?). Still, I dare to predict that someday, somewhere, somehow, Science shall observe either photons or neutrinos or anti-neutrinos suddenly issue from, or disappear into, a barycenter out in space between two heavenly bodies.

What now is implied if Earth's barycenter is a second multi-h-combo? To start the answer, let's give the name Terry Sr. to the leading form of this second one of Earth's multi-h-combos. The leading form of Earth's first multi-h-combo remains Terry, or Terry Jr., if you prefer.

Since the Moon orbits Terry Sr., it must be that such is possible because, at the center of the Moon, is a multi-h-combo whose carrying generator is using the u-spaces provided to it by a rotating, rational form concentric with Terry Sr.. How should we describe that captive form?

I'm disinclined to spell out every aspect of what's involved here; so, I shall limit myself to the following: It's usually said the Moon's maximum distance from Earth = 406,700 km = 252,711.663 mi., and its orbit around Earth takes 27 days, 7 hrs., 43 min., 11.5 sec. = 114,371.5 sec. total = 1.5880983745 x 10^{100} K.

Concentric with the leading form Terry Senior is a captive rational form we'll call *Jerry* Sr. and whose current OD is:

553,571,354,635,599,960,397,502,567,110,240,596,196,485,055,488,000,000,000 = OD5.54 x 10^{56}. Note: $2^{188} = 3.9 \times 10^{56}$ & $2^{189} = 7.85 \times 10^{56}$.

Its native, rational OD =

14,344,116,266,184,232,989,187,762,390,265,888,308,945,441 =
The square of: 3,787,362,705,918,754,883,729 = 3.787 x 10^{21}. Note: $2^{72} = 4.7 \times 10^{21}$.

Multiply the native by the current OD and double that. The duration of each cycle on the part of the captive Jerry Sr. in Terry Sr. is 1.5880983745 x 10^{100} K = 114,371.5 sec. = the duration of the Moon's orbit.

The radius of Jerry Senior's actuality = its current OD = 5.54 x 10^{56} Φ = 406,699.999 km. = 252,711.66388 mi. = the Moon's maximum distance from Earth. The captive Jerry Sr., then, accounts for the duration of the Moon's orbit and its maximum distance from Earth. I deem it superfluous to work out any further details. Besides, there's a far more important calculation at hand.

Which is Earth's primary multi-h-combo: the one led by Terry Sr. or the one led by Terry Jr.? Let's say it's the former; and so, Terry Sr. is the leading form currently in the 197th generic zone somewhere between OD2^{196} and 2^{197}. Let's see what that entails.

If Terry Sr. leads Earth's primary multi-h-combo, then it is *Terry Senior's* carrying generator which is in potency to the form we called Gene. In case you've forgotten, that's the captive rational form: (1) which is concentric with some leading form in the Sun, (2) which is currently at, or slightly above, OD2^{197}, and (3) which, by rotating, is carrying the Earth around the Sun as—I'm now saying—Terry Senior's carrying generator uses the u-spaces provided by the captive Gene. That, of course, means it is Terry Sr. which is currently at an OD somewhere slightly above OD2^{196} and having a rectilinear velocity of c. 709.56 MPH as the captive Gene, thru his rotating u-spaces, carries Terry Sr. around the

Sun at 66,616 MPH average. If Terry *Senior's* carrying generator is in potency to the Sun's captive Gene, then what is the role of the generator carrying Terry *Junior's* multi-combo?

Terry *Junior's* carrying generator is in potency to some one of the captive forms included in the roughly 2^{196} captive forms logically concentric with Terry *Senior*. Which one?! It's the same captive rational form to which the Moon's carrying generator is in potency. We've called that rational form, captive in Terry Sr., Jerry Sr. What must be the current OD of the leading form Terry *Junior* for his carrying generator to be in potency to the Jerry Sr. captive in Terry *Senior*?

To answer, we must first determine how far the captive Jerry Sr. is above OD2^{188}. To do that, we subtract 2^{188} from Jerry Senior's current OD given above as 5.54×10^{56}. That gives us this:

553,571,354,635,599,960,397,502,567,110,240,596,196,485,055,488,000,000,000 minus
392,318,858,461,667,547,739,736,838,950,479,151,006,397,215,279,002,157,056 =
161,252,496,173,932,412,657,765,728,159,761,445,190,087,840,208,997,842,944.

Jerry Senior's number of steps above OD2^{188} is thus $1.61252496 \times 10^{56}$. Since 2^{188} = 3.923189×10^{56} and 2^{189} = 7.8463772×10^{56}, the captive Jerry Sr., at 5.53571×10^{56}, is still less than half-way between OD2^{188} and OD2^{189}.

For Terry Junior's carrying generator to be in potency to the captive Jerry Sr. in Terry Sr., the leading form Terry Junior must currently be a certain number of steps above OD2^{187}, and that number of steps must—according to formula #11-1 on page 91—be one-half the captive Jerry Senior's steps above OD2^{188}. We calculate that half so:

161,252,496,173,932,412,657,765,728,159,761,445,190,087,840,208,997,842,944÷2 =
80,626,248,086,966,206,328,882,864,079,880,722,595,043,920,104,498,921,472.

Terry Junior's current OD is thus $2^{187} + 8.0626248087 \times 10^{55}$. That give us this:

196,159,429,230,833,773,869,868,419,475,239,575,503,198,607,639,501,078,528
+ 80,626,248,086,966,206,328,882,864,079,880,722,595,043,920,104,498,921,472 =
276,785,677,317,799,980,198,751,283,555,120,298,098,242,527,744,000,000,000.

The current OD of the leading form Terry Junior is thus $2.767856773178 \times 10^{56}$. The number of positions in each rotation by that form is 2 times its current OD so:

276,785,677,317,799,980,198,751,283,555,120,298,098,242,527,744,000,000,000 x 2 =
553,571,354,635,599,960,397,502,567,110,240,596,196,485,055,488,000,000,000.

That gives us $5.53571354636 \times 10^{56}$ positions per rotation by the leading form Terry Jr.. If, like the leading form Terry Sr., Terry Jr. is a form native to the square of $2^{64} - 1$, then the duration of each rotation by Terry Jr. = $(2^{64} - 1)^2$ x the number of positions in each of Terry Junior's rotations. The result is:

340,282,366,920,938,463,426,481,119,284,349,108,225 x
553,571,354,635,599,960,397,502,567,110,240,596,196,485,055,488,000,000,000 =
188,370,570,815,032,175,088,593,557,567,427,365,472,270,509,989,915,694,911,637,654,
068,962,183,479,552,442,188,800,000,000,000 K.

Each rotation on the part of Terry Jr. thus = $1.88370570815 \times 10^{95}$ K. Divide that by the number of K per second (***i.e.:*** $1.388543802\text{x}10^{95}$), and the duration of each of his rotations is only 1.3566 seconds. That's much too fast. Again, though, that would be the duration of what's at the center of Earth and only $7.34683969\text{x}10^{-47}$ cm. in Geometry speak. What, then, accounts for an Earth rotating 360^{0} in 86,164.098903691 seconds?

That takes us back to the answer already given in the description of a captive rational form called Jerry and whose native and current OD's are the same at:

77,344,239,284,521,637,719,212,996,502,657,086,968,088,464,480,164 = $7.7344\text{x}10^{49}$ =
the square of 8,794,557,367,174,407,016,872,358 = $8.79\text{x}10^{24}$.

Here, we need only change Jerry's name to Jerry Jr. and recall: (1) that his every rotation lasts 86,164.0999997899 sec., and (2) that he imposes that rate of rotation upon all the irrational forms between him and his leading form Terry Jr. at $8.063 \text{ x } 10^{55}$ steps above $OD2^{187}$. All that's changed is the number of forms—all of which are irrational—between the two juniors: captive Jerry Jr. and his captor Terry Jr.. That number of forms is:

Captor Terry Junior's OD Minus Captive Jerry Junior's OD =
276,785,677,317,799,980,198,751,283,555,120,298,098,242,527,744,000,000,000
–77,344,239,284,521,637,719,212,996,502,657,086,968,088,464,480,164 =
276,785,599,973,560,695,677,113,564,342,123,795,441,155,559,655,535,519,836 =
$2.767856\text{x}10^{56}$ forms each at a discrete OD and 4th dimension level.
Remember: $2^{188} = 3.923189\text{x}10^{56}$.

I suppose it should also be mentioned that at, $OD2^{187}$, the radius of the form is 2^{187} Φ, and that translates into a radius of c. 89,549 miles.

✡ ✞ ✡ ✞ ✡ ✞ ✡ ✞ ✡ ✞ ✡ ✞ ✡ ✞ ✡ ✞ ✡ ✞ ✡ ✞ ✡ ✞ ✡ ✞ ✡ ✞ ✡ ✞

> The most unsatisfactory side of this system . . . . lay in its description of light, which Newton also conceived, in accordance with his system, as composed of material points. Even at that time the question, What in that case becomes of the material points of which light is composed, when the light is absorbed?, was already a burning one.
> ——**ALBERT EINSTEIN: *Maxwell's Influence On The Evolution Of The Idea Of Physical Reality***, first published, 1931 in ***James Clerk Maxwell: A Commemoration Volume***, Cambridge University Press, and found on page 267 of ***Ideas And Opinions*** in the hardbound book published by Bonanza Books; New York, 1954. ©1954 by Crown Publishers, Inc. (I've yet to find Science's answer to this "burning" question which remains whether light is a material point or a wave. I'm particularly anxious to hear String Theory explain how strings absorb and radiate light without internal change. With its concept of the 4th dimension, Esoptrics easily explains what changes the multi-combos undergo as they absorb and radiate light: Accelerating, they up the number of their 4th dimension levels and thus gain "slots" in which to park concentric to themselves the newly acquired light or neutrinos or anti-neutrinos, or what have you; decelerating, they dump some of those levels and thus radiate the light or whatever else had been parked in those forfeited "slots". That must entail at least tiny rises and falls [often too tiny to detect] in the particle's interior and mass; but, since they're limited to one u-space's 10^{-47} cm., the particle remains a *physically* dimensionless, single, non-composite dot.).

Another error hath proceeded from too great a reverence, and a kind of adoration of the mind and understanding of man; by means whereof, men have withdrawn themselves too much from the contemplation of nature, and the observations of experience, and have tumbled up and down in their own reason and conceits. Upon these intellectualists, which are notwithstanding commonly taken for the most sublime and divine philosophers, Heraclitus gave a just censure, saying, "Men sought truth in their own little worlds, and not in the great and common world [Sextus Empiricus, *Against the Logicians*, I. 133]"; for they disdain to spell, and so by degrees to read in the volume of God's works: and contrariwise by continual meditation and agitation of wit do urge and, as it were, invocate their own spirits to divine and give oracles unto them, whereby they are deservedly deluded.
——**FRANCIS BACON** (1561-1626)**:** ***Advancement Of Learning***, First Book, Sec. 5: Par. 6, as given at the top of the left-hand column of page 16 of Vol. 30 of ***Great Books Of The Western World*** as published by Encyclopædia Britannica, Inc.; Chicago, 1952 (Shades of Aristotle's complaint on my page 30! Thus, c. even 150 years before Aristotle [384-322 BC], Heraclitus [c.540-c.480BC] was effectively praising the scientific over the philosophical method. But, tell me, Mr. Bacon: Were you alive today and seeing what this author has achieved by "too great a reverence, and a kind of adoration of the mind" and a withdrawal "too much from the contemplation of nature, and the observations of experience," [though not from the empirical observation of the tripartite structure of what's "working-over" "the raw material of the senses"] would you still harbor your above opinion?)

To pass on: in the first event or occurrence after the fall of man, we see (as the Scriptures have infinite mysteries, not violating at all the truth of the story or letter) an image of the two estates, the contemplative state and the active state, figured in the two persons of Abel and Cain, and in the two simplest and most primitive trades of life; that of the shepherd (who, by reason of his leisure, rest in a place, and living in view of heaven, is a lively image of a contemplative life), and that of the husbandman: where we see again the favour and election of God went to the shepherd, and not to the tiller of the ground.
——**FRANCIS BACON:** op. cit., Par. 7, page 18 left-hand column (Well, Mr. Bacon! Will you please make up your mind, sir? But then, at the bottom of the left-hand column of page 16, you use the phrase "the two ways of contemplation"; and so, perhaps the target of your complaint is the kind of *fixated* contemplation which *never* looks outward to "the contemplation of nature" to any extent whatsoever. Is there such a kind of contemplation?! Maybe more to the point: If Esoptrics is correct, does it not prove a lot more time should have been spent on *fixated* contemplation of intra-mental things before the turn outward to the extra-mental ones? Who, before the full knowledge and mastery of self, can achieve any significant knowledge or mastery of anything else?)

Chapter Thirteen:

GENERIC ZONES AMONG THE PLANETS:

Is there any evidence to support Esoptrics' contention that the Universe contains generic zones? I don't imagine I can produce *conclusive proof* of such; but, I do dare to speculate there is some very interesting evidence in the distances from the Sun exhibited by the bodies which orbit it. That evidence is presented in chart #13-1 so:

PLANET	PERIHELION IN MILES	APHELION IN MILES	GENERIC ZONES' MAX & MIN RADII 1 MI. = 2.190529854607x10^{51}Φ	
			IN Φ	IN MILES
Mercury	28,583,702	43,382,549	2^{195}~2^{196}	22,924,505~45,849,011
Venus	66,782,600	67,693,910	2^{196}~2^{197}	45,849,010~91,698,022
Earth	91,402,640	94,509,460	2^{197}~2^{198}	91,698,020~183,396,041
Mars	128,409,597	154,865,853	2^{197}~2^{198}	91,698,020~183,396,041
The Belt	191,488,963	303,965,490	2^{198}~2^{199}	183,396,041~366,792,082
Jupiter	460,237,112	507,040,016	2^{199}~2^{200}	366,792,082~733,584,164
Saturn	838,741,509	934,237,322	2^{200}~2^{201}	733,584,164~1,467,168,328
Uranus	1,699,449,110	1,868,039,489	2^{201}~2^{202}	1,467,168,328~2,934,336,657
Neptune	2,771,162,073	2,819,185,846	2^{201}~2^{202}	1,467,168,328~2,934,336,657
Pluto	2,756,872,958	4,583,311,152	2^{202}~2^{203}	2,934,336,714~5,868,673,429

CHART #13-1: SCIENCE VS. ESOPTRICS ON THE DISTANCE OF THE SUN FROM THE BODIES ORBITING IT

At the risk of insulting the "know-it-alls", I'll take pity on those ignorant of Astronomy and clue them in on the meanings of "perihelion" and "aphelion". "Perihelion" indicates the number of miles from the center of the Sun to a particular orbiting body when that latter is at its *shortest* distance from the Sun. "Aphelion" indicates the number of miles when that latter is at its *longest* distance. Thus, every orbit on the part of a body circling the Sun takes that body from its aphelion to its perihelion over and over again.

Note, between Mars and Jupiter, the term "The Belt". It refers to what is usually called "the main asteroid belt". Strictly speaking, the terms "perihelion" and "aphelion" do not apply to it. More exactly, the figures given describe the "width" of the belt according to Wikipedia's assurance the belt lies between 2.06 and 3.27 AU and that the AU (***i.e.:*** astronomical unit = avg. Earth-Sun distance) is typically given as 92,955,807.267 miles.

Examining the chart, you can rather readily see that Mercury's orbit falls within the boundaries of what Esoptrics calls the 196th generic zone. Venus then falls within the boundaries of the very next one. Earth and Mars then both fall within the very next one, though by no means at the same end of it. The main asteroid belt then falls within the very next one, Jupiter in the very next; Saturn in the very next, Uranus and Neptune both in the very next zone at opposite ends of it; and Pluto mostly in the very next.

It's curious that Earth & Mars, like Uranus & Neptune, fall in a generic zone beginning with an *irrational* generic OD. Does that so affect the generic zone's parameters (***ex. gr.:*** making them run from the first rational OD below its base to the first below the next zone) as to mean Earth & Mars are in separate ones and so too Uranus & Neptune?

Yes, I'm fully aware Pluto is no longer classed a planet. But, what the heck?! Give it a break. Discovered in 1930, it was called a planet for 75 years. I am also well aware that its perihelion falls slightly below the lower limit of the 203rd generic zone. We've already seen, however, how that can happen due to a planet's carrying generator being in potency to a Sun-centered form with a *native* OD above 2^{128} but a *current* one even much higher.[1]

Some may hasten to point out that not all the planets, asteroids, etc., rotate around the Sun in the same plane. That's easy enough to explain: It is by no means necessary for every duo-combo concentric with the Sun's multi-h-combo to be oriented the same way or to rotate "clockwise" rather than "counter-clockwise". In a multi-h-combo with an ontological depth of roughly 2^{227} (***i.e.:*** $2.157\text{x}10^{68}$), there can be a lot of variety among the duo-combos logically concentric with one another.

In the above chart, there lurks what some very acute thinkers might perceive and, in doing so, consider it far more difficult than the one just mentioned. It boils down to this: How can a smaller heavenly body be further away from the Sun than a larger one? For, isn't the distance from the Sun proportional to the ontological depth of the heavenly body's multi-h-combo and the size of the heavenly body proportional to that depth? How, for example, can Pluto be 4.5 billion miles from the Sun, unless, at its center, is a multi-h-combo with an ontological depth between $\text{OD}2^{201}$ and $\text{OD}2^{202}$ and, thereby, having a leading form making its carrying generator in potency to a Sun centered form between $\text{OD}2^{202}$ and $\text{OD}2^{203}$? But, how can Pluto with such a multi-h-combo at its center be smaller and have roughly one-twentieth the mass Jupiter has despite the fact Jupiter has about one-eighth the ontological depth Pluto would have to possess?

My mind suggests to me two possible answers. The first answer goes like this: Some heavenly bodies have one multi-h-combo, some more than one, and some none at all. If a heavenly body has more than one, only one of them can be at its center or barycenter, and all the others—drawn to it by its force of gravity—must be clustered around it, and must have an ontological depth smaller than that central one has. They are thus what we might call *subordinate* multi-h-combos. Naturally, increase the number of such subordinate multi-h-combos in a given heavenly body, and you increase the mass and force of gravity exerted by that heavenly body and, thereby, increase its size. Naturally, depending upon how many subordinate multi-h-combos are clustered around the central one—and depending upon how great is the difference between their and the central one's ontological depth—that's what shall determine the density of the heavenly body. In bodies such as the Sun, the number of subordinate multi-h-combos may be so great, as to cause some to be thrown from the Sun and thus to become the progenitors of orbiting planets. The same principle may hold with planets which then throw off moons.

---

[1] In the above statement, I'm assuming—and perhaps mistakenly so (See footnote #1 on page 78.)—that minor forms are at D_N in each cycle's 1st phase and then D_C in the 2nd. If it's D_C in the 1st and D_N in the 2nd, then how far a planet goes beneath the lower limit of its generic zone is determined solely by gravity rather than, in part, by any immense contraction on the part of the form providing the planet's carrying generator with u-spaces. That, the planet can indeed do without its carrying generator having to switch from one form to another for its supply of u-spaces. That's because the form supplying u-spaces all the way out to $2^{203}\ \Phi$ from the Sun's center also provides u-spaces all the way to the Sun's center.

According to the above paragraph's chain of thought, it is now quite easy to explain why Pluto is much further from the Sun than its much larger "cousins". At Pluto's core is a multi-h-combo with an ontological depth much greater than that of the multi-h-combos at the core of its fellow "planets"; but, it's the only one there.

The second answer deals with what we can perhaps call explosively *truncated* vs. explosively *eliminated* multi-h-combos. We'll take them in that order.

It may be that, for some K, there was at Pluto's center a multi-h-combo with an ontological depth of 2^{201} and thus having a carrying generator in potency to a form logically concentric to the Sun's multi-h-combo and currently somewhere between $OD2^{202}$ and $OD2^{203}$. After some number of K, though, an explosion suddenly reduced Pluto's multi-h-combo to an ontological depth of between 2^{193} and 2^{194}. One might then expect Pluto to be about half the size of Mercury and to be orbiting the Sun at half Mercury's distance. As I see it, though, Esoptrics says this: To whatever form Pluto's carrying generator was in potency at the time of the explosion, that's the form to which the generator shall continue to be in potency, until some other influence forces a change. Of course, if that's what happened in Pluto's case, one would expect to find in the solar system, and perhaps in Pluto's vicinity, bits and pieces of what had once been a part of Pluto. This day, March 23, 2013, and on the Internet at solarsystem.nasa.gov/news/display, Johns Hopkins University Applied Physics Laboratory tells me there is such debris in Pluto's vicinity.

That, then, is what I mean by an explosively *truncated* multi-h-combo. What is meant by an explosively *eliminated* multi-h-combo?

Suppose there had been, for some K, a planet between Mars and Jupiter we can call Marupiter. Its multi-h-combo had an ontological depth of somewhere between 2^{197} and 2^{198} thus making its carrying generator in potency to a Sun-centered form currently at somewhere between $OD2^{198}$ and 2^{199}. Suddenly, Marupiter exploded so violently, the largest piece of it left was not quite 600 miles in diameter. If not a one of those pieces of Marupiter has a multi-h-combo at its center, why are they all still in the 199th generic zone?

First of all, why are there such large chunks if there's no multi-h-combo in any of them? Esoptrics answers: Categorical forms currently in the MT generic zone, and not OD1 types, attract one another, and what's in each of those chunks is a great many atoms, and, in the center of each of those atoms' nucleus, is a multi-A-combo. In each atom, then, is a combo whose leading form is native to the categorical $OD2^{128}$ and is currently somewhere in the MT generic zone. Necessarily, then, whatever groups of atoms came out of the explosion in close proximity to one another remain so because of the attractive gravitational force between $OD2^{128}$ categorical forms in the MT zone.

But why are there carrying generators still in potency to one or more forms in the 199th generic zone? Esoptrics answers: To whatever form the generators carrying Marupiter's atoms were in potency at the time of the explosion, that's the form to which those generators shall continue to be in potency, until some other influence forces a change.

Yes, usually, such atom-carrying generators receive their u-spaces from the one or more multi-h-combos at the center of the heavenly body their atoms compose. That, however, does not mean they cannot—upon the sudden, catastrophic loss of those multi-h-combos—fall back upon the u-spaces a Sun centered form previously provided to those now eliminated multi-h-combos. Remember, too, how tiny is the velocity which makes an atom's carrying generator in potency to a form at $OD2^{198}$. That, then, is what is implied by "explosively eliminated multi-h-combos".

✡✝✡✝✡✝✡✝✡✝✡✝✡✝✡✝✡✝✡✝✡✝✡

Few of the specific facts used here are cited, even significant data points like distances between planets and the rate of Earth's rotation.
——**ANNA CALL: *Clarion Review*** (Has pedantry no limit???!!! Must one not even say the name of the current president of the United States is Barack Obama without citing an external, respected source confirming it???!!! This is outrageous almost to the point of being malevolent. Even grammar school children are so adept at using the Internet, they can easily find dozens of web sites confirming data points like planetary distances and the daily rotation of the Earth. Why, then, should one waste time, money, energy, and paper on citing *every data point whatsoever*???!!! How frustrating it is to spend a thousand dollars or more on a book review and then to be presented with one doing little more than to reveal the reviewer's astonishing levels of pedantry and gross inability to grasp what the book is saying!!!)

. . . . it is not as serious a flaw as the rare in-text citations, which usually don't include page numbers and dates.
——**ANNA CALL: *Clarion Review*** (Would to God Ms. Call had listed them! If they are "rare", could she not have listed each one with its page number?)

Ever since the early 80's, a competing theory has been struggling for acceptance. Known as MOND, for Modified Newtonian Dynamics, it posits that dark matter's main effect . . . . isn't caused by extra stuff, but instead by a change in how gravity works under certain conditions. . . . Stacy McGaugh of Case Western Reserve University . . . . . himself was completely dismissive about MOND when he first heard about it. "Who wants to waste their time hearing about that crap," he recalls thinking when MOND's creator, the Israeli astrophysicist Mordehai Milgrom, showed up to give a presentation many years ago. McGaugh went to listen anyway. His reaction afterward? "This is crazy talk."
——**MICHAEL LEMONICK: *Cosmic Fuggedaboudit: Dark Matter May Not Exist At All*** (On April 3, 2014, I found this on the Internet at the web site: science.time.com/2013/02/26/ cosmic-fuggedaboudit-dark-matter-may-not-exist-at-all/. It says I'm not the only one targeted for gutter level slurs from the ranks of the intellectually superior, utterly objective, open-minded and unbiased members of the mono-directional empirical observation crowd. It seems hurling gutter level insults at one's opponents is often a part of being a scientist. I try to keep my jabs more sophisticated and elegant. LOL. What's also interesting about this quote is its indication I'm not the only one denying there's any such thing as dark matter. As we'll eventually see: In Esoptrics, "dark matter" is so *relative* to us, but is by no means absolutely dark.).

Chapter Fourteen:

THE GREAT ACCELERATION ABOVE MT[1], THE BIRTH OF THE 8 REVERSE CATEGORIES, & ANTI-GRAVITY:

#	8 CATEGORIES BELOW MT	VS.	8 REVERSE CATEGORIES ABOVE MT
1st	OD1 = 8^{256} Alpha duos[2]	VS.	$OD2^{256}$ = 1 Omega duo[3]
2nd	$2^1(1)$ = OD2~OD3	VS.	$2^{-1}(2^{256})$ = $OD2^{255}$~$OD2^{256}$-1
3rd	$2^2(1)$ = OD4~OD15	VS.	$2^{-2}(2^{256})$ = $OD2^{254}$~$OD2^{255}$-1
4th	$2^4(1)$ = OD16~OD255	VS.	$2^{-4}(2^{256})$ = $OD2^{252}$~$OD2^{254}$-1
5th	$2^8(1)$ = OD256~$OD2^{16}$-1	VS.	$2^{-8}(2^{256})$ = $OD2^{248}$~$OD2^{252}$-1
6th	$2^{16}(1)$ = $OD2^{16}$~$OD2^{32}$-1	VS.	$2^{-16}(2^{256})$ = $OD2^{240}$~$OD2^{248}$-1
7th	$2^{32}(1)$ = $OD2^{32}$~$OD2^{264}$-1	VS.	$2^{-32}(2^{256})$ = $OD2^{224}$~$OD2^{240}$-1
8th	$2^{64}(1)$ = $OD2^{64}$~2^{128}-1	VS.	$2^{-64}(2^{256})$ = $OD2^{192}$~$OD2^{224}$-1

CHART #14-1: 8 CATEGORIES VS. 8 REVERSE CATEGORIES

It's already been said that the original categorical forms are those native to OD's 2^0, 2^1, 2^2, 2^4, 2^8, 2^{16}, 2^{32}, 2^{64}, 2^{128}, and 2^{256}. The first 8 of those 10 are *below* the MT. Focus on that, and the logic of the *mirror* should quickly suggest that there must be 8 categories *above* the MT. Notice how that works in the above chart: The first category below MT is 2^0 times the <u>bottom</u> of the epoch's ontological depth, OD1, and the first *reverse* category above MT is 2^0 times the <u>top</u> of that depth, $OD2^{256}$. The second category below MT is 2^1 times OD1 and the first reverse category above MT is 2^{-1} times Omega. Then it's 2^2 vs. 2^{-2}; then 2^4 vs. 2^{-4}; then 2^8 vs. 2^{-8}; then 2^{16} vs. 2^{-16}; then 2^{32} vs. 2^{-32}; and, finally, 2^{64} x 2^{-64} times OD1 vs. Omega respectively. Hereafter and for economy's sake, let's refer to these 8 reverse categories as RC#1, RC#2, RC#3, RC#4, RC#5, RC#6, RC#7, and RC#8.

What are reverse categories, and what produces them? As for the latter, a vast number of major forms—if not prevented by the limit of 2^{128} times their native OD—rise to a current OD far above both their native OD and $OD2^{128}$. A huge number of them stop within RC#8's boundaries shortly after a smaller number stop within RC#7's boundaries shortly after a smaller number stop within RC#6's boundaries, and so on as I surmise my reader can foresee. As a rise in the 4th dimension at 1OD/K, this all occurs in $(2^{256}–2^{128})$K = c. 10^{-19}

[1] The term "MT" = the mirror threshold = OD 2^{128} = the square root of 9th epoch's ontological depth = the square root of how many 4th dimension levels are possible to a 9th epoch u-space.

[2] The term "Alpha duo" = any duo-combo whose piggyback form is native to OD1.

[3] The term "Omega" = the one duo-combo whose piggyback form is native to $OD2^{256}$.

sec.. Because I suspect it shall make it easier for my reader to follow the details, I will start with RC#2 and work my way down from there.

Here's how the reverse category at $OD2^{-1}$ Omega (***i.e.:*** RC#2) comes to be: Some number of MT duos (***i.e.:*** duo-combos each with a piggyback form native to $OD2^{128}$) accelerate to at least $OD2^{255}$ +1 but no further than $OD2^{256}$-1. For many years, I speculated that number is 6, because, for many years, I speculated God created 6 forms native to each of the OD's of 2^{255} thru 2^{256}–1. These days, now that I've changed that to 8, I speculate 8 MT duos are accelerated to at least $OD2^{255}$+1. I say that because, with only 8 forms native to each of the OD's from 2^{255} to, and including, 2^{256}–1, no more than 8 multi-g-combos can make concentric to themselves forms native to, and currently at, $OD2^{255}$ or higher.

The result is the first kind of multi-g-combo and the 8 most massive of the multi-g-combos (***i.e.:*** black holes?) possible to the Universe. Each of these cosmic multi-combos is composed of at least 2^{255} +1 (***i.e.:*** $5.7896\text{x}10^{76}$+1 at $OD2^{255}$ +1) duo-combos each concentric with all the rest and each at a discrete OD and level of the 4th dimension. Each of the 8 thus has a mass at least $(2^{255} +1)^2$ (***i.e.:*** c. $3.4\text{x}10^{153}$) times that of a single duo-combo. Hereafter, let's refer to each of these multi-g-combos as an RC#2 multi-combo, and the term "ontological depth of the multi-combo" shall refer to the number of duo-combos made concentric in the multi-combo. Be sure not to confuse "ontological depth of the multi-combo" with "ontological depth of the Universe's 9th epoch". The latter is 2^{256}; the former varies.

The "ontological depth" of an RC#2 multi-combo at $OD2^{255}$+1 is 2^{255}+1, and the ontological depth of an RC#2 multi-combo at $OD2^{256}$-1 is 2^{256}-1. The latter is because Omega can never be made concentric to any other form. Every RC#2 multi-combo's ontological depth includes forms currently at OD's 2^1, 2^2, 2^4, 2^{16}, 2^{32}, 2^{64}, 2^{128} and maybe 2^0; but, none of those is *native* to the given OD, because—though every OD1 form can make logically concentric to itself other OD1 forms—no higher form can ever make logically concentric to itself a categorical form.[1] What's current, then, at those OD's are forms not native to them and either accelerated or decelerated to them in order to fill the gap. What's currently at OD2 is probably a form native to OD3 and decelerated to OD2. What's currently at OD4 is probably a form native to OD3 and accelerated to OD4. If there's a form at OD1 (I'm still vacillating on that issue.), it's a form native to OD3 and decelerated to OD1.

Here, let's pause briefly to consider an important issue: Each of the RC#2 multi-g-combos holds concentric to itself one duo-combo currently at each of the OD's of at least 2 thru at least 2^{255}+1. Why are the OD's 2 thru 2^{128} there? In accelerating from $OD2^{128}$ to $OD2^{255}$+1, the MT duo serving as the lead form did not pass thru any OD below 2^{128}. Esoptrics answers that whatever an accelerating MT duo accomplishes *above* the MT, it accomplishes *below* the MT in a kind of *reverse* influence. That, after all, is what one would expect when dealing with the logic of a *mirror*. Therefore, just as each of the MT duos accelerating *above* MT to $OD2^{255}$ +1 makes concentric to itself 127 generic OD's above MT (***viz.:*** 2^{129}, 2^{130}, 2^{131}, etc., to, and including, 2^{255}) along with their species, so also do each of those MT duos make concentric to itself 127 generic OD's below MT (***viz.:*** 2, 2^2, 2^3, etc., to, and including, 2^{127}) along with their species.[2]

More than that, the MT duo's acceleration up from 2^{128} to 2^{255}+1 and down from 2^{128} to OD2 is an action so powerful it spawns a huge number of atomic multi-combos each having an ontological depth of at least 2^{128} (if none has a logically concentric form at OD1)

[1] That prohibition has major consequences for the structure of the atomic nucleus. They'll be spelled out in a later chapter.

[2] Such a principle militates strongly against the notion that the RC#2 multi-g-combos have a form at OD1 among the forms logically concentric with them.

and each having a leading form native to $OD2^{128}$. Because every OD1 form's carrying generator gets its u-spaces from an OD2 form, the fusion within the nucleus (and an effect on the OD2 form to be detailed later) causes, in each atom, an OD1 form to accelerate. The result is thus a huge number of atoms each with a small nucleus[1] orbited by a duo-combo native to OD1 and holding logically concentric to itself c. 10^{37} OD1 duo-combos.

In each of these RC#2 multi-combos, the accelerated MT duo's piggyback form is the leading form, and its carrying generator serves as the carrying generator for the entire RC#2 multi-combo. It receives its field of locomotion (***i.e.:*** its u-spaces) from Omega. It thus receives a field of locomotion composed of $(2x2^{256})^3 = 2^{771} = 2.484x10^{232}$ u-spaces able to be converted into microstates of excitation and each serving as a location in space logically separate from, and outside of, all the others.[2] If an RC#2 multi-combo is exactly at $OD2^{255}$, its carrying generator is "leapfrogging" thru its field of locomotion at the speed of light (***i.e.:*** $2^{128}K/\Phi = C$). If an RC#2 multi-combo is at $OD2^{255}+1$, its carrying generator is moving thru its field of locomotion at $2^{-255}C$. At $OD2^{255}+2$, the velocity shifts to $2(2^{-255})C$. At $OD2^{255}+3$, the velocity is $3(2^{-255})C$. I take it you follow the progression and, thus, anticipate that, at $OD2^{255}+2^{254}$, the velocity is .5C using the u-spaces provided by Omega, and, at $OD2^{255} + 2^{255} -1$ the velocity is virtually light speed.

Each of these RC#2 multi-combos serves as the anchor for (***i.e.:*** is at the center of) one of the 8 main regions of the universe, and each now races toward what geometry-speak would call 1 of the 8 "corners" of Omega. Why is that? That brings us to Esoptrics' two-fold version of anti-gravity. Its two-fold version is expressed by what Esoptrics calls:

THE FIRST LAW OF THE PRESERVATION OF THE CATEGORIES

EVERY CATEGORICAL FORM LEADING A MULTI-G-COMBO IN POTENCY TO ANOTHER CATEGORICAL FORM IS REPELLED BY THAT OTHER CATEGORICAL FORM AND DRIVEN THEREBY TO THE OUTERMOST REACHES OF THAT OTHER'S U-SPACES ENVELOPE.

THE SECOND LAW OF THE PRESERVATION OF THE CATEGORIES

ALL CATEGORICAL FORMS LEADING MULTI-G-COMBOS OPERATING IN THE SAME GENERIC ZONE REPEL ONE ANOTHER.

If a multi-g-combo's leading form is a *categorical* one whose carrying generator is in potency to (***i.e.:*** gets its field of locomotion from) another *categorical* form, such as Omega, then that multi-g-combo moves inexorably toward the outermost limits of the categorical form to which its carrying generator is in potency. In the case of an RC#2 multi-combo in potency to Omega, that outermost limit means, for Esoptrics, 2^{256} successive microstates of excitation (***i.e.:*** $2^{256}\Phi$) away from the center of the Universe. For our sensation dependent minds, it means roughly 9 trillion light years.

The second law above is basically a repeat of the first one. For, from what's been

[1] Its precise structure shall be described in a later chapter.

[2] I repeat: No *single* one of those c. 10^{232} u-spaces is to any extent or in any way whatsoever a square, a cube, a circle, or a sphere save in the imagination of those seeking an effective teaching aid; nevertheless, *collectively* they *actually* produce *in real effect* a sphere. Our senses do not fool us about the shape of the Universe or any of its material occupants; they merely do not—and, as is blatantly obvious, *cannot*—make manifest to our *senses* the ultra-microscopic *why* behind those shapes.

said, multi-g-combos operating in the same generic zone (***i.e.:*** performing a macrostate of excitation at an OD within the OD range of that generic zone) have carrying generators getting their u-spaces from the next highest categorical form; and so, they are each being driven from the center of that next highest categorical form. What the second law adds to that is this: In addition to being driven away from the center of the u-spaces envelope (***i.e.*** form) providing their carrying generators with a field of locomotion, they are repelling each other. In short, there's a "double barreled" anti-gravity force driving them away from one another, and that suggests the rate at which they are fleeing one another should steadily increase. It comes as no surprise, then, when, on page 299 of his book, ***The Fabric Of The Cosmos***, Prof. Greene informs us of two groups concluding that, since c. 7 billion years old, the Universe's rate of expansion has been accelerating.

In RC#2, if the leading form of a multi-A-combo (***i.e.:*** atomic nucleus) is exactly at MT, its entire atom is moving at the speed of light with its carrying generator utilizing the field of locomotion provided by the leading form of the RC#2 multi-combo toward which it's drawn, or, if possible, it will utilize the field of locomotion provided by Omega—a logically problematic switch on which I'll not elaborate. If the leading form of a multi-A-combo is at MT+1, the velocity of it and its atom is 2^{-128} times light speed, C, and its carrying generator is utilizing the field of locomotion provided by that one of the RC#2 multi-combo's forms which is currently at OD2^{128}. It thus can't go very far. At MT+2, the velocity of it and its atom is $2(2^{-128})$C, and its carrying generator is utilizing the field of locomotion provided by that one of the RC#2 multi-combo's forms which is currently at OD$2(2^{128})$ = OD2^{129}. At MT+3, the velocity of it and its atom is $3(2^{-128})$C, and its carrying generator is utilizing the field of locomotion provided by that one of the RC#2 multi-combo's forms which is currently at OD$3(2^{128})$. If you follow the progression, then you foresee that, at MT $+2^{127}$, it and its atom are moving at .5C, and its carrying generator is utilizing the field of locomotion provided by that one of the RC#2 multi-combo's concentric forms currently at OD2^{255} (***i.e.:*** 2^{128} x 2^{127}). What if the multi-A-combo's velocity calls for a form lower than OD2^{256} but higher than the OD of the RC#2's leading form? In that case, the generator utilizes the u-spaces provided by whatever form in the RC#2 is the highest multiple of 2^{128}.

Let's turn to what's implied by the third reverse category, which is to say the one running from OD2^{254} thru and including OD2^{255}-1. Here again, we're talking about acceleration on the part of a certain number of MT duos. I suspect the number is 64-8=56 or, at least, no greater than 56. I say that because only 8^2 duo-centered combos were created native to each of the OD's 2^{254} thru and including 2^{255}-1, and, probably, 8 of those were ingested by the 8 MT duos which accelerated into RC#2.[1]

If 8 is the number of MT duos in RC#2; and 56 the number in RC#3—then the MT duos accelerating into RC#3 divide into 8 groups of 7 each, and each group of 7 is in potency to a different one of the 8 RC#2 multi-combos. For example, the RC#2 multi-combo traveling toward Omega's "upper, left-front corner" (Let's call it Cal.) is presenting each of 7 different RC#3 multi-combos with u-spaces. Each of that 7—in keeping with the law of the preservation of the categories—shall now travel outward seeking the outermost boundaries of Cal. For Esoptrics, that means each of those 7 RC#3 multi-combos can go no fur-

[1] As I write the above this Tuesday, Feb. 5, 2013, I find my mind suggesting the above is incorrect. I recall often telling myself that an accelerating form can fill in the gaps by accelerating a form to whatever OD would otherwise go begging. Accept that, and it follows that there can be 8 RC#2 multi-combos, 8^2 RC#3 multi-combos, 8^4 RC#4 multi-combos, 8^8 RC#5 multi-combos, 8^{16} RC#6 multi-combos, 8^{32} RC#7 multi-combos, and 8^{64} RC#8 multi-combos. Such a progression is more aesthetically appealing than the one mentioned in the above and following paragraphs. Notice I said "can be". Here and there, those numbers may not be realized for some reason.

ther than 2^{255} logically sequential microstates of excitation away from the center of Cal. For sensation dependent minds, it means $2^{255}\Phi$ and, therefore, roughly 4.5 trillion light years.[1]

As each of the 7 RC#3 multi-combos travels away from Cal's center, what's the source of its field of locomotion? Therein lies a crucial bit of information. How so?!

Omega is not a multi-combo. There are no duo-combos concentric to it. Its ontological depth is 1. It, therefore, presents only one field of locomotion to each and every one of the 8 RC#2 multi-combos in potency to it. Such is far from being true of what each of the 8 RC#2 multi-combos is doing for each of the 7 RC#3 multi-combos in potency to it, which is to say looking to it for a field of locomotion (***i.e.:*** set of u-spaces). The ontological depth of each RC#2 multi-combo, if at OD2^{255}+1, is at least 2^{255}, which is to say at least 2^{255} duo-combos—each at a different one of the OD's (***i.e.:*** 4th dimension's levels) from at least 2^1 thru and including 2^{255}+1—are concentric in it. That means at least 2^{255} duo-combos each of which—thru its piggyback form—is offering one of 2^{255} different fields of locomotion (***i.e.:*** u-spaces envelopes). Each field is concentric with all the others at a point which—for sensation dependent minds—is only 10^{-47} cm. at the level of its 2^{255} concentric generators. One of these duo-combos (***i.e.:*** the one currently at OD2) has a form offering a field of locomotion consisting of only 4x4x4 u-spaces (***i.e.:*** for sensation dependent minds, 64 different places within an area only 4Φ in dia.). The one currently at OD3 is offering only 6x6x6 u-spaces (***i.e.:*** for sensation dependent minds, 216 different places within an area only 6Φ in dia.). The one currently at OD4 is offering only 8x8x8 u-spaces; at OD5, 10x10x10, and so on, until, at OD2^{255}, we have a form offering a field of locomotion consisting of $(2x2^{255})^3$ u-spaces (***i.e.:*** the cube of twice the OD) and offering the ability to travel 4.5 trillion light years away from its center. To which one of these roughly 10^{77} different fields of locomotion would an RC#3 combo's carrying generator turn for the u-spaces which shall allow it to engage in rectilinear locomotion?

Remember: Above the MT zone, every categorical form repels every categorical form in potency to it, and—because every RC#3 multi-g-combo is in the same generic zone (***i.e.:*** the 255th one ranging from OD2^{254} to OD2^{255}–1)—they, too, repel each other. That means each RC#3 multi-combo must, in a discrete direction, go an immense distance from the center of the RC#2 multi-combo to which it is in potency. To do that, its carrying generator must get from the RC#2 multi-combo a field of locomotion allowing it to travel many light years away from the center of the RC#2 multi-combo. Esoptrics, then, says: If the RC#3 multi-combo is at OD2^{254}+1 (***i.e.:*** if that's the current OD of its leading form), its generator gets its field of locomotion from what, in the RC#2 multi-combo, is currently at OD2^{255}+2 and thus offering a field of locomotion 4.5 trillion light years in radius. If the RC#3 multi-combo is at OD2^{254}+2, its generator receives its field of locomotion from what, in the RC#2 multi-combo, is currently at OD2^{255}+4. If at OD2^{254}+3, the field is from OD2^{255}+6; at OD2^{254}+4, OD2^{255}+8, and so on, until, at OD2^{254}+2^{254}-1, the field of locomotion is coming from what's currently at OD2^{256}–2 and is 9 trillion light years in radius.

So then, the ontological depth of an RC#2 multi-combo is at least 2^{255}, since it must be at least at OD2^{255}+1, but never greater than 2^{256}-2 (That's at OD2^{256}-1 and if it has no form logically concentric to it at OD1.), and the ontological depth of an RC#3 multi-combo is at least 2^{254}, since it must be at least at OD2^{254}+1, but never greater than 2^{255}-2 (That's at

[1] Suppose Cal moves to the outermost reaches of Omega and an RC#3 multi-combo moves to the outermost reaches of Cal. Won't the RC#3 multi-combo's center now be 9 trillion + 4.5 trillion = 13.5 trillion light years away from the center of Omega, which is to say the center of the Universe? Yes, ***if*** the direction of the RC#3 multi-combo's locomotion away from Cal's center is the same as Cal's locomotion away from the center of Omega. But, let's not bother with that for now. It's not relevant enough to what I'm mainly trying to clarify at the moment.

$OD2^{255}$-1 and if it has no form logically concentric to it at OD1.). That thus gives us 2 distinct kinds of multi-g-combos: kind #1 in RC#2, and kind #2 in RC#3. At the head of each of these is a duo-combo whose piggyback form is a categorical one native to $OD2^{128}$. The leading form of each is thus an MT duo. In the RC#2 multi-combos, an MT duo has accelerated to at least $OD2^{255}$+1 to avoid light speed. In the RC#3 multi-combos, an MT duo has accelerated to at least $OD2^{254}$+1 to avoid light speed.

The acceleration producing the second kind of multi-g-combos also complicates the structure of the atomic nucleus. How so?! We'll deal with that later in another chapter. For now, then, let us deal with the other reverse categories. Chart #14-2 below graphs in advance what I'm about to spell out in detail.

#	RC#	MIN. OD	MAX OD	#	RC#	MIN OD	MAX OD	#	RC#	MIN OD	MAX OD
1	2	2^{255}+1	2^{256}-1	**23**	7-10	2^{233}+1	2^{234}-1	**45**	8-20	2^{211}+1	2^{212}-1
2	3	2^{254}+1	2^{255}-1	**24**	7-9	2^{232}+1	2^{233}-1	**46**	8-19	2^{210}+1	2^{211}-1
3	4-2	2^{253}+1	2^{254}-1	**25**	7-8	2^{231}+1	2^{232}-1	**47**	8-18	2^{209}+1	2^{210}-1
4	4-1	2^{252}+1	2^{254}-1	**26**	7-7	2^{230}+1	2^{231}-1	**48**	8-17	2^{208}+1	2^{209}-1
5	5-4	2^{251}+1	2^{252}-1	**27**	7-6	2^{229}+1	2^{230}-1	**49**	8-16	2^{207}+1	2^{208}-1
6	5-3	2^{250}+1	2^{251}-1	**28**	7-5	2^{228}+1	2^{229}-1	**50**	8-15	2^{206}+1	2^{207}-1
7	5-2	2^{249}+1	2^{250}-1	**29**	7-4	2^{227}+1	2^{228}-1	**51**	8-14	2^{205}+1	2^{206}-1
8	5-1	2^{248}+1	2^{249}-1	**30**	7-3	2^{226}+1	2^{227}-1	**52**	8-13	2^{204}+1	2^{205}-1
9	6-8	2^{247}+1	2^{248}-1	**31**	7-2	2^{225}+1	2^{226}-1	**53**	8-12	2^{203}+1	2^{202}-1
10	6-7	2^{246}+1	2^{247}-1	**32**	7-1	2^{224}+1	2^{225}-1	**54**	8-11	2^{202}+1	2^{203}-1
11	6-6	2^{245}+1	2^{246}-1	**33**	8-32	2^{223}+1	2^{224}-1	**55**	8-10	2^{201}+1	2^{202}-1
12	6-5	2^{244}+1	2^{245}-1	**34**	8-31	2^{222}+1	2^{223}-1	**56**	8-9	2^{200}+1	2^{201}-1
13	6-4	2^{243}+1	2^{244}-1	**35**	8-30	2^{221}+1	2^{222}-1	**57**	8-8	2^{199}+1	2^{200}-1
14	6-3	2^{242}+1	2^{243}-1	**36**	8-29	2^{220}+1	2^{221}-1	**58**	8-7	2^{198}+1	2^{199}-1
15	6-2	2^{241}+1	2^{242}-1	**37**	8-28	2^{219}+1	2^{220}-1	**59**	8-6	2^{197}+1	2^{198}-1
16	6-1	2^{240}+1	2^{241}-1	**38**	8-27	2^{218}+1	2^{219}-1	**60**	8-5	2^{196}+1	2^{197}-1
17	7-16	2^{239}+1	2^{240}-1	**39**	8-26	2^{217}+1	2^{218}-1	**61**	8-4	2^{195}+1	2^{196}-1
18	7-15	2^{238}+1	2^{239}-1	**40**	8-25	2^{216}+1	2^{217}-1	**62**	8-3	2^{194}+1	2^{195}-1
19	7-14	2^{237}+1	2^{238}-1	**41**	8-24	2^{215}+1	2^{216}-1	**63**	8-2	2^{193}+1	2^{194}-1
20	7-13	2^{236}+1	2^{237}-1	**42**	8-23	2^{214}+1	2^{215}-1	**64**	8-1	2^{192}+1	2^{193}-1
21	7-12	2^{235}+1	2^{236}-1	**43**	8-22	2^{213}+1	2^{214}-1				
22	7-11	2^{234}+1	2^{235}-1	**44**	8-21	2^{212}+1	2^{213}-1				

CHART #14-2: THE 64 TYPES OF MULTI-G-COMBOS

RC#4 is produced when perhaps as many as 8^4 (***i.e.:*** 4,096) –64 MT duos accelerate to at least $OD2^{252}$+1 but no further than $OD2^{254}$-1. I subtract 64 because that is possibly the combined number of MT duos accelerated to RC#2 + RC#3 (***i.e.:*** 8 RC#2 + 56 RC#3).

Notice how between $OD2^{252}$ (***i.e.:*** the floor of RC#4) and $OD2^{254}$ (***i.e.:*** the floor of RC#3), there is the generic $OD2^{253}$. That gives us both a 3rd and a 4th kind of multi-g-combo. One kind has a minimum ontological depth of 2^{252} at $OD2^{252}$+1 and a maximum of 2^{253}-2 at $OD2^{253}$-1. The other's such depth runs from 2^{253} to 2^{254}–2. Each has, as its leading form, an MT duo. The latter can be referred to as an RC#4-2 multi-g-combo and the former an RC#4-1 multi-g-combo. How many of the RC#4 multi-g-combos are RC#4-1 vs. RC#4-2 types? It's up to God.

Whatever the number, each RC#4-1 type's carrying generator gets its u-spaces from the duo-combos logically concentric in some one of the RC#4-2 multi-g-combos, while the latter type's carrying generators get their u-spaces from the RC#3 multi-g-combos. Because the leading form of each is a categorical one in potency to a categorical one, each RC#4-1 multi-g-combo is steadily driven away from the center of the RC#4-2 multi-g-combo to which it is in potency, even as each RC#4-1 repels its fellows, since they're all in the same generic zone. By the same token, each RC#4-2 multi-g-combo is steadily driven away from the center of the RC#3 multi-g-combo to which it is in potency, even as each RC#4-2 repels its fellows, since they're all in the same generic zone. I see no reason why RC#4-1 types can't accelerate above $OD2^{253}$ and become RC#4-2 types.

RC#5 is produced when perhaps as many as 8^8 (***i.e.:*** 16,777,216) minus perhaps 4,096 MT duos are accelerated to $OD2^{248}+1$ but not beyond $OD2^{252}-1$. Notice that, between $OD2^{248}$ (***i.e.:*** the start of RC#5) and $OD2^{252}$ (***i.e.:*** the start of RC#4), there are the generic OD's of 2^{249}, 2^{250}, and 2^{251}. This gives us the 5th, 6th, 7th, & 8th kinds of multi-g-combos. We can call them RC#5-1, RC#5-2, RC#5-3, & 5-4 respectively. It should be rather obvious what ontological depths are involved for each of these RC#5 types. If not, see chart #14-2 on page 116. How many of the RC#5 multi-g-combos are RC#5-1 vs. -2, etc.? It's up to God.

Each RC#5 type has, as its leading form, an MT duo. Each can be referred to as an RC#5 multi-g-combo, and each one's carrying generator gets its field of locomotion from the RC#5 type above it, or—in the case of the RC#5-4 types—from the RC#4-1 types. Because the leading form of each is a categorical one in potency to a categorical one, each RC#5 type is steadily driven away from the center of the RC#5 type above it or—in the case of the RC#5-4 types—from the center of the RC#4-1 type above it. Here again, of course, each RC#5-1 repels its fellows as each RC#5-2 repels its fellows, etc., because each such type is operating in the same generic zone.

RC#6 is produced when perhaps as many as 8^{16} (***i.e.:*** 281,474,976,710,656) minus perhaps 16,777,216 MT duos are accelerated to $OD2^{240}+1$ but not beyond $OD2^{248}-1$. Notice that, between $OD2^{240}$ (***i.e.:*** the start of RC#6) and $OD2^{248}$ (***i.e.:*** the start of RC#5), there are the 7 generic OD's of 2^{241} thru 2^{247}. That opens the door to the 9th thru the 16th kinds of multi-g-combos. We can refer to them as the RC#6–1, –2, –3, –4, –5, –6, -7 and –8 kinds of multi-g-combos respectively. Collectively, they are RC#6 multi-g-combos. It should be rather obvious what ontological depths are involved for each of these RC#6 types. If not, see chart #14-2 on page 116. How many of the RC#6 multi-g-combos are RC#6-1 vs. -2, etc.? It's up to God.

Each RC#6 type has as its leading form an MT duo. Each one's carrying generator gets its field of locomotion from the RC#6 type above it, or—in the case of the RC#6-8 types—from the RC#5-1 types. Because the leading form of each is a categorical one in potency to a categorical one, each RC#6 multi-g-combo is steadily driven away from the center of the RC#6 type above it or—in the case of the RC#6-8 types—from the center of the RC#5-1 type above it. Here again, of course, each RC#6-1 repels its fellows as each RC#6-2 repels its fellows, etc., because each such type is operating in the same generic zone.

RC#7 is produced when perhaps as many as 8^{32} (***i.e.:*** 7.9228×10^{28}) minus perhaps 8^{16} MT duos are accelerated to $OD2^{224}+1$ but not beyond $OD2^{240}-1$. Notice that, between $OD2^{224}$ (***i.e.:*** the start of RC#7) and $OD2^{240}$ (***i.e.:*** the start of RC#6), there are the 15 generic OD's of 2^{223} thru 2^{239}. That opens the door to the 17th thru the 32nd types of cosmic multi-g-combos. We can call them the RC#7–1, –2, –3, –4, –5, –6, –7, . . . –16 kinds of multi-h-combos respectively. Collectively, they are RC#7 multi-g-combos. The ontological depths of each should be rather obvious. If not, see chart #14-2 on page 116. How many of the RC#7 multi-g-combos are RC#7-1 vs. -2, etc.? It's up to God.

Each RC#7 type has as its leading form an MT duo. Each one's carrying generator gets its field of locomotion from the RC#7 type above it, or—in the case of the RC#7-16 types—from the RC#6-1 types. Because the leading form of each is a categorical one in potency to a categorical one, each RC#7 multi-g-combo is steadily driven away from the center of the RC#7 type above it or—in the case of the RC#7-16 types—from the center of the RC#6-1 type above it. Here again, of course, each RC#7-1 repels its fellows as each RC#7-2 repels its fellows, etc., because each such type is operating in the same generic zone.

RC#8 is produced when perhaps as many as 8^{64} (***i.e.:*** 6.277×10^{57}) minus perhaps 8^{32} MT duos are accelerated to $OD2^{192}+1$ but not beyond $OD2^{224}-1$. Notice that, between $OD2^{192}$ (***i.e.:*** the start of RC#8) and $OD2^{224}$ (***i.e.:*** the start of RC#7), there are the 31 generic OD's of 2^{193} thru 2^{223}. That opens the door to the 33rd thru the 64th kinds of multi-g-combos. We can refer to them as the RC#8–1, –2, –3, –4, –5, –6, –7, . . . –32 types of multi-g-combos respectively. Collectively, they are RC#8 multi-g-combos. The ontological depths of each should be rather obvious. If not, see chart #14-2 on page 116. How many of the RC#8 multi-g-combos are RC#8-1 vs. -2, etc.? It's up to God.

Each RC#8 type has as its leading form an MT duo. Each one's carrying generator gets its field of locomotion from the RC#8 type above it, or—in the case of the RC#8-32 types—from the RC#7-1 types. Because the leading form of each is a categorical one in potency to a categorical one, each RC#8 multi-g-combo is steadily driven away from the center of the RC#8 type above it or—in the case of the RC#8-32 types—from the center of the RC#7-1 type above it. Here again, of course, each RC#8-1 repels its fellows as each RC#8-2 repels its fellows, etc., because each such type is operating in the same generic zone.

How do the multi-h-combos (***i.e.:*** stars, planets, etc.) fit into the above picture? Consider first the 8 RC#2 multi-g-combos. Having an ontological depth of at least 2^{255}, each of them holds logically concentric to itself duo-combos and thus forms—and, thus, u-spaces envelopes—currently at each and every one of the OD's (***i.e.:*** 4th dimension's levels) from at least OD2 to at least $2^{255}+1$. Suppose a form native to $OD2^{126}$ is accelerated to $OD2^{225}$. Let's call it Linus. Its acceleration shall be well within the limit of $2^{126} \times 2^{128}$, and its carrying generator can thereby get its u-spaces from a rational form which is currently at $OD2^{226}$ and logically concentric to one of the RC#2 multi-g-combos. Linus will thus be a multi-h-combo serving as the core of a rather massive heavenly body orbiting the center of its RC#2 multi-g-combo at a radius of $2^{226}\Phi$, which is to say c. 16,749 light years. It could be doing the same around any RC#3, RC#4, RC#5, RC#6, or RC#7-3 and above, multi-g-combo. For, each of those has logically concentric to itself a duo-combo and thus form and u-spaces envelope currently at $OD2^{226}$. To confirm that, refer to chart #14-2 on page 116.

Pardon me if I now decline to say more about multi-h-combos. I opine that, in the above paragraph, I've said enough for my reader to figure out the rest on his or her own.

Let's now round out this chapter with some concrete examples of what it's saying about the Universe's macro-structure. To that end, we start by saying an MT duo has accelerated into RC#7 (***i.e.:*** $OD2^{224}$ to $OD2^{240}-1$) and is currently at somewhere between $OD2^{228}$ (***i.e.:*** 5.5×10^{70}) and $OD2^{229}$. It thus becomes an RC#7-4 multi-g-combo at the center of a galaxy with a diameter somewhere between 2^{229} and $2^{230}\Phi$, which is to say somewhere between 66,996.7 and 133,933.4 light years.

The diameter of the Milky Way is generally given as 100,000 light years. If that's correct, we are perhaps here describing the RC#7 multi-combo (black hole?) serving as the anchor for the Milky Way. Let's call this multi-g-combo Whitey.

Next, a duo-combo native to some *generic* OD such as perhaps $OD2^{112}$ has been accelerated by Whitey to such an extent that it is a cosmic multi-h-combo with a leading form somewhere between $OD2^{226}$ & 2^{227} and, therefore, in potency to (***i.e.:*** getting its field of locomotion from) that one of Whitey's concentric forms which is somewhere between $OD2^{227}$ & OD^{228}. Let's call Whitey's victim Sonny. Whatever is exactly the form in Whitey giving Sonny his field of locomotion (Let's call it Whitey Jr..), Sonny is at the outermost reaches of that field of locomotion. Because we're now talking about a radius rather than a diameter, we're saying Sonny is somewhere between 2^{227} and $2^{228}\Phi$ from the captor Whitey's center. That means Sonny is somewhere between 16,749.2 and 33,498.4 light years from Whitey's center. Since the cosmic multi-h-combo at Sonny's center is a *generic* one, Sonny, unless he accelerates, goes no further away from, or closer to, the captor Whitey's center than what is caused by the captive Whitey Jr. pulsing the way all forms do. If we speak in terms dear to sensation dependent minds, Sonny's leading form extends its influence over an "area" somewhere between 16,749 to 33,498 light years in diameter as he orbits captor Whitey at a radius of between 16,749 and 33,498 light years.

On October 28, 2011, I read on Wikipedia that the Sun's distance from the center of the Milky Way is 24,900 ± 1,000 light years. If that's correct, we are perhaps here describing the cosmic multi-h-combo at the center of our Sun and holding it together.

Next, since the leading form in every cosmic multi-h-combo is native to a generic OD, one or more other cosmic multi-h-combos should be orbiting Sonny somewhere between, say, 2^{194} and 2^{195} Φ, then one or more between 2^{195} and 2^{196} Φ, then one or more between 2^{196} and 2^{197} Φ, and so forth. See chart #13-1 on page 107.

As you can see there, Mercury, Venus, Jupiter, Saturn, and Uranus all fall nicely into the generic zones suggested by Esoptrics. Earth does not quite do so, and Pluto is far out of step with Esoptrics. Mars might seem to be so; but, it falls well within the generic zone set for Earth, and nothing in Esoptrics forbids 2 planets to fall within the same generic zone. In chart #13-1, then, the generic zone associated with Mars does not apply to it; but, it does perhaps apply to the asteroid belt, since Wikipedia tells me the asteroid belt is a region found roughly between the orbits of Mars and Jupiter. I've yet to find anything about its perihelion or aphelion. Neptune, too, seems out of step with Esoptrics. It's not. It falls well within the generic zone set for Uranus, and, again, nothing in Esoptrics forbids 2 planets to fall within the same generic zone.

Before closing this chapter, let me draw my reader's attention to what seems another interesting bit of agreement between Esoptrics and Science. This day, August 23, 2013, I viewed a web page at http://astronomy.swin.edu.au/cosmos/D/dwa. At that site, the Swinburne Center for Astrophysics & Supercomputing says "Dwarf galaxies are the most abundant type of galaxy in the universe". Look again at chart #14-2 on page 116, and you should be able to notice this: The chart tells you that most multi-g-combos fall in RC#8; and so, they must produce galaxies whose diameters, at best, would be 2^{225}Φ (***i.e.:*** c. 4,187 light years) and, at worst, only 2^{294}Φ (***i.e.:*** c. 2,865,563 miles). For Esoptrics, then, as for the Swinburne Center, "dwarf galaxies are the most abundant type of galaxy in the universe".

In what's been said in the course of the last few pages, Esoptrics gives what I dare say is a very simple explanation of why—though the stars and planets remain at rather fixed distances from the centers of their galaxies (because the center of each star and planet is a multi-***h***-combo with a *generic* leading form)—the galaxies themselves are racing away from one another (because the center of each galaxy is a multi-***g***-combo with a *categorical* leading form) within clusters of galaxies each racing away from its fellow clusters (because the center of each cluster is a multi-***g***-combo with a *categorical* leading form) within clusters of clusters of galaxies each racing away from its fellow clusters of clusters (because the center of each cluster of clusters is a multi-***g***-combo with a *categorical* leading form). In addressing why the galaxies are racing apart while their component stars and planets are not, does Science give an explanation anywhere near as simple as Esoptrics' is? In other words, Esoptrics gives not merely an explanation but a very simple one of how there is anti-gravity between the galaxies, the clusters of galaxies and the clusters of clusters and, yet, none between the heavenly bodies occupying the galaxies. But, in addressing the issues of what anti-gravity is and why it's found at the level of the galaxies but not at the level of their occupants, does Science give an explanation anywhere near as simple as Esoptrics' is?

2013 ESOPTRICS UPDATE

Esoptrics also gives what I dare propose is a very simple explanation of why the galaxies might be moving apart at *increasing* velocities—namely: It's reasonable to suspect that, if—as every multi-g-combo's categorical leading form repels the categorical leading forms in potency to it—the latter also repel all their kind in the same generic zone, then the galaxies should not only flee from one another but, rather, should do so at a steadily increasing velocity. In addressing the issues of why that acceleration might be increasing, does Science give an explanation anywhere near as simple as Esoptrics' is?

Pondering those questions, be sure to keep in mind what Prof. Greene says in his book ***The Fabric Of The Cosmos***. At the bottom of page 285 and continuing on to 286, he describes the Big Bang as *supposedly* having occurred at "time zero" and thus being "the creation event". On pages 286 & 318, though, he insists inflation's burst "is best thought of" as having occurred in a "preexisting universe" instead of as the creation event. On page 302, he then states we have no convincing answer to the question of whether or not there was a pre-inflation era and what it might have been like. Understandably, on page 337, he then contends "we still have no insight into what happened at the very beginning of the universe." On page 322, he again confesses our lack of knowledge regarding what conditions there might have been in the pre-inflation era.

What, now, should one say of Science's explanations of gravity, anti-gravity, and galaxies fleeing one another with accelerating velocity, when, by Prof. Greene's own admission, Science knows nothing of what preceded that behavior? Perhaps we should say: How can an explanation admittedly beginning with a giant, empty hole be anything but a joke? See, though, how different are Esoptrics' explanations: In great but simple detail, it perfectly ties them in with a thorough and graphic illustration of what God, in the creation of the Universe, did in the pre-inflation era of the 1st thru the 8th epochs (***i.e.:*** according to chart #10-2 on pg. 83, c. 10^{-38}sec. from time zero) before—in creating the 9th epoch's duo-combos—God caused the Universe's diameter to expand from half that of an atom to trillions of light years in 10^{-18} sec. in our manner of measuring.

For all of that, scientists and philosophers, too, almost everywhere turn a profoundly deaf ear to Esoptrics, if, that is, they're not busy scornfully dismissing Esoptrics as "crap" from the brainless head of a muckraking lunatic. Oh, history! If, in your future, Esoptrics proves to be basically correct, what a black eye you shall give to them!

Chapter Fifteen:

ESOPTRICS' 2 KINDS OF INFLATION:

> We will get to the issue of the origin of the universe shortly, but first a few words about the first phase of the expansion. Physicists call it inflation. . . . [Skip! . . . . according to even conservative estimates, during this cosmological inflation, the universe expanded by a factor of 1,000,000,000,000,000, 000,000,000,000,000 [10^{30} – ENH] in .0000000000000000000000000000000 0001 [10^{-35} – ENH] second. It was as if a coin 1 centimeter in diameter suddenly blew up to ten million times the width of the Milky Way. That may seem to violate relativity, which dictates that nothing can move faster than light, but that speed limit does not apply to the expansion of space itself.
>
> The idea that such an episode of inflation might have occurred was first proposed in 1980, based on considerations that go beyond Einstein's theory of general relativity and take into account aspects of quantum theory. Since we don't have a complete quantum theory of gravity, the details are still being worked out, and physicists aren't sure exactly how inflation happened. But according to the theory, the expansion caused by inflation would not be *completely* uniform, as predicted by the traditional big bang picture.
>
> ——**STEPHEN HAWKING & LEONARD MLODINOW:** ***The Grand Design***, pages 129-130 of the paperback edition by Bantam Books Trade Paperbacks, New York, 2012. ©2010 Stephen Hawking & Leonard Mlodinow.

If one has adequately understood chapter 10's bit on creation and chapter 14's about the great acceleration and the reverse categories, then one should probably be able to understand rather readily how Esoptrics' two kinds of inflation fit with the above words from Profs. Hawking and Mlodinow. Esoptrics' first kind—for the first c. 10^{-19} sec.—involved some previously created generators moving at 2^{128} times light speed as the result of being driven out of place—in a way not "completely uniform"—by the most recently created generators. Esoptrics' second kind involves not "the expansion of space itself", but does involve u-spaces envelopes (***i.e.:*** the multi-g-combos) fleeing each other. Let's go over them again.

At the start of the Universe's ninth epoch, God first created the one form which is native to OD2^{256} and which, throughout the 9th epoch, never varies from that OD. As said earlier, that form is called Omega. We can also call it the primary u-spaces envelope, because it offers more u-spaces than can any other form.

One K later, God created 8 duo-combos each with a piggyback form native to OD2^{256}–1 and each with a generator in one of the 8 u-spaces clustered around Omega's center. Remember the 8 building blocks stacked in a cube 2 tiers deep and each tier 2 rows of 2 blocks each? They are the 8 u-spaces contiguous to Omega's center, and each is briefly the microstate of excitation for one of the 8 generators just created in the 8 duo-combos.

At 2K into the ninth epoch, God created 8 duo-combos with piggyback forms native to $OD2^{256}$ –2 and each with a generator in one of the 8 u-spaces clustered around Omega's center. What?! Was not each already hosting a generator? Yes indeed! That, then, necessarily means this: The second set of 8 must force each generator in the first set of 8 to move to one of the u-spaces logically adjacent to the first u-space in which it was created.

To picture what that means, take 64 of the cube shaped building blocks dear to kiddies, and use them to construct a cube 4 tiers deep and each tier 4 rows of 4 blocks each. We've now added 56 u-spaces to the original 8 for a total of 64 (***i.e:*** 4^3). Under the influence of the creation of the second set of 8 duo-combos, the first 8 generators shall be scattered among the 56 u-spaces logically adjacent to the original 8.

Wait a minute! That means the first 8 generators have each moved 1Φ only 1K after being created, and that's 2^{128} (***i.e.:*** $3.4x10^{38}$) times what we've said is light speed for generators engaging in rectilinear locomotion (***i.e.:*** $1\Phi/2^{128}K$). Is that colossal increase in speed going to continue? For c. 10^{-19} sec., yes, but not uniformly! Why so?!

At 3K into the ninth epoch, God created 8 duo-combos with piggyback forms native to $OD2^{256}$–3 and each with a generator in one of the 8 u-spaces clustered around Omega's center. What?! Was not each already housing a generator? Yes indeed! That, then, necessarily means this: The *third* set of 8 must force each generator in the *second* set of 8 to move to one of the 7 u-spaces logically adjacent to the first u-space in which it was created. We now have 16 generators trying to use the 56 logically adjacent to the original 8. Such crowding means one or more but not all of the second set of 8 are bound to collide with one or more of the first set of 8 and, thereby, cause 1 or more but not all of them to again move 1 Φ per K. Some, then, of the Universe's occupants shall now be racing outward at as much as 2^{128} times the speed of light; but, not all shall be doing so.

At 4K into the 9th epoch, God created 8 duo-combos with forms native to $OD2^{256}$–4 and each in the act of using one of the 8 u-spaces around Omega's center—thus displacing the prior set of 8. Again, that must mean some duo-combos moving at 2^{128} times light speed, but not all, since 24 generators in 56 u-spaces = only *some* collisions probable.

At 5K, God created 8 duo-combos with forms native to $OD2^{256}$–5 and their generators filling Omega's 8 central u-spaces. At 6K, God created 8 duo-combos with forms native to $OD2^{256}$–6 and their generators filling Omega's 8 central u-spaces; at 7K, 8 with forms native to $OD2^{256}$–7; at 8K, 8 with forms native to $OD2^{256}$–8. Over and over God does that, until, at $2^{255}K$ (***i.e.:*** $5.7x10^{76}K$ = c. 10^{-19} sec.) into the 9th epoch, God has created 8 duo-combos native, thru their piggyback forms, to each of the OD's of 2^{255} thru 2^{256}–1. In that $2^{255}K$ (***i.e.:*** c. 10^{-19} sec.), the inflation—caused by such repeated filling and emptying of the 8 u-spaces around Omega's center—may have caused some duo-combos to be billions of light years away from Omega's center. If, however, each duo-combo were pushed away from Omega's center uniformly, the result would be 8 "strings" each of 2^{255} duo-combos stretching from Omega's center to a point 2^{255} Φ from that center. Geometry-speak would describe that as 4.5 trillion light years from that center. That's a *possible* result, but not a very *probable* one; and so, Esoptrics says the Big Bang's inflation was probably not uniform (***i.e.:*** not all duo-combos of the same age shall be equidistant from Omega's center).

At 2^{255} K into the 9th epoch, inflation has come to an end *with regard to the primary u-spaces envelope called Omega*. The 2^{258} duo-combos (***i.e.:*** 8 for each of 2^{255} OD's) to which Omega is supplying u-spaces are now traveling at a maximum of light speed =$1\Phi/2^{128}K$.

At 2^{255} +1K, God now shifts to creating duo-combos by means of the 8 duo-combos God created native to $OD2^{255}$. Let's call each of those 8 a Psi. At the center of each of those 8 Psis—and at 2^{255} +1K into the 9th epoch—God creates 8 duo-combos with forms native to $OD2^{255}$–1 for a total of 64 native to $OD2^{255}$–1. At the center of each of those 8 Psis—and at

2^{255} +2K into the 9th epoch—God creates 8 duo-combos each native to OD2^{255}–2 for a total of 64 native to OD2^{255}–2. At the center of each of those 8 Psis—and at 2^{255} +3K into the 9th epoch—God creates 8 duo-combos each native to OD2^{255}–3 for a total of 64 native to OD2^{255}–3. For 2^{254} K, that's the way it goes, until God has created 64 forms native to each of the OD's 2^{254} thru and including 2^{255}–1. For each of the 8 Psis, that's a total of 2^{254} sequential groups-of-8 at 1 per K and the members of each group pushing their predecessors away from the center of a Psi at as much as 2^{128} times light speed. Here again, though, the expansion (***i.e.:*** inflation) won't be uniform, because not all duo-combos of the same age shall be equidistant from the center of its Psi.

The above paragraph describes what we can call the 2nd episode of the first kind of inflation. In it, there are only 2^{254} successive acts of creation relative to each of the 8 Psis, instead of the 2^{255} successive acts which occurred relative to Omega. That's because, thru Omega, God created 8 piggyback forms native to each of the OD's 2^{255} thru and including 2^{256}-1; but, thru each of the 8 Psis, God created 8 piggyback forms native to each of the OD's 2^{254} thru and including 2^{255} -1. Again, though, as in the episode of inflation relative to Omega, the episode of inflation around each of 8 Psis means each newly created set of 8 must shove the previously created set of 8 away from the center of the form thru which creation is being pursued.

In sum then: At (2^{255} + 2^{254})K, there have been two episodes of the first kind of inflation. One was relative to Omega, started at 1K, and ceased at 2^{255}K into the ninth epoch. The other was relative to each of the 8 Psis, started at 2^{255} +1K, and ceased at (2^{255} + 2^{254})K into the ninth epoch. Now commences a third episode.

At (2^{255} + 2^{254})K into the 9th epoch, God has just created 64 duo-combos (***i.e.:*** 8 per each of the 8 Psis) each of which contains a piggyback form native to OD2^{254}. Call each of those OD2^{254} forms a Chi. At the center of each of those 64 Chis—and at 1+(2^{255} + 2^{254})K into the 9th epoch—God creates 8 duo-combos each with a form native to OD2^{254}–1 and their generators filling a Chi's 8 central u-spaces. At the center of each of those 64 Chis—and at 2+(2^{255} + 2^{254})K into the 9th epoch—God creates 8 duo-combos each with a form native to OD2^{254}–2 and their generators. That fills each Chi's 8 central u-spaces, and causes the prior 8 to move at 1Φ/1K. At the center of each of those 64 Chis—and at 3+(2^{255} + 2^{254})K into the 9th epoch—God creates 8 duo-combos each for a form native to OD2^{254}–3 and their generators. Again, that fills each Chi's 8 central u-spaces, thus causing some of the members of the 2 prior sets of 8 to move at many times light speed.

On it goes just as before in relation to Omega, except that, this go around, there are only 2^{253} successive acts of creation relative to each of the 64 Chis, instead of the 2^{255} successive acts which occurred relative to Omega and the 2^{254} successive acts which occurred relative to each of the 8 Psis.

If you've followed that, you perhaps see this: For Esoptrics, the first kind of inflation lasts for the first 2^{256} K (***i.e.:*** 1.1579×10^{77} K = 8.339102×10^{-19} sec.) into the 9th Epoch and involves 256 episodes. The first lasts 2^{255} K and is 2^{255} successive waves at intervals of 1K and each consisting of 8 duo-combos issuing from Omega's 8 central u-spaces. The second lasts 2^{254} K and is 2^{254} successive waves at intervals of 1K and each consisting of 8 groups of 8 duo-combos each and each group of 8 issuing from the 8 central u-spaces of 1 of the 8 forms native to OD2^{254}. The third lasts 2^{253} K and is 2^{253} successive waves at intervals of 1K and each consisting of 8^2 groups of 8 duo-combos each and each group of 8 issuing from the 8 central u-spaces of 1 of the 64 forms native to OD2^{253}. The fourth lasts 2^{252} K and is 2^{252} waves at intervals of 1K and each consisting of 8^3 groups of 8 duo-combos each and each group of 8 issuing from the 8 central u-spaces of 1 of the 512 forms native to OD2^{252}.

Hopefully, that's enough to enable you to figure out the other 252 episodes for yourself. That, then, is what the 256 episodes of the first kind of inflation mean for Esoptrics.

Let's now turn to that second kind in which—to use the erroneous phrase given us by Profs. Hawking & Mlodinow (***i.e.:*** H&M for short)—"space itself" expands as a result of the great acceleration and the production of the reverse categories. Maybe the drawing below can allow us to describe that second kind in graphic detail.

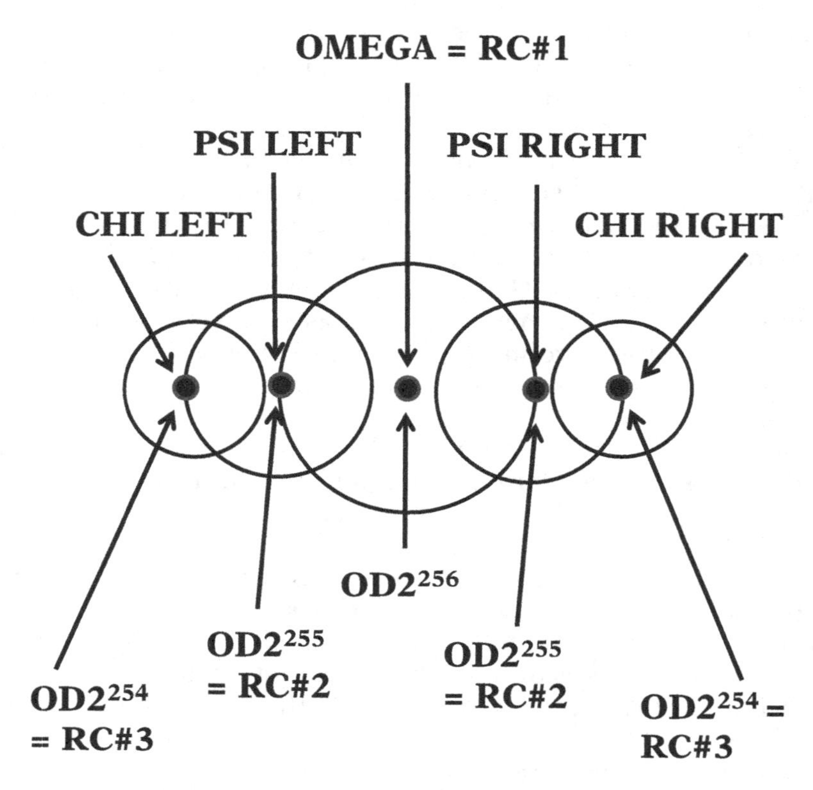

DRAWING #15-1: HOW "SPACE ITSELF" EXPANDS

First of all, bear in mind that the above is merely a figurative expression and an exceedingly truncated one at that. There should be 8 Psis moving away from Omega's center in a three-dimensional pattern and 8 Chis moving away from the center of each of those 8 Psis. Who in his right mind would imagine I can illustrate that? Furthermore, instead of 2 as in the above drawing, there should be 7 circles to each side of Omega and diminishing in size with each step away from Omega's center. Only then would RC#4 thru RC#8 be shown.

The term "space itself" is a misnomer and a very misleading one at that. Those who use it indicate they have no idea regarding what space really is. In the above, instead of seeing "space itself" expand, we're seeing the *u-spaces envelopes* of 4 multi-g-combos expanding away from one another. Let's elaborate on that so:

Psi left and Psi right are each a multi-g-combo with a leading MT form currently at somewhere between $OD2^{255}$ +1 and $OD2^{256}$ -1. As such, each is a u-spaces envelope offering at least $(2x2^{255})^3$ u-spaces to any generator which might be in potency to it. Omega, also, is a u-spaces envelope and is one offering $(2x2^{256})^3$ u-spaces to each Psi's generator. Using Omega's u-spaces, each of those carrying generators is moving its Psi away from Omega's center at a steadily increasing velocity. As each generator does that, the u-spaces envelopes offered by Psi left and Psi right are *themselves* moving away from Omega's center and from one another; and so, the 2 *u-spaces envelopes themselves*—as opposed to H&M's "space itself"—are moving outward but without their u-spaces *expanding*. To be sure, an increase in its generator's rectilinear velocity shall cause a Psi to move to a higher current OD and, thus, expand in dia. by upping the number of u-spaces it's offering to a Chi's generator. But, don't misinterpret that: The increase only adds u-spaces *to the outer perimeter of the envelope*. It doesn't cause the previously present ones to expand or to move away from their envelope's center. What u-spaces were there previously still have the same "size" and distance from their envelope's center they had before. Only the envelope is moving away from its fellow u-spaces envelopes and, if accelerating, expanding in diameter.

Let's imagine a multi-h-combo arrives on the scene. Call its carrying generator Jay Left. Jay Left is using Psi Left's u-spaces as Jay Left, by accelerating, moves to the left away from the center of Psi Left. Jay Left is moving at .5C by switching from one u-space to one next "in line to the left" every 2^{129}K. That's how fast Jay Left is moving *relative to Psi Left's center*. How fast is he moving relative to *Omega's* center?

It depends upon how fast Psi Left, as a u-spaces envelope, is moving away from Omega's center. Psi Left, thru his generator, may well be racing away at light speed. If that's true, then Jay Left is moving away from Omega's center at 1.5C. Is he, then, exceeding light speed? By no means! He himself is moving at only .5C. His additional speed is not his doing; it's the doing of the u-spaces he's using, and, as H&M told us on page 121, light speed's limit "does not apply to the expansion of space itself." Naturally, Esoptrics changes that to: Light speed's limit does not apply *to the u-spaces themselves*.

Let's make the scenario even more interesting by introducing a second multi-h-combo. Call its carrying generator Jay Right. Jay Right is using Psi Right's u-spaces as Jay Right, by accelerating, moves to the right away from the center of Psi Right. Jay Right is moving at .5C by switching from one u-space to one next "in line to the right" every 2^{129}K. That's how fast Jay Right is moving *relative to Psi Right's center*. How fast is he moving *relative to Jay Left*? Like Jay Left, Jay Right is moving away from Omega's center at 1.5C; but, they are doing that in opposite directions; and so, their velocity relative to one another is 3C, despite the fact neither one of them exceeding 2^{128}K/Φ.

Now, let's really up the ante. Drawing #15-1 depicts only 3 of the 8 reverse categories. Let's add in the other 5. To do that, return to drawing #15-1, and imagine this: To the left of Chi Left, add in 5 circles we'll call Mu Left, Nu Left, Roe Left, Tau Left, and Lambda Left. To the right of Chi right, add in 5 circles we'll call Mu Right, Nu Right, etc. thru Lambda Right. The carrying generator of each RC is moving at light speed, 2^{128}K/Φ. All the RC's to the left are moving left at light speed, and all the RC's to the right are moving right at light speed. What, though, is the velocity of Lambda Left relative to Lambda Right? It's 14C, because the velocity of each relative to Omega is 7C.

In potency to Lambda Left, imagine a multi-h-combo's generator fleeing Lambda Left's center at .5C to the left. Its name is Kay Left. In potency to Lambda Right, imagine a multi-h-combo's generator fleeing Lambda Right's center at .5C to the right. Its name is Kay Right. What's Kay Right's velocity relative to Kay left? It's 15C because the velocity of each relative to Omega is 7.5C in opposite directions.

For Esoptrics, the upshot of it all is this: There is no such thing as "space itself" expanding; but, the logical diameter of the Universe ***is***, because the RC combos, as u-spaces envelopes, are fleeing one another in compliance with the 2 laws of categorical preservation. Since each may be doing that at light speed—or at a speed steadily climbing to light speed—some may be fleeing Omega's center at several times light speed thereby making the Universe's diameter expand at several times light speed.[1] For all of that, after the first 10^{-18} seconds into creation, no generator ever changes u-spaces faster than 2^{128}K/Φ.

---

[1] On page 523 of ***The Fabric Of The Cosmos***, Prof. Greene describes space's regions as fleeing each other at greater than light speed. He thus agrees with what Esoptrics says above.

Such, then, is what Esoptrics means by the second kind of inflation. To the scientific community (and probably the philosophical one too), it shall mean nothing save, perhaps, a reason to laugh themselves half to death at the thought of how great a fool must be the man idiotic enough to come up with such a "merely silly" brand of Theoretical Physics. I dare to suggest that those who do so are able to do so only because, on the one hand, they know nothing about what space *really* is and, on the other hand, know nothing of the *real* reason why the speed of light is the same throughout all inertial systems. In turn, that latter ignorance of theirs stems from the fact that they know nothing about what *really* distinguishes one inertial system from another—a topic handled in the next chapter.

✡✞✡✞✡✞✡✞✡✞✡✞✡✞✡✞✡✞✡✞✡✞✡✞✡

> Thus, the origin of the outward motion is *not* an explosion that took place within space. Instead, the outward motion arises from the relentless outward swelling of space itself. (pg. 231)
>
> . . . .
>
> By attributing the observed motion of galaxies to the swelling of space, general relativity provides an explanation that not only treats all locations in space symmetrically, but also accounts for all of Hubble's data in one fell swoop. It is this kind of explanation, one that elegantly steps outside the box (in this case, one that actually *uses* the "box"—space that is) to explain observations with quantitative precision and artful symmetry, that physicists describe as almost being too beautiful to be wrong. There is essentially universal agreement that the fabric of the [Sic!] space is stretching.
>
> ——**BRIAN GREENE:** ***The Fabric Of The Cosmos***, pages 231 & 233 of the 2005 paperback edition by Vintage Books, a division of Random House, Inc., New York. ©2004 by Brian Greene (This passage from Prof. Green is a very revealing example of just how "universal" is the physicists' total ignorance of what space *really* is. There is no such thing as "space itself", because space is not a single reality but, rather, is a vast number of envelopes each made up of a definite number of microstates of excitation. Therefore, there can't possibly be any such thing as "space itself" either "swelling" or "stretching". In its place, there are only u-spaces [***i.e.:*** microstates] envelopes which—where they are the ones provided by multi-***g***-combos—are racing away from one another because of their multi-g-combos' compliance with the first and second laws of the preservation of the categories. Here, again, we have an example of scientists offering an explanation—in this case, regarding the way space behaves—when, by their own admission, their explanation starts with a giant, empty hole. See pages 6, 335, 471, 486, and 493 of his book where Prof. Greene, one way or another, admits scientists have not the foggiest notion regarding what the "ultramicroscopic makeup" of time and space might be. When, in one breath, scientists openly confess such ignorance and then, in the very next, tell us what "space itself" is doing, how is that not merely a clown act?).

Chapter Sixteen:

ESOPTRICS VS. NEWTON & EINSTEIN REGARDING ACCELERATION, TIME, SPACE & SIMULTANEITY:

I should start this chapter with an attempt to make sure I've not left my reader in the dark regarding an important point. It's what happens as a result of being in the habit of unconsciously assuming others see in one's statements the same implications one sees without listing them. And who is not at least sometimes given to such a bad habit?

In distinguishing between rectilinear and curvilinear locomotion, I do not thereby mean to leave out the distinction between curvilinear locomotion *strictly speaking* and rectilinear motion in a circle. What is that distinction?

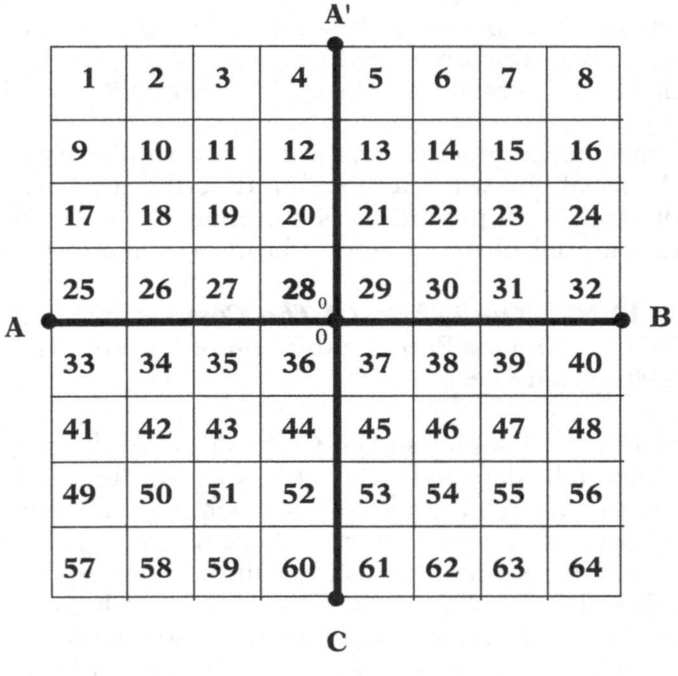

A'

1	2	3	4	5	6	7	8
9	10	11	12	13	14	15	16
17	18	19	20	21	22	23	24
25	26	27	28_0	29	30	31	32
33	34	35	^{0}36	37	38	39	40
41	42	43	44	45	46	47	48
49	50	51	52	53	54	55	56
57	58	59	60	61	62	63	64

A — B

C

DRAWING #16-1: CURVILINEAR VS. CIRCULAR RECTILINEAR MOTION

Imagine a generator is at "square" #25. After some number of K, it's suddenly at 26, then 27, then 28, 29, 30, 31, 32. That's rectilinear motion in a straight line. Imagine that, after some number of K at #25, its suddenly at 33, then 34, then 26, then 25, then 33, then 34, then 26, then 25, and so on over and over again. That is rectilinear motion in a circle—an oddly shaped one perhaps but circular nevertheless. The same would be true of the sequence 25, 33, 41, 42, 43, 35, 27, 26, 25. Despite that fact, neither of those sequences is an instance of curvilinear motion strictly speaking. What, then, is the latter?

Imagine a generator is at "square" #1 (***i.e.:*** is performing that state of excitation). While it's continuously there, the entire overall "square" rotates 360° an astronomical number of times. That is an instance of curvilinear locomotion *strictly speaking*. In case you don't readily see it, the distinction here is this: In circular rectilinear locomotion, the generator is the *agent* of its motion; in curvilinear locomotion *strictly speaking*, the generator is the *patient* of its motion. In circular rectilinear locomotion, the generator moves in a quasi-circle *by changing thru a series of u-spaces*; in curvilinear locomotion strictly speaking, the generator moves in a circle *without changing from one u-space to another*. It can do that, because it's solely the u-spaces themselves which—being supplied by a form rotating around its own center—are circling their providing form's center.

Perhaps the most important difference between curvilinear and rectilinear motion in a circle is the speed limit. According to Esoptrics, the famous light speed limit imposed by Einstein's Special Relativity applies only to rectilinear velocity, which is to say it applies only where generators—after the Universe's first 10^{-18} sec.—are changing from u-space to logically contiguous u-space; it does not apply to curvilinear motion, which is to say it does not apply to the speed at which the u-spaces themselves are carried around the center of the rotating form producing them. What, then, is that speed limit? In the case of rational forms rotating in the primary and secondary planes, that limit is an astronomically huge multiple of light speed. As we shall eventually see, that's what accounts for the chaos Quantum Mechanics detects in the sub-atomic world.

To get the clearest grasp of what I've just said, contrast it with this:

> But, quantum mechanics practitioners argue, probability waves are *not* like water waves. A probability wave, although it describes matter, is not a material thing itself. And, such practitioners continue, the speed-of-light barrier applies only to material objects, things whose motion can be directly seen, felt, detected.
>
> ——**BRIAN GREENE: *The Fabric Of The Cosmos***, page 503 of the 2005 paperback edition by Vintage Books, a division of Random House, Inc., New York, 2005. ©2004 Brian Greene.

That's not exactly what Esoptrics says. Generators are the only "things whose motion can be *directly*, seen, felt, detected." Still, the speed-of-light barrier applies to them only when they are engaged in *rectilinear* motion, which is to say only when they are shifting from u-space to u-space. It does not apply to them when they are stationary in moving u-spaces traveling in a curvilinear manner as a result of being supplied by a form rotating around its center. As Quantum Mechanics has discovered, such curvilinear motion on the part of the generators can be detected. What Quantum Mechanics has not discovered is that, in detecting generators seeming to break the speed-of-light barrier, what they are actually doing is *in*directly detecting the *curvilinear* motion of the *forms* supplying the generators with the u-spaces which make *rectilinear* motion possible to generators.

If all of that is sufficiently clear, let's move on to the issue of acceleration.

The nature of acceleration seems to be a formidable problem for many, if not all, members of the scientific community from Newton to Einstein to today's scientists. To illustrate that, consider these words:

The procedure just described overcomes a deficiency in the foundations of mechanics which had already been noticed by Newton and was criticized by Leibnitz[1] and, two centuries later, by Mach: inertia resists acceleration, but acceleration relative to what? Within the frame of classical mechanics the only answer is: inertia resists acceleration *relative to space*. This is a physical property of space—space acts on objects, but objects do not act on space. Such is probably the deeper meaning of Newton's assertion *spatium est absolutum* (space is absolute). But the idea disturbed some, in particular Leibnitz, who did not ascribe an independent existence to space but considered it merely a property of "things" (contiguity of physical objects). Had his justified doubts won out at that time, it hardly would have been a boon to physics, for the empirical and theoretical foundations necessary to follow up his idea were not available in the seventeenth century.
——**ALBERT EINSTEIN:** ***Ideas And Opinions***, page 348 of the hardbound book by Bonanza Books, New York, 1954. ©1954 Crown Publishers (Einstein died April 18, 1955 – ENH).

It is characteristic of Newtonian physics that it has to ascribe independent and real existence to space and time as well as to matter, for in Newton's law of motion the concept of acceleration appears. But in this theory, acceleration can only denote "acceleration with respect to space." Newton's space must thus be thought of as "at rest," or at least as "unaccelerated," in order that one can consider the acceleration, which appears in the law of motion, as being a magnitude with any meaning. Much the same holds with time, which of course likewise enters into the concept of acceleration.
——**ALBERT EINSTEIN** page 360 of the hardbound book quoted immediately above this quote.

And so, with these developments we learn that *geometrical shapes of trajectories in spacetime provide the absolute standard that determines whether something is accelerating*. Spacetime, not space alone, provides the benchmark. (pg. 61)
. . . .
According to general relativity, the benchmarks for all motion, and accelerated motion in particular, are freely falling observers—observers who have fully given in to gravity and are being acted on by no other forces. Now, a key point is that the gravitational force to which a freely falling observer acquiesces arises from all the matter (and energy) spread throughout the cosmos. (pg. 73) . . . . Thus, in general relativity, when an object is said to be accelerating, it means the object is accelerating with respect to a benchmark determined by matter spread throughout the universe. (pg. 74)
——**BRIAN GREENE:** ***The Fabric Of The Cosmos***, pages 61, 73 & 74 of the 2005 paperback edition by Vintage Books, a division of Random House, Inc., New York. ©2004 Brian Greene.

Esoptrics agrees to a major extent with the venerable Prof. Einstein's view that space and time together constitute the "benchmark" and—in Prof. Greene's words—*"the*

---

[1] Yes, that's how he spells it, and that's how I spelled it for years and, for that reason, still prefer to spell it. Oh, and how it aggravated me when most switched to Leibniz instead Leibnitz!

absolute standard that determines whether something is accelerating." It does not by any means *fully* agree, and that's for several reasons.

First of all, what determines whether or not a *carrying generator* is accelerating is not the same as what determines whether or not a *piggyback form* is accelerating. Secondly, as far as Einstein's notions of time and space are from Newton's, that's how far Esoptrics' notions of time and space are from Einstein's, if not further. Thirdly, Esoptrics does not to any extent agree that acceleration is "with respect to a benchmark determined by matter spread throughout the universe." I will elaborate on the first two qualifications in the order given. I'll not bother to elaborate on the third. It's not worth it.

In every generator, acceleration *increases* its frequency of change from u-space to logically contiguous u-space and, so, *decreases* the number of successive units of u-time, K, spent per u-space. *Decreasing* K per a generator's stay in a u-space then *increases* the frequency with which K affects it and, so, causes its i-time's frequency to accelerate (***i.e.:*** in the ratio u/i, u's value diminishes). Thus, how it moves thru both time and space is part of what determines if a generator is accelerating. Since scientists know nothing about either: (1) u-time and how it relates to i-time, or (2) u-spaces and their ability to be logically outside of one another in logical sequence—such an explanation shall likely be meaningless to them. May God grant further elaboration shall remedy that to some extent.

In every form, acceleration involves 2 factors: (1) an increase in its current OD, and (2) by *increasing* the *ratio* of successive u-times per unit of the form's i-time: (a) an *increase* in the number of successive units of u-time spent in each of the two phases of its every cycle, and (b) a decrease in the frequency with which u-time affects it (***i.e.:*** in the ratio u/i, u's value *increases*). Here again, then, time is part of what says whether or not a form is accelerating; but, movement thru space is not. In place of *space*, the other determining factor here is change in *ontological distance*. Yes, change in OD is movement thru space as movement thru space's 4th dimension; but, it's not such as change from one u-space to a logically contiguous one. No matter how great the form's acceleration in OD, that movement thru the 4th dimension is still contained within the c. 10^{-47} cm. of a single u-space.

In every form, acceleration does involve contraction in the apparent "size" of the u-spaces. That's because, as a form's current OD increases, so does the dia. of its actuality's u-spaces envelope. As u-spaces envelopes, a first form at OD2^{155} has a dia. of $2^{156}\Phi$, and, relative to it, a second form at OD2^{128}—with a dia. of $2^{129}\Phi$—has a dia. 2^{-27} smaller. If the first form accelerates to 2^{156}, its dia. is then $2^{157}\Phi$, and, relative to it, the second form now has a dia. 2^{-28} smaller. For the first form, that means the dia. of the MT has contracted 50%, and so has the dia. of every u-space, since the dia. of the latter is 2^{-129} times that of the MT. I have no idea to what extent Esoptrics' kind of spatial contraction matches Einstein's.

Note the contrast between the2 kinds of acceleration and the connection between them from time's standpoint: For accelerating *generators*, i-time speeds up as they spend less K in stasis; for accelerating *forms*, i-time slows down as they spend more K in stasis, and forms are far more important to us than generators. For, though generators provide us with objects and—in the units of tension they generate—give us sense images of those objects, it's forms which enable us to feel those objects and sense images. Sensory consciousness is a power limited to forms and to no extent possible to generators. That's why Einstein was correct about how clocks work for us. For, we, *as forms*, experience in each the result of it somehow having leading forms at the same OD at which ours reside.[1]

[1] The above is an oversimplification for economy's sake. Piggyback forms are limited to mono-directional empirical observation's fixation on sensations; and so, no piggyback form can be a *human* (as opposed to vegetative or animal) soul, which is to say that kind of immaterial reality able to be aware of abstract concepts because it's capable of bi-directional empirical observation (a power sci-

CHAPTER SIXTEEN

How does it come about that time slows for the forms as it speeds up for the generators? What's the connection? Remember what formula #11-3 told us on page 92: Every generator's velocity is determined by its piggyback form's current OD, and, for that velocity to increase, the generator's piggyback form must increase its current OD and, so, its K per act. The two kinds of acceleration are thus inseparable. You can't have one without the other. As I just said, though, what matters for us is what happens from the standpoint of the forms; and so, that's what even scientists fixate on and, thereby, miss the other half of what's involved—namely: Though acceleration causes i-time to slow down for the forms, it causes i-time to speed up for the generators actually producing rectilinear locomotion. That's understandable when you realize they know nothing about forms vs. generators and how they go together and can't know anything about such matters as long as they remain ignorant of God's 7 *a priori* coordinates.[1]

How is it possible for time to accelerate and decelerate? As far as I am aware, scientists have not yet given us even so little as the slightest trace of how that might be possible. That's understandable, since even they themselves readily admit they have no idea what either time or space *really* are.

> Space and time capture the imagination like no other scientific subject. For good reason. They form the arena of reality, the very fabric of the cosmos. Our entire existence—everything we do, think, and experience—takes place in some region of space during some interval of time. Yet science is still struggling to understand what space and time actually are.
> ——**BRIAN GREENE: *The Fabric Of The Cosmos***, page IX of the 2005 paperback edition by Vintage Books, a division of Random House, Inc., New York. ©2004 Brian Greene.

> If one takes the positivist position, as I do, one cannot say what time actually is. All one can do is describe what has been found to be a very good mathematical model for time and say what predictions it makes.

---

entists seem never to use.). Such "*reflection* capable" consciousness empowers the human soul—in a kind of mystifying rebound action—to introspect upon the acts of will whereby it focuses its power of attention outwardly or inwardly and, thereby, to define directly and/or reflexively observed targets in the terms of the acts of will whereby it focused its power of attention upon them. The "stuff" of abstract concepts is memory of which acts of will focused the power of attention on which target. What reality has such power? As Catholic Doctrine explains: It's a reality created by God at the instant of conception and added to—as Esoptrics describes it—the human body's leading piggyback form. Esoptrics then speculates that this "at–conception–created" reality is either a heavenly benchmark or heavenly sextet pertaining solely to the person to whom God joins it. Every human is thus the combination of a *heavenly* form with a leading *piggyback* form plus the latter's subordinate piggyback forms and all the multi-A-combos "corralled" by the latter. The number of the latter depends upon how many generic zones lie between MT and the OD of the body's leading form.

[1] Despite how obnoxious I'm being on this point, I repeat: *Un*aware of the coordinates which precede time and space, scientists deal with those coordinates suggested to them by what they observe of time and space. *Aware* of the coordinates which precede time and space, Esoptrics deals with the kinds of time and space suggested by those *a priori* coordinates. It's at the very start of it all, then, that the vast difference between Esoptrics and all other cosmological theories originates.

> . . . .the least initial departure from the truth later multiplies a thousand times over.
> ——**ARISTOTLE: *On The Heavens***, Book 1: Chap. 5: 271^b 9.

——**STEPHEN HAWKING:** ***The Universe In A Nutshell***, page 31 of the hardbound edition by Bantam Books, New York, 2001. ©2001 Stephen Hawking.

In the theory of relativity there is no unique absolute time, but instead each individual has his own personal measure of time that depends on where he is and how he is moving. (pg. 34)

. . . .

The point is that the theory of relativity says that there is no unique measure of time that all observers will agree on. Rather, each observer has his or her own measure of time. (pg. 162)

——**STEPHEN HAWKING:** ***A Brief History Of Time***, pages 34 & 162 of the 10th anniversary paperback edition by Bantam Books, New York, 1998. ©1998 Stephen Hawking.

Esoptrics, however, claims it can tell you exactly what time is and, doing so, can present everyone everywhere in the Universe with a unique, absolute time which all will define the same, identical way, if they are rightly informed individuals, which is to say: if they are well aware of the logic of the mirror and God's 7 *a priori* coordinates. That time, Esoptrics calls u-time. In addition to being ultimate time, it is universal time because all rightly informed individuals will define it the same way. Why and how so?!

All rightly informed individuals will admit that the absolute duration of this the 9th epoch of the Universe is 2^{385} (***i.e.:*** 7.88×10^{115}) regularly occurring, consecutive instances of u-time. Therefore, let U_T = the absolute duration of this the 9th epoch of the universe, and—for all rightly informed individual always and everywhere—the absolutely invariable and universal definition of u-time *relative to the duration of this 9th epoch of the Universe* is:

That which occurs in a durationless instant every $2^{-385}U_T$.

Therefore, for all rightly informed individuals, u-time relentlessly marches on at the pace of one step every $2^{-385}U_T$. As a result, whatever happens in this 9th epoch starts and, if it ends, ends at a definite number of K into the 9th epoch. That means every rightly informed individual can figure out what is or is not simultaneous, and thereby perishes one of Relativity's most cherished assertions.

. . . . Clocks that are moving relative to each other fall out of synchronization and therefore give different notions of simultaneity. (pg. 55)

. . . .

. . . . Again, although it's hard to accept at a gut level, there is no paradox here: *observers in relative motion do not agree on simultaneity—they do not agree on what things happen at the same time.* (pg. 57)

——**BRIAN GREENE:** ***The Fabric Of The Cosmos***, pages 55 & 57 of the 2005 paperback edition by Vintage Books, a division of Random House, Inc., New York. ©2004 Brian Greene.

Yes, individuals limited to *looking at clocks* "do not agree on simultaneity"; but, rightly informed ones able *to calculate in the terms of the logic of the mirror and God's 7 a priori coordinates* will agree on *"what things happen at the same time"*, because they'll be able to calculate perfectly well the difference between u-time and their own i-time.

I said all rightly informed individuals will define u-time in the same identical way when defining it *relative to the duration of this 9th epoch of the Universe.* For all of that, they will not all define u-time the same identical way *relative to themselves*. Why is that?

As many a philosopher said for thousands of years: Time is the measure of *change*. At least some scientists, too, admit that:

> No one has as yet found the definitive, fundamental definition of time, but, undoubtedly, part of time's role in the makeup of the cosmos is that it is the bookkeeper of change. . . .
>
> . . . . things in the universe must *change* from moment to moment for us even to define a notion of *moment to moment* that bears any resemblance to our intuitive conception. If there were perfect symmetry between how things are now and how they were then, if the change from moment to moment were of no more consequence than the change from rotating a cue ball, time as we normally conceive it wouldn't exist.
>
> ——**BRIAN GREENE: *The Fabric Of The Cosmos*,** pages 225 & 226 of the 2005 paperback edition by Vintage Books, a division of Random House, Inc., New York. ©2004 Brian Greene.

If time is the measure of change, then—as Prof. Greene admits—it is not, *strictly speaking,* the measure of duration in an *internally changeless* state of excitation. For Esoptrics, then—since every generator's & form's every state of excitation is internally changeless for as long as it lasts—the history of this the 9th epoch of the Universe is written in the terms of two radically different and alternating kinds of moments—namely: absolutely durationless instants of time, and timeless instants of *undetectable* duration. Notice I said "*undetectable*" and did not say "*incalculable*". What's the distinction?

Where time is the measure of change, there is—*for any given combo*—no such thing as a moment of time *for it,* save in the durationless instant in which either: (1) a generator changes from one temporarily changeless microstate of excitation to another (***i.e.:*** from one u-space to a logically contiguous one), or (2) where a form changes from one macrostate of excitation to another. As long as generators or forms are engaged in a *changeless* act of excitation, they undergo no kind of internal change whatsoever; they do not age; they do not pass from what is now past to what was previously future. The march of u-time thus has no effect upon them whatsoever, does not affect them in any way whatsoever, and does not register with them in any way whatsoever. For that reason: As long as a given generator or form is performing a temporarily changeless state of excitation, the duration of that state—*from the standpoint of that given generator or form's internality*—is *effectively* zero i-time no matter what its duration is *from the standpoint of u-time*.

To every sensing form, consciousness seems a continuous, unbroken flow of change from past to future despite the fact that its *continuous existence* is very far from being a *continuous flow of change.*[1] Next, you don't have to be a genius to figure out that, if the

---

[1] Naturally, that will prompt some to snarl: "If that's so, consciousness is a maximally deceitful thing proving that, if it's divinity's creation, that divine author is more deceitful than any finite mind can describe." That would be true were it not for the fact that he who writes here knew even before the sixth year of grammar school that time, space, change, and consciousness cannot possibly be continuous. Ask my fellow grammar school students how vociferous I was about it. Furthermore, look around, and you'll find I am far from being the only one to have done that either in this or any other century. If, then, there are those deceived by consciousness, it's not because God is vicious but, rather, because they are intellectually dishonest enough to *want* to be deceived. Why?! This is not the place to discuss that, save to say this: Instinctively, *competition addicts* are utterly intolerant of any level of knowledge sure to eliminate every environment in which competition can occur and leave

experienced duration of a *changeless* state of excitation is zero i-time, then it is such no matter what the duration of that changeless state of excitation might be *from the standpoint of u-time as change-measuring time*. As a result, as an increase in the OD of one's form produces—from the standpoint of u-time—an increase in the duration of that form's *changeless* state of excitation and, thereby, slows one's progress thru change, one has not the slightest *sensible, observable, detectable, or manifest* indication of either: (1) the fact that one's awareness of change-measuring time has slowed, or (2) why it has slowed.

That's what *undetectable* duration and i-time mean. That, though, has nothing to do with *calculable* duration. A rightly informed individual—knowing how to think in the terms of the logic of the mirror and God's 7 *a priori* coordinates—knows: The *calculable* duration of his form's every act in K (***i.e.:*** in number of successive instants of u-time) = the square of his form's current OD. Knowing that, he automatically know that—from the standpoint of his form (as opposed to from the standpoint of U_T)—the frequency of u-time varies as the inverse of the square of his form's current OD. He also knows that—if he sets up an arbitrary unit of i-time such as a second—then the duration of that unit is a definite number of acts of consciousness, which is to say a definite number of acts (***i.e.:*** changeless states of excitation) on the part of his form. The result of that arbitrary unit of i-time is this: For his form, the frequency of u-time varies as the inverse of the square of his form's current OD divided by the duration of his second in number of acts of consciousness.[1]

Is that too complex for you? In that case, here's an example of how, for our planet's human occupants, the frequency of u-time (***i.e.:*** the alphakronons a/k/a K) is almost exactly once every 7.201789375 x 10^{-96} sec.. I hope you love large numbers as much as I do.

An irrational form—probably one native to the irrational OD2^{127}—accelerates to a much higher OD. Because it's irrational, it does that without producing a multi-combo, which is to say it causes no duo-combos to become *concentric* with it; but, because it's generic, it does produce what we might call a compound multi-combo, which is to say one in which many multi-combos work in concert with one another. We'll say more about that when discussing molecules. The much higher OD to which it accelerates is:

58,918,244,245,183,424,346,609,661,111,934,869,931,738,863,090 =
5.8918244245183 x 10^{46}.

That's an irrational OD between 2^{155} and 2^{156}. Square that OD, and the result is:

3,471,359,504,935,089,700,933,088,746,630,243,083,438,841,885,485,525,027,314,456,378,
935,922,012,513,479,412,245,764,348,100 = 3.471359504935089700933 x 10^{93}

Divide the number 1 by the above, and the result is:

.000,000,000,000,000,000,000,000,000,000,000,000,000,000,000,000,000,000,000,000,000,
000,000,000,000,000,000,000,000,000,000,288071575006374573723772217199464 =
2.8807157500637457 x 10^{-94}.

---

some victors and other losers, unless, of course, one is the ultimate super-competitor super-hungry for the kind of cruel victory which forever destroys for all others any chance to outdo him.

[1] For irrational forms, the formula for determining when instants of u-time occur appeals to the inverse of the square of the irrational form's current OD. For rational forms, the formula must appeal to the inverse of the rational form's current OD times its native OD. Since the former are more important to us and outnumber the latter 2^{128} to 1, the latter are of less interest to us, save when explaining why Quantum Mechanics detects what it does.

That tells us how often a u-time occurs relative to how often there's a change of act on the part of the above form whose current OD squared is abbreviated as:

$$3.47135950493508970093 \times 10^{93}$$

That then raises the question of how many acts such a form performs per second. Esoptrics say 20 cycles of 2 phases each for a total of 40 acts of consciousness' form per second. I'll explain later why Esoptrics says that. Multiply by 40, then, the number given above in full and abbreviated as: $3.47135950493508970093 \times 10^{93}$. The result is:

138,854,380,197,403,588,037,323,549,865,209,723,337,553,675,419,421,001,092,578,255, 157,436,880,500,539,176,489,830,573,924,000 = $1.38854380197403588 \times 10^{95}$.

Divide the number 1 by that, and the result is:

.000,000,000,000,000,000,000,000,000,000,000,000,000,000,000,000,000,000,000,000,000, 000,000,000,000,000,000,000,000,000,000,007,201789375159364343094305429986. = $7.2017893751593643431 \times 10^{-96}$.

That is the *frequency* of the alphakronons, K, *for us*. It tells us that, *for us*, a durationless instant of u-time inexorably occurs once every 7.2×10^{-96} of our seconds. It tells us u-time is the result and indication of how often forms at OD1 always change from one changeless macrostate of excitation to another. It tells us that's always and everywhere the fastest rate of change ever to be found anywhere in this or any other universe. Because that's so, that rate of change can be used to calculate and state what—relative to u-time's rate of change—is the frequency of change for any and every generator or form changing at a lesser rate. U-time is thus the result and measure of the rate of change at OD1, and i-time the result and measure of the rate of change at all OD's greater than 1. I-time is, therefore, calculated and defined in the terms of the ratio, u/i, of the number of consecutive instants of u-time per each instant of this-or-that generator's or form's individual time.

U-time can also tell us the *duration* of every act by any generator or form not changing at u-time's rate. Doing so, though, it tells us only what the duration of those others is *relative to the duration of acts at OD1*. It cannot tell us their *absolute* duration because it cannot tell us the absolute duration of acts at OD1. As for duration, one can only argue in a circle so: "The duration of acts by forms at OD10 is 100 times that of acts at OD1 (***i.e.:*** u/i =100/1), and the duration of acts at OD1 is a hundredth that of acts at OD10 (***i.e.:*** i/u =1/100)." Only God and infinity's other occupants can tell you the duration of acts at OD1 without relating it to some other duration itself defined in the terms of duration at OD1. For example the duration of an act at OD1 is 2^{-385} that of the Universe's 9th epoch; but, that latter duration is known only as 2^{385} times the duration of each act at OD1. In sum, then, duration, i-time and u-time's *definition* are relative to the observer; but, u-time itself is absolutely and forever 1K per act at OD1 and 2^{385} K per cycle of the Universe's 9th epoch.

Turn now to what, these days, some scientists are saying about time as a purely subjective phenomenon with no counterpart in the real world outside the human mind.

> Just as we envision all of space as *really* being out there, as *really* existing, we should also envision all of time as *really* being out there, as *really* existing, too. Past, present, and future certainly appear to be distinct entities. But, as Einstein once said, "For we convinced physicists, the distinction between past, present, and future is only an illusion, however persistent." The only

> thing that's real is the whole of spacetime.
>
> . . . .
>
> In this way of thinking, events, regardless of when they happen from any particular perspective, just *are*. They all exist. They eternally occupy their particular point in spacetime. There is no flow. . . . [Skip] . . . . if we stare intently at this familiar temporal scheme and confront it with the cold hard facts of modern physics, its only place of refuge seems to lie within the human mind.
>
> ——**BRIAN GREENE: *The Fabric Of The Cosmos*,** page 139 of the 2005 paperback edition by Vintage Books, a division of Random House, Inc., New York. ©2004 Brian Greene.

Prof. Green et al seem to think that's a new viewpoint. It's far from it. Over 740 years ago, basically the same assessment was voiced by a famous Catholic philosopher-theologian, Doctor of the Church, and canonized saint who had a thorough grasp of what is meant when God is said to be omniscient and omnipresent. Thus, we read:

> For, since time is not without change (*motum*), eternity, which is completely without change, in no way pertains to time. Furthermore, since the eternal never fails to be what it is, the eternal, as an ever current presence, is present to any time or instant of time whatsoever. We may see a kind of example of that in the case of a circle. Consider a particular point on the circle's perimeter. Although it be indivisible, it does not simultaneously occupy any other point's position, since it is the chain of positions which produces the continuity of the perimeter. On the other hand, the center of the circle, which is no part of the perimeter, is directly opposite any particular point on the perimeter. Hence, whatever is in any part of time coexists with what is eternally present to it, although with respect to some other time it is past or future. Something can be present to the eternal only by being present to the whole of it, since the eternal does not have succession's kind of duration. The divine intellect, therefore, sees as currently present to the whole of its eternity whatever takes place throughout the whole course of time. And yet what takes place in a certain part of time was not always existent. It stands, then, that God has knowledge of those things which according to the march of time do not yet exist.
>
> ——**ST. THOMAS AQUINAS** (1225 or 27 to 1274 AD)**: *Summa Contra Gentiles*,** Book I: Chap. 66: Par. 7 (My translation – ENH).

To get a clear picture of what St. Thomas was saying, look at the circle on the next page, and imagine, on the perimeter (***i.e.:*** "circumference", if you prefer) a dot at the 3 o'clock position and another at the 9 o'clock position. The former represents the Universe at high noon on this day, Sunday, June 9, 2013. The latter represents the Universe at high noon a trillion millennia from now. From the standpoint of the former, the latter is far off in time. From the standpoint of God, each is equally present and always so. That's symbolized by the fact that the distance from the circle's center to each point on its perimeter is the same. No matter how many points one imagines on the perimeter,[1] every one of them is always equally distant from the circle's center, which is the metaphor's way of saying every

---

[1] It would be a number including every one of the 2^{385} u-times of this 9th epoch of the Universe plus all the u-times of whatever epochs are yet to come.

one of them is always as present to God as any of the others are; and yet, God also sees exactly where each point falls in logical sequence for those limited to experiencing the "dots" one per temporarily changeless state of excitation. When St. Thomas and Prof. Green "envision all of time as *really* being out there," the only difference is that, for St. Thomas, that's the way it is from *God's* standpoint; whereas, for Prof. Greene, that's the way it is from the standpoint of the *Universe itself* without any help from God. Surprise! Surprise! Surprise!

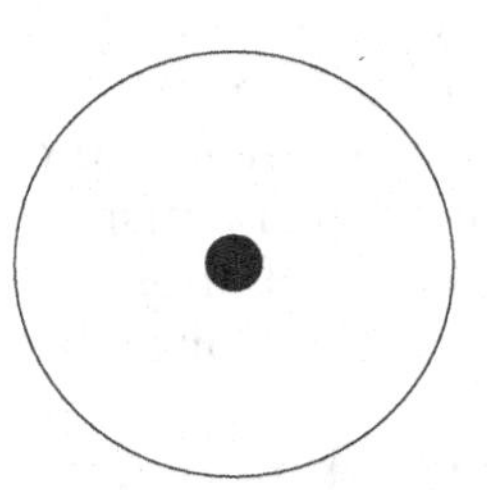

DRAWING #16-2: ST. THOMAS' METAPHORICAL EXPRESSION OF HOW EVERY INSTANT IN THE UNIVERSE'S HISTORY IS ALWAYS EQUALLY PRESENT TO GOD.

Some, of course, do far more than imagine the whole of *this* universe's time is really out there. For them, what's really out there is the whole time of this and all other possible universes.

> We will discuss duality and M-theory further in Chapter 5, but before that we turn to a fundamental principle upon which our modern view of nature is based: quantum theory, and in particular the approach to quantum theory called alternative histories. In that view, the universe does not have just a single existence or history, but rather every possible version of the universe exists simultaneously in what is called a quantum superposition.
> ——**STEPHEN HAWKING and LEONARD MLODINOW: *The Grand Design***; pages 58 & 59 of the paperback edition by Bantam Books Trade Paperbacks; New York, 2012. ©2010 Stephen Hawking & Leonard Mlodinow.

Here, again, that's nothing new, since St. Thomas would readily admit that God's omniscience includes seeing as equally present to the whole of God's eternity every time and instant of time pertaining to every *possible* universe. To express figuratively how that could be, imagine the circle is a sphere with an *equatorial* perimeter sporting an astronomical number of points. Crossing thru each of those points, we fancy an astronomical number of perimeters each at a discrete angle and sporting an astronomical number of points thru which an astronomical number of perimeters are crossing, and so on and on and on for as far as only The Infinitely Informed can tell. Again, since all of these perimeters and their points are on the outer surface of the sphere, each is equally distant from the sphere's center—a picture metaphorically expressing the fact that each and every one of this astronomically astronomical number of time instants always was, is, and shall be equally present to God and, therefore, each currently existent *relative to God* no matter what they are *relative to one another* and *the march of time.*[1]

[1] Some delight in asking at what point in time God created the Universe. My answer is: at no time and at all times. For, if you introduce the notion of one or more instants of time preceding creation, then they are merely more points on the circle or sphere. As such, they are—from God's standpoint

As I said: The only difference here is that, for we Theists, that's the way it is from *God's* standpoint; whereas, for the Atheists, that's the way it is from the standpoint of the *Universe itself* without any input from God. Just let them be aware that Theists thought of all that long, long before modern Science's Atheists did. And isn't it amazing how Theists, who are, without exception, idiots (according to the Atheists, that is) could do that long before the Atheists who (according to their unbiased opinion of themselves, that is) are, without exception, indescribably far above every Theist in intellectual acumen and, therefore, never anywhere nearly as much an idiot as every Theist always is?[1]

Perhaps the more rapturous point here is this: What is true for God's view of creation is also true for the view of all those angels and saints in infinity (***i.e.:*** heaven) with God. They, too, are ceaselessly aware of every instant of their own and everyone else's life and all the possible variations upon each of those instants and all the possible variations upon each of those variations and on and on for as far as only infinity's inhabitants can tell.

> This is schematically illustrated in Figure 5.1, but the perspective should make you scratch your head. The "outside" perspective of the figure, in which we're looking at the whole universe, all of space at every moment of time, is a fictitious vantage point, one that none of us will ever have. We are all *within* spacetime. Every experience you or I ever have occurs at some location in space at some moment of time. And since Figure 5.1 is meant to depict all of spacetime, it encompasses the totality of such experiences—yours, mine, and those of everyone and everything. If you could zoom in and closely examine all the comings and goings on planet earth, you'd be able to see Alexander the Great having a lesson with Aristotle, Leonardo da Vinci laying the final brushstroke on the Mona Lisa, and George Washington crossing the Delaware, as you continued scanning the image from left to right, you'd be able to see you grandmother playing as a little girl, you father celebrating his tenth birthday, and your own first day at school; looking yet farther to the right in the image, you could see yourself reading this book, the birth of your great-great-granddaughter, and, a little farther on, her inauguration as President.
> ——**BRIAN GREENE: *The Fabric Of The Cosmos***, pages 130 & 131 of the 2005 paperback edition by Vintage Books, a division of Random House, Inc., New York. ©2004 Brian Greene.

Yes, for those *without* God in their lives, the above "zoom in" is indeed "a fictitious vantage point" which none of *them* can ever enjoy. Quite the opposite is true for those *with*

(***i.e.:*** from infinity's point of view) neither more nor less than what each and every one of the other points is—namely: what always was, always is, and always will be equally present to the whole of God's eternity. Relative to one another, instants of time precede and follow one another in existence; but, relative to God, all are simultaneously existent. For God, there is no time when creation's every instant of time is not yet really there—whether "creation" means one universe or "every possible version of the universe". Eternity is the "no time" eternally including all times.

[1] The originator of the Big Bang Theory is a Catholic priest from Belgium by the name of Fr. Georges LeMaitre (1894-1966). According to the Atheists, that does not change the indisputable fact that, because he was a Theist, he was, *ipso facto*, intellectually—and grossly so—inferior to every Atheist. Let their view of Fr. LeMaitre indicate just how intellectually superior they *really* are. Incidentally, the originator of the science of genetics is a Fr. Gregor Mendel (1822-1884)—like myself, another idiot because a Theist.

God in their lives. The above "zoom in" is a vantage point they will enjoy with God in infinity for all time simultaneously past, present, and to come.

Pondering time, some scientists (if not all) are greatly puzzled by what they call "the arrow of time". Let's call it the "t-arrow" for short.

> And among the features of common experience that have resisted complete explanation is one that taps into one of the deepest unresolved mysteries in modern physics—the mystery that the great British physicist Sir Arthur Eddington called the *arrow of time*. (pgs. 12 & 13)
>
> . . . . .
>
> . . . . we continually experience an obvious directionality to the way things unfold in time but the laws themselves treat what we call forward and backward in time on an exactly equal footing. As there is no arrow within the laws of physics that assigns a direction to time, no pointer that declares, "Use these laws in this temporal orientation but not in the reverse," we were led to ask: If the laws underlying experience treat both temporal orientations symmetrically, why are the experiences themselves so temporally lopsided, always happening in one direction but not the other? Where does the observed and experienced directionality of time come from? (pg. 159)
>
> ——**BRIAN GREENE:** ***The Fabric Of The Cosmos***, pages 13 & 159 of the 2005 paperback edition by Vintage Books, a division of Random House, Inc., New York. ©2004 Brian Greene.

That's right. No law of *Physics* (Including entropy, Prof. Greene!) can account for why our t-arrows are "always happening in one direction". That's because *Physics* (Including entropy, Prof. Greene!) has nothing to do with the direction of our t-arrows. It is, rather, Theology alone which has to do with it. More specifically, it is God's love of free will—as God's infinitely informed standards define it—which alone has to do with it. How so?!

According to God's infinitely informed standards, free will does not mean the *absence* of *all predetermination*; rather, it means the *presence* of *self* predetermination. "Self", though, does not have the same meaning for God's infinitely informed standards as it has for most, if not all, of us. For us, "self" means one's *partial* self; for God, The Infinitely Informed, it means one's *complete* self. What does that mean?

The *partial* self is the self each individual knows in the latest instant of his or her march toward a time which is *not yet* as far into the future as one can go this side of infinity. As such, the partial self is that self which is the sum total of all one's past but only remembered experiences without including those still to come this side of infinity.

The *complete* self is the self each individual knows in the last instant of his or her march toward a time which *is* finally as far into the future as one can go this side of infinity. As such, the complete self is that self which is the sum total of all the experiences one has had and ever shall have this side of infinity. What's the significance of that distinction?

I dare say all of us, as we move thru time, change our opinions regarding at least some of our past choices and wish we had—at such-and-such a time—made such-and-such a choice rather than the one we did make. One's complete self will be no different, except that, being at its future's end, its choices—choices to have chosen such-and-such at such-and-such a time—will be final and forever unchanged. No, the complete self will not lack the *power* to change; it will lack any *reason* to do so. That's because, the complete self's knowledge of itself shall be so complete as to leave it self-determined: (1) to have every reason to stick to each of its final choices, (2) to have no reason to abandon any of its final choices, and (3) to have every reason to reject wholly and entirely and forever every reason to abandon any of its final choices. It's called self-preservation of the self dearest by far to one's self.

Only to God, The Infinitely Informed, are each complete self's final and forever irrevocable choices as present as every partial self's choices are to it. As many a theologian has expressed it: God knows us far better than we know ourselves. Therefore, knowing with infinitely informed perfection what a truly *self*-determined free will means (***i.e.:*** what one's *complete self* determines for itself should have been its t-arrow's final goal and direction at every instant of its partial self's march thru time); and loving with infinite intensity such a kind of free will—God insures that every instant of every t-arrow's direction shall be as its owner's *complete self* determines for itself that direction should be. For all always and everywhere this side of infinity, every iota of every experience is exactly as one's own *complete self* determines for itself that experience should be. God insures every t-arrow's direction *shall* be as it is; each of us—thru God's foreknowledge of one's complete self and acting as its proxy by that complete self's choice—determines what his or her own t-arrow's direction *should* be. Every t-arrow's direction is thus evidence for a loving God's existence.

Understandably, some will protest: "How can that be possible where billions of people are simultaneously living on this planet and often at odds with one another?" It can be possible because an omnipotent, omnipresent God easily knows which t-arrows—however great their number—perfectly complement one another no matter how much it might seem to them that such is far from being the case.

Some will add: "So then, if someone commits an injustice against me, it's not really an injustice, since my assailant is merely giving me what my complete self calls upon that assailant to give me." I reply: Who consents to suffer an *injustice* suffers an *injustice*. Every injustice thus remains an *injustice* and every unjust assailant an *unjust assailant*.

It's senseless to dwell any further on this aspect of time, since it has nothing to do with Theoretical Physics. I'll return to it later. For now, back on the track!

It goes without saying that there can be no such thing as a thorough grasp of acceleration, unless one first has a thorough grasp of velocity. After all, acceleration, by definition, is a change in velocity. However vigorously they might deny it, scientists are far off base when it comes to velocity. Consider this:

> Maxwell's equations dictate that electromagnetic waves travel at a speed of about 300,000 kilometers a second, or about 670 million miles per hour. But to quote a speed means nothing unless you specify a frame of reference relative to which the speed is measured. That's not something you usually need to think about in everyday life. When a speed limit sign reads 60 miles per hour, it is understood that your speed is measured relative to the road and not the black hole at the center of the Milky Way. But even in everyday life there are occasions in which you have to take into account reference frames. For example, if you carry a cup of tea up the aisle of a jet plane in flight, you might say your speed is 2 miles per hour. Some on the ground, however, might say you were moving at 572 miles per hour. Lest you think that one or the other of those observers has a better claim to the truth, keep in mind that because the earth orbits the sun, someone watching you from the surface of that heavenly body would disagree with both and say you are moving at about 18 miles per second, not to mention envying your air-conditioning.
>
> —**STEPHEN HAWKING and LEONARD MLODINOW: *The Grand Design***; pages 92 & 93 of the paperback edition by Bantam Books Trade Paperbacks; New York, 2012. ©2010 Stephen Hawking & Leonard Mlodinow.

What if one did measure the tea carrier's velocity relative to "the black hole at the center of the Milky Way"? Some say the velocity of our galaxy thru the Universe is roughly 630 kilometers per sec. = 2,268,000 KPH = 1,409,270 MPH. In that case, is the tea carrier's velocity his 2 MPH + the plane's 570 MPH + Earth's velocity around its axis (***i.e.:*** c. 1,000 MPH) + Earth's velocity around the Sun (***i.e.:*** c. 66,616 MPH) + the Sun's velocity thru the Milky Way (***i.e.:*** 504,000 MPH) + the Milky Way's velocity thru the Universe? And to what is the Milky Way's velocity relative? Is it relative to something which also has a velocity relative to something else? What, then, is the tea carrier's velocity?

Today's scientists can't tell you, because they know nothing of Esoptrics and, thus, nothing of the *ultramicroscopic makeup* of space and rectilinear vs. curvilinear motion as set by God's 7 *a priori* coordinates. Those familiar with such know rectilinear velocity is relative to the rate of K per change among some particular u-spaces envelope's u-spaces and to *nothing else*. If, then, a jetliner's velocity is 570 MPH (***i.e.:*** 1Φ per 4.003493×10^{44}K = c. $\Phi/10^{-55}$ sec.), it's such for all *knowledgeable* observers everywhere. For, in each of the jet's many atoms is a multi-A-combo whose carrying generator is shifting from u-space to logically contiguous u-space at that rate. As for the tea carrier walking up and down the aisle at 2 MPH, his rectilinear velocity is 572 MPH regardless of his direction. Whether walking toward the nose or the tail, his carrying generators—using the same u-spaces envelope as the jetliner—are changing u-spaces 2 MPH more frequently than the jet's.

Earth's rotation around its axis at 1,000 MPH has nothing to do with the *rectilinear* velocity of either the jetliner's or the tea carrier's atoms. It's solely the *curvilinear* velocity of the *u-spaces* the jetliner's and the tea carrier's atoms are using. The *curvilinear* velocity of each *u-space* in the tertiary plane is relative solely to 2 *wholly local* factors: (1) its distance in Φ's from its envelope's center, and (2) the number of K (***i.e.:*** successive units of u-time) per tertiary plane cycle on the part of its envelope (***i.e.:*** that one of the forms logically concentric with Earth's multi-h-combo which is providing the u-spaces envelope). For each u-space, the *distance* traveled per cycle on the part of its envelope = (8Δ)–4 where Δ = the u-space's distance from its envelope's center in Φ's. The *time* in which that distance is covered (***i.e.:*** the other factor in determining velocity) is relative solely to the number of K (***i.e.:*** consecutive u-times) per cycle on the part of the envelope (***i.e.:*** the providing form) containing the u-space.

The *rectilinear* velocity of Earth's orbit around the Sun is relative solely to the rate at which its multi-h-combo's carrying generator is changing from u-space to logically contiguous u-space, and that rate is relative solely to the number of K per change. As pointed out earlier, the vast majority of Earth's curvilinear velocity around the Sun is the curvilinear velocity of the *u-spaces* Earth's carrying generator is using.

Likewise, as Earth orbits the Sun, the curvilinear velocity of the u-spaces being used by Earth's carrying generator is relative solely to their distance in Φ's from their envelope's center and its number of K per cycle. Neither the Sun's velocity thru the galaxy, nor the galaxy's velocity thru something else, adds anything either to the rectilinear velocities of the jetliner, the tea carrier, and the Earth or to the curvilinear velocities of the u-spaces being used by those three.

Thus, whether one is trying to calculate either the rectilinear velocity of this-or-that object or the curvilinear velocity of the u-spaces this-or-that body's carrying generators are using, it is unnecessary to take into consideration either the rectilinear velocity of any other body whatsoever or the curvilinear velocity of anything but the form providing the u-spaces under consideration. I repeat, though: That's the way it is, if one is sufficiently versed in Esoptrics and what is the *ultramicroscopic makeup* of space and rectilinear vs. curvilinear locomotion as determined by God's 7 *a priori* coordinates. Here are more examples of the contrast between being and not being sufficiently versed:

> . . . . imagine two events that take place at the same spot but at different times, in a jet aircraft. To an observer on the jet there will be zero distance between those two events. But to a second observer on the ground the events will be separated by the distance the jet has traveled in the time between the events. This shows that two observers who are moving relative to each other will not agree on the distance between two events. (pg. 96)
>
> , , , ,
>
> If you bounce a ball on a jet, an observer aboard the plane may determine that it hits the same spot each bounce, while an observer on the ground will measure a large difference in the bounce points.
> ——**STEPHEN HAWKING and LEONARD MLODINOW: *The Grand Design***; pages 96 & 97 of the paperback edition by Bantam Books Trade Paperbacks; New York, 2012. ©2010 Stephen Hawking & Leonard Mlodinow (The second set of words are given under a picture on pg. 97.).

Yes, those knowing nothing of Esoptrics would disagree on what the bounce points were; but, those sufficiently versed in Esoptrics would be able to calculate and to agree on exactly which u-spaces—whether by the atoms in the observer on the plane or in the observer on the ground or in the jetliner or in the ground or in the bouncing ball—were being used at which given number of u-times into the Universe. It's the difference between relying on one's senses and Geometry vs. relying on God's input and Algebra.

> Now suppose the two observers observe a pulse of light traveling from the tail of the aircraft to its nose. Just as in the above example, they will not agree on the distance the light has traveled from its emission at the plane's tail to its reception at the nose. Since speed is distance traveled divided by the time taken, this means that if they agree on the speed at which the pulse travels—the speed of light—they will not agree on the time interval between the emission and the reception.
>
> What makes this strange is that, though the two observers measure different times, they are watching the *same physical process*. Einstein didn't attempt to construct an artificial explanation for this. He drew the logical, if startling, conclusion that the measurement of the time taken, like the measurement of the distance covered, depends on the observer doing the measuring. That effect is one of the keys to the theory in Einstein's 1905 paper, which has come to be called special relativity
>
> We can see how this analysis could apply to timekeeping devices if we consider two observers looking at a clock. Special relativity holds that the clock runs faster according to an observer who is at rest with respect to the clock. To observers who are not at rest with respect to the clock, the clock runs slower. If we liken a light pulse traveling from the tail to the nose of the plane to the tick of a clock, we see that to an observer on the ground the clock runs slower because the light beam has to travel a greater distance in that frame of reference. But the effect does not depend on the mechanism of the clock; it holds for all clocks, even our own biological ones.
> ——**STEPHEN HAWKING and LEONARD MLODINOW: *The Grand Design***; pages 96, 97 & 98 of the paperback edition by Bantam Books Trade Paperbacks; New York, 2012. ©2010 Stephen Hawking & Leonard Mlodinow.

Yes, if one's knowledge is limited to the *measurements* scientists are accustomed to make either with their naked senses or their instrument aided senses (***i.e.:*** with *mono*-directional empirical observation), then, for the observer on the plane, the light travels solely the distance from the tail to the nose; whereas, for the observer on the ground the light travels the distance from: (1) the tail when it and the nose were at one set of locations, to (2) the nose when it and the tail are at a set of locations further along in space. Again, if one's knowledge is limited to the images impressed on one's senses by a clock, time has one length for an observer at rest and another for an observer in motion. Oppositely, if one's knowledge extends to Esoptrics in sufficient detail, then—whether the observer on the plane or the observer on the ground or the observer at rest with a clock or the observer in motion with a clock—all can still calculate exactly which u-spaces the light traversed and exactly the number of K into the 9th epoch at which it started and ended its journey.

The gist of it all is this: Scientists are so ignorant of the God-given ultramicroscopic makeup of time, space, and rectilinear vs. curvilinear locomotion, they cannot begin to understand the real reason why light speed is the same in every inertial system—namely: What distinguishes inertial systems from one another is that, in each system, the rate at which rectilinearly moving generators change from one microstate of excitation to another is different; but, in all inertial systems, that rate of change is the same for generators at light speed. At least, so it is for those versed in Esoptrics' algebraic calculations rather than reliant solely on sense experience. It may help to understand how that's so, if we recast the formula $1\Phi/2^{128}K$. To that end, let Ω_D = Omega's dia. of $2^{257}\Phi$ and $\forall$ = 1 act of a form at OD1. We can then say that, for all inertial systems and all observers always and everywhere, light speed = $2^{-257}\Omega_D$ per $2^{128}\forall$. The difference comes about because what $2^{-257}\ \Omega_D$ and $2^{128}\forall$ are *to an observer's sense experience* depends on the current OD of the observer's form. For an observer at OD1 with a dia. of 2Φ and called Alphon, the ratio of Ω_D to his dia. is 2^{256} to 1; and so, $2^{-257}\ \Omega_D$ = ½ Alphon's dia. = 1Φ for Alphon. But, for an observer at $OD2^{155}$ with a dia. of $2^{156}\Phi$ and called Bogon, the ratio of Ω_D to his dia. = 2^{101} to 1; and so, for him, $2^{-257}\ \Omega_D = 2^{-156}$ his dia. of $2^{156}\Phi$, and $2^{-156} \times 2^{156}\ \Phi = 1\Phi$ for Bogon. If, then, at $OD2^{155}$ Bogon's dia. is 2^{27} times that of an atom seeming to his senses $5\text{x}10^{-8}$ cm. in dia. at $OD2^{128}$, then, the size of an Φ another 2^{129} (***i.e.:*** $2^{156} \div 2^{27}$) lower in OD = $7.3\text{x}10^{-47}$cm.; and yet, if Bogon knows Esoptrics, he can easily calculate the difference between what 1Φ is for him at his OD vs. for Alphon at OD1 and Omega at $OD2^{256}$—namely: For Alphon, it's 2^{155} bigger than it is for him and, for Omega, is 2^{101} smaller than it is for him. Also, knowing the ratio of $\forall$ to each of his acts at $OD2^{155}$ is $(2^{155})^2$ to 1, he can easily calculate what the frequency of $\forall$ is for each of Alphon's acts at OD1 and each of Omega's acts at $OD2^{256}$—namely: For Alphon, it's 1/1 and, for Omega, it's $2^{512}/1$. He thus also knows the ratio of Alphon's i-time to his (***viz.:*** $2^{310}/1$) and the ratio of his to Omega's (***viz.:*** $2^{202}/1$). For Bogon, sense experience may *observe* different times and distances; but, Esoptrics' masters will all, relative to $\forall$, *calculate* the same times—whether u-times or i-times—and, relative to Omega, the same distances for every one of the 4th dimension's 2^{256} OD levels.[1]

[1] What a prison and kind of searing torment OD1 would be! How so?! Suppose Alphon is a soul at OD1. For him, sense experience would alternately be of 2 vs. 4Φ in dia. and 8 vs. 64 in volume. His sense images would thus alternate between 8 and 64 "pixels" in resolution and each effectively for him what's for us c. 3.4 cm. in dia.. Over and over again for 2^{128} (***i.e.:*** $3.4\text{x}10^{38}$) times (***i.e.:*** for us, c. $2.4\text{x}10^{29}$ years of 40 acts of awareness per sec. and thus c. $1.26\text{x}10^{9}$ per yr.), he'd experience the same 2 alternating but frozen frames before a shift of c. 3.4 cm. occurred. But, what's 3.4 cm. for him and 10^{-47} cm. for us is, for all always everywhere, $2^{-257}\Omega_D$ even if one is too carnal minded to calculate and be aware of it. Likewise, what's c. 1/40 sec. for him and c. 10^{-96} sec. for us is, for all always everywhere, $2^{-385}U_T$. And what's the time length of each of his acts? None! For, there is no time

> Without any calculation or detailed analysis, we realize that the uniformity of the physical environment, as evidenced by the uniformity of the microwave background radiation and the uniform distribution of galaxies throughout space, allows us to infer uniformity of time.
> ——**BRIAN GREENE:** ***The Fabric Of The Cosmos***, page 234 of the 2005 paperback edition published by Vintage Books, a division of Random House, Inc., New York. ©2004 Brian Greene.

Yes, Prof. Greene! But, as with those who used what their outwardly looking minds observed to justify *inferring* a geocentric Universe or infinitely divisible time, space, matter, locomotion, and change, you *infer* wrongly. Why is that? It's because of how gross and widespread throughout the scientific community is ignorance of the 4th dimension and the resulting fact that there is nothing even remotely close to uniformity in the way generators and forms change from one state of excitation to another. Hopefully, in future pages, you'll come to realize that what your outwardly looking eyes observe is not the *reality* of it all but, rather, the uniformity of the *stroboscopic effect* the molecular forms have upon the 4th dimension's 2^{256} levels.

✡✝✡✝✡✝✡✝✡✝✡✝✡✝✡✝✡✝✡✝✡✝✡✝

> When I first encountered this idea many years ago, it was a bit of a shock. Up until that point, I had thought I understood the concept of entropy fairly well, but the fact of the matter was that, following the approach of textbooks I'd studied, I'd only ever considered entropy's implications for the future. And, as we've just seen, while entropy applied toward the future confirms our intuition and experience, entropy applied toward the past just as thoroughly contradicts them.
> ——**BRIAN GREENE:** ***The Fabric Of The Cosmos***, page168 of the 2005 paperback edition by Vintage Books, a division of Random House, Inc., New York. ©2004 Brian Greene (The fact of the matter is that I've never been able to make any sense out of the concept of entropy and still can't make any sense out of it. For one thing, it seems to me that whether or not this-or-that condition is one of order [***i.e.:*** low entropy] vs. disorder [***i.e.:*** high entropy] is strictly in the eye of the beholder. For another, it seems to me that what's order vs. disorder depends upon the final tally of all the consequences of the condition under consideration. And who but One Who is Infinitely Informed could possibly know such a tally? On page 152 of his book, Prof. Greene tells us high entropy means there are many ways a particular physical condition can be realized, and low entropy means few ways. Here again, though, who but The Infinitely Informed can be sure of such an assessment? In the final analysis, then, the concept of entropy strikes me as nothing more than a Rube Goldberg contrivance.).

---

where there is no change. All we can say is: Each of his acts endures ¼ as long as they would at OD2; 1/9 as long as at OD3; 1/16 as long as at OD4, etc. all the way to 2^{-512} as long as at Omega.

Chapter Seventeen:

ACTION AT A DISTANCE & ENTANGLEMENT:

For many (if not most) scientists, action at a distance is a major puzzle. For Einstein, it was "spooky" and "ludicrous".[1] For Esoptrics, it is no puzzle at all, since it is exactly the result called for by Esoptrics' explanation of what space *really* is—namely: an astronomical number of "u-space envelopes", each of which is the actuality (***i.e.:*** macrostate of excitation) of a form which, for every generator in potency to it, is a definite number of u-spaces which are only *logically* distinct from, and outside of, one another.

> . . . . the majority view leads to the conclusion that *measurement on 1 produces instantaneous change at 2, a change that depends precisely on exactly what is measured at 1*. In other words, there is some counterintuitive togetherness-in-separation between 1 and 2; action at 1 produces immediate consequences for 2 and the consequences are different for different actions at 1. This is usually called the *EPR effect*.[2] The terminology is somewhat ironic since Einstein himself refused to believe in such a long-range connection, regarding it as an influence that was too 'spooky' to be acceptable to a physicist. (pg. 79).
>
> . . . .
>
> The EPR effect's implication of deep-seated relationality present in the fundamental structure of the physical world is a discovery that physical thinking and metaphysical reflection have still to come to terms with in fully elucidating all its consequences. As part of that continuing process of assimilation, it is necessary to be as clear as possible about what is the character of the entanglement that EPR implies. One must acknowledge that a true case of action at a distance is involved, and not merely some gain in additional knowledge. Putting it in learned language, the EPR effect is ontological and not simply epistemological (pg. 80).
>
> ——**SIR JOHN POLKINGHORNE:** ***Quantum Theory, A Very Short Introduction***, pages 79 & 80 of the paperback edition by the Oxford University

---

[1] See pages 11 and 80 of Prof. Greene's book ***The Fabric Of The Universe.*** On Nova's web site at www.pbs.org/wghb/nova/physics/spooky-action-distance.html, he has an article entitled "Spooky Action At A Distance" in which he repeats what he says in his book. If the article is no longer there, rest assured it was there today this June 12, 2013.

[2] Actually, it's called the EPR *paradox*. The term "EPR *effect*" is a medical one referring to the Enhanced Permeability and Retention effect for lipid and macromolecular agents in solid tumors. For those not aware, the initials EPR refer to Albert Einstein and his two colleagues Boris Podolsky and Nathan Rosen who worked with Einstein to produce a thought experiment which, they hoped, would reveal in Quantum Mechanics a flaw serious enough to discredit it. It didn't.

Press, Oxford, 2002. ©2002 John Polkinghorne.

If there is space between two objects—if there are two birds in the sky and one is way off to your right and the other is way off to your left—we can and do consider the two objects to be independent. We regard them as separate and distinct entities. Space, whatever it is fundamentally, provides the medium that separates and distinguishes one object from another. That is what space does. Things occupying different locations in space are different things. Moreover, in order for one object to influence another, it must in some way negotiate the space that separates them. . . . . [Skip!] . . . . (pg. 79)

Physicists call this feature of the universe *locality*, emphasizing the point that you can directly affect only things that are next to you, that are local. Voodoo contravenes locality, since it involves doing something over here and affecting something over there without the need for anything to travel from here to there, but common experience leads us to think that verifiable, repeatable experiments would confirm locality. And most do.

But a class of experiments performed during the last couple of decades has shown that something we do over here (such as measuring certain properties of a particle) *can* be subtly entwined with something that happens over there (such as the outcome of measuring certain properties of another distant particle), without anything being sent from here to there. . . . . [Skip!] . . . . This sounds like voodoo; Einstein, who was among the first physicists to recognize—and sharply criticize—this possible feature of quantum mechanics, called it "spooky". . . . .

Nevertheless, these results, coming from both theoretical and experimental considerations, strongly support the conclusion that the universe admits interconnections that are not local. Something that happens over here can be entwined with something that happens over there even if nothing travels from here to there—and even if there isn't enough time for anything, even light, to travel between the events. This means that space cannot be thought of as it once was: Intervening space, *regardless of how much there is,* does not ensure that two objects are separate . . . . [Skip!] . . . . According to quantum theory and the many experiments that bear out its prediction, the quantum connection between two particles can persist even if they are on opposite sides of the universe. From the standpoint of their entanglement, notwithstanding the many trillions of miles of space between them, it's as if they are right on top of each other.

Numerous assaults on our conception of reality are emerging from (pg. 80) modern physics . . . . [Skip!] . . . . But of those that have been experimentally verified, I find none more mind-boggling than the recent realization that our universe is not local. (pg. 81)

——**BRIAN GREENE:** ***The Fabric Of The Cosmos***, pages 79, 80 & 81 of the 2005 paperback edition by Vintage Books, a division of Random House, Inc., New York. ©2004 Brian Greene.

That's the all-important gist of it all: The Universe is *not local*; it has *no locality*, because space is not and cannot be anything like what it was long thought to be. So far is space from the popular notion of it, no amount of it standing between two objects ensures they are separate. That's what I have been trumpeting since grammar school. More importantly, except in Esoptrics, there is nowhere else in Theoretical Physics the slightest trace of an explanation of *how* a universe devoid of locality could, nevertheless, appear—to our senses and our measuring instruments—to have locations at various distances from one another and particles—without any leaps—traversing those distances over the course of some time span. Oppositely, Esoptrics gives far more than a mere trace of an explanation. It gives a very graphic, mathematically precise and highly detailed one which I continue to publish before the world at my expense.[1]

Esoptrics is not out to explain how there can be direct interaction between two *spatially* separated objects. It's out to explain how there can be direct interaction between two objects only *logically* separated. Esoptrics is not out to explain how there can be direct interaction between two separated objects without *any kind* of *intermediating* agent. It's out to explain how—between two *logically* separated objects—there can be direct interaction by means of an intermediating agent which, like them, is not *spatially* extended and *not to any extent traveling between them*. How does Esoptrics do that?

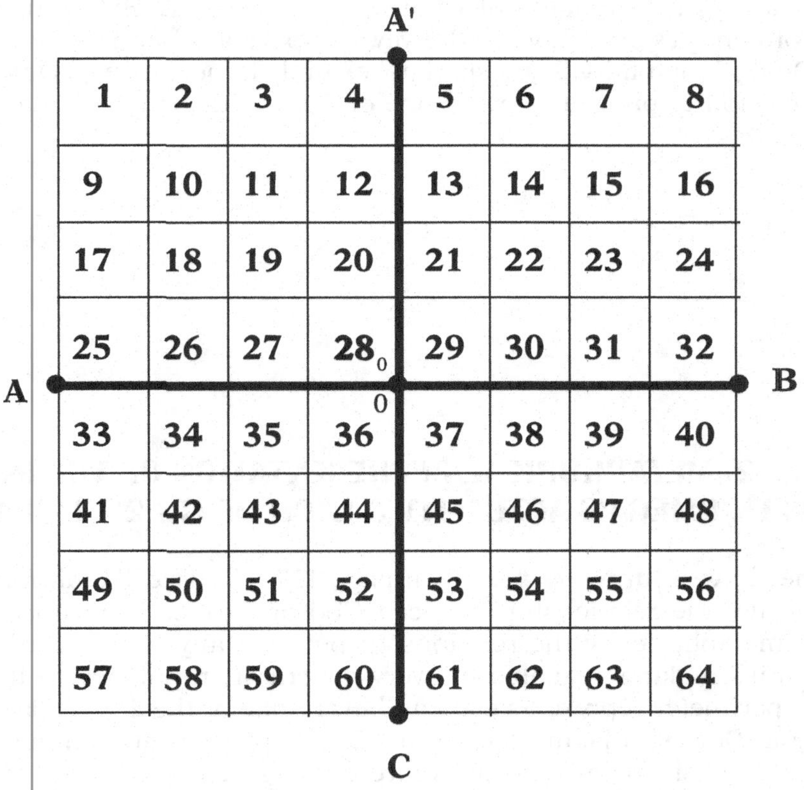

A'

1	2	3	4	5	6	7	8
9	10	11	12	13	14	15	16
17	18	19	20	21	22	23	24
25	26	27	28_0	29	30	31	32
33	34	35	^{0}36	37	38	39	40
41	42	43	44	45	46	47	48
49	50	51	52	53	54	55	56
57	58	59	60	61	62	63	64

A B

C

DRAWING #17-1: THE NATURE OF ACTION AT A DISTANCE.

[1] That's currently in the form of 20 self-published books in print, and this one, #21, shall soon join them. If the world shall never listen to any of what I have said, so be it. It's no skin off my back. At 77, my physical, mental, and financial health are so far superior to that of the vast majority of others, I cannot justly go to my grave with anything on my mind but extreme gratitude to God for a life far, far more rewarding than what is given to 99.999999% of humans.

Needless to say, I cannot draw either a cube composed of 2^{771} cubes or a square composed of 2^{257} squares. Be tolerant, then, and try to imagine the above is such a cube (or such a square, if you prefer) and, as such, is a metaphor for Omega, the one form native to, and currently at, $OD2^{256}$.

Next, imagine a Universe in which there are no *multi*-combos and, besides Omega, only 2 other *duo*-combos. Call them Huey and Dewey. Each one's generator is receiving its u-spaces from Omega. Omega is presenting those 2 generators with 2^{771} (***i.e.:*** the cube of 2^{257} = c. 10^{232}) u-spaces to be used 1 per state of excitation. Huey is currently using the u-space represented by the square marked #32 in drawing #17-1 on the prior page. Dewey, on the other hand, is currently using the u-space represented by the square marked #25. That's a figurative way of saying that Huey and Dewey are at opposite extremities of Omega's actuality. For Esoptrics, that means the distance between Huey and Dewey = 2^{257} u-spaces (***i.e.:*** Φ's). For the Geometry-speak of sensation limited minds, that means the distance between them is c. 18 trillion light years.

Despite that fact, Huey and Dewey are continuously and instantly in contact with one other by means of the actuality of Omega. That's because Huey's u-space is not one particular *area* of Omega's actuality and Dewey's u-space a different *area* of Omega's actuality. Because Omega's actuality is not spatially extended, the whole of Omega's actuality is equally and continuously present to every one of the 2^{771} u-spaces it's making available to Huey and Dewey.

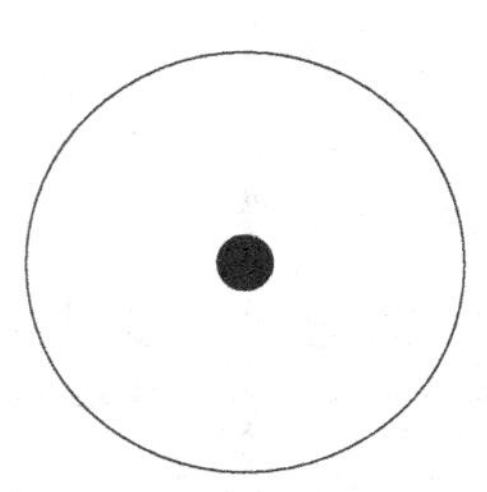

DRAWING #17-2: METAPHORICAL REPRESENTATION OF THE RELATIONSHIP BETWEEN OMEGA'S ACTUALITY & EACH OF ITS 2^{771} U-SPACES.

Remember how in drawing #16-2, on page 137, we used the above drawing to express metaphorically the relationship between God and time. The same drawing can be used to express metaphorically the relationship between any form (in this case, Omega) and the u-spaces it's making available to every generator in potency to it. In the current case, the circle's perimeter represents each and every one of the 2^{771} u-spaces Omega is offering to Huey and Dewey. Imagine each of the 2^{771} u-spaces is at a unique location on the perimeter, and that shall metaphorically represent the fact that—from their own standpoint—each of the u-spaces is logically outside of, and distinct from—all the others, and a u-space at 12 o'clock on the perimeter is further from one at the 6 o'clock position than from one at the 3 o'clock position. Still, because each point on the perimeter is equidistant from the dot at the circle's center, that metaphorically expresses the fact that the whole of Omega's actuality is *equally* present to, and *wholly* in contact with, each and every one of its 2^{771} u-spaces. That's possible because, though logically *divided* 6 ways, Omega's actuality is in no way *extended* whether logically or spatially. Its OD is logically extended in the sense that $OD2^{256}$ is 1OD higher than $2^{256}-1$, 2OD's higher than $2^{256}-2$, etc.; but, that has nothing to do with what Omega's actuality is within itself, despite what it is from the standpoint of the generators dividing that actuality into 2^{771} u-spaces.

CHAPTER SEVENTEEN

That, then, is how whatever happens to Huey is instantly presented to Dewey and vice versa, and every such exchange is by means of an intermediary. Still, there is no intermediary *in between* them—at least, not in the sense of something which *moves* from one to the other. On the contrary, it's more accurate to say the intermediary (***i.e.:*** Omega's actuality) is in each of them wholly, equally, and continuously. It's merely a matter of grasping the difference between: (1) what's *physically* "right on top of each other" (Prof. Greene's phrase) vs. what's not *logically* so; and (2) the nature of the u-spaces vs. that of their cause. Huey & Dewey are not one and the same thing; but, what provides logically separate u-spaces to each of their carrying generators is invariably the same one thing in its entirety.

What, though, causes two particles to be entangled? I'm far from sure I can answer definitively. For now, nothing comes to my mind save this: Due to how rational forms rotate in the *primary* and *secondary* planes (planes logically related to one another in a ratio bearing way which can be metaphorically expressed as perpendicular to one another), the Universe's every u-space executes an elaborate multi-level weaving pattern an astronomical number of times per c. 10^{-18} sec. in our terminology. For example, every 10^{-18} sec., and by means of the *2 upper planes*: (1) every u-space offered by the leading form in Earth's multi-h-combo circles Earth an astronomical number of times at an astronomical number of discrete "angles" as (2) the leading form in the Sun's multi-h-combo causes them to circle the Sun an astronomical number of times at an astronomical number of angles as (3) the leading form in the multi-g-combo at the Milky Way's center (***i.e.:*** one in the 8th reverse category) causes them to circle that center an astronomical number of times at an astronomical number of angles as (4) a leading form in the multi-g-combo in the 7th reverse category's center causes them to circle that center an astronomical number of times at an astronomical number of angles as (5) a leading form in the multi-g-combo in the 6th reverse category's center causes them to circle that center an astronomical number of times at an astronomical number of angles as (6) a leading form in the multi-g-combo in the 5th reverse category's center causes them to circle that center an astronomical number of times at an astronomical number of angles as (7) a leading form in the multi-g-combo in the 4th reverse category's center causes them to circle that center an astronomical number of times at an astronomical number of angles as (8) a leading form in the multi-g-combo in the 3rd reverse category's center causes them to circle that center an astronomical number of times at an astronomical number of angles as (9) a leading form in the multi-g-combo in the 2nd reverse category's center causes them to circle that center an astronomical number of times at an astronomical number of angles as (10) Omega causes them to circle the Universe's center an astronomical number of times at an astronomical number of angles.

Why do I say c. 10^{-18} sec.? At OD2^{256}, Omega spends 2^{128} K making one *primary* plane rotation in each of the 2^{128} positions in its *secondary* plane. That's a total of 2^{256}K per Omega's every rotation in its *secondary* plane and 2^{256}K (***i.e.:*** c. 10^{-18} sec.) per position in the *tertiary* plane. Note, though, that, at the center of every galaxy in the 2nd thru the 8th reverse categories, every multi-g-combo is one native to OD2^{128}. That means this: At the level of the 2nd thru the 8th reverse categories, each multi-g-combo's every rotation in its secondary plane lasts only 2^{128} K (***i.e.:*** c. 10^{-58} sec.). Every c. 10^{-18} sec., then, every multi-g-combo completes 2^{128} rotations in its secondary plane per one rotation by Omega in its secondary plane. By simple math, that's one rotation in the secondary plane per 10^{-58} sec. for each multi-g-combo and 2^{128} such rotations per the 10^{-18} sec. it takes Omega to complete one rotation in its secondary plane. Thus the Universe's every u-space not provided by Omega executes a fantastically elaborate weaving pattern 2^{128} times per 10^{-18} seconds.[1]

[1] The weaving pattern may be far more elaborate than the above. It may be it occurs at every one of the generic levels in RC's 4 thru 8. That would add another 56 steps to the above 10.

Naturally, all of that has to be taking place at a velocity which is an astronomical number of times greater than light speed. That, though, does not violate the speed limit of $1\Phi/2^{128}$K. For, as said many times, that speed limit applies to the rectilinear velocity of the generators changing from u-space to u-space. That has nothing to do with what was just said in the prior two paragraphs, because what's being described there is the curvilinear velocity of the u-spaces as their providers rotate in the primary and secondary planes. Yes, because a vast number of those u-spaces are being used by a vast number of generators, those generators and the particles they serve are necessarily involved in that astronomically speedy and elaborate pattern of weaving. Still, no matter how wildly the atoms and their sub-atomic particles go whirling throughout the Universe, their carrying generators are not changing from u-space to u-space at greater than $1\Phi/2^{128}$K.

Notice I said that's a pattern repeated c. every 10^{-18} sec. for Omega and c. every 10^{-58} sec. for the other 7 reverse categories. That implies a home position in the tertiary plane for each u-space. In turn, that means each multi-combo has a *home place* in the tertiary plane, and, in most cases, that's going to mean home is some particular atom. Suppose, then, home for two particular particles (***i.e.:*** multi-combos) is the same one atom. In that case, if, in the course of the wild gyrations described above, they become further separated than they usually are in their atom, they shall still be predisposed to maintain between themselves the balance they maintained in their atom. That, then, is what entanglement means to me at this stage of Esoptrics' development.

✡✞✡✞✡✞✡✞✡✞✡✞✡✞✡✞✡✞✡✞✡✞✡✞

In physics, the smallest doll [He means smallest ultimate constituent of the universe – ENH] is called the Planck length. To probe to shorter distances would require particles of such high energy that they would be inside black holes. We don't know exactly what the fundamental Planck length is in M-theory, but it might be as small as a millimeter divided by a hundred thousand billion billion billion [10^{-32} mm. = 10^{-33} cm. – ENH]. We are not about to build particle accelerators that can probe to distances that small. They would have to be larger than the solar system, and they are not likely to be approved in the present financial climate.
——**STEPHEN HAWKING:** ***The Universe In A Nutshell***, pg. 178 as published in hardbound copy by Bantam Books, New York, November 2001. ©2001 by Stephen Hawking. (Here, in addition to Prof. Greene [See my pages 54&174.], we have another scientist—perhaps the most famous one alive—telling us Science cannot probe what's at 10^{-33} cm., let alone smaller. From that, it necessarily follows Science cannot *observe* whether or not Esoptrics is correct when it says the makeup of the Universe's ultramicroscopic constituents is found at c. 10^{-47}cm. in "diameter". After all, it takes us another 14 zeroes beyond the 33 already requiring a particle accelerator larger than the solar system.)

Chapter Eighteen:

ESOPTRICS VS. QUANTUM MECHANICS:

These days, authors writing about Quantum Mechanics devote many of their words to the description of what they call "double slit experiments". This they do, it seems, because the results of these experiments are mind boggling to put it mildly. Thus, we read:

> In the Austrian experiment, opening the second gap did indeed increase the number of molecules arriving at some points on the screen—but it decreased the number at others, as in the figure below. In fact, there were spots where no buckyballs landed when both slits were open but where balls did land when only one or the other gap was open. That seems very odd. How can opening a second gap cause fewer molecules to arrive at certain points?
>
> We can get a clue to the answer by examining the details. In the experiment, many of the molecular soccer balls landed at a spot centered halfway between where you would expect them to land if the balls went through either one gap or the other. A little farther out from that central position very few molecules arrived, but a bit farther away from the center than that, molecules were again observed to arrive. This pattern is not the sum of the patterns formed when each gap is opened separately, but you may recognize it from Chapter 3 as the pattern characteristic of interfering waves. The areas where no molecules arrive correspond to regions in which waves emitted from the two gaps arrive out of phase, and create destructive interference; the areas where many molecules arrive correspond to regions where the waves arrive in phase, and create constructive interference. (pg. 66)
>
> . . . .
>
> In the 1940s Richard Feynman had a startling insight regarding the difference between the quantum and Newtonian worlds. Fenyman was intrigued by the question of how the interference pattern in the double-slit experiment arises. (pg. 74). . . .
>
> According to Newtonian physics—and to the way the experiment would work if we did it with soccer balls instead of molecules—each particle follows a single well-defined route from its source to the screen. There is no room in this picture for a detour in which the particle visits the neighborhood of each slit along the way. According to the quantum model, however, the particle is said to have no definite position during the time it is between the starting point and the endpoint. Feynman realized one does not have to interpret that to mean that particles take *no* path as they travel between source and screen. It could mean instead that particles take *every* possible

> path connecting those points. This, Feynman asserted, is what makes quantum physics different from Newtonian physics. The situation at both slits matters because, rather than following a single definite path, particles take every path, and they take them all *simultaneously*! That sounds like science fiction, but it isn't. Feynman formulated a mathematical expression—the Feynman sum over histories—that reflects this idea and reproduces all the laws of quantum physics (pg. 75). . . .
>
> In the double-slit experiment Feynman's ideas mean the particles take paths that go through only one slit or only the other; paths that thread through the first slit, back out through the second slit, and then through the first again; paths that visit the restaurant that serves that great curried shrimp, and then circle Jupiter a few times before heading home; even paths that go across the universe and back. This, in Feynman's view, explains how the particle acquires the information about which slits are open—if a slit is open, the particle takes paths through it. When both slits are open, the paths in which the particle travels through one slit can interfere with the paths in which it travels through the other, causing the interference. It might sound nutty, but for the purposes of most fundamental physics done today—and for the purposes of this book—Feynman's formulation has proved more useful than the original one. (pgs. 75 & 76)
>
> ——**STEPHEN HAWKING and LEONARD MLODINOW:** ***The Grand Design***; as found on the noted pages of the paperback published by Bantam Books Trade Paperbacks; New York, 2012. ©2010 Stephen Hawking and Leonard Mlodinow.

Yes! Even "paths that go across the universe and back"! The instant I read that, I knew: (1) what—though they have not the slightest inkling of it—they're *really* observing in their double slit experiments, and (2) why they neither do nor can have any inkling of it. *What* they were observing are the effects of the distinction between: (1) that *curvilinear* motion imposed on the *u-spaces* by the way their providing forms rotate in 3 planes, and (2) the rectilinear motion the generators execute by means of those whirling u-spaces. Every generator thus has 2 paths intertwined with one another—namely: the path of its *own rectilinear* motion and the path of *the u-spaces* it uses for its rectilinear motion.

To repeat what I just said on pages 149 & 150: In the course of c. every 10^{-58} sec., every multi-A-combo's carrying generator is: (1) using u-spaces provided by a form in Earth which causes those u-spaces to circle Earth a huge number of times at a huge number of angles while (2) its own carrying generator is using u-spaces provided by a form in the Sun which causes those u-spaces to circle the Sun a huge number of times at a huge number of angles while (3) its own carrying generator is using u-spaces provided by a form in the 8th reverse category which causes those u-spaces to circle the 8th reverse category a huge number of times at a huge number of angles while (4) its own carrying generator is using u-spaces provided by a form in the 7th reverse category which causes those u-spaces to circle the 7th reverse category a huge number of times at a huge number of angles while (5) its own carrying generator is using u-spaces provided by a form in the 6th reverse category which causes those u-spaces to circle the 6th reverse category a huge number of times at a huge number of angles while (6) its own carrying generator is using u-spaces provided by a form in the 5th reverse category's which causes those u-spaces to circle the 5th reverse category a huge number of times at a huge number of angles while (7) its own carrying generator is using u-spaces provided by a form in the 4th reverse category which causes those u-spaces to circle the 4th reverse category a huge number of times at a huge number of angles while (8) its own carrying generator is using u-spaces provided by a form in the 3rd reverse category which causes those u-spaces to circle the 3rd reverse category a huge number of times at a huge number of angles while (9) its own carrying generator is using u-spaces provided by a form in the 2nd reverse category's which causes those u-spaces to circle the 2nd reverse category a huge number of times at a huge number of angles. Under the influence of Omega, that 9 level pattern—completed once per c. 10^{-58} sec.—is then repeated 2^{128} times c. every 10^{-18} sec. as no generator's rectilinear motion exceeds 1Φ per pattern of c. 10^{-58} sec. in duration. I call attention to that latter fact just in case one has forgotten that a generator moving at the light speed limit of $1\Phi/2^{128}$K is changing from u-space to logically contiguous u-space at the rate of c. $1\Phi/10^{-58}$ seconds.

Each 9 level pattern is such that the whirling u-spaces—without any generator changing u-spaces more than once and most not at all—take the generators (and thus the combos they're carrying) along a vast number of paths at speeds so great, virtually every particle in the Universe can "go across the universe and back" in what is, for us, too small a fraction of a second to detect—namely: c. 10^{-58} sec. for our way of speaking.

> In Chapter 4 we saw how particles of matter fired at a screen with two slits in it could exhibit interference patterns just as water waves do. Feynman showed that this arises because a particle does not have a unique history. That is, as it moves from its starting point A to some point B, it doesn't take one definite path but rather simultaneously takes every possible path connecting the two points. From this point of view, interference is no surprise because, for instance, the particle can travel through both slits at the same time and interfere with itself.
> ——**STEPHEN HAWKING and LEONARD MLODINOW: *The Grand Design***; pages 135 & 136 of the paperback by Bantam Books Trade Paperbacks; New York, 2012. ©2010 Stephen Hawking and Leonard Mlodinow.

If that's what he says, Prof. Feynman (1918-1988) is way off the mark. Contrary to his contention, particle A *does* take one definite *rectilinear* path—one in the tertiary plane; but, as it does that using a generator shifting from u-space to logically sequential u-space, it's using u-spaces which—as the product of a form's *rotating* actuality—are plying an astronomical number of *curvilinear* paths thru 2 planes intersecting the tertiary one. In the higher of those 2, every form invariably executes 360^{0} paths involving one shift per K. At such a velocity, every RC form but Omega can execute at least 2^{128} curvilinear paths in 10^{-58} sec. and repeat them at least 2^{128} times in 10^{-18} sec.. Those paths are thus taking place "rather simultaneously" (Feynman's phrase in the above quote) because they occur in a time span so small, it is almost—if not totally—impossible to detect that they are not occurring "simultaneously" as Hawking and Mlodinow relate on page 75 of ***The Grand Design***.[1] In short, it's the *u-spaces* and *not the particles* taking "every possible path".

It should take little effort, then, to see how strenuously Esoptrics would reject this:

> According to quantum physics, no matter how much information we obtain or how powerful our computing abilities, the outcomes of physical processes cannot be predicted with certainty because they are not *determined* with certainty. Instead, given the initial state of a system, nature determines

---

[1] Notice, on page 75, they describe Feynman as saying, in italics, the particles take all the paths "*simultaneously*", but, above, describe him as saying they take all the paths "rather simultaneously" and, there, use no italics or comma after "rather". Esoptrics agrees with "rather simultaneously".

> its future state through a process that is fundamentally uncertain. In other words, nature does not dictate the outcome of any process or experiment, even in the simplest of situations. Rather, it allows a number of different eventualities, each with a certain likelihood of being realized. It is, to paraphrase Einstein, as if God throws the dice before deciding the result of every physical process.
> —**STEPHEN HAWKING & LEONARD MLODINOW: *The Grand Design*** page 72 of the paperback by Bantan Books Trade Paperbacks, New York, 2012. ©2010 Stephen Hawking & Leonard Mlodinow.

According to Esoptrics, Profs. Hawking and Mlodinow are far closer to the truth when, on page 74 of their book, they—though referring to a different issue—write:

> Our use of probabilistic terms to describe the outcome of events in everyday life is therefore a reflection not of the intrinsic nature of the process but only of our ignorance of certain aspects of it.

Esoptrics, though, says: *Even in the microscopic world of the atom and below*, the use of "probabilistic terms" by "quantum physics" is "a reflection not of the intrinsic nature of the process but only of" how ignorant Quantum Mechanic's proponents are "of certain aspects of" what's really happening in the ultramicroscopic realm. Indeed, their use of "probabilistic terms" even reflects how ignorant they are of their ignorance. For, how could anyone with even so little as half a grain of common sense insist that, "*no matter how much* information we obtain", we cannot predict "the outcomes of physical processes" "with certainty"? Would any but the densest of the dense fancy they see with certitude that there is nowhere any information able turn the tables on what "quantum physics" says about the predictability of "physical processes" in the world of tiny things?

> Putting the matter in a nutshell, classical probabilities correspond to ignorance and they combine by simple addition. Quantum probabilities combine in an apparently more elusive and unpicturable way. There then arises the question: Would it, nevertheless, be possible to understand quantum probabilities as also having their origin in the physicist's ignorance of all the detail of what is going on, so that the underlying basic probabilities, corresponding to inaccessible but completely detailed knowledge of what was the case, would still add up classically?
> —**REV. DR. JOHN POLKINGHORNE: *Quantum Theory, A Very Short Introduction***, page 42 of the paperback edition by the Oxford University Press, Oxford, 2002. ©2002 by John Polkinghorne.

If Esoptrics is correct, the answer to the reverend doctor's question is unmistakably affirmative. And what a blow that shall be to all those incessantly insisting most adamantly that the answer cannot possibly be yes!

> We shall see later that a deterministic interpretation of quantum theory is possible in which probabilities arise from ignorance of details. However, we shall also see that the theory that succeeds in this way has other properties that have made it seem unattractive to the majority of physicists.
> . . . . .
> In 1954, David Bohm published an account of quantum theory that was fully

deterministic, but which gave exactly the same experimental predictions as those of conventional quantum mechanics. In this theory, probabilities arise simply from ignorance of certain details. This remarkable discovery led John Bell to re-examine von Neumann's argument stating that this was impossible and to exhibit the flawed assumption on which this erroneous conclusion had been based.
——**REV. DR. JOHN POLKINGHORNE** on page 53 of the above work.

So, the impossible is not so impossible after all! Why, then, all this incorrigibly obstinate clinging to "Quantum *probabilities*"? You don't have to be a genius to see what it's out to accomplish. It's an attempt to exclude God from the picture. It's an attempt to create so much uncertainty and confusion (***i.e.:*** a modern "Tower of Babel" and more accurately a tower of *pseudo-scientific babble*), one can convince even the most critical of thinkers that belief in God is far from rational and intellectually sound. Thus, in their book, we find Profs. Hawking and Mlodinow *repeatedly* struggling to push God aside.

We will describe how M-theory may offer answers to the question of creation. According to M-theory, ours is not the only universe. Instead, M-theory predicts that a great many universes were created out of nothing. Their creation does not require the intervention of some supernatural being or god. Rather, these multiple universes arise naturally from physical law. They are a prediction of science. (pages 8 & 9)

Over the centures [Sic! "centures" for "centuries"-ENH] many, including Aristotle, believed that the universe must have always existed in order to avoid the issue of how it was set up. Others believed the universe had a beginning, and used it as an argument for the existence of God. The realization that time behaves like space presents a new alternative. It removes the age-old objection to the universe having a beginning, but also means that the beginning of the universe was governed by the laws of science and doesn't need to be set in motion by some god. (page 135)

Some would claim the answer to these questions is that there is a God who chose to create the universe that way. It is reasonable to ask who or what created the universe, but if the answer is God, then the question has merely been deflected to that of who created God. In this view it is accepted that some entity exists that needs no creator, and that entity is called God. This is known as the first-cause argument for the existence of God. We claim, however, that it is possible to answer these questions purely within the realm of science, and without invoking any divine beings. (page 172)[1]

Because there is a law like gravity, the universe can and will create itself

[1] Answer *Infinite* God, and the question has not "been deflected to that of who created God." For, by definition, The *Infinite* has no limits—least of all a need to be created. Oppositely, a self-created *finite* reality screams the question of how a *finite* reality could have a self-sufficiency that fantastic. Which makes more sense: that the *infinite* has it or that the *finite* does? H&M would have us believe that, if the *finite* can't have it, then neither can the *infinite*; and so, we must concede it to the finite to avoid a regression *ad infinitum*. They fail to note that to grant self-sufficiency to God ***is*** a one step regression *ad Infinitum*.

> from nothing in the manner described in Chapter 6. Spontaneous creation is the reason there is something rather than nothing, why the universe exists, why we exist. It is not necessary to invoke God to light the blue touch paper and set the universe going. (page 180)

Therein you see what makes it possible for supposedly great minds to stoop to dogmatically insisting no one can ever possibly garner enough information to overturn Quantum Theory's assurance that all is merely an endless series of probabilities in an endless series of universes governed by an endless series of different collections of laws (***The Grand Design***, pg. 143). For, what you see therein is a mind crippled by a predilection—not for whatever thinking explains what we observe in the extra-mental world—but, rather, for whatever thinking explains the observed in a way which excludes God from the picture.

What, then, is precisely the crippling effect of that exclusion? This: Quantum Mechanics' probabilities are merely the predictable result of not knowing the difference between the rectilinear motion of the generators vs. the curvilinear motion of the u-spaces, and that's a difference unknowable without first knowing what *God-produced* space is like.

Speaking of David Bohm (1917-1992), I am greatly fascinated by this:

> Bohm imagined that the wavefunction of a particle is another, *separate element of reality*, one that exists *in addition to the particle itself*. It's not particles *or* waves, as in Bohr's complementarity philosophy; according to Bohm, it's particles *and* waves. Moreover, Bohm posited that a particle's wavefunction interacts with the particle itself—it "guides" or "pushes" the particle around —in a way that determines its subsequent motion. While this approach agrees fully with the successful predictions of standard quantum mechanics, Bohm found that changes to the wavefunction in one location are able to immediately push a particle at a distant location, a finding that explicitly reveals the nonlocality of his approach.
>
> ——**BRIAN GREENE: *The Fabric Of The Cosmos***, page 206 of the 2005 paperback edition by Vintage Books, a division of Random House, Inc., New York. ©2004 by Brian Greene.

That's marvelously close to Esoptrics. Perhaps, one need only add: One must distinguish between: (#1) a form as the "wavefunction" of the particles (***i.e.:*** generators) looking to it for their u-spaces, and (#2) a form as the "wavefunction" of its carrying generator. For, every #2 form "guides" and "pushes" its carrying generator in one manner while every #1 form moves its "in potency" generators differently. Every #1 form determines the generators' subsequent motions by spinning the u-spaces they use in their rectilinear motions. Every #2 form determines from which form its carrying generator shall seek its u-spaces.

As for Bohm's take on "nonlocality", add this on how "changes to the wavefunction in one location are able to immediately push a particle at a distant location": The wavefunction changes pushing a particle at a distant location are not distant from that particle (***i.e.:*** generator). For, the wavefunction belongs either to the generator's piggyback form or to the form providing the generator with u-spaces. In either case, the whole of the form is in immediate contact with that generator whether it's the carrier for a duo- or a multi-combo. The sensation addicts' take on distance and locality is sadly mistaken.

Chapter Nineteen:

OF MOLECULES & MACROSCOPIC ORDER VS. MICROSCOPIC CHAOS:

Just as Esoptrics paints the region of the u-spaces as one of dizzying gyrations, so also does Quantum Mechanics describe the realm of the sub-atomic particles. They do so, though, knowing nothing about u-spaces, which is to say knowing nothing about what space *really* is—namely: an immense number of u-spaces envelopes many of which are logically concentric with one another in a multi-combo. Still, regardless of how each accounts for the mind-boggling hurly-burly down in that realm of the extremely tiny, each soon enough compels inquisitive minds to see in that supreme turbulence the same puzzle:

> Classical theories such as Newton's are built upon a framework reflecting everyday experience, in which material objects have an individual existence, can be located at definite locations, follow definite paths, and so on. Quantum physics provides a framework for understanding how nature operates on atomic and subatomic scales, but as we'll see in more detail later, it dictates a completely different conceptual schema, one in which an object's position, path, and even its past and future are not precisely determined. . . . .
>
> Can theories built upon a framework so foreign to everyday experience also explain the events of ordinary experience that were modeled so accurately by classical physics? They can, for we and our surroundings are composite structures, made of an unimaginably large number of atoms, more atoms than there are stars in the observable universe. And though the component atoms obey the principles of quantum physics, one can show that the large assemblages that form soccer balls, turnips, and jumbo jets—and us—will indeed manage to avoid diffracting through slits. So though the components of everyday objects obey quantum physics, Newton's laws form an effective theory that describes very accurately how the composite structures that form our everyday world behave.
>
> . . . . In the case of quantum physics, physicists are still working to figure out the details of how Newton's laws emerge from the quantum domain. What we do know is that the components of all objects obey the laws of quantum physics, and the Newtonian laws are a good approximation for describing the way macroscopic objects made of those quantum components behave.
>
> The predictions of Newtonian theory therefore match the view of reality we all develop as we experience the world around us. But individual atoms and molecules operate in a manner profoundly different from that of

> our everyday experience. Quantum physics is a new model of reality that gives us a picture of the universe. It is a picture in which many concepts fundamental to our intuitive understanding of reality no longer have meaning.
>
> . . . . but as we've said, in general, the larger the object the less apparent and robust are the quantum effects.
>
> —**STEPHEN HAWKING & LEONARD MLODINOW:** ***The Grand Design*** pages 67-68 of the paperback edition by Bantam Books Trade Paperbacks, New York, 2012. ©2010 Stephen Hawking & Leonard Mlodinow.

That's their rather elaborate description of what's much more succinctly set forth by another famous scientist who is also a Fellow of the Royal Society and—believe it or not—an Anglican priest. He puts the above so:

> One aspect of the problems we are considering in this chapter can be phrased in terms of asking how it can be that the quantum constituents of the physical world, such as quarks and gluons and electrons, whose behavior is cloudy and fitful, can give rise to the macroscopic world of everyday experience, which seems so clear and reliable.
>
> —**REV. DR. JOHN POLKINGHORNE:** ***Quantum Theory, A Very Short Introduction***, page 43 of the paperback edition by the Oxford University Press, Oxford, 2002. ©2002 John Polkinghorne.

I dare suggest Profs. Hawking and Mlodinow are begging the question. Basically, they're saying that, because there are stable macroscopic realities, the mad cap behavior patterns of the microscopic realities don't show up in the former. That, though, is precisely the question: How is it possible to have stable macroscopic realities ultimately composed of nothing but the most colossal chaos conceivable everywhere in the Universe at the microscopic level? Well, they do admit physicists are still trying to figure out the answer; and so, it's perhaps true they were fully aware that they were begging the question.

How is this contrast—between the microscopic domain of "cloudy and fitful" behavior vs. the macroscopic domain of behavior "so clear and reliable"—not a puzzle for Esoptrics? It's because Esoptrics easily explains the latter as the result of the stroboscopic effect of the molecular forms—major but irrational generic ones accelerated to somewhere between the MT zone's top at $OD2^{129}$ and the 8th reverse categories bottom at $OD2^{192}$. Consider the following example of what a molecular form is.

Let's imagine a particular duo-combo we shall call Robbie. Robbie's piggyback form is native to a generic but irrational OD such as 2^{127}, 2^{125}, 2^{123}, or the like. Robbie, being an irrational form, does not rotate in the upper planes and rotates oddly in the tertiary. Perhaps more importantly, if accelerated, Robbie does not produce a multi-combo, which is to say does not create a combo in which 2 or more duo-combos become *concentric* with one another. He does, however, produce what we can perhaps call a pyramid or p-combo.[1] If you prefer, we can also call it an m-combo as in molecular combo. Because Robbie's native OD is generic, he exerts a kind of attraction which we can perhaps call a weak version of gravity and which causes 2, 4 or 6 duo-combos to be in potency to him (***i.e.:*** to look to him to provide their carrying generation with their u-spaces). Why is that?

Looking at drawing #19-1 on the next page, let's imagine it represents Robbie. Next, imagine Robbie's A'C axis alone rotates 45° forward around the AB axis and parks there.

[1] Using the image of a pyramid may be misleading. For, Robbie may be somewhere in the interior of one part of an irregularly shaped object such as inside a human being's brain.

Robbie's A'C axis it thus slanted relative to his B'C' axis; and so, Robbie's A'C axis—because it's 1 of 3—exerts a "charge" of +1/3 at each end. To remedy that, 2 other irrational but generic duo-combos must come along in which the opposite is true. That means this: In each of them, and around the AB axis, the A'C axis has slanted 45° in the opposite direction. One of them—as its carrying generator uses the u-spaces provided to it by Robbie—moves to the A' end of Robbie's A'C axis, and the other does the same with regard to the C end of Robbie's A'C axis. The m-combo is now 2 layers deep and composed of 3 combos not concentric to one another. This is the simplest of the m-combos.[1]

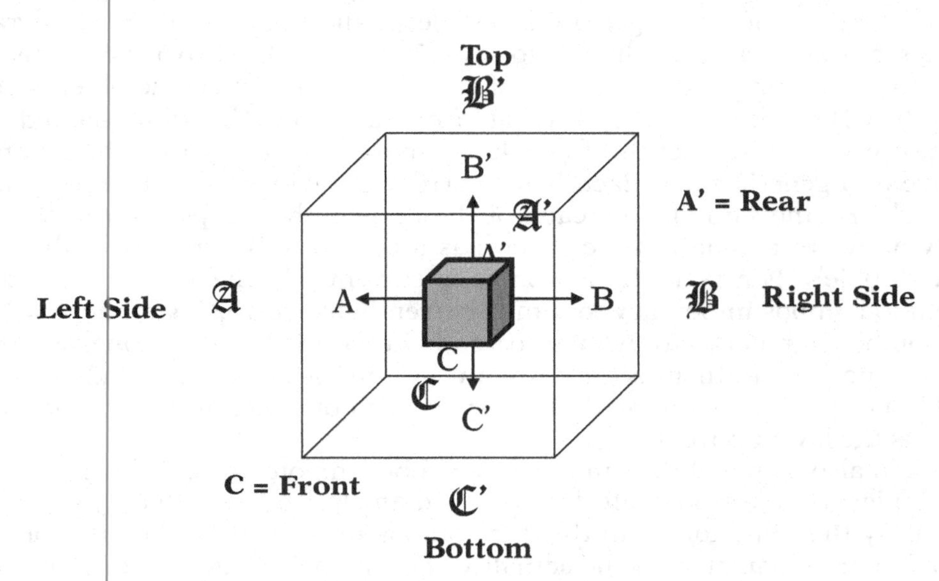

DRAWING #19-1: ROBBIE'S 3 AXES BEFORE ANY OF THEM ARE TILTED & LOCKED IN PLACE

What if, in each of Robbie's 2 adherents, the B'C' axis is slanted to the A'C axis. Each shall draw 2 duo-combos in which the B'C' axis has slanted around the A'C axis but in the opposite direction. The m-combo is now 3 layers deep.

How many layers can there be? The answer is how many *generic* zones (***i.e.:*** powers of 2) there are between Robbie's current OD and the mirror threshold, $OD2^{128}$. If Robbie is currently at or slightly above $OD2^{129}$, the m-combo can be only 2 generic layers deep, and the bottom layer consists of 2 atoms rather than 2 duo-combos. If Robbie is currently at or

[1] Atomic *multi*-combos can produce a similar result but in a different manner. As a non-rotating, irrational, but generic form, Robbie produces a charge by *slanting* an axis. In the atomic multi-combos, the leading form—as a rational, rotating one—produces a charge by the way it rotates. Because multi-combos include many forms concentric with one another; but duo-combos like Robbie do not—the former add the force of gravity to their charges, and the latter do not. As a result, in the world of sub-atomic *multi*-combos, charged particles (except for the electron which stands aloof because no multi-combo not an electron can capture forms native to OD1), pull tightly together; whereas, in the world of *duo*-combos like Robbie, the attracted duo-combos fan out to the limits of the form which is attracting them and providing them with u-spaces. That's another important part of Esoptrics' explanation of how the madly chaotic microscopic world can serve as the underpinning of a macroscopic world nowhere near that chaotic. The big problem here, of course, is that Robbie is a kind of reality of which the scientific world has not the slightest inkling.

slightly above $OD2^{140}$, the m-combo is 12 generic layers deep, and the bottom layer consists of 2^{12} atoms. If Robbie is currently at or slightly above $OD2^{152}$, the m-combo is 24 generic layers deep, and the bottom layer consists of 2^{24} atoms (***i.e.:*** 16,777,216).

Look again at drawing #19-1 on the prior page, and imagine this: Robbie's A'C axis has rotated to an intermediate stop around his AB axis; his B'C' axis has rotated to an intermediate stop around his A'C axis; and his AB axis has rotated to an intermediate stop around his B'C' axis. All 3 of Robbie's axes now exert a charge at each end. As a result, Robbie will now capture 6 duo-combos which may capture 6 each, which may capture 6 each, etc..[1] If, additionally, Robbie is currently at or slightly above $OD2^{152}$, Robbie will now be at the top of an m-combo 24 generic layers deep, and the bottom of that pyramid may consist of as many as 6^{24} atoms which is to say 4,738,381,338,321,616,896 atoms forming a rather large organic composite. It's a way of saying every macroscopic object's complexity depends on the OD of the m-combo's leading form and the number of its slanted axes.

In case you missed it, each of the duo-combos in potency to Robbie is more than merely native to a generic zone. Each has a current OD which is in the next generic zone below Robbie's. By the same token, each of the duo-combos in potency to them is more than merely native to a generic zone. Each has a current OD which is in the 2nd generic zone below Robbie's. For example: If Robbie is currently in the 153rd generic zone, then each of the duo-combos in potency to him is currently in the 152nd generic zone, and each of the duo-combos in potency to them is currently in the 151st generic zone, and each of the duo-combos in potency to them is currently in the 150th generic zone, and so on, until the result is a kind of pyramid composed of 24 layers each of which is either 2 or 4 or 6 times as extensive as the layer above it.

It should also be noted that, in the duo-combos in potency to Robbie, the "diameter" of each one's piggyback form is half that of the "diameter" of the actuality of Robbie's piggyback form. By the same token, in the duo-combos in potency to the duo-combos in potency to Robbie, the "diameter" of the actuality of each one's piggyback form is one-quarter that of the "diameter" of the actuality of Robbie's piggyback form. In the pyramid's third layer, the "diameter" of the actuality of each of that layer's piggyback forms is one-eighth that of the actuality of Robbie's piggyback form, then one-sixteenth, then one-thirty-second and so on, until, at the 24th layer, the "diameter" of each actuality is 2^{-23} that of Robbie's actuality.

At least, those are the *basic* "diameters". If we wish to be more specific, then we must say this: In the first layer below Robbie, the "diameter" may range from ½ to only a hair less than Robbie's diameter. In the second layer below Robbie, the "diameter" may range from ¼ to only a hair less than ½ of Robbie's "diameter". In the third layer below Robbie, the "diameter" may range from 1/8 to only a hair less than ¼ of Robbie's "diameter". I assume my reader follows the progression; and so, I will go no further with it.

What, now, is meant by Robbie's stroboscopic effect upon his environment? Remember how a strobe light works: We shine it upon the rapidly spinning blades of a fan and then fiddle with the rate at which the strobe light flashes. Adjust that rate just right, and, to our eyes, the blades cease to spin. Instead of a blur of motion making it impossible to count how many blades there are, we see each blade standing completely still and, therefore, can now count them quite easily. If we prefer, we can adjust the strobe light a little further, and the blades shall now rotate, but rotate so slowly, we can still easily count the

[1] Needless to say, I do not even begin to be intellectually competent enough to talk about the various ways in which such branching can progress. At best, all I can say is: The various species of crystals, plants, fishes, birds, animals, and what have you, illustrate how many and complex are the ways in which such branching can progress.

number of blades. Perhaps an even more revealing insight goes like this: Suppose, without making any adjustments to the strobe light, we point first at a fan blade spinning at 1500 RPM and then point at blades spinning 1525 RPM. The same one strobe light, in the first case, produces fan blades standing still and, in the second case, produces blades spinning but doing it so slowly, we can still count the blades rather easily.

In what sense, then, is Robbie such a strobe light? As an irrational form currently at c. $OD2^{152}$, Robbie's every cycle has 2 phases each with a duration of c. $(2^{152})^2 = 2^{304}$K. In other words: Robbie is a strobe light with a "flash rate" of 2^{304}K per flash. For Robbie, then, every form with the same, identical "flash rate" is stationary while those with a slightly higher or lower rate are—from his standpoint—moving slightly faster or slightly slower.

But, what has that got to do with the chaos in the sub-atomic world? Note more precisely what the strobe light is doing when pointed at the fans: It's causing us to see the spinning blades as occupying per minute a small fraction of the number of positions they actually occupy per minute. For the strobe light (***i.e.:*** for the stroboscopic effect of the molecular form), the first fan is occupying only 1 out of 1500 and the other 25 out of 1525. That's the same result Robbie has upon the chaos in the microscopic world: His "flash rate" causes it to appear that per 2^{312}K the u-spaces and their occupants are occupying an astronomically small fraction of the positions they actually occupy per 2^{312}K.

In its description of the molecular forms and their stroboscopic effect, Esoptrics is presenting its never before heard answer to what some scientists dub "the quantum measurement problem". It involves what's called the stage one vs. the stage two formulation of quantum theory, and the observed fact that, in stage one, Erwin Schrödinger's (1887-1961) famous 1926 equation remains "the basic equation of quantum mechanics"[1] as it continues to account accurately for the "wavefunction" of the microscopic domain's occupants such as electrons, protons, atoms, and the like. For all of that, when, in stage two, experimenters probe the quantum world, the Schrödinger equation's "ironclad power over quantum processes"[2] collapses. One scientist spells out the problem so:

> The stage one/stage two formulation of quantum theory, even though stage two has remained mysterious, predicts probabilities for measuring one outcome or another. . . . [Skip!] . . . . The fantastic experimental success of this approach has far outweighed the discomfort of not having a precise articulation of what actually happens in stage two.
>
> But the discomfort has always been there. And it is not simply that some details of wavefunction collapse have not quite been worked out. The *quantum measurement problem*, as it is called, is an issue that speaks to the limits and the universality of quantum mechanics. It's simple to see this. The stage one/stage two approach introduces a split between what's being observed (an electron, or a proton, or an atom, for example) and the experimenter who does the observing. Before the experimenter gets into the picture wave-functions happily and gently evolve according to Schrödinger's equation. But, then, when the experimenter meddles with things to perform a measurement, the rules of the game suddenly change, Schrödinger's equation is cast aside and stage-two collapse takes over. Yet, since there is no difference between the atoms, protons, and electrons that make up the experimenter and

[1] The above phrase comes from page 928 of the 1995 hardbound edition of Merriam Webster's ***Biographical Dictionary***.

[2] The phrase is from page 202 of Prof. Greene's ***The Fabric Of The Cosmos***.

> the equipment he or she uses, and the atoms, protons, and electrons that he or she studies, why in the world is there a split in how quantum mechanics treats them? If quantum mechanics is a universal theory that applies without limitations to *everything*, the observed and the observer should be treated in exactly the same way.
>
> Niels Bohr disagreed. He claimed that experimenters and their equipment *are* different from elementary particles. Eventhough they are made from the same particles, they are "big" collections of elementary particles and hence governed by the laws of classical physics. Somewhere between the tiny world of individual atoms and subatomic particles and the familiar world of people and their equipment, the rules change because the sizes change. The motivation for asserting this division is clear: a tiny particle, according to quantum mechanics, can be located in a fuzzy mixture of here and there, yet we don't see such behavior in the big, everyday world. But exactly where is the border?
>
> ——**BRIAN GREENE:** ***The Fabric Of The Cosmos***, pages 202 & 203 of the 2005 paperback edition by Vintage Books, a division of Random House, Inc., New York. ©2004 Brian Greene.

That's right! Scientists can't tell you *where the border is*; but, Esoptrics can tell you *exactly* where it is—namely: at that mirror threshold invariably found at $OD2^{128}$ throughout the course of this the 9th epoch of the Universe. Niels Bohr (1885-1962) was right: "Experimenters and their equipment *are* different from elementary particles." They contain something *more* than "the atoms, protons, and electrons that he or she studies". He, however, did not have the foggiest notion regarding *what* that something *more* is. That's because, like all scientists living and dead, he had no knowledge whatsoever of even so little as a single, solitary, one of the principles set forth by Esoptrics—let alone what it says about major forms accelerated from below to above the MT in OD (***i.e.:*** level of the 4th dimension).

Despite what Prof. Greene & co. say, the difference between our realm and the quantum one is not mainly a matter of size; it's mainly a matter of the difference between the kind of major forms still *at* or *below* the MT vs. those irrational generic ones accelerated *above* it but still below the 8th reverse category's floor. The former have many duo-combos logically concentric with them; the latter do not. The former rotate at fantastic speeds in the primary and secondary planes; the latter do not rotate at all in those planes. The former relentlessly rotate programmatically in the tertiary plane; the latter, if they rotate in that plane, do so only variably.[1] Above all, the latter persistently impose a stroboscopic effect upon the forms at or below the MT; whereas, those at or below the MT never, to any extent, impose such an effect upon any forms above the MT.

For Esoptrics, then, the answer to the "quantum measurement problem" is quite simple: Wholly unlike the MT and *sub*-MT particles, the experimenter has, additionally, a *supra*-MT, hierarchical structure of *molecular forms* whose *stroboscopic effect* is interfering with the very different behavior of forms of a very different ilk.

✡✞✡✞✡✞✡✞✡✞✡✞✡✞✡✞✡✞✡✞✡✞✡✞✡✞

[1] Probably, inorganic molecular forms rotate in the tertiary plane only to the extent they are *externally* moved to do so; whereas, organic molecular forms rotate in the tertiary plane to the extent they are *self* moved to do so.

Chapter Twenty:

DARK MATTER & DARK ENERGY:

For Esoptrics, there is no such thing as matter which *really* is dark. There is only matter which is dark *for us*, because we can't detect the light "dark matter" is emitting. It's dark for us because its *emitted energy* is dark for us. But, how is there light and other kinds of emitted energy we can't detect? The answer lies in a large difference between: (1) the ontological depth of (***i.e.:*** the 4th dimension levels in) the multi-h-combo providing the observer with u-spaces, and (2) the OD (***i.e.:*** level of the 4th dimension) at which the emitted energy and light are traveling thru the Universe. How's that?

First of all, remember what's been said repeatedly: In its 4th dimension, every u-space—while remaining c. 10^{-47} cm. in "size"—may be from 1 to 2^{256} –1 layers thick. The number of 4th dimension layers in a given u-space depends upon how many duo-combos are concentric at it and/or how many forms' u-spaces are overlapping it. The latter depends upon how far it is from the centers of nearby multi-combos. Consider an example—one ignoring overlaps due to nearby multi-combos.

Suppose a particular multi-g-combo in the 2nd reverse category (***i.e.:*** OD's 2^{255} thru 2^{256}–1) has an ontological depth of $2^{255}+2^{254}$. Let's call it Zeon. Zeon's carrying generator is using the u-spaces provided by Omega (***i.e.:*** the OD2^{256} form). At some particular number of K into the 9th epoch, Zeon's generator is using some particular one of Omega's u-spaces. Call it Zero. As long as Zeon's generator is at Zero, Zero's 4th dimension has $2^{255}+2^{254}$ levels, because that many of Zeon's forms and generators are logically concentric at Zero. For all u-spaces 1Φ from Zero, their 4th dimension has $(2^{255}+2^{254})$ –1 levels, because all but 1 of Zeon's forms overlap them. For all u-spaces 2Φ from Zero, their 4th dimension has $(2^{255}+2^{254})$ –2 levels, because all but 2 of Zeon's forms overlap them. For all u-spaces 3Φ from Zero, their 4th dimension has $(2^{255}+2^{254})$ –3 levels, because all but 3 of Zeon's forms overlap them. I assume my reader catches the drift.

The negative ciphers of minus 1, 2, and 3 apply to the lower end of the OD's involved. That's because: For the u-spaces 1Φ from Zero, there is no overlap at OD1; for the u-spaces 2Φ form Zero, there are no overlaps at OD's 1 or 2, etc.. That, of course, remains true only as long as Zeon remains at Zero. When Zeon moves 1Φ from Zero, it is then Zero's 4th dimension which lacks the overlap at OD1.

Suppose Zeon, at Zero, emits one or more duo-combos at OD2^{128} at light speed. Call that collection Epsilon. Climbing the "ladder" of Zero's 4th dimension at the rate of one "rung" per K, Epsilon can, in c. 10^{-18} sec., reach the OD of Zeon's leading form. At that OD of $2^{255}+2^{254}$, Zeon's leading form offers Epsilon a u-spaces envelope c. 6.75 trillion light years in radius. Epsilon is thus free to navigate the Universe at light speed for 6.75 trillion of our years. That, though, is what it's doing at an OD of $2^{255}+2^{254}$. So what?!

At Earth's center, says Esoptrics, is a multi-h-combo whose maximum ontological depth is c. 2^{197}. How can such a combo intercept—and make available to those using one of its u-spaces envelopes—light or energy crossing the Universe at $OD2^{255}+2^{254}$? Though they pass right thru it, there's no way it can do so, because they're passing thru at an OD and 4th dimension level astronomically far above the highest such in Earth's multi-h-combo.

But, what about the Sun?! If, as Esoptrics opines, the ontological depth of the multi-h-combo at it center is c. 2^{227}, can it intercept—and make available to those using one of its u-spaces envelopes—light or energy crossing the Universe at $OD2^{255}+2^{254}$? Here again, it cannot, because they're passing thru the Sun at an OD and 4th dimension level vastly far above the highest OD and 4th dimension level found in the Sun's multi-h-combo.

What about the Milky Way?! Esoptrics suggests the multi-g-combo at its center has an ontological depth of 2^{229} thereby giving it a leading form whose actuality has a dia. of c. 133,000 light years. Can such a multi-g-combo intercept—and make available to those using some one of its u-spaces envelopes—light or energy crossing the universe at $OD2^{255}+2^{254}$? No, because the energy is passing thru it at an OD and 4th dimension level far above the highest OD and 4th dimension level in the Milky Way's multi-g-combo.

The notion of ontological distance—and, thus, four dimensional u-spaces—easily explains dark matter and energy. It also rather easily explains why the universe we observe has a diameter much smaller than the one attributed to it by Esoptrics. How so?!

In its article, ***Universe***, Wikipedia gives the age of the Universe as 13.7 billion years plus or minus 120 million. That would lead one to believe it's *radius* should be 13.7 billion light years and its *diameter* 27.4 billion light years. Not so! In its article, ***Observable Universe***, Wikipedia tells us the observable universe is 93 billion light years in diameter, but also mentions 78 billion light years as the diameter suggested by the cosmic microwave background radiation. As you might expect, it explains the difference between 27.4 billion and the two larger diameters is due to the expansion of "space itself".

#	8 CATEGORIES BELOW MT	VS.	8 REVERSE CATEGORIES ABOVE MT
1st	$OD1 = 8^{256}$ Alpha duos[1]	VS.	$OD2^{256} = 1$ Omega duo[2]
2nd	$2^{1}(1) = OD2 \sim OD3$	VS.	$2^{-1}(2^{256}) = OD2^{255} \sim OD2^{256}-1$
3rd	$2^{2}(1) = OD4 \sim OD15$	VS.	$2^{-2}(2^{256}) = OD2^{254} \sim OD2^{255}-1$
4th	$2^{4}(1) = OD16 \sim OD255$	VS.	$2^{-4}(2^{256}) = OD2^{252} \sim OD2^{254}-1$
5th	$2^{8}(1) = OD256 \sim OD2^{16}-1$	VS.	$2^{-8}(2^{256}) = OD2^{248} \sim OD2^{252}-1$
6th	$2^{16}(1) = OD2^{16} \sim OD2^{32}-1$	VS.	$2^{-16}(2^{256}) = OD2^{240} \sim OD2^{248}-1$
7th	$2^{32}(1) = OD2^{32} \sim OD2^{264}-1$	VS.	$2^{-32}(2^{256}) = OD2^{224} \sim OD2^{240}-1$
8th	$2^{64}(1) = OD2^{64} \sim 2^{128}-1$	VS.	$2^{-64}(2^{256}) = OD2^{192} \sim OD2^{224}-1$

In the above chart from page 111, the fifth reverse category includes the OD's of 2^{248} to $2^{252}-1$; and so, multi-g-combos moving in the 5th reverse category have ontological depths of anywhere from 2^{248} to $2^{252}-1$. A multi-g-combo with an ontological depth of 2^{248} would have a leading form whose actuality has a diameter of c. 70.25 billion light years (***i.e.:*** $2^{249}\Phi$). A multi-g-combo having an ontological depth of 2^{249} would have a leading form whose actuality has a diameter of c. 140.5 billion light years (***i.e.:*** $2^{250}\Phi$).

Suppose, in the 5th reverse category, there is a super cluster of galaxies, and each of the clusters has at its center a multi-g-combo with an ontological depth of 2^{248} plus some number between 2^{248} and $2^{249}-1$. Call each a Kappa class cluster. In each of these Kappa

---

[1] The term "Alpha duo" = any duo-combo whose piggyback form is native to OD1.

[2] The term "Omega" = the one duo-combo whose piggyback form is native to $OD2^{256}$.

class clusters, the leading form's carrying generator is receiving its u-spaces from a multi-g-combo in the next highest generic zone, which, in this case, is one with an ontological depth of 2^{249}. As just pointed out, the leading form of such a multi-g-combo would have a diameter of c. 140.5 billion light years. That being the case, the Kappa class cluster can expand to a diameter of 140.5 billion light years, thus putting us within the range of the diameters presented by Wikipedia.

To what extent could these Kappa class clusters intercept—and offer those using their u-spaces—light or energy emitted by multi-h and multi-g-combos in higher generic zones and reverse categories? Above its first generic zone (***i.e.:*** 2^{248} thru $2^{249}-1$) the 5th reverse *category* includes the 3 generic zones of: (1) OD2^{249} to $2^{250}-1$, (2) OD2^{250} to $2^{251}-1$, and (3) OD2^{251} to $2^{252}-1$. Light and energy emitted in any of those zones could be crossing the Universe at OD$2^{252}-1$ at least. For, there is no law saying they can't find a way to climb to the 3rd or 2nd reverse categories. At an OD between 2^{248}& $2^{249}-1$, can any of the Kappa clusters' leading forms intercept—and offer those using their u-spaces—energy crossing the Universe at an OD and 4th dimension level anywhere from 2^{250} to $(2^{256}-2^{250})$ levels higher than their own? If not, then nothing in those Kappa class clusters or any of the generic zones and reverse categories below them can ever receive any of the light or energy from the generic zones and reverse categories above them. Therefore, to all observers residing on some planet in either the 6th, the 7th, or the 8th reverse categories, the Universe must appear to have a current diameter somewhat less than, but expanding toward, a diameter of c. 140.5 billion light years. Thus does Esoptrics' take on dark matter and energy explain how the universe we observe is much smaller than what Esoptrics suggests.[1]

> *This is a remarkable number*. If it's correct, then not only does ordinary matter—protons, neutrons, electrons—constitute a paltry 5 percent of the mass/energy of the universe, and not only does some currently unidentified form of dark matter constitute at least *five times* that amount, but also the *majority* of the mass/energy in the universe is contributed by a totally different and rather mysterious form of dark energy that is spread throughout space.
> ——**BRIAN GREENE:** ***The Fabric Of The Cosmos***, pages 300 & 301 of the 2005 paperback edition by Vintage Books, a division of Random House, Inc., New York. ©2004 Brian Greene.

No, Prof. Greene! Dark matter is not particles other than those we know; it's the same ones, but emitting light and energy traversing the Universe at OD's far above the OD's of the leading forms in our observable universe's multi-g and multi-h-combos. So to speak, our "antennas" ("ladders") are too "short" to reach as far up into the 4th dimension as they need to reach in order to intercept what you mistakenly call "dark energy" from "dark matter". You are closer to the truth when, on page 295, you state that dark matter's "identity" is still "a major, looming mystery". Still, that wins you no gold star. For, you fail to realize that, though it's still a great mystery for you and your fellow scientists, it is by no means such for Esoptrics due to what it knows of God's 7 *a priori* coordinates.

✡✞✡✞✡✞✡✞✡✞✡✞✡✞✡✞✡✞✡✞✡✞✡✞✡✞

[1] On pages 285 & 293, Prof. Greene states the observable universe is a tiny fraction of the whole—thus perhaps deeming the Universe even larger than Esoptrics says it is.

Einstein found he could have his cake and eat it too: he could maintain all the appealing experimentally confirmed features of general relativity while basking in the eternal serenity of an unchanging cosmos, one that was neither expanding nor contracting.

With this result, Einstein no doubt breathed a sigh of relief. How heart-wrenching it would have been if the decade of grueling research he had devoted to formulating general relativity resulted in a theory that was incompatible with the static universe apparent to anyone who gazed up at the night sky. But, as we have seen, a dozen years later the story took a sharp turn. In 1929, Hubble showed that cursory skyward gazes can be misleading. His systematic observations revealed that the universe is *not* static. It *is* expanding.

——**BRIAN GREENE: *The Fabric Of The Cosmos***, page 279 of the 2005 paperback edition by Vintage Books, a division of Random House, Inc., New York. ©2004 Brian Greene (Here again we have a vividly telling example of how tricky is the ground one treads when relying upon what's *inferred* by what the extra-mental world presents to the outward looking mind. It's what Prof. Greene himself admits when, on page 5 of his book, he writes that "The overarching lesson" of "the last century" of "scientific inquiry" is that "human experience is often a misleading guide to the true nature of reality." Even hundreds of years before Christ, many a writer warned of that.).

(Experience) . . . . represents at best only a stage on the way to real understanding in terms of universals and is thus by most ancient writers despised as a makeshift and uncertain form of knowledge (pg. 156). . . . .

The experience from which the empiric draws his conjectures is, of course, the homely and substantial experience of a world of public objects, which forms for all sane and unreflective persons the basis of ordinary life. It has been regularly insisted, however, since the earliest times, that experience in this sense is nothing ultimate: the alleged paradoxes of motion and change and the more familiar facts of perceptual error and illusion are enough (it is thought) to show that it cannot be straightforwardly identified with the real. Hence, in addition to the rejection of habit learning as a road to knowledge, there arises that further prejudice against the deliverances of the senses and in favor of necessary reasoning from first principles, of which the Parmenidean distinction between the "ways" of truth and opinion is an early and famous example.

——**H. L. HEATH: *Experience***, pages 156 & 157 in volume 3 of the 1972 reprint of ***The Encyclopedia Of Philosophy*** published by the Macmillan Publishing Co., Inc. & The Free Press. ©1967 Macmillan, Inc. (Yes, "since the earliest times", it's been widely known and admitted that relying on what sense imagery *infers* is risky business. Why?! New tools reveal previously hidden details which, when observed, cause sense imagery to infer very differently from what they inferred previously. Hubble is an ideal example. Prior to his observations, celestial gazing inferred a *static* Universe; after his, it inferred an *expanding* one. *Inferred* conclusions always run the risk of being exploded by newly discovered details. As the old saying goes: The devil is in the details. It's doubly true for *inferred* conclusions—whether those of mono- or bi-directional empirical observation.).

Chapter Twenty-One:

ESOPTRICS' MONONS, DIONS & SEXTONS VS. SCIENCE'S LEPTONS & QUARKS:

Every form native to OD1 is a monon. I use the term "monon" because it applies to only one categorical OD: $OD2^0$ (***i.e.:*** OD1). When a monon accelerates to become the leading form in a sub-atomic multi-combo (one necessarily composed solely of monons), the resulting particle is also a monon or "multi-monon" or "monon particle" vs. "duo-monon", whichever you prefer. I equate it with Science's electron.

Every form native to either OD2 or OD4 is a dion. I use the term "dion", because it applies to those two categorical OD's only. If a form native to OD2 accelerates to become the leading form in a sub-atomic multi-combo, the resulting particle is also a dion or "multi-dion", etc.. If a form native to OD4 accelerates to become the leading form in a sub-atomic multi-combo, the resulting particle is also a dion or "multi-dion", etc..

Every form native to either $OD2^4$ or 2^8 or 2^{16} or 2^{32} or 2^{64} or 2^{128} is a sexton. I use the term "sexton" because it applies to those 6 categorical OD's. Wherever one of these is the leading form in a sub-atomic multi-combo, the result is a sexton particle, etc..

Monon and dion particles can be looked upon as 2 species within a genus called "sub-sextons". I deem Esoptrics' "sub-sextons" equivalent to Science's leptons. As with Science's leptons, all monon and dion particles are necessarily three-thirds non-neutral, which is to say they have a charge of 1 and either positive or negative. Why is that?

In chart #8-3, pg. 62, the categorical forms native to OD's 1, 2, or 4 have either no rotations at all in either the primary or the secondary planes or no *definitive* ones. They, then, cannot achieve neutrality in either of those 2 planes. That, in turn, implies they never achieve neutrality in the tertiary plane either and, thus, are always 3-thirds non-neutral and, as such always have a charge of 1 whether positive or negative.

For forms native to OD1, not even the tertiary plane has a *definitive* rotation at OD1. That leads me to speculate all forms native to OD1 rotate the same way in the tertiary plane no matter what their current OD. If so, all are not merely non-neutral on all 3 axes; additionally, each has a charge of 1 which, in every one of them, is of the same one polarity.

For me, chart #8-3 also implies this: Some forms native to OD2 or 4 rotate and others counter-rotate in the tertiary plane. Some, then, would have a charge of +1, others –1.

In contrast to the sub-sextons, the sextons either: (1) are neutral in all 3 planes and, thereby, have no charge, or (2) are neutral in 2 of the 3 planes and, thereby, have a charge of either plus 1/3 or minus 1/3, or (3) are neutral in only one of the 3 planes and, thereby, have a charge of either plus 2/3 or minus 2/3. That's due in part to the fact that, unlike the sub-sextons, sextons can either rotate or counter-rotate in the primary and secondary planes. It's also due in part to a complex issue with which I have wrestled at great length. Indeed, to this day, I'm not all that confidant I've resolved it. What is that issue? Drawing #21-1 on the next page shall help us to answer that question.

2013 ESOPTRICS UPDATE

Drawing #21-1 is a copy of drawing #2-6 on page 17. I can no longer remember how long ago I first drew it; but, it probably goes back over 30 years. Its algebraic version goes back to 1961. That's a well documented assertion, since I have a hard copy of a book I wrote in 1961, and, on its page 242, one finds the expression $(+A) \otimes (-A) \rightarrow 0$.

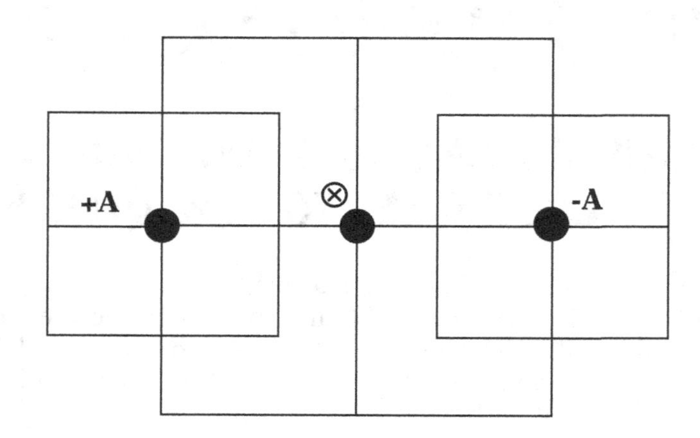

DRAWING #21-1: GEOMETRICAL EXPRESSION OF ESOPTRICS' FIRST PRINCIPLE.

For many years, I suspected the above figure was giving me a key to the structure of the hydrogen atom; and yet, I persistently resisted what it rather clearly implied. How is that? Over and over again, it seemed to be telling me that neutrality is achieved at the center, ⊗, because +A and –A are rotating in opposite directions around ⊗. I rejected that as impossible because, for it to be possible, ⊗ would have to be the center of two forms simultaneously at the same OD in the same one duo- or multi-combo and rotating in opposite directions. But, two forms simultaneously at the same OD in the same combo???!!!

For some time, though, I found myself repeatedly musing: Suppose ⊗ is a rational form rotating *clockwise* around the B'C' axis and, thereby, persistently executing the pattern: A → A' → B → C → A in the tertiary plane. What if ⊗ is also a rational form rotating *counter*-clockwise around the B'C' axis and, thereby, persistently executing the pattern: A → C → B → A' → A in the tertiary plane? Is that enough of a difference for each to maintain its individuality despite the fact both are simultaneously at the same OD in the same combo? And what if each is native to the same OD as they would have to be for +A and –A to rotate in opposite directions at the same velocity? In 2009, after perhaps years of toying with the idea, I decided to answer yes.[1]

[1] Some readers may object that +A and –A must collide. That they would do, if their curvilinear motions were *continuous*; but, because they are discontinuous, they shall—*figuratively* speaking—leap past one another. More accurately, at two stages in each revolution, they shall merely *exchange* logically contiguous "locations" (relationships to a heavenly sextet would be more accurate) in the tertiary plane. I don't remember where I read it many years ago; but, if memory serves me correctly, one of the observable phenomena which puzzles nuclear physicists is the observable fact of two particles executing orbits which should produce a head on collision; and yet, no collision occurs. For Esoptrics, there's nothing puzzling about that at all.

Almost immediately, my mind puzzled: But, that would produce neutrality on the B'C' axis only. What about the other two? If only 1 out of the 3 axes is neutral, doesn't that leave ⊗ two-thirds non-neutral? As is often the case with my mind, a question asked is a question soon answered; and so, I soon found myself insisting: At least for every sexton (***i.e.:*** forms native to OD's 2^4, 2^8, 2^{16}, 2^{32}, 2^{64}, & 2^{128}), there must be 6 kinds of forms. More specifically, there must be 2 modes for each of 3 types. Each type is distinguished by which axis is the axis of its rotation in the tertiary plane, and each mode is distinguished by which way it rotates around the axis of its tertiary plane. The first result was:

NAME	TERTIARY PLANE'S AXIS OF ROTATION	SPIN METHOD
+Aleph	AB	A' → B' → C → C' → A'
–Aleph	AB	A' → C' → C → B' → A'
+Beth	A'C	A → B' → B → C' → A
–Beth	A'C	A → C' → B → B' → A
+Daleth	B'C'	A → A' → B → C → A
–Daleth	B'C'	A → C → B → A' → A

CHART #21-1: THE 3 TYPES OF FORMS NATIVE TO OD's 2^4, 2^8, 2^{16}, 2^{32}, 2^{64} & 2^{128} TOGETHER WITH THE 2 MODES OF EACH.

Hereafter, then, if a sexton's axis of rotation in the tertiary plane is AB, it shall be called an Aleph sexton. If it rotates around that axis clockwise, it shall be called a positive Aleph or +Aleph for short. If it rotates around that axis counter-clockwise, it shall be called a negative Aleph or –Aleph for short.

If a sexton's axis of rotation in the tertiary plane is A'C, it shall be called a Beth sexton. If it rotates around that axis clockwise, it shall be called a positive Beth or +Beth for short. If it rotates around that axis counter-clockwise, it shall be called a negative Beth or a –Beth for short.

If a sexton's axis of rotation in the tertiary plane is B'C', it shall be called a Daleth sexton. If it rotates around that axis clockwise, it shall be called a positive Daleth or +Daleth for short. If it rotates around that axis counter-clockwise, it shall be called a negative Daleth or –Daleth for short.

For a while, my mind was satisfied with the thinking expressed in the above chart. Soon enough, though, a disturbing thought arose: Chart #21-1 implies that ⊗ can be as many as 6 forms simultaneously at the same OD in the same multi-combo; but, drawing #21-1, page 168, clearly suggests only 2. Furthermore, is drawing #21-1 saying neutrality at ⊗ is possible only if ⊗ is circled by two multi-combos which are mirrored images of one another, or is it saying the neutrality at ⊗ is the result of the fact that, at ⊗, there are 2 rational forms which are mirrored images of one another? Soon enough, my mind suggested each of the 3 basic types must have 6 modes rather than merely 2. That speculation produced chart #21-2 on the next page.

"*Simple* types" implies *compound* types in each of which 2 of the simple types are combined as called for by drawing #21-1 on page 168. The possible combinations and their results are given on page 171 in chart #21-3. In that chart, neutrality with regard to some one of the three axes becomes neutrality with regard to some one of the three planes. Neutral in a plane means 2 forms are logically concentric to one another at the same OD and are rotating in opposite directions in the same plane. To say that a *plane* has a charge of 1/3 means it's one of the 3 planes and is not neutral. To say that a *combo* has a charge of 2/3 means 2 of its 3 planes are not neutral. To say a *combo* has a charge of zero means that, in all three of its planes, 2 forms are rotating in opposite directions.

	AXIS AND DIRECTION OF ROTATION IN		
NAME	TERTIARY PLANE	SECONDARY PLANE	PRIMARY PLANE
$Aleph_1$ = A1	AB+	A'C+	B'C'+
$Aleph_2$ = A2	AB+	A'C+	B'C'-
$Aleph_3$ = A3	AB+	A'C-	B'C'-
$Aleph_4$ = A4	AB-	A'C-	B'C'-
$Aleph_5$ = A5	AB-	A'C-	B'C'+
$Aleph_6$ = A6	AB-	A'C+	B'C'+
$Beth_1$ = B1	A'C+	B'C'+	AB+
$Beth_2$ = B2	A'C+	B'C'+	AB-
$Beth_3$ = B3	A'C+	B'C'-	AB-
$Beth_4$ = B4	A'C-	B'C'-	AB-
$Beth_5$ = B5	A'C-	B'C'-	AB+
$Beth_6$ = B6	A'C-	B'C'+	AB+
$Daleth_1$ = D1	B'C'+	AB+	A'C+
$Daleth_2$ = D2	B'C'+	AB+	A'C-
$Daleth_3$ = D3	B'C'+	AB-	A'C-
$Daleth_4$ = D4	B'C'-	AB-	A'C-
$Daleth_5$ = D5	B'C'-	AB-	A'C+
$Daleth_6$ = D6	B'C'-	AB+	A'C+

CHART #21-2: 3 SIMPLE TYPES OF SEXTONS & THE 6 BASIC MODES OF EACH.

In chart #21-3, page 171, I've dealt only with the Alphas listed in the above chart. I see no need to spend the time and effort it would take to repeat the above chart in so far as it pertains to the Betas and the Deltas listed in the above chart. After all, I think you can see for yourself that, though the arrangement of the letters would be different, the results would be the same—namely: 3 neutral combos and 12 with charges of either +1/3, -1/3, +2/3, or -2/3. Since that's true of each of the 3 sets of Alpha, Beta, and Delta types, the end result is 36 compound types each with one of the mentioned fractional charges.

That, as you may or may not know, agrees rather nicely with what the physicists say about the kinds of quarks which form the nucleus of the atom. The big difference here is this: They see the quarks as ultimate particles; whereas, Esoptrics has always and consistently maintained that even leptons and quarks break down into an astronomical number of duo-combos. In my current thinking, duo-combos are either photons (***i.e.:*** 2 or perhaps more duo-combos *linked* as opposed to *concentric*) or neutrinos and anti-neutrinos (***i.e.:*** 1 duo-combo each).

Let's again touch on Esoptrics' explanation of why none of the above pertains to the monons and dions (***i.e.:*** what Esoptrics calls sub-sextons and equates with Science's leptons). The OD's 2^0, 2^1, and 2^3 do not have integral square roots, and the fact that their native OD is a power of 2 is here meaningless. For, it's impossible for 2^0 and 2^1 to have any distinct rotations in either of the upper 2 planes and impossible for 2^3 to have *only* distinct rotations in both of those planes. For OD1, distinct rotations are impossible even in the tertiary plane, until forms native to that OD accelerate beyond it. $OD2^2$ has an integral square root, but cannot have a single, solitary distinct rotation in either the primary or the secondary planes.

COMBO	3rd PLANE	2nd PLANE	1st PLANE	CHARGE
A1+A2	AB+/AB+ = +1/3	A'C+/A'C+ = +1/3	B'C'+/B'C– = 0	+2/3
A1+A3	AB+/AB+ = +1/3	A'C+/A'C– = 0	B'C'+/B'C'– = 0	+1/3
A1+A4	AB+/AB– = 0	A'C+/A'C–= 0	B'C'+/B'C– = 0	0
A1+A5	AB+/AB– = 0	A'C+/A'C– = 0	B'C'+/B'C'+ = +1/3	+1/3
A1+A6	AB+/AB– = 0	A'C+/A'C+ = +1/3	B'C'+/B'C'+ = +1/3	+2/3
A2+A3	AB+/AB+ = +1/3	A'C+/A'C– = 0	B'C–/B'C'– = -1/3	±2/3?[1]
A2+A4	AB+/AB– = 0	A'C+/A'C– = 0	B'C'–/B'C'– = –1/3	–1/3
A2+A5	AB+/AB– = 0	A'C+/A'C'– = 0	B'C'–/B'C'+ = 0	0
A2+A6	AB+/AB– = 0	A'C+/A'C+ = +1/3	B'C'–/B'C+ = 0	+1/3
A3+A4	AB+/AB– = 0	A'C–/A'C– = –1/3	B'C–/B'C'– = –1/3	–2/3
A3+A5	AB+/AB– = 0	A'C–/A'C– = –1/3	B'C'–/B'C'+ = 0	–1/3
A3+A6	AB+/AB– = 0	A'C–/A'C+ = 0	B'C'–/B'C'+ = 0	0
A4+A5	AB–/AB– = –1/3	A'C–/A'C– = –1/3	B'C'–/B'C'+ = 0	–2/3
A4+A6	AB–/AB– = –1/3	A'C–/A'C+ = 0	B'C–/ B'C'+ = 0	–1/3
A5+A6	AB–/AB– = –1/3	A'C–/A'/C+ = 0	B'C'+/B'C'+ = +1/3	±1/3?[2]

CHART #21-3: 12 CHARGED & 3 NEUTRAL COMPOUND TYPES POSSIBLE TO THE SIMPLE ALPHA TYPE PIGGYBACK FORMS

As I see it, then, none of the forms native to one of the 4 OD's, 2^0, 2^1, 2^2, and 2^3, could ever be neutral in any of the 3 planes; and so, their charge—whether minus or plus, I cannot yet decide—would have to be 3/3 = 1. The same would be true of any particle (***i.e.:*** multi-combo) they might form by accelerating to a current OD higher than their native ones. This, too, I contend agrees rather nicely with what the physicists say about the leptons.

I claim that the above agrees rather nicely with what the physicists say about the kinds of leptons and quarks which constitute the atom. What, then, do the physicists say?

In Israel, there is a theoretical physicist named Haim Harari. Born in 1940, he is still living as of today, June 27, 2013. In the April 1983 edition of the ***Scientific American***, vol. 248: #4, there's an article by him entitled ***The Structure Of Quarks And Leptons.*** There, he says there are 6 leptons: electrons, electron type neutrinos, muons, muon type neutrinos, taus, and tau type neutrinos. For each of these 6, he adds, there's an anti-particle. Wikipedia, in its article ***Lepton***, gives the resulting list of 12 as: electron / antielectron (***i.e.:*** positron), muon / antimuon, tau / antitau, electron neutrino / electron antineutrino, muon neutrino / muon antineutrino, and tau neutrino / tau antineutrino.

Of forms native to OD1, Esoptrics suggests this to me: As long as a form native to OD1 is still at OD1 or moving as a duo-combo only at some higher current OD, it's an electron type neutrino. If a form native to OD1 accelerates to be the lead form in a multi-combo, the result is an electron. Yes, an OD1 form is irrational; but, it's also categorical, and that fact overrides the fact it's irrational.

Of forms native to OD2, Esoptrics suggests this to me: Despite what the physicists

---

[1] Since there's far more activity in the primary than in the tertiary plane, this should probably be a minus rather than a plus 2/3.

[2] Again, since there's far more activity in the primary than in the tertiary plane, this should probably be a plus rather than a minus 1/3.

say, if a form native to OD2 accelerates to be the lead form in a multi-combo, it's a positron and, therefore, the positron is not truly an anti-electron, because it does not and cannot contain a single, solitary form native to OD1. This error on the part of the physicists is due to the fact that, if equally accelerated, the difference between a form native to OD1 and one native to OD2 is a single duo-combo out of as many as 10^{38}. That, I dare suggest, is a difference in mass too insignificant for current instrumentality to detect. Yes, an OD2 form is irrational; but, it's also categorical, and that fact overrides the fact it's irrational.

If forms native to OD2 are still at OD2 or moving at some higher OD as a duo-combo only, they are a kind of neutrino or anti-neutrino not yet recognized by Science as not really an electron kind. That's understandable when you realize how little difference there is between forms native to OD1 and those native to OD2.

Of forms native to OD4, Esoptrics suggests this to me: As long as a form native to OD4 is still at OD4 or moving at some higher OD as a duo-combo only, it's a muon type neutrino or anti-neutrino depending upon which way it's rotating in the tertiary plane. If a form native to OD4 accelerates to be the lead form in a multi-combo, the result is either a muon or anti-muon depending upon which way the lead is rotating in the tertiary plane.

Of forms native to OD8, Esoptrics suggests this to me: As long as a form native to OD8 is still at OD8 or moving as a duo-combo only, it's a tau type neutrino or anti-neutrino depending upon which way it's rotating in the tertiary plane. If a form native to OD9 accelerates to be the lead form in a multi-combo, the result is either a tau or anti-tau depending upon which way the lead is rotating in the tertiary plane. Yes, I switched from OD8 to OD9, because OD8, as neither rational nor categorical, cannot be the lead form in a multi-combo. As rational, an OD9 form can, and the difference between it and an OD8 form would be too little for Science to note any difference between a tau and a tau type neutrino.

Wikipedia tells me muons and taus decay into electrons. Esoptrics clearly declares that impossible, since no muon or tau ever includes even so little as a single form native to OD1. I dare suggest that, under very unusual and short lived circumstances, they may *appear* to be electrons; but, that's only because of the little number of duo-combos involved in the difference. Equally accelerated, muons have only 3 and the taus only 7 fewer duo-combos than the electrons do. As said above, that's generally a difference of 3 and 7 out of 10^{38}, and thus a difference in mass too small for current technology to detect.

Turning next to the quarks, Prof. Harari again lists 6: bottom (***a/k/a:*** *b*), charmed (***a/k/a:*** *c*), down (***a/k/a:*** *d*) strange (***a/k/a:*** *s*), top (***a/k/a:*** *t*), and up (***a/k/a:*** *u*). He says the quarks *c*, *t*, and *u* each exhibit an electrical charge of +2/3; whereas, each of the other 3 (***i.e.:*** *b*, *d*, and *s*) exhibit a charge of -1/3. Additionally, each quark may exhibit one of 3 kinds of color which Prof. Harari dubs, blue, red, and yellow. In the lower *left*-hand column of page 57, he warns us these colors have nothing to do with what we usually call colors. It's just the name given to one of the 3 long-range forces. Returning to the lower *right*-hand column of page 57, Prof. Harari tells us this: If each quark of a certain color is considered a distinct kind of quark, then 3 colors times 6 quarks gives us a grand total of 18 kinds of quarks. He gives special note to the fact that, though each of the quarks has both a color and an electrical charge, none of the leptons has color. He then adds that there is an anti-particle for each of the leptons and quarks already described. Each particle and its anti-particle have the same mass; but, their colors and electrical charges are reversed. For example, the positron is the anti-particle of the electron, and, whereas the electrical charge of the latter is *minus* 1, the electrical charge of the former is *plus* 1. By the same token, the red *up* quark has an electrical charge of *plus* 2/3; whereas, its opposite (***i.e.:*** the antired *anti-up* quark) has a charge of *minus* 2/3. If, as he says, there are 18 quarks each with an anti-particle, then 18 quarks + 18 anti-quarks = 36 quarks.

CHAPTER TWENTY-ONE

Switch now to the lower part of the right-hand column on page 60 of the said article. There. Prof. Harari puzzles over a very intriguing pattern of electrical charges found in the leptons and quarks—namely: Going from -1 to +1 in intervals of 1/3, we find some particle exhibits one of the resulting charges. Why are there not other values such as +4/3 and -5/3? Then too, every particle which has an *integral* charge (***ex. gr.:*** +1 or -1) has no color; whereas, every particle which has a *fractional* charge has color. He puzzles over whether that implies a connection between electrical charge and color and, therefore, between leptons and quarks.

Note what Prof. Harari gives us as the two main differences between the leptons and the quarks: On the one hand, the quarks have *fractional* electrical charges of 2/3 and 1/3; whereas, the leptons have only <u>non</u>-fractional electrical charges of 1. On the other hand, the quarks have color and the leptons do not. That causes him to puzzle over whether there might be some connection between the electrical charge of a particle and its color.

Let us make two assumptions: (1) that what science calls an electrical charge corresponds to what I have called the various states of non-neutrality; and (2) that what science calls color corresponds to what I have called the 3 different ways Aleph, Beth and Daleth forms rotate in the tertiary plane (***i.e.:*** which of the 3 axes they use to do so). If it be proper to make those two assumptions, then there is a rather startling similarity between what Prof. Harari says about leptons versus quarks and what Esoptrics says about sub-sextons versus sextons. What is even more startling is that Esoptrics does in fact delineate a very clear relation between the electrical charge and color. For, it explains in complete detail how the color found in the sextons (***i.e.:*** their ability to rotate in 3 planes whereas sub-sextons cannot) is precisely what allows them to produce the *fractional* charges.

Perhaps the *most* astonishing point of all, though, is this: Prof. Harari wonders why nature has favored the charges of 1, 1/3, and 2/3. Esoptrics spells out a possible answer in complete detail right down to how an intrinsically triune infinity causes it all.

Thus, in distinguishing between the 3 planes and the difference between: (1) the way forms native to OD's above OD8 can rotate in the primary and secondary planes, and (2) they way forms native to the 4 OD's 1, 2, 4, and 8 cannot, Esoptrics accounts for the 36 kinds of quarks and the 12 kinds of leptons and, to boot, explains why the latter have integral charges and the former fractional ones.

Having said all of that, I must admit there may be a formidable difference between what Science and Esoptrics say about quarks (***a/k/a*** sextons). Science says 36, and, in the above, Esoptrics seems to agree wholeheartedly. I'm not sure it actually does do so. For, in any given quark, the leading form may be any of the 6 rational categorical forms of 2^4, 2^8, 2^{16}, 2^{32}, 2^{64}, and 2^{128}. If, as the leading form in a quark, each of those 6 can take on any of the 36 roles described above, doesn't that mean there are actually 36x6 = 216 kinds of quarks according to Esoptrics? If that is in fact what Esoptrics is claiming, how would Science go about proving or disproving it? Is it possible for Science to determine whether or not every quark has a leading form native to the same OD? Others more gifted than I am shall have to answer that question. I cannot.

Before closing this chapter, I should perhaps deal with a question raised by what I've said about the 36 kind of quarks (***a/k/a*** sextons). That question—as some readers shall almost certainly perceive quite readily on their own—has to do with creation. Chapter 10 described God as creating such-and-such a number of duo-combos native to such-and-such an ontological distance. Do the numbers given in chapter 10 still hold true? I seriously doubt it. It seems to me one must multiply by 36 the numbers given for the six categorical OD's of 2^4, 2^8, 2^{16}, 2^{32}, 2^{64}, and 2^{128}. That's easily done because all 36 could be simultaneously created at the same u-space. For, each rotates in a manner so unique, their individuality is not compromised by temporarily occupying the same u-space. I dare suspect, though, that they spent no more than a few alphakronons so united. Under what circumstances, then, might they re-unite? I have no idea for now.

✡✞✡✞✡✞✡✞✡✞✡✞✡✞✡✞✡✞✡✞✡✞✡✞✡✞✡✞

But bear in mind that no one has ever seen a string and, except for some maverick ideas discussed in the next chapter, it is likely that even if string theory is right, no one ever will. Strings are so small[1] that a direct observation would be tantamount to reading the text on this page from a distance of 100 light years: it would require resolving power nearly a billion billion times [10^{18} – ENH] finer than our current technology allows. Some scientists argue vociferously that a theory so removed from direct empirical testing lies in the realm of philosophy or theology, but not physics.

I find this view shortsighted, or, at the very least, premature. While we may never have technology capable of seeing strings directly, the history of science is replete with theories that were tested experimentally through indirect means.

——**BRIAN GREENE:** ***The Fabric Of The Cosmos***, page 352 of the 2005 paperback edition by Vintage Books, a division of Random House, Inc., New York. ©2004 Brian Greene (I, of course, agree with what those "scientists argue vociferously". It's the same as what Esoptrics has been rather thunderously proposing for over half a century. More importantly, Esoptrics has done far more than merely "argue vociferously". It has presented to the world so many and mathematically precise details as to leave little doubt that—at least in the search for the ultramicroscopic makeup of time and space—Philosophy and Theology have indeed done a far better job in the field of Theoretical Physics than the physicists have. As for Prof. Green's bit about "theories that were tested experimentally through indirect means", for thousands of years, 99.9999999% of even the most brilliant minds believed it "tested experimentally through indirect means" that the continuum is real and, therefore, there's no limit to the smallness of the smallest segments of time and space. Even today, probably at least 80% are still convinced of it. Einstein was (pg. XXIII). For c. 1,400 years, almost no one doubted the geocentric theory is one so solidly "tested experimentally through indirect means", no sane person would question it; and yet, today, it's one of history's most debunked theories. And what of the static Universe prior to Hubble?! Here, too, "the history of science is replete with theories" which were *thought* to be "tested experimentally through indirect means", but which turned out to be mistaken. The gist of it all is: No matter what the number or the dogmatism of the scientists trumpeting the correctness of this or that *inferred* theory, watch out! Drawing conclusions based on what sensory experience *infers* is risky business. See the quotes on pages 166 & 246. Recall on my page 54 Prof. Greene's other words about "extremely small scales".).

---

[1] On page 345, he gives the measurement as 10^{-33} cm.; and so, in the above, he's telling us 10^{-33} cm. is 10^{18} times smaller than what "current technology allows" us to observe—making the latter 10^{-15} cm... What, then, of Esoptrics assertion that carrying generators operate at 10^{-47} cm.? How much further is that into "the realm of philosophy or theology, but not physics"?

Chapter Twenty-Two:

THE GREAT ACCELERATION'S EFFECT UPON THE HYDROGEN ATOM'S STRUCTURE:

From the very beginning, this issue has always been—and to this day still is—the one puzzling me the most. Several times, I thought I had it nailed down only to see my solution evaporate into thin air.

Fortunately, that's far from true with regard to *every* part of the hydrogen atom's structure. From the very beginning and to this day, I have been firmly convinced (I do not thereby claim *absolute certitude*.) that every atom's every multi-A-combo is an atomic multi-combo which: (1) has a leading form native to $OD2^{128}$ (***i.e.:*** the mirror threshold = MT for short) and currently somewhere between $OD2^{128}$ and $OD2^{129}$, which is to say no lower than $OD2^{128}$ +1 and no higher than $OD2^{129}$–1,[1] (2) has logically concentric with its leading form one duo-combo currently at each of the OD's from 2^1 thru one OD below the leading form's current OD,[2] and (3) has a carrying generator serving as the means by which the whole of the atom moves thru this-or-that u-spaces envelope.

What now follows, then, is my latest *attempt* to describe the structure of the hydrogen atom. To that end, we start with the structure of the 8 reverse categories as produced by the great acceleration. Omega, at $OD2^{256}$, serves as the first reverse category. For the 7 below Omega, the floor of each is 2^{-1}, 2^{-2}, 2^{-4}, 2^{-8}, 2^{-16}, 2^{-32}, and 2^{-64} times Omega respectively, which is to say $OD2^{255}$, $OD2^{254}$, $OD2^{252}$, $OD2^{248}$, $OD2^{240}$, $OD2^{224}$, and 2^{192} respectively.

I long supposed that structure repeated in the hydrogen atom; and so, the MT's generic zone (***i.e.:*** the one including the OD's of 2^{128} to and including 2^{129} –1) was then the first *atomic accelerated* category, and, for the 7 atomic accelerated categories below MT, the floor of each was again, 2^{-1}, 2^{-2}, 2^{-4}, 2^{-8}, 2^{-16}, 2^{-32}, and 2^{-64} times MT respectively, which is to say OD^{127}, $OD2^{126}$, $OD2^{124}$, $OD2^{120}$, $OD2^{112}$, $OD2^{96}$, and $OD2^{64}$, respectively. I now strongly suspect that's mistaken. Why?! A much more appealing answer dawned on me this day, Tuesday, July 2, 2013, and I'll get to the reason behind it in a moment.

My current answer is illustrated on the next page in chart #22-1. It is divided into two halves. The first half pertains to the reverse categories. The second half pertains to the accelerated categories created in the hydrogen atom by the great acceleration which concurrently produced both sets of 8 accelerated categories each. Remember: MT = 2^{128}.

FROM THE LEFT: Column 1 is marked "C#" because it gives us *both* the numbers of the reverse categories far above MT *and* the numbers of the accelerated categories at and

---

[1] Well, I have at times found myself wondering if it might, for some reason, currently be between 2^{127} and 2^{128} –1; but, that speculation never lasts for long before it gives way to the above range.

[2] As mentioned earlier, I have recently begun to suspect that each multi-A-combo may include logically concentric with itself a duo-combo currently at OD1; but, no such duo-combo could have a form *native* to OD1. It would probably be native to OD3.

below MT. Column 2 is marked "OD_N" because it gives us the native OD of the forms accelerated into each of the reverse categories in order to serve as the multi-g-combos anchoring whatever galaxies, clusters of galaxies and clusters of clusters are included in the given category. As said earlier, the native OD of each is 2^{128}. Column 3 is marked "x↑", because it tells us the multiple of acceleration imposed upon the MT forms (***i.e.:*** ones native to $OD2^{128}$) in column 2 in order to move them to the floor of the given category. Column 4 is marked "xΩ", because it gives us the multiple of Omega which establishes the floor of the given category. Column 5 is marked "OD's", because it tells what OD's are logically concentric in every leading categorical form in the given category.

REVERSE CATEGORIES					HYDROGEN CATEGORIES			
1	**2**	**3**	**4**	**5**	**6**	**7**	**8**	**9**
C#	**OD_N**	**x↑**	**xΩ**	**OD's**	**OD_N**	**x↑**	**xMT**	**OD'S**
1	2^{256}	2^0	$2^0 = 2^{256}$	2^{256}	2^{128}	2^0	$2^0 = 2^{128}$	$2^1\sim2^{129}-1$
2	2^{128}	2^{127}	$2^{-1} = 2^{255}$	$2^1\sim2^{256}-1$	2^{64}	2^{63}	$2^{-1} = 2^{127}$	$2^{64}\sim2^{128}-1$
3	2^{128}	2^{126}	$2^{-2} = 2^{254}$	$2^1\sim2^{255}-1$	2^{32}	2^{94}	$2^{-2} = 2^{126}$	$2^{32}\sim2^{127}-1$
4	2^{128}	2^{124}	$2^{-4} = 2^{252}$	$2^1\sim2^{254}-1$	2^{16}	2^{109}	$2^{-3} = 2^{125}$	$2^{16}\sim2^{126}-1$
5	2^{128}	2^{120}	$2^{-8} = 2^{248}$	$2^1\sim2^{252}-1$	2^8	2^{116}	$2^{-4}= 2^{124}$	$2^8\sim2^{125}-1$
6	2^{128}	2^{112}	$2^{-16} = 2^{240}$	$2^1\sim2^{248}-1$	2^4	2^{119}	$2^{-5} = 2^{123}$	$2^4\sim2^{124}-1$
7	2^{128}	2^{96}	$2^{-32} = 2^{224}$	$2^1\sim2^{240}-1$	2^2	2^{120}	$2^{-6} = 2^{122}$	$2^2\sim2^{123}-1$
8	2^{128}	2^{64}	$2^{-64} = 2^{192}$	$2^1\sim2^{224}-1$	2^1	2^{120}	$2^{-7} = 2^{121}$	$2^1\sim2^{122}-1$

CHART #22-1: EFFECT OF THE GREAT ACCELERATION UPON THE UNIVERSE'S 8 REVERSE CATEGORIES & THE HYDROGEN ATOM'S 8 ACCELERATED CATEGORIES

FROM THE LEFT: In column 6, the chart has switched from describing the 8 reverse categories relative to Omega to describing the 8 accelerated categories relative to MT. That sixth column is marked "OD_N", because it gives us the native OD of the forms accelerated into each of the given categories. Column 7 is marked "x↑", because it tells us the multiple of acceleration imposed upon the categorical forms in column 6 in order to move them to the floor of the given category. Column 8 is marked "xMT", because it gives us the multiple of the mirror threshold which establishes the floor of the given category. Column 9 is marked "OD's", because it tells us the OD's of the forms logically concentric with the leading form if its current OD is at the uppermost reach of the given category.

Reverse the order of the exponents in column 3 so: 0, 64, 96, 112, 120, 124, 126, and 127. Next, place them alongside the exponents in column 7. The result is: 64-1=63; 96-2=94; 112-3=109; 120-4=116; 124-5=119; 126-6=120; 127-7=120. That's the relationship we'll explain shortly by setting forth the insight which dawned on me today, July 2, 2013—the date, from 1263-1969, of the feast of the Blessed Virgin's visitation with St. Elizabeth.

Chart #22-2 on the next page repeats some of what's given above in the right half of chart #22-1. To that data, it adds a problematic detail: the number of forms native to a particular OD which are accelerated into a given one of the accelerated categories as the leading form of an atomic multi-combo. From what principle are those numbers derived?

That question is answered by *drawings* #22-1 and #22-2 on the following 2 pages. They've been seen before as metaphorical depictions of the first and second principles of Esoptrics. Being that fundamental, it is only logical to expect that they should stand as keys to the structure of the hydrogen atom. If so, then, #22-1 describes the structure of the first accelerated category's multi-A-combo with the second accelerated category's two multi-a-combos in potency to the former. Drawing #22-2 then describes the structure just described plus the third accelerated category's four multi-a-combos in potency to the second accelerated category's two multi-a-combos. Yes, that means two of the four are in potency to one of the second accelerated category's multi-a-combos, while the other two are in potency to the second one of that second category's multi-a-combos.

ACCELERATED CATEGORICAL ZONE #	OD's IN EACH CATEGORICAL FORM	OD_N OF THE ACCELERATED FORMS	NUMBER OF FORMS PER OD_N ACCELERATED
1	$2^1 \sim 2^{129}-1$	2^{128}	1 quark
2	$2^{64} \sim 2^{128}-1$	2^{64}	2 quarks
3	$2^{32} \sim 2^{127}-1$	2^{32}	4 quarks
4	$2^{16} \sim 2^{126}-1$	2^{16}	8 quarks (or 4?)
5	$2^8 \sim 2^{125}-1$	2^8	16 quarks (or 4?)
6	$2^4 \sim 2^{124}-1$	2^4	32 quarks (or 4?)
7	$2^2 \sim 2^{123}-1$	2^2	64 leptons (or 4?)
8	$2^1 \sim 2^{122}-1$	2^1	128 leptons (or 4?)
TOTAL FORMS ACCELERATED =			255 (or 27?)

CHART #22-2: NUMBER OF FORMS PER CATEGORICAL OD ACCELERATED INTO A GIVEN ONE OF THE HYDROGEN ATOM'S ACCELERATED CATEGORIES

In drawing #22-1 below, ⊗ is a nexus representing the multi-A-combo which: (1) has 2 mutually opposed leading forms native to the categorical $OD2^{128}$ and currently at some OD above 2^{128} but no higher than $2^{129}-1$, and (2) has the carrying generator serving as the means by which the entire atom travels thru this-or-that u-spaces envelope. It is the most massive of the atom's multi-combos and is perhaps what scientists call the "top quark".

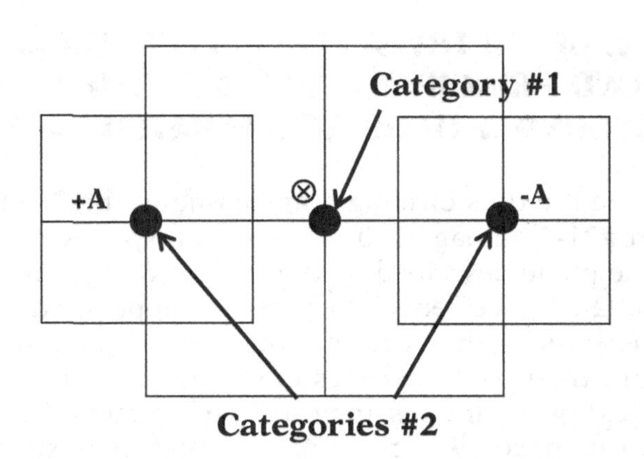

DRAWING #22-1: GEOMETRY-SPEAK'S ALLEGORICAL DEPICTION OF ESOPTRICS' FIRST PRINCIPLE APPLIED TO THE HYDROGEN ATOM'S FIRST & SECOND ACCELERATED CATEGORIES

In #22-1 above, +A and –A represent the two multi-combos each of which: (1) has 2 mutually opposed leading forms native to the categorical $OD2^{64}$ and currently at some OD above 2^{127} but no higher than $2^{128}-1$, (2) has a carrying generator serving as the means by which it moves thru one of the 2 u-spaces envelopes provided by the nexus ⊗, and (3) is rotating around ⊗ in a direction which is the opposite of its "sibling's" direction. Because the 2 leading forms of each are native to $OD2^{64}$, each is perhaps what scientists call a quark.

In drawing #22-2 below, ⊗, $+A_1$ (***a/k/a*** $\otimes_2$), and $-A_1$ (***a/k/a*** $\otimes_2$) signify ⊗, +A, and –A in drawing #22-1 respectively. Each of the points marked A_2 signify a multi-combo which: (1) has 2 mutually opposed leading forms native to the categorical $OD2^{32}$ and currently at some OD above 2^{126} but no higher than $2^{127}-1$, (2) has a carrying generator serving as the means by which it moves thru the u-spaces envelope provided by one of the two nexuses marked $\otimes_2$, and (3) is rotating around that $\otimes_2$ in a direction which is the opposite of its "sibling's" direction. Because the 2 leading forms of each are native to $OD2^{32}$, each is perhaps what scientists call a quark.

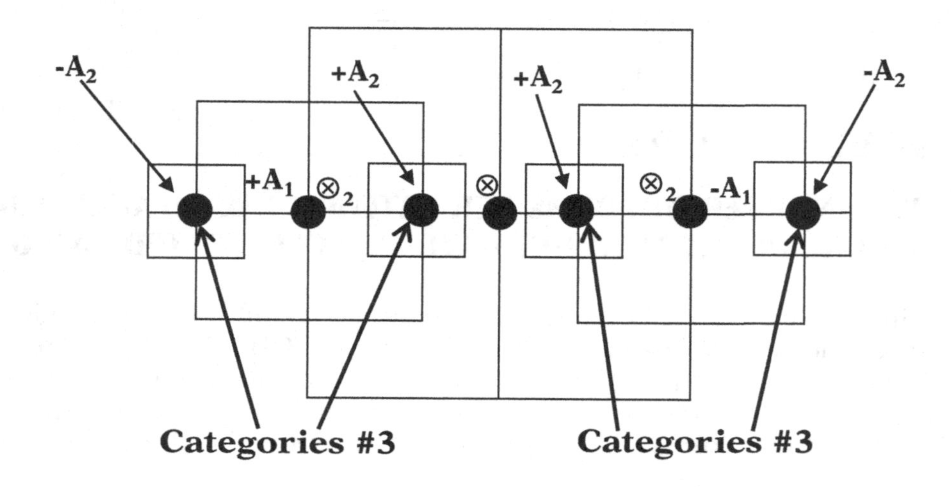

DRAWING #22-2: GEOMETRY-SPEAK'S ALLEGORICAL DEPICTION OF ESOPTRICS SECOND PRINCIPLE APPLIED TO THE HYDROGEN ATOM'S FIRST, SECOND & THIRD ACCELERATED CATEGORIES

In what manner are the dots circling a given one of the 3 nexuses? If the nexus is a Daleth sexton (***vid:*** chart #21-2 on page 170), then its 2 u-spaces envelopes are—in opposite directions—circling in the plane described by the surface of this sheet of paper. If the nexus is a Beth sexton, then its 2 u-spaces envelopes are—in opposite directions—circling in the plane described by an arc passing thru the left and right edges of this sheet of paper. If the nexus is an Aleph sexton, then its 2 u-spaces envelopes are—in opposite directions—circling in the plane described by an arc passing thru the top and bottom edges of this sheet.

How far out from its nexus is a given dot circling its nexus? As I have said before: Categorical forms operating above the MT's generic zone repel one another;[1] but, when operating in or below the MT's generic zone, they attract one another. As a result, the

[1] Don't forget: Whether above or below the MT's generic zone, all accelerated *generic* forms always attract one another as they attract, and are attracted by, all accelerated categorical forms. Among accelerated *rational* generic forms, the attraction is much stronger than that among accelerated *irrational* generic forms. I sometimes suspected this accounts for the difference between what scientists call the strong vs. the weak nuclear force; but, a better explanation shall follow later.

atom's 8 accelerated categorical zones pull together to produce a kind of compound combo with a "diameter" no greater than what is needed to allow the 255 (27?) atomic multi-combos to execute their patterns of rotation without any 2 generators trying to perform at once the same one state of excitation at the same OD within the same 1 of the 3 planes.

That stipulation raises a question I've yet to resolve. Notice how in the extreme right-hand column of *chart* #22-2 on page 177, I placed a series of "or 4" notations followed by a "or 27" notation at the bottom of the column. Why?!

Look at drawing #22-2 on the prior page. It shouldn't take long to see that, if we insert two more dots to the opposite sides of each of the 4 smallest squares, things are getting awfully crowded. That suggests this: It may be that, as one moves from the 3rd to the 4th accelerated category, only the 2 outermost, smallest squares can each support 2 dots. Drawing #22-3 below illustrates the meaning of that. Notice how only the extreme left and right hand large dots each has 2 small dots on opposite sides of those 2 large dots. Each of the 4 tiny dots represents a multi-combo in the atom's 4th accelerated category.

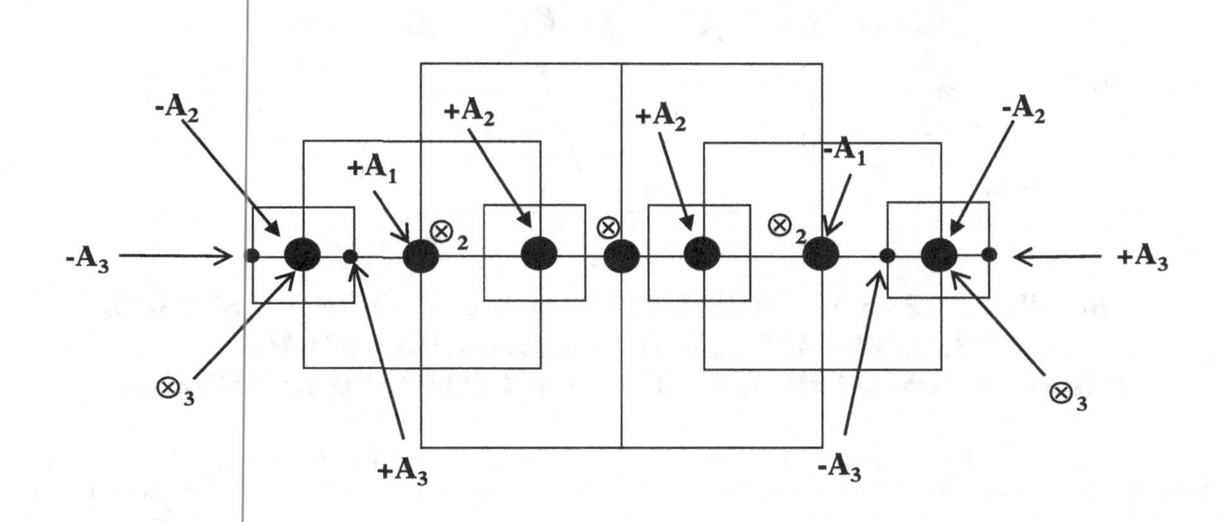

DRAWING #22-3: ALTERNATE VERSION #1 OF ESOPTRICS' THIRD PRINCIPLE APPLIED TO THE HYDROGEN ATOM'S FIRST, SECOND, THIRD, & FOURTH ACCELERATED CATEGORIES

If the remedy expressed in the above drawing solves the overcrowding problem, then it seems it should be applied when going from the 4th to the 5th accelerated category, and from the 5th to the 6th, and from the 6th to the 7th, and from the 7th to the 8th. If so, then the progression is: 1+2+4+4+4+4+4+4=27 rather than 1+2+4+8, etc., = 255. But, *does* it solve the problem? And what is the source of the problem in the first place?

If each dot in the above represents a sexton of the *Daleth* type, then each of the dots in the above drawing is 2 u-spaces envelopes one of which is rotating and the other counter-rotating in the plane of the sheet's surface. What, though, if each of the dots at the center of the 4 smallest squares is each a sexton of the *Aleph* type? That would produce the result illustrated below in drawing #22-4. Notice that each of the 4 smallest squares is sporting two tiny dots for a total of 8 tiny dots. In each of the 4 smallest squares, the tiny dots are no longer to the left and right of the large dot at the square's center; instead, one is toward the sheet's top edge and the other to the sheet's bottom edge. That's designed to illustrate this principle: Each of the 4 smallest squares is two u-spaces envelopes one of which is rotating and the other counter-rotating in a plane perpendicular to the sheet's surface and circling the sheet from its top edge to its bottom edge and, thereby, causing the 4 tiny dots to rotate and counter-rotate in a plane perpendicular to the plane of rotation common to the 2 mid-sized squares. That avoids the overcrowding and maintains the progression of 1+ 2 + 4 + 8.

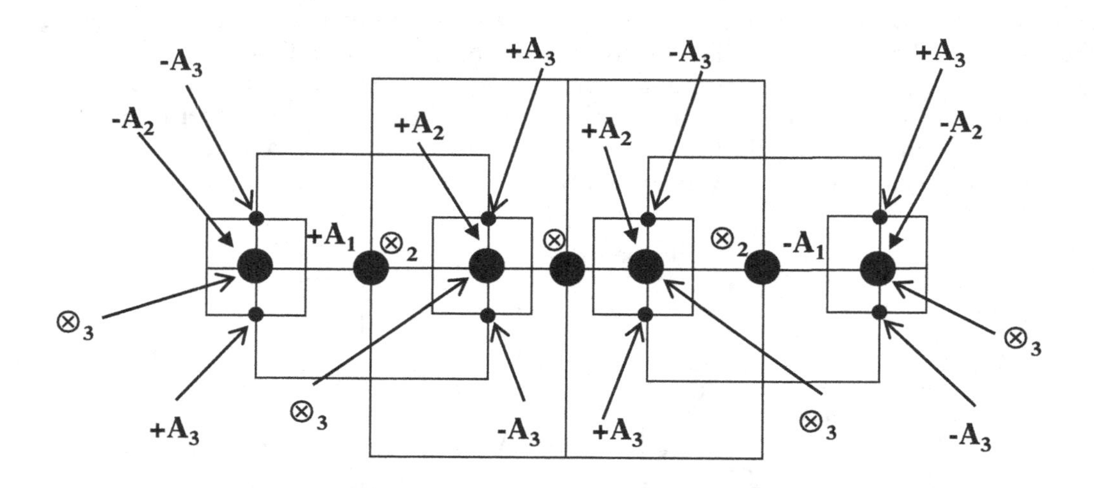

DRAWING #22-4: ALTERNATE VERSION #2 OF ESOPTRICS' THIRD PRINCIPLE APPLIED TO THE HYDROGEN ATOM'S FIRST, SECOND, THIRD, & FOURTH ACCELERATED CATEGORIES

What if each of those 4 smallest squares is a sexton of the *Beth* type? In that case, each of them is two u-spaces envelopes one of which is rotating and the other counter-rotating in a plane perpendicular to the sheet's surface and circling the sheet from its left edge to its right edge thereby causing each of the 8 tiny dots to rotate around its own center. That avoids the overcrowding and maintains the progression of 1+ 2 + 4 + 8 +16. As I said, we're here dealing with a question I've yet to resolve; and so, it leaves me wondering if perhaps the exact structure of the hydrogen atom depends upon how the 3 types of sextons (***i.e.:*** the 3 "colors" of quarks?) are arranged within it.

Exactly how does the great acceleration produce the atomic structure just described? Here, now, is the answer given to me July 2, 2013:

The great acceleration does not take place in 7 *successive* stages. It's not a case of some MT forms accelerating all the way to $OD2^{255}$ before any MT forms accelerate all the way to $OD2^{254}$ before any MT forms accelerate all the way to $OD2^{252}$ and so forth. On the contrary, whatever their final destination, the MT forms all start climbing simultaneously. In a manner of speaking, there's a single "wave" with 7 "layers" each containing a definite number of MT forms—the number of which is greatest in the lowest layer (***i.e.:*** layer #1), smaller in the next layer (***i.e.:*** layer #2), even smaller in the next layer (***i.e.:*** layer #3), etc., until it's only 8 in the top layer (***i.e.:*** layer #7). As the whole wave with its astronomical number of forms native to $OD2^{128}$ accelerates upward, the first layer—after accelerating 2^{64} times above 2^{128}—stops at $OD2^{192}$ (***i.e.:*** floor of the 8th reverse category). As the acceleration upward continues, the smaller number of MT forms in the second layer—after accelerating

2^{96} (***i.e.:*** 2^{64}x2^{32}) times above 2^{128}—stop at OD2^{224} (***i.e.:*** floor of the 7th reverse category). As the acceleration upward continues, the smaller number of MT forms in the <u>*third*</u> layer—after accelerating 2^{112} (***i.e.:*** 2^{64}x2^{32}x2^{16}) times above 2^{128}—stop at OD2^{240} (***i.e.:*** floor of the 6th reverse category). As the acceleration upward continues, the smaller number of MT forms in the <u>*fourth*</u> layer—after accelerating 2^{120} (***i.e.:*** 2^{64}x2^{32}x2^{16}x2^{8}) times above 2^{128}—stop at OD2^{248} (***i.e.:*** floor of the 5th reverse category). As the acceleration upward continues, the smaller number of MT forms in the <u>*fifth*</u> layer—after accelerating 2^{124} (***i.e.:*** 2^{64}x2^{32}x2^{16}x2^{8} x2^{4}) times above 2^{128}—stop at OD2^{252} (***i.e.:*** floor of the 4th reverse category). As the acceleration upward continues, the smaller number of MT forms in the <u>*sixth*</u> layer—after accelerating 2^{126} (***i.e.:*** 2^{64}x2^{32}x2^{16}x2^{8}x2^{4}x2^{2}) times above 2^{128}—stop at OD2^{254} (***i.e.:*** floor of the 3rd reverse category). As the acceleration upward continues, the 8 MT forms in the <u>*seventh*</u> layer—after accelerating 2^{127} (***i.e.:*** 2^{64}x2^{32}x2^{16}x2^{8}x2^{4}x2^{2}x2^{1}) times above 2^{128}—stop at OD2^{255} (***i.e.:*** floor of the 2nd reverse category).[1]

Now then, what an MT form does above the MT, it does below the MT in a *reverse* fashion, which is to say what it does above the MT is *mirrored* below the MT. Therefore, as a multitude of MT forms accelerate 2^{64} times to reach OD2^{192}—and, thereby, make logically concentric to themselves all the forms in the 1st 64 generic zones <u>*above*</u> MT (The result is a multi-g-combo including the 65 generic zones of the 129th thru the 193rd)[2]—an astronomical number of MT forms each makes logically concentric to itself all the forms in the 1st 64 generic zones <u>*below*</u> MT (The result is multi-A-combos including the 65 generic zones of the 64th thru the 129th.). Be sure to note carefully what that means: They do it without changing their *current* OD. In other words, they do it without either accelerating or decelerating; they do it by virtue of the power of the MT to produce a *mirrored* reaction.

Ah, but! Doing that, each of those MT forms tries to make logically concentric to itself a *categorical* form native to OD2^{64}. That, however, is a first class no-no. Except in the case of forms native to OD1, no categorical form can ever be made one of the captured forms logically concentric to a leading form. Thus, the attempt to capture the forms native to OD2^{64} produces a violent reaction mirroring what has just been done by a *downward* seizure: In each of the atoms, a form native to OD2^{64} accelerates *upward* in an attempt to make concentric to itself the 1st 64 generic zones above it. That, of course, would put the MT form logically concentric to it, and that, again, is a first class no-no. As a result, in every atom, the accelerating categorical form native to OD2^{64} stops short of the MT form and takes on a current OD of no less than $2^{127}+1$ (***i.e.:*** 1 + the floor of the 2nd accelerated category) and no more than $2^{128}-1$. It thus accelerates 2^{63} rather than 2^{64} times. Note that the prior exponent is the latter minus 1.

So then, numerous MT forms have stopped at OD2^{192} (***i.e.:*** the floor of the 8th reverse category) or slightly above; but, a lesser number continue for another 32 generic zones and to stop at OD2^{224} (***i.e.:*** the floor of the 7th reverse category) as a result of accelerating 2^{96} times 2^{128}. Again, what's done *above* MT is repeated in reverse order *below* MT; and so, the multi-A-combos reach *downward* to make logically concentric to themselves—

---

[1] If you prefer, think of a telescope with 7 (64?) cylinders. With the outermost cylinder in place and the other 6 retracted down inside of it, the 2nd cylinder emerges with the other 5 retracted down inside of it. When the 2nd cylinder is fully extended, the 3rd cylinder emerges with the other 4 retracted down inside of it. When the 3rd cylinder is fully extended, the 4th cylinder emerges with the other 3 retracted down inside of it. Need I describe the remaining 3 steps?

[2] If you're as slow in math as I am, you may think the number should be 64 generic zones. Yes, the 129th generic zone plus 64 generic zones takes in the 193rd generic zone. One must, however, then add the 129th zone into the calculation for a total of 65. For example: 1+9=10; but, the total number of digits therein is 10—namely: 1, 2, 3, 4, 5, 6, 7, 8, 9, and 10. Count them for yourself.

in addition to the 64 generic zones already captured—the 1st 32 generic zones below $OD2^{64}$. Doing that, though, each seeks to make logically concentric to itself one of the categorical forms native to $OD2^{32}$. That, of course, is the same old first class no-no; and so, each of those $OD2^{32}$ forms reacts violently by trying to do in reverse what the 2nd set of accelerating MT forms has just done *downward* in the atom—namely: to accelerate *upward* 2^{96} times. That, of course, would put each of those categorical $OD2^{32}$ forms at $OD2^{128}$ and, thereby, enable them to make logically concentric to themselves the categorical $OD2^{128}$ forms and the $OD2^{64}$ ones now at $OD2^{127}$. Since that's not allowed, each of the categorical forms native to $OD2^{32}$ stops short of capturing either of those 2 categorical forms and, thereby, comes to rest at an OD no less than $2^{126}+1$ (***i.e.:*** 1 + the floor of the 3rd accelerated category) and no greater than $2^{127}–1$. Each thus accelerates 2^{94} rather than 2^{96} times. Note that the prior exponent is the latter minus 2. The sequence –3, –4, –5, –6, –7 shall now follow.

So then, numerous MT forms have stopped at $OD2^{224}$ (***i.e.:*** the floor of the 7th reverse category); but, a fewer number continue for another 16 generic zones to stop at $OD2^{240}$ (***i.e.:*** the floor of the 6th reverse category) after accelerating 2^{112} x 2^{128}. Again, what's done *above* MT is repeated in reverse order *below* MT; and so, the multi-A-combos reach downward to make logically concentric to themselves—in addition to the 96 generic zones already captured—the 1st 16 generic zones below $OD2^{32}$. Doing that, though, each seeks to make logically concentric to itself one of the categorical forms native to $OD2^{16}$. That, of course, is the same old first class no-no; and so, each of those forms native to $OD2^{16}$ reacts violently by trying to do what the 3rd set of accelerating MT forms has just done—namely: to accelerate 2^{112} times. That, of course, would put each of those categorical forms at $OD2^{128}$ and, thereby, enable them to make logically concentric to themselves the categorical forms currently at $OD2^{128}$, $OD2^{127}$, and $OD2^{126}$. Since that's not allowed, each of the categorical forms native to $OD2^{16}$ stops short of capturing any of those 3 categorical forms and, thereby, comes to rest at an OD no less than $2^{125}+1$ (***i.e.:*** 1 + the floor of the 4th accelerated category) and no greater than $2^{126}–1$. Each thus accelerates 2^{109} rather than 2^{112} times. Note that the prior exponent is the latter minus 3.

Do you catch the drift? If so, let's jump to the end result of it all: By the time the great acceleration above MT puts 8 MT forms at $OD2^{255}$ (***i.e.:*** the floor of the 2nd reverse category), the <u>upward</u> seizure—by those 8 MT forms—of the 127 generic zones <u>above</u> MT shall have produced—on the part of a great many MT forms—a <u>downward</u> seizure of the 127 generic zones <u>below</u> MT. In the interval, there have been—besides the 3 downward seizures we've already described—an additional 4 downward seizures which first sought to capture categorical forms native to $OD2^{8}$, then categorical forms native to $OD2^{4}$, then categorical forms native to $OD2^{2}$, and then, lastly, categorical forms native to $OD2^{1}$. Those 4 no-no's shall produce 4 violent reactions which shall see: (1) forms native to $OD2^{8}$—after trying to accelerate 2^{120} to $OD2^{128}$—accelerate only 2^{116} to $OD2^{124}$ (***i.e.:*** the floor of the atom's 5th accelerated category),[1] in order to leave unmolested the categorical forms at OD's 2^{128}, 2^{127}, 2^{126}, and 2^{125}; (2) forms native to $OD2^{4}$—after trying to accelerate 2^{124} to $OD2^{128}$—accelerate only 2^{119} to $OD2^{123}$ (***i.e.:*** the floor of the atom's 6th accelerated category),[2] in order to leave unmolested the categorical forms currently at OD's 2^{128}, 2^{127}, 2^{126}, 2^{125}, and 2^{124}; (3) forms native to $OD2^{2}$—after trying to accelerate 2^{126} to $OD2^{128}$—accelerate only 2^{120} to $OD2^{122}$ (***i.e.:*** the floor of the atom's 7th accelerated category),[3] in order to leave

---

[1] Note that the exponent 116 = the exponent 120 –4.

[2] Note that the exponent 119 = the exponent 124 –5.

[3] Note that the exponent 120 = the exponent 126 –6.

unmolested the categorical forms currently at OD's 2^{128}, 2^{127}, 2^{126}, 2^{125}, 2^{124}, and 2^{123}; (4) forms native to $OD2^1$—after trying to accelerate 2^{127} to $OD2^{128}$—accelerate only 2^{120} to $OD2^{121}$ (***i.e.:*** the floor of the atom's 8th accelerated category),[1] in order to leave unmolested the categorical forms currently at OD's 2^{128}, 2^{127}, 2^{126}, 2^{125}, 2^{124}, 2^{123}, and 2^{122}.

During all of the above, what's happening to the forms native to $OD2^0$? Every OD1 form's generator is in potency to an OD2 form. Therefore, when an OD2 form accelerates to $OD2^{121}$ +1, the OD1 form in potency to it shall also accelerate to $OD2^{120}$ +1—the generic zone below that of the OD2 form. Doing so, it becomes a multi-e-combo making concentric to itself 2^{120} duo-combos each, like itself, native to $OD2^0$. Watch what that entails.

According to Esoptrics, the mass of a multi-combo relative to that of a single duo-combo taken as a mass of 1 = the square of the number of duo-combos logically concentric in that multi-combo. Therefore, for a multi-e-combo having 2^{120} duo-combos concentric to itself, its mass = $(2^{120}+1)^2$. For economy's sake we'll round that to the square of 2^{120} and say the mass of such a multi-e-combo is 2^{240} (***i.e.:*** 1.7668×10^{72}) times that of 1 duo-combo.

Compare that to the mass of the multi-A-combo. With somewhere between 2^{128} to 2^{129}-2 duo-combos logically concentric with it, its mass is 2^{256} to 2^{258} times that of a single duo-combo. What, now, is its mass compared to that of the multi-e-combo just described? To answer, we first divide 2^{256} by 2^{240}, and the answer is 2^{16} (***i.e.:*** 65,536) as the minimum multiple. To calculate the maximum multiple, we divide 2^{258} by 2^{240}, and the answer is 2^{18} (***i.e.:*** 262,144). That thus gives us a range of 65,536 to 262,144 times the mass of the multi-e-combo. That's the range, though, only where the multi-e-combo is in its lowest energy state, which is to say where its current OD is at 2^{120}+1 making its carrying generator in potency to an OD2 form at $OD2^{122}$ +2. Some scientists (***ex. gr.:*** pg. 263 of Prof. Greene's ***The Fabric Of The Cosmos***) tell me the mass of the top quark is roughly 350,000 times that of the electron. I leave it to the reader to decide if I'm here saying anything of value.

Speaking of electrons, Wikipedia, in its article ***Electron Shell***, tells me there are 7 such shells and that Science calls them either K, L, M, N, O, P, and Q or 1, 2, 3, 4, 5, 6 and 7. Shell K (***i.e.:*** #1) is the innermost shell; shell L (***i.e.:*** #2) is one from the innermost shell; shell M (***i.e.:*** #3) is two from the innermost shell; and so forth. Each shell then breaks down into anywhere from 1 to 7 sub-shells called s, p, d, f, g, h, and i. The breakdown is so:

SHELL	SUB-SHELLS PER SHELL
K = #1	s only
L = #2	s & p
M = #3	s, p & d
N = #4	s, p, d & f
O = #5	s, p, d, f & g
P = #7	s, p, d, f, g & h
Q = #8	s, p, d, f, g, h & i

CHART #22-3: THE 7 SHELLS & 7 SUB-SHELLS OF ELECTRONS.

Each of the sub-shells can house up to a definite number of electrons. Chart #22-4 on the next page gives the maximum number of electrons per sub-shell according to Wikipedia. The left column gives the name of the sub-shell, and the right column gives the maximum number of electrons per sub-shell.

[1] Note that the exponent 120 = the exponent 127 –7.

SUB-SHELL	ELECTRONS
s	2
p	2+4=6
d	6+4=10
f	10+4=14
g	14+4=18
h	18+4=22
i	22+4=26

CHART #22-4: MAXIMUM NUMBER OF ELECTRONS PER SUB-SHELL.

Because the K shell has only the s sub-shell, the K shell houses a maximum of 2 electrons. Because the L shell has both the s and the p sub-shells, the L shell houses a maximum 2+6 = 8 electrons. Because the M shell has both the s, the p, and the d sub-shells, the M shell allows for 2+6+10 = 18 electrons, and so forth. Chart #22-5 below gives the whole picture in detail.

SHELL	MAX ELECTRONS PER SHELL $=2N^2$
K = #1	2
L = #2	2+6 = 8
M = #3	2+6+10 = 18
N = #4	2+6+10+14 = 32
O = #5	2+6+10+14+18 = 50
P = #6	2+6+10+14+18+22 = 72
Q = #7	2+6+10+14+18+22+26 = 98

CHART #22-5: MAXIMUM NUMBER OF ELECTRONS PER SHELL.

Look at chart #22-5 carefully, and you'll notice this: The maximum number of electrons per shell = $2(N^2)$ where N = the number of the shell. Does that formula sound familiar? It's one of Esoptrics most fundamental formulas, and it's one I've been well aware of for at least 48 years. See formula #8-9 on page 61.

SHELL # =	ACCELERATED ATOMIC CATEGORY #
1	7
2	6
3	5
4	4
5	3
6	2
7	1

CHART #22-6: WHICH SHELL CORRESPONDS TO WHICH ONE OF THE HYDROGEN ATOM'S ACCELERATED CATEGORIES.

Chart #22-6 above correlates Esoptrics accelerated categories with science's electron shells. Where did accelerated category #8 go? It got snubbed by the electrons. Why?!

Accelerated category #8 comes about when categorical OD2 forms accelerate to somewhere between $OD2^{121}$ and $OD2^{122}$-1. Unless the rectilinear velocity of their carrying generators is tremendous, they have stopped at an OD well below $OD2^{122}$-1. Let's call each of these accelerated OD2 forms an Octavian.

In keeping with the fact its basic nature is to be in potency to an OD2 form unless accelerated past all such, the OD1 form accelerates to at least $OD2^{120}$ +1 thus forming a multi-e-combo (***i.e.:*** electron) whose carrying generator seeks its u-spaces from an Octavian at $OD2^{121}$ +2. But, suppose the electron accelerates to $OD2^{120} + 2^{119}$. Its carrying generator shall now seek its u-spaces from a form currently at $OD2^{121} + 2^{120}$. Is it possible none of the Octavians has logically concentric to its leading form one currently at $OD2^{121} + 2^{120}$? It's not only possible but even very likely. Thus, none of the Octavians gives the electron _all_ the u-spaces envelopes it needs for its carrying generator to have access to the full range of the rectilinear velocities possible within its own generic zone (***i.e.:*** that 121st one including the OD's of 2^{120} thru 2^{121}–1). What, then, is the electron's carrying generator going to do? It's going to snub all the Octavians and, thereby, render the 8th accelerated category _not_ one of the electron shells. How does the electron snub the Octavians?

Focus on the multi-combos in the atom's 7th accelerated category. Whether 4 or 64, each shall be called a "Septimus". Each Septimus has a leading form native to $OD2^2$ and currently accelerated to somewhere between $OD2^{122}$ and 2^{123}–1. Even if a Septimus is accelerated no further than $OD2^{122}$+1, its leading form has logically concentric to itself one form currently at each of the OD's from 2^{121} to 2^{122}. Each Septimus thus *unfailingly* offers to the electron in the 121st generic zone every one of the u-spaces envelopes it needs to give its carrying generator access to all the velocities possible within the 121st generic zone. In short, for the electron, the 7th accelerated category is the 1st electron shell, because it, not the atom's 8th accelerated category, is the first one able to supply an electron with a *reliable* supply of forms at each of the OD's required by the electron if it is to enjoy all the velocities possible to it. The number of electrons able to occupy that 1st shell = $2(1^2) = 2$.

What, though, if an electron's leading form accelerates to slightly above $OD2^{121}$ and thus into that 122nd generic zone (***a/k/a:*** atom's 7th accelerated category) including the OD's of 2^{121} thru 2^{122}–1? It shall then need a reliable supply of forms (***i.e.:*** u-spaces envelopes) at each of the OD's ranging from 2^{122} to 2^{123}–1. That, no Septimus can supply, since—unless accelerated to near light speed—each Septimus has logically concentric to its leading form no form higher than much below $OD2^{123}$–1. Therefore, accelerated to slightly above $OD2^{121}$, the electron must—to enjoy all the possible velocities—seek the needed u-spaces envelopes from the next highest accelerated category, which is to say the 6th accelerated one.

Focus, then, on the multi-combos in the 6th accelerated category. Whether 4 or 32, each shall be called a "Sextus". Each Sextus has a leading form native to $OD2^4$ and currently accelerated to somewhere between $OD2^{123}$ and 2^{124}–1. Even if a Sextus is accelerated no further than $OD2^{123}$+1, its leading form has logically concentric to itself one form currently at each of the OD's from 2^{122} to 2^{123}–1. Each Sextus thus *unfailingly* offers to the electron in the 122nd generic zone every one of the u-spaces envelopes it needs to give its carrying generator access to all the velocities possible within the 122nd generic zone. In short, for the electron, the 6th accelerated category is the 2nd electron shell, and the number of electrons able to occupy that 2nd shell = $2(2^2) = 8$.

What if an electron's leading form accelerates to slightly above $OD2^{122}$ and thus into that 123rd generic zone (***a/k/a:*** atom's 6th accelerated category) including the OD's 2^{122} thru 2^{123}–1? To enjoy all its possible velocities, it shall then need a reliable supply of forms (***i.e.:*** u-spaces envelopes) at each of the OD's ranging from 2^{123} to 2^{124}–1. That, no Sextus can supply, since—unless accelerated to near light speed—each Sextus has logically concentric to its leading form no form higher than much below $OD2^{124}$–1. Therefore, accelerated to slightly above $OD2^{122}$, the electron must seek the needed u-spaces envelopes from the next highest accelerated category, which is to say the atom's 5th accelerated one.

Are you following the drift? If so you can perhaps readily see that, for Esoptrics, the relationship between the 7 electron shells and 7 sub-shells can be charted so:

SUB-SHELLS	SHELLS (Min & Max OD Ranges In Parentheses)			
	#8 Min (2^1~2^{121}) Max (2^1~2^{122}–1) M=2^{121} E=2^{120}	#7 = #1 Min (2^2~2^{122}) Max (2^2~2^{123}–1) M=2^{121} E=2^{120}	#6 = #2 Min (2^4~2^{123}) Max (2^4~2^{124}–1) M=2^{122} E=2^{121}	#5 = #3 Min (2^8~2^{124}) Max (2^8~2^{125}–1) M=2^{123} E=2^{122}
s		M~2M–1 E~2E–1	M~1.5M–1 E~1.5E–1	M~4/3M–1 E~4/3E–1
p			1.5M~2M–1 1.5E~2E–1	4/3M~5/3M–1 4/3E~5/3E–1
d				5/3M~2M–1 5/3E~6/3E–1

SUB-SHELLS	SHELLS (Min & Max OD Ranges In Parentheses)			
	#4 = #4 Min (2^{16}~2^{125}) Max (2^{16}~2^{126}–1) M=2^{124} E=2^{123}	#3 = #5 Min (2^{32}~2^{126}) Max (2^{32}~2^{127}–1) M=2^{125} E=2^{124}	#2 = #6 Min (2^{64}~2^{127}) Max (2^{64}~2^{128}–1) M=2^{126} E=2^{125}	#1 = #7 Min (2^1~2^{128}) Max (2^1~2^{129}–1) M=2^{127} E=2^{126}
s	M~5/4M–1 E~5/4E–1	M~6/5M–1 E~6/5E–1	M~7/6M–1 E~7/6E–1	M~8/7M–1 E~8/7E–1
p	5/4M~6/4M–1 5/4E~6/4E–1	6/5M~7/5M–1 6/5E~7/5E–1	7/6M~8/6M–1 7/6E~8/6E–1	8/7M~9/7M–1 8/7E~9/7E–1
d	6/4M~7/4M–1 6/4E~7/4E–1	7/5M~8/5M–1 7/5E~8/5E–1	8/6M~9/6M–1 8/6E~9/6E–1	9/7M~10/7M–1 9/7E~10/7E–1
f	7/4M~2M–1 7/4E~2E–1	8/5M~9/5M–1 8/5E~95E–1	9/6M~10/6M–1 9/6E~10/6E–1	10/7M~11/7M–1 10/7E~11/7E–1
g		9/5M~2M–1 9/5E~2E–1	10/6M~11/6M–1 10/6E~11/6E–1	11/7M~12/7M–1 11/7E~12/7E–1
h			11/6M~2M–1 11/6E~2E–1	12/7M~13/7M–1 12/7E~13/7E–1
i				13/7M~2M–1 13/7E~2E–1

CHART 22-7: OD RANGE OF THE U-SPACES ENVELOPES OFFERED BY EACH SHELL'S SUB-SHELLS TO EACH ELECTRON OF A GIVEN ACCELERATION. M = MINIMUM OD OF THE U-SPACES ENVELOPES OFFERED TO ELECTRONS. E = THE MINIMUM OD OF AN ELECTRON'S ACCELERATED LEADING FORM.

There may be some for whom the above chart is self-explanatory. For others as slow as myself, here are some expanded notes: With entries such as #7=#1 and #6=#2, the chart is saying that what Esoptrics calls shell #7 is the same as what Science calls shell #1, and what Esoptrics calls #6 is the same as what Science calls #2, and so forth. In shell #4, the notation, Min (2^{16}~2^{125}), tells us this shell is the result of multi-combos each of which has an OD2^{16} leading form currently at an OD no less than 2^{125}. The notation, Max (2^{16}~2^{126}–1), tells us that, were the leading form accelerated to the maximum possible in that shell, then its current OD would be 2^{126}–1. Such a leading form would then have logically concentric with itself one form (***i.e.:*** u-spaces envelope) at each of the OD's of 2^{16} thru 2^{126}–1.

Continuing with shell #4, the notation, M = 2^{124}, tells us that—to the electrons seeking a u-spaces envelope in this shell—the multi-combos in this shell make available u-spaces envelopes having a current OD no less than 2^{124}. The notation, E = 2^{123}, tells us that the electrons seeking a u-spaces envelope from this shell have a current OD of no less than 2^{123} and, for that reason, seek their u-spaces envelopes from an OD range starting at 2^{124}, which is to say starting in the next highest generic zone.

Continuing with shell #4, the notations, M~5/4M–1 & E~5/4E–1, describe for us the relationship between: (1) the u-spaces envelopes supplied by the multi-combos in shell #4, and (2) the single pair of electrons in sub-shell s of shell #4. They tells us that, in sub-shell s of shell #4, 2 electrons have leading forms currently—depending upon their level of acceleration—somewhere between OD2^{123} and OD2^{123}+2^{121} and, so, receiving their u-spaces envelopes from forms currently somewhere between OD2^{124} and (OD2^{124}+2^{122}), which is to say receiving those envelopes from forms currently in the next highest generic zone.

Continuing with shell #4, the notations, 5/4M~6/4M–1 & 5/4E~6/4E–1, describe for us the relationship between: (1) the u-spaces envelopes supplied by the multi-combos in shell #4, and (2) the 3 pairs of electrons in sub-shell p of shell #4. They tells us that, in sub-shell p of shell #4, 6 electrons have leading forms currently—depending upon their level of acceleration—somewhere between OD2^{123}+2^{121} and OD2^{123}+2^{121}+2^{121} and, therefore, receiving their u-spaces envelopes from forms currently somewhere between OD2^{124}+2^{122} and OD2^{124}+2^{122}+2^{122}, which is to say receiving those envelopes from forms currently in the next highest generic zone.

Continuing with shell #4, the notations, 6/4M~7/4M–1 & 6/4E~7/4E–1, describe for us the relationship between: (1) the u-spaces envelopes supplied by the multi-combos in shell #4, and (2) the 5 pairs of electrons in sub-shell d of shell #4. They tells us that, in sub-shell d of shell #4, 10 electrons have leading forms currently—depending upon their level of acceleration—somewhere between OD2^{123}+2^{121}+2^{121} and OD2^{123}+2^{121}+2^{121}+2^{121} and, therefore, receiving their u-spaces envelopes from forms currently somewhere between OD2^{124}+2^{122}+2^{122} and OD2^{124}+2^{122}+2^{122}+2^{122}, which is to say receiving those envelopes from forms currently in the next highest generic zone.

Continuing with shell #4, the notations, 7/4M~2M–1 & 7/4E~2E–1, describe for us the relationship between: (1) the u-spaces envelopes supplied by the multi-combos in shell #4, and (2) the 7 pairs of electrons in sub-shell f of shell #4. They tells us that, in sub-shell f of shell #4, 14 electrons have leading forms currently—depending upon their level of acceleration—somewhere between OD2^{123}+2^{121}+2^{121}+2^{121} and OD2^{123}+2^{121}+2^{121}+2^{121}+2^{121} and, therefore, receiving their u-spaces envelopes from forms currently somewhere between OD2^{124}+2^{122}+2^{122}+2^{122} and OD2^{124}+2^{122}+2^{122}+2^{122}+2^{122}, which is to say receiving those envelopes from forms currently in the next highest generic zone.

Surmising that's enough to make chart #22-7 sufficiently clear even for those as slow as I am, I will not bother to give to the other shells the same elaborate treatment I've just given to shell #4 and its sub-shells. Still, I should take pains to insure my reader is well alerted to some facts pertinent to more than just any single one of the shells.

First of all, on page 183, I mentioned that, in its article ***Electron Shell***, Wikipedia described shell K (***i.e.:*** #1) as the innermost shell; shell L (***i.e.:*** #2) as one from the innermost shell; shell M (***i.e.:*** #3) is two from the innermost shell, and so forth. Chart #22-7 fully agrees with that. For, as you can see for yourself: (1) in what is shell #7 for Esoptrics, and shell #1 for Science, the electrons receive their u-spaces from envelopes whose radius varies from 2^{121} to $2^{122}-1\Phi$; (2) in what is shell #6 for Esoptrics, and shell #2 for Science, the electrons receive their u-spaces from envelopes whose radius varies from 2^{122} to $2^{123}-1\Phi$; (3) in what is shell #5 for Esoptrics, and shell #3 for Science, the electrons receive their u-spaces from envelopes whose radius varies from 2^{123} to $2^{124}-1\Phi$; and so forth until (4) in what is shell #1 for Esoptrics, and shell #7 for Science, the electrons receive their u-spaces from envelopes whose radius varies from 2^{127} to $2^{128}-1\Phi$. That latter figure, of course, is what leads to the conclusion that the "diameter" of the Φ is 2^{-129} times the dia. of the hydrogen atom with its electron in its outermost orbit.

Secondly, some of the fractions of M and E must result in fractional OD's. As repeatedly said, there are no fractional OD's; and so, wherever the result is a fractional OD, it must be rounded up or down to the nearest whole number.

Thirdly, electrons getting their u-spaces from a *rational* form use ones rotating in all three planes. What, though, of electrons getting their u-spaces from an <u>ir</u>rational form? I currently speculate each such form and its u-spaces rotate only in the tertiary plane and in the same manner as does the first rational form below each such form.

In its article ***Electron shell***, Wikipedia says outer shell Electrons travel farther from the nucleus because the pull of the atom's nucleus upon them is weaker and, therefore, more easily broken. Esoptrics says how far an electron travels from the nucleus has nothing to do with any "pull" from the atom's nucleus. It has to do with the facts: (1) the electron's leading form, as native to OD1, repels every form in the nucleus, and (2) that the electron's leading form causes its carrying generator to be in potency to whatever form has the current OD demanded by the current OD of the electron's leading form. The electron then—though its native OD drives it from the nucleus—stays where it is, because its carrying generator cannot go beyond the outer limits of the form (***i.e.:*** u-spaces envelope) to which the electron's leading form has limited its generator. Then, too, bear in mind that, like all forms, the form, to which the electron's generator is looking for its u-spaces (***i.e.:*** to which it is in potency), shall expand and contract at regular intervals. The electron, then, must alternately, and at regular intervals, dive toward the center of the nucleus and swing away from it. If memory serves me correctly, that's been observed.

To go beyond its current distance from the nucleus, the electron's leading form must cause its carrying generator to become in potency to a form *with a larger diameter to its u-spaces envelope*. That means the electron's leading form must cause its carrying generator to become in potency to a form *with a current OD higher than the OD of the form which held the electron closer to the nucleus*. To have such an effect, the electron's leading form must increase its own current OD. For each increase in its current OD, the electron's leading form ingests one duo-combo, and emits one per OD decrease. Esoptrics suggests this involves absorbing and emitting neutrinos or anti-neutrinos (***i.e.:*** single duo-combos) or photons (***i.e.:*** a few chained duo-combos). Though internet sources speak of beta decay causing the radiation of electrons and anti-neutrinos or positrons and neutrinos, they speak only of *light* being absorbed and emitted; but, Esoptrics suggests every photon is a small number of chained neutrinos or chained anti-neutrinos; and so, photons would fill the bill wherever increases and decreases in OD come in small handfuls.

How do electrons or any lepton or quark absorb and emit anything? As said on pages 44 & 105: Acceleration adds 4th dimension levels at which what's absorbed can be parked; deceleration reduces the levels thereby expelling what had been parked in them. That's Esoptrics' simple explanation. I've yet to see an intelligible one from Science.

Chapter Twenty-Three:

GRAVITY + ESOPTRICS VS. THE HIGGS FIELD:

When the subject is gravity, what becomes paramount by far is what's commonly called "the inverse square law". On page 424 of his book, ***The Fabric Of The Cosmos***, Prof. Greene assures us that, so far (He copyrighted his book in 2004 and seems to have written it in 2003.), no deviations from that law have been detected in 3 dimensional space. As many books (Wikipedia, too, in its article ***Inverse Square Law***) explain: Referred to the gravitational force between 2 bodies, that law means this: Double the distance between 2 objects, and that force's intensity drops to one-quarter of its intensity at the prior distance, because $2^2 = 4$, and the inverse = ¼. At triple the distance, it's one-ninth; at 4 x, $1/4^2$; etc.. That's said true, though, only in a 3 dimensional realm. Thus, we read:

> Also, in more than three dimensions the gravitational force between two bodies would decrease more rapidly than it does in three dimensions. In three dimensions the gravitational force drops to ¼ of its value if one doubles the distance. In four dimensions it would drop to $^1/_8$, in five dimensions it would drop to $^1/_{16}$, and so on.
> ——**STEPHEN HAWKING and LEONARD MLODINOW: *The Grand Design***; page 161 of the paperback by Bantam Books Trade Paperbacks; New York, 2012. ©2010 Stephen Hawking & Leonard Mlodinow.

Esoptrics contradicts that and maintains that—though in *four* dimensions (Ontological distance is the 4th.)—gravity still decreases in accordance with the inverse square law. How is that possible? It's possible because Esoptrics' two explanations of the gravitational force (I've not yet decided which is correct.) are each radically different from any other explanation advanced. What are those explanations? Curiously, each says this:

ESOPTRICS' INVERSE SQUARE LAW APPLIED TO GRAVITY

THE INTENSITY OF THE GRAVITATIONAL FORCE BETWEEN TWO MULTI-COMBOS DIMINISHES AS THE INVERSE OF THE SQUARE OF THE LOWEST OD LEVEL AT WHICH THEIR LOGICALLY CONCENTRIC FORMS OVERLAP DURING BOTH PHASES OF THEIR CYCLES.

Why is that? Onward to the details regarding one way the above maybe applies!

Esoptrics says any given *multi*-combo's *mass* equals its ontological depth (***i.e.:*** the number of duo-combos logically concentric in it) squared. That mass is the generators' doing; but: (1) only the multi-combo's forms enable them to produce that mass, since only the presence of each form at a discrete level of the 4th dimension enables the generators to be concentric without colliding; and (2) it is thru the actuality of the multi-combo's forms that the gravitational force of that mass is instantly exerted throughout the whole of each form's actuality. Because the gravitational force is the product of the mass and the mass equals the square of the multi-combo's ontological depth, the *gravitational force* of any given multi-combo relative to a single duo-combo equals the square of its ontological depth. For example, if a multi-combo's leading form is at OD2^{255}, its ontological depth is 2^{255}, and its gravitational force is $(2^{255})^2 = 2^{510}$ (***i.e.:*** $3.35\text{x}10^{153}$)G, where G = the gravitational force of a single duo-combo. Again, if a multi-A-combo's leading form is at OD2^{128}, its ontological depth is 2^{128}, and its gravitational force is $(2^{128})^2 = 2^{256}$ (***i.e.:*** $1.1579\text{x}10^{77}$)G.

Why, now, does that intensity diminish as the inverse of the distance (Say "radius", if you prefer.) squared? The answer most readily suggesting itself is this: At $2^{255}+1\Phi$ away from a multi-combo at OD$2^{255}+1$, the intensity of its gravitational force should be the bare minimum. That would expectedly be so, if the intensity of its gravitational force diminishes as the inverse of the distance squared. Thus, at a distance of $2^{255}\Phi$, the intensity of its gravitational force diminishes to $1/(2^{255})^2 \text{ x } 2^{510}=1/2^{510}\text{x}2^{510}$ = 1G, and = zero at 1Φ or more beyond that. That, in turn, implies this: At 2Φ, the intensity of its gravitational force is 1/4 x 2^{510}G; at 3Φ = 1/9 x 2^{510}G; at 4Φ = 1/16x2^{510}G; and so on for 2^{255} such steps of 1Φ each.

Exactly what brings about this reduction? How might I spell out the answer to that in the most understandable manner possible? My puny mind can think of nothing but to fall back upon a drawing composed of squares. As I do so, be sure to bear in mind what I myself too easily fail to bear in mind—namely: The squares are merely a kind of "figure of speech" and *metaphor* trying to serve as the most understandable way of explaining what is purely logical and, therefore, highly abstract in nature.[1] Thus, the microstates of excitation represented by the squares are by no means squares with one measurement from side to side and another from corner to corner. Indeed, the microstates have no kind of *physical* (***i.e.:*** spatial) measurement whatsoever. All each has—as a result of God's *a priori* coordinates—is a logical location relative to the other microstates included in the same u-spaces envelope. If the fourth dimension of the whole is the focus of our attention, then all each microstate has is the current OD of the form providing it.[2] If the other three dimensions of

---

[1] Though I've already stressed this point *ad nauseam*, I'll still do it some more now. On page 352 of ***The Fabric Of The Cosmos***, Prof. Greene states that—in what is called "the standard model of particle physics"—all the basic particles are considered points, which is to say "a dot of no size" (pg. 349). Also on page 349, he describes what he calls "the key to string theory's success". That "key", he says, is that its basic particles have "spatial extent". Esoptrics, of course, says both viewpoints are wrong. Nothing has "spatial extent" except in collective effect, and no ultimate particle is *strictly* "a dot of no size". Ultimates *strictly* such are *in no way* either divisible or extended within themselves and leave one wholly unable to explain how they could be separate from one another and add up to an extended whole. Esoptrics' basic particles are very different. *Logically*, each is very much extended and divisible and separate from all others in a whole extended in a way which cannot so far make any sense to the physicists, because they know nothing of how God's *a priori* coordinates influence "the ultramicroscopic makeup of space, time, and matter" (Greene pg. 493). On my pages 54&174, see my quotes from, and answers to, Prof. Greene's bit about: "Shrinking smaller than the Planck scale", may entail "concepts of space and time segue into notions for which 'shrinking smaller' is as meaningless as asking whether the number nine is happy." Esoptrics goes far below the Planck scale to give history's only account of that "ultramicroscopic makeup", and let's see if history judges it "as meaningless as asking whether the number nine is happy."

[2] On page 426 of his book, Prof. Green writes "no other serious proposal" "would so thoroughly shake the foundation of physics" than evidence for more than the familiar 3 dimensions. On 367, he adds: Prior to string theory (On 382, this goes back to the 70's.), "no theory said anything at all about the number of spatial dimensions". *No* theory said *anything at all*?! Esoptrics said 4 well before

the whole are the focus of our attention, then each microstate has only some positive vs. negative percentage of some 1, 2, or 3 of the available 6 potencies. "Logically a square or cube" means only: The best way to teach the lesson at hand is to use pictures of squares which imagination can then elevate to the status of cubes.

Mindful of that, we turn to the following series of 4 drawings. In each, "OMEGA" signifies the one categorical form native to, and currently at, $OD2^{256}$. In each, the arrow pointing down to the dot between squares 1 & 2 is pointing to Omega's center.

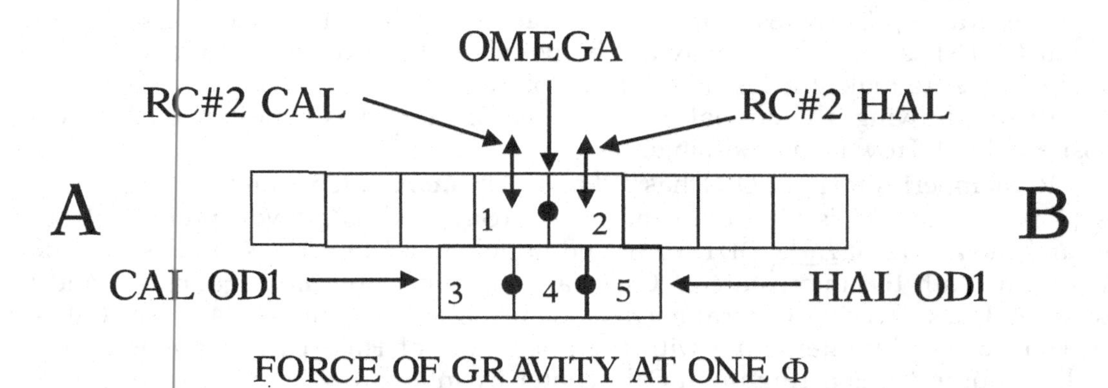

DRAWING #23-1: HOW GRAVITY IS EXERTED AT 1Φ APART

Cal & Hal are each a multi-g-combo currently in the second reverse category (***i.e.:*** OD's 2^{255} thru $OD2^{256}–1$).[1] The dot between squares 3 & 4 represents the center of Cal and, therefore, the center of Cal's carrying generator. The dot between squares 4 & 5 represents the center of Hal and, therefore the center of Hal's carrying generator. The upper row of squares represents a tiny part of the field of locomotion (***i.e.:*** u-spaces envelope) which Omega is presenting to Cal and Hal. Cal's carrying generator is performing the microstate of excitation represented by square #1, and, therefore, the dot between squares 3 & 4 must be *imagined* as centered on square #1. Hal's carrying generator is performing the microstate represented by square #2 and, therefore, the dot between squares 4 & 5 must be *imagined* as centered on square #2. The centers of Hal and Cal are thus 1Φ apart for sensation dependent minds; but, for minds not so dependent, they are 1 microstate of excitation away from one another (***i.e.:*** Cal's & Hal's microstates are logically contiguous.). The 2 rows are depicted as adjacent because, if, in the drawing, we were to impose the lower on the upper, the resulting confusion would be insurmountable.

---

string theory. Dismiss it as not a "serious proposal"; but, don't say *no* theory ever said *anything at all* about 4 dimensions. On pages 360 & 367, he mentions Theodor Kaluza's (1885-1954) 1919 appeal to 4 space dimensions but, dismisses it as an *assumption* having, unlike String Theory, no equations predicting Kaluza's 4th dimension. Is String Theory not *theory*, and does *Theoretical* Physics make no *assumptions*?! See his page 322 listing parts of it which "depend" "on the assumptions we make." "Said no equations predicting" is not the same as "said nothing at all". Clearly, in writing no theory ever said *anything at all* about the number prior to the 70's, he's contradicting himself big time.

[1] Don't get confused. Saying the 2nd reverse category = the OD's 2^{255} thru $OD2^{256}–1$, does not say Cal & Hal are multi-g-combos whose logically concentric forms are one per $OD2^{255}$ thru $2^{256}–1$. Their logically concentric forms are at least one per $OD2^{1}$ thru 2^{255}. Listing the 2nd reverse category's OD's merely gives the range of the OD's to which Cal's & Hal's leading forms can be current. It also tells us the zone (***i.e.:*** Omega's) to which their carrying generators will turn for their u-spaces.

The two-pointed arrow bisecting square #1 (***i.e.:*** cube in imagination) indicates that Cal, thru his generator, occupies that square at no less than 2^{255} OD levels thus giving that u-space a 4th dimension of 2^{255} OD levels within its 10^{-47} cm.. That's because Cal is a multi-g-combo whose logically concentric duo-combos are at least one per $OD2^1$ thru 2^{255}.

The two-pointed arrow bisecting square #2 indicates that Hal, thru his generator, is occupying that square at no less than 2^{255} OD levels thus giving that u-space a 4th dimension of 2^{255} OD levels within its 10^{-47} cm... That's because Hal is a multi-g-combo whose logically concentric forms are at least one per $OD2^1$ thru 2^{255}.

The drawing indicates that Cal and Hal each hold concentric to itself a form currently at OD1. That, of course, may not be possible;[1] but, we shall pretend the drawing is correct in order to make the lesson easier to follow.

In the drawing, the crucial point is this: Square #4 is simultaneously two squares opposite in kind. How is that possible?

Remember: A form at OD1 has a dia. of 2Φ (***i.e.:*** 2 microstates) and, therefore, a radius of 1Φ. Square #4 is thus both the B arm (Say "radius", if you prefer.) of <u>*Cal's*</u> OD1 form and the A arm of <u>*Hal's*</u> OD1 form, and notice how #4 overlaps (***i.e.:*** is logically interfacing with) both the microstate of Cal's carrying generator (***i.e.:*** square #1) and the microstate of Hal's carrying generator (***i.e.:*** square #2). Thru square #4, then, Cal is acting upon Hal's center by interacting with the microstate of Hal's carrying generator, even as Hal, thru square #4, is acting upon Cal's center by interacting with the microstate of Cal's carrying generator. For Esoptrics, then, it is at OD1—*and only at OD1!*—that Hal and Cal each exert upon the other's carrying generator the intensity of their gravitational force. What is that intensity? It is $(1/1^2)\text{x}2^{510}=(1/1)\text{x}2^{510}=2^{510}$G.

Now, consider this:

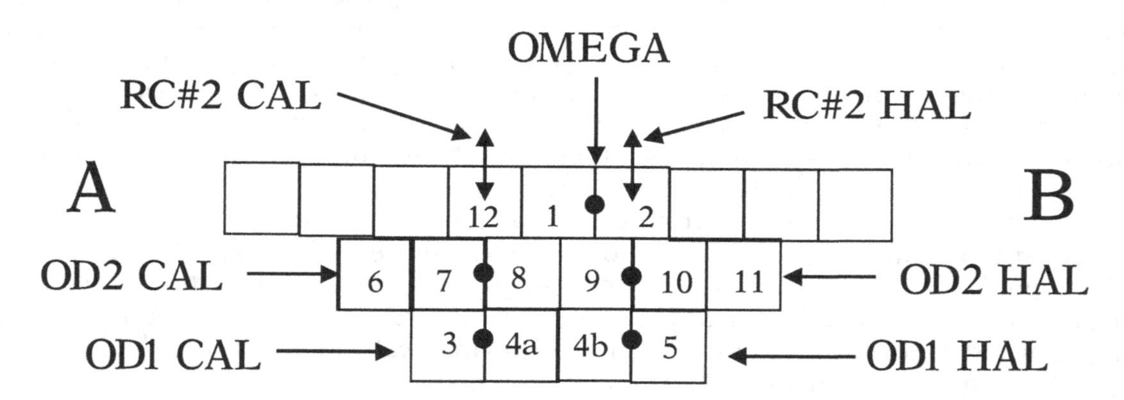

DRAWING #23-2: HOW GRAVITY IS EXERTED AT 2Φ APART

In drawing #23-2, Cal's carrying generator is performing the microstate indicated by square 12, and Hal's carrying generator is still performing the microstate indicated by square #2. As a result, the centers of Cal and Hal are 2Φ (***i.e.:*** 2 logically sequential microstates) apart. What, in drawing #23-1, were 2 instances of square #4 has now split into the 2 squares 4a and 4b. That's because Cal and Hal are now 2Φ apart; and so, the B arm of Cal's OD1 form no longer interacts with the A arm of Hal's OD1 form. As a result, Cal & Hal no longer exert a gravitational force upon one another <u>*at OD1*</u>.

[1] It certainly can't be a form *native* to OD1, but may be a form native to OD3 decelerated to OD1.

They are, however, exerting a gravitational force upon one another *at OD2*. As you can see in drawing #23-2: Because at OD2, each arm is 2 states of excitation in length, squares 8 & 9 constitute both the B arm of Cal and the A arm of Hal, and, thru square #9, Cal is interacting with the microstate of Hal's carrying generator (***i.e.:*** square #2) while, thru square #8, Hal is interacting with the microstate of Cal's carrying generator (***i.e.:*** square #12). Therefore, at that OD2—*and only at OD2!*—each is exerting upon the other the intensity of its gravitational force. What is that intensity? It is $(1/2^2)x2^{510}=(1/4)x2^{510}=2^{508}G$.

Turn to drawing #23-3 below on this page. Cal's generator is performing the microstate represented by square #11, as Hal's generator continues to perform the one represented by square #2. That puts them 3 logically sequential u-spaces apart. On line OD2, notice how that affects Hal's and Cal's OD2 forms. Squares 4c and 4b are the A arm of Hal's OD2 form, and squares 4a and 4c are the B arm of Cal's OD2 form. Because neither of the 2 "parts" of Hal's A arm (***i.e.:*** squares 4c & 4b) overlap the innermost part of Cal's B arm (***i.e.:*** square 4a); and because neither of the 2 "parts" of Cal's B arm (***i.e.:*** squares 4a & 4c) overlap the innermost part of Hal's A arm (***i.e.:*** square 4b)—Cal & Hal no longer exert a gravitational force upon one another *at OD2*.

They are, however, exerting a gravitational force upon one another *at OD3*. As you can see in drawing #23-3: Because at OD3, each arm is 3 states of excitation in length, squares 4a, 4c & 4b constitute both the B arm of Cal and the A arm of Hal, and, thru square #4a, Hal is interacting with the microstate of Cal's carrying generator (***i.e.:*** sq. #11), while, thru square #4b, Cal is interacting with the microstate of Hal's carrying generator (***i.e.:*** sq. #2). Therefore, at that OD3—*and only at OD3!*—each is exerting upon the other the intensity of its gravitational force. That intensity is: $(1/3^2)x2^{510}=(1/9)x2^{510}G$.

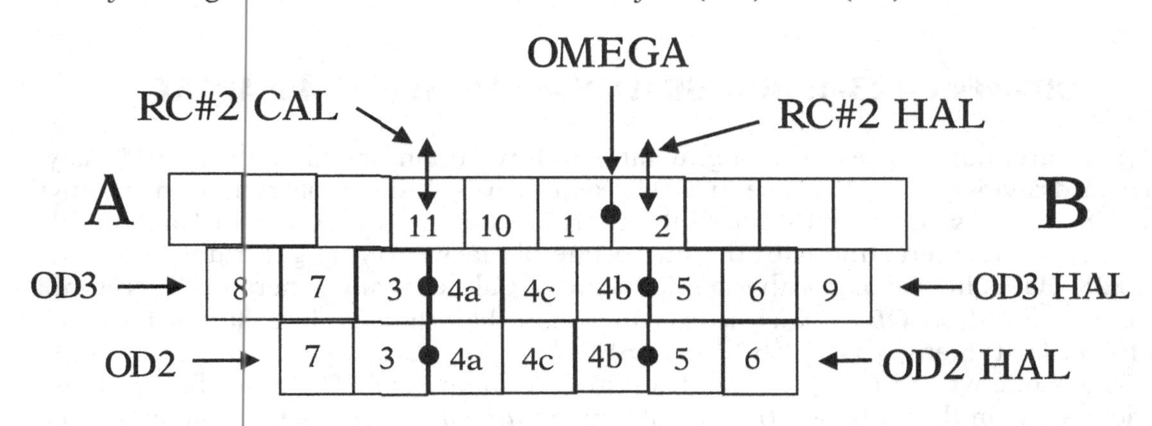

DRAWING #23-3: HOW GRAVITY IS EXERTED AT 3Φ APART

Turn to drawing #23-4 on the next page. Cal's generator is performing the microstate represented by square #36, as Hal's generator continues to perform the one represented by square #2. That puts them 4 logically sequential u-spaces apart. On line OD2, notice how that affects Hal's and Cal's OD2 forms. Squares 9 and 9a are the A arm of Hal's OD2 form, and squares 8 and 8a are the B arm of Cal's OD2 form. Because neither of the 2 "parts" of Hal's A arm (***i.e.:*** squares 9 & 9a) overlap the innermost part of Cal's B arm (***i.e.:*** square 8); and because neither of the 2 "parts" of Cal's B arm (***i.e.:*** squares 9 & 9a) overlap the innermost part of Hal's A arm (***i.e.:*** sq. #9)—Cal & Hal no longer exert a gravitational force upon one another *at OD2*.

The same condition prevails on line OD3. On it, squares 18, 19 & 20 are the A arm of Hal's OD3 form, and squares 17, 18 & 19 are the B arm of Cal's OD3 form. Because none of the 3 "parts" of Hal's A arm (***i.e.:*** squares 18, 19 & 20) overlap the innermost part of Cal's B arm (***i.e.:*** square 17); and because neither of the 3 "parts" of Cal's B arm (***i.e.:*** squares 17, 18 & 19) overlap the innermost part of Hal's A arm (***i.e.:*** square 20)—Cal & Hal no longer exert a gravitational force upon one another *at OD3*.

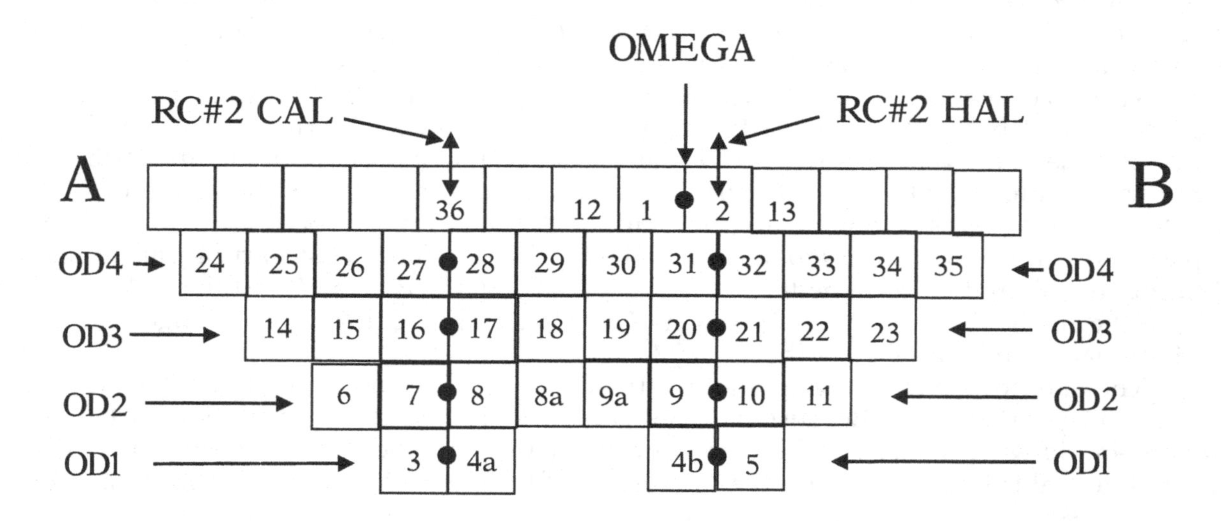

FORCE OF GRAVITY AT FOUR Φ

DRAWING #23-4: HOW GRAVITY IS EXERTED AT 4Φ APART

They are, however, exerting a gravitational force upon one another *at OD4*. As you can see in drawing #23-4: Because at OD4, each arm is 4 states of excitation in length, squares 28, 29, 30 & 31 constitute both the B arm of Cal and the A arm of Hal, and, thru square #31, Cal is interacting with the microstate of Hal's carrying generator, while, thru square #28, Hal is interacting with the microstate of Cal's carrying generator. Therefore, at that OD4—*and only at OD4!*—each is exerting upon the other the intensity of its gravitational force. That intensity is $(1/4^2)x2^{510}=(1/16)x2^{510}=2^{506}G$.

That's one way to interpret Esoptrics' inverse square law. The other has to do with Esoptrics assertion that—though the rate at which *rational* forms change from act to act is determined by their current times their native OD—the rate at which <u>ir</u>rational forms change from act to act is determined by their current OD squared. For example, in drawing #23-1 on page 191 where Hal and Cal are only 1Φ apart, Hal's OD1 form is pulsing against Cal's OD1 form once per K, and Cal's OD1 form is pulsing against Hal's OD1 form once per K. But in drawing #23-2 on page 192 where Hal and Cal are 2Φ apart, Hal's OD2 form is pulsing against Cal's OD2 form once per 4K, and Cal's OD2 form is pulsing against Hal's OD2 form once per 4K, and that rate = both the square of their distance from one another in Φ's and the square of the lowest OD level at which they contact one another. Likewise, in drawing #23-3 on page 193 where Hal and Cal are 3Φ apart, Hal's OD3 form is pulsing against Cal's OD3 form once per 9K, and Cal's OD3 form is pulsing against Hal's OD3 form once per 9K, and that rate = both the square of their distance from one another in Φ's and the square of the lowest OD level at which they contact one another. Again, in drawing #23-4 on page 194 where Hal and Cal are 4Φ apart, Hal's OD4 form is pulsing against Cal's OD4 form once per 16K, and Cal's OD3 form is pulsing against Hal's OD4 form once per 16K, and that rate = both the square of their distance from one another in Φ's and the square of the lowest OD level at which they contact one another. Thus: (1) When Hal and Cal are 2Φ apart, the frequency with which they act upon one another at OD2 is one-fourth the frequency at which they acted upon one another at OD1, and that one-fourth = the inverse of the square of their distance from one another at 2Φ; (2) when Hal and Cal are 3Φ apart, the frequency with which they act upon one another at OD3 is one-ninth the frequency at which they acted upon one another at OD1, and that one-ninth = the inverse of the square of their distance from one another at 3Φ; and (3) when Hal and Cal are 4Φ apart, the frequency with which they act upon one another at OD4 is one-sixteenth the frequency at which they acted upon one another at OD1, and that one-sixteenth = the inverse of the square of their distance from one another at 4Φ.

Here again, then, the intensity of the gravitational force between them diminishes as the inverse of the square of the lowest OD level at which their component, logically concentric forms overlap during both phases of their cycles. It's just that, in this version of Esoptrics' inverse square law, the law has to do with the difference in frequency of change from act to act rather than a mere change in the lowest OD level at which contact occurs. This version perhaps explains more satisfactorily *why* the inverse square law correlates with a change in the lowest OD level of contact.

Some readers may here perceive two problems seeming to be implied by the above explanations. The first goes like this: As the forms change from act to act, the length of their arms should alternately expand and contract. As a result, whereas—at 2Φ apart—Hal's and Cal's OD1 forms may not contact one another in the contracted phase of their cycles, they would contact one another in the expanded phase of their cycles.

Such an objection stems from a failure to remember Esoptrics' inverse square law refers to what happens during *both* phases of each cycle. At 2Φ apart, Hal's and Cal's OD1 forms may overlap in *one* phase of each cycle; but, they can't overlap in *both* phases.

The second problem goes like this: It may be that—among the duo-combos logically concentric to their leading forms—Hal or Cal or both have one or more *rational* forms currently at an OD to which they are not native. If so, there must be intervals at which the frequency of impact of one upon the other is not the square of the lowest OD level of contact.

That's true; but, one must bear in mind that the ratio of irrational to rational forms is 2^{128} to 1—making this discrepancy's occurrence an astronomically infrequent one. Also, whenever it does occur, it's only at some particular u-space whose distance in Φ from such a form's center = that form's current OD. Thus, to detect that discrepancy, one would have to observe what's occurring at a location c. 10^{-47} cm. in "diameter" and occurring there for perhaps an astronomically small number of K. That may be within the purview of what a *philosopher* or theologian can *deduce*; but, it's far from being within the purview of what a *physicist* can *observe*. In short, it's a discrepancy of no significance whatsoever.

For economy's sake, in the above, my explanation appeals to two multi-g-combos in potency to Omega; but, describing *gravity*, one should appeal either to a multi-h- vs. a multi-g-combo or to two multi-h-combos. That's harder to do, and I'm reckless enough to assume my reader can readily make any needed substitutions. Why is there such a need?

For Esoptrics, the gravitational force between Hal and Cal is an instance of *anti-gravity*, because each is a multi-g-combo produced by an $OD2^{128}$ categorical form accelerated far above the MT zone. In keeping, then, with the law of the conservation of the categories, they must repel one another until the distance between them in Φ's exceeds their ontological depth. For teaching aid's sake, we've pretended the opposite is true.

I'm also inclined to think that, according to Esoptrics, Hal and Cal also exert their anti-gravitational force against the center of Omega. If so, each will continue moving away from the center of Omega until each—in opposite directions from one another—reach the outermost limits of Omega's u-spaces envelope at $2^{256}\Phi$ from Omega's center. Once there, they will stay there, since: (1) throughout this the 9th epoch of the Universe, there is no form currently at an OD higher than $OD2^{256}$ and, therefore, no u-spaces envelope with a radius greater than $2^{256}\Phi$;[1] and (2) contrary to what some theoretical physicists speculate, the law of the conservation of the categories never allows either Hal or Cal to turn back to the center of Omega. We do not live in an "accordion Universe" destined to collapse back to where it started. Furthermore, since no duo- or multi-combo can travel beyond the limits of the u-spaces envelope being utilized by its carrying generator (limits determined by the OD of the form providing that envelope),[2] neither is our Universe doomed to an unending expansion sure to leave its every occupant—whether galaxy, atom or what have you—so far apart from one another, nothing can ultimately remain save a kind of universal and eternal death eliminating forever any and every kind of life-form.[3]

To be sure, at 2^{512}K (***i.e.:*** $[2^{256}]^2$) since the 9th epoch's start, Omega's dia. shall double allowing Hal & Cal to resume their flight from Omega's center until each is $2^{257}\Phi$ from it. Another 2^{512}K later, Omega's dia. shall return to $2^{257}\Phi$, and Cal & Hal retreat to $2^{256}\Phi$ from its center. It's not much of an accordion effect and can't occur until long after the 9th epoch—at 2^{385}K from its start—yields to a 10th epoch starting with another Big Bang.

That's quite distinct from the theory of the accordion Universe which imagines that

---

[1] On page 30 of his book ***The Fabric Of The Cosmos***, Prof. Greene muses over whether the entire Universe has a place in space, a speed thru it, and whether it can move "ten feet to the left or right". Rightly put, the question is: Does Omega have a place in space, a speed thru it, and the ability to move some distance left, right, or what have you? Omega's position is not in space; it's in OD, and that position is describable as: "the highest level of intensity God currently makes available to the forms". "Edge of the Universe" does not mean "edge of space"; it means the current limit to the number of intensities (***i.e.:*** OD & 4th dimension levels) possible to the forms in their macrostates of excitation. Omega has no velocity or ability to move some distance from the center of the Universe because—with no currently available intensity greater than $OD2^{256}$—there is no higher u-spaces envelope to which Omega's carrying generator can turn for microstates of excitation able to give that generator a means by which to carry Omega's center away from where it—throughout this the Universe's 9th epoch—is nailed. That shall change, though, once the Universe's 10th epoch commences. For, our Omega shall then be a categorical form above which the next highest categorical form (as opposed to *reverse* categorical form) is the one at $OD2^{512}$. To put it another way: "*Outermost* limits of the Universe" does not mean the "*outermost* limits of the three dimensional u-spaces envelopes"; it means the *uppermost* limit of the OD levels currently made available to the forms (***i.e.:*** the entities providing the u-spaces envelopes). The confusion arising here is wholly and entirely the result of the failure to understand how the Universe can—even without taking time into consideration—be *four* rather than *three* dimensional. Alas! Like virtually every human being who has ever walked the face of this planet, the theoretical *physicists* (as opposed to the theoretical *philosopher-theologian*) are so intellectually crippled by sense imagery's infantile take on space, it remains to be seen if they can ever achieve even the slightest ability to understand what Esoptrics is saying.

[2] Compare the above to what Prof. Greene says on pages 388-389 of ***The Fabric Of The Cosmos***. There he says Joe Polchinski and his co-workers in the field of String Theory "showed" that the "endpoints" of the strings would be "stuck" or "trapped", as he says, "within certain regions". Pardon me, if I dare suggest Prof. Polchinski & Co. come nowhere near as close as Esoptrics does to explaining what those regions are and how they limit the movements of those ultimates they call "strings" and Esoptrics calls "carrying generators performing microstates of excitation".

[3] On page 301 of his book, Prof. Greene says that, if inflationary cosmologists are right, then in c. 100 billion years, the Universe shall be "a vast, empty, and lonely place." Isn't a godless Universe wonderful?!

this Universe created by the last Big Bang must end by collapsing back to what it was like just before that last Big Bang and then—in an identical repeat of it—produce an exact copy of it. In other words, all who lived and died in the course of the prior Big Bang will—in the course of the next one—duplicate their life stories down to the finest details. Esoptrics agrees there shall indeed be another Big Bang; but, thru it, God will create a Universe vastly different from the prior one and including in itself—as an insignificantly tiny occupant relative to itself—that prior one.

Among those able to grasp the above well (assuming any of the above shall ever be *read* by anyone at all, let alone *understood*), some may perceive it presents us with two very different scenarios with regard to gravity. What are those 2? The difference has to do with variations in ontological depth and distance in Φ's. How so?!

Let's say Hal and Cal have the same ontological depth of $2^{255}+32$. Even if Hal and Cal are $2^{254}\Phi$ (***i.e.:*** c. 2.25 trillion light years) apart, each still contacts the other at $2^{254}+32$ OD levels (***i.e.:*** $[2^{255}+32] - 2^{254}$). Until Hal and Cal are further apart than $2^{255}+32\Phi$, each one's gravitational force is equal to the other's, and each is impacting the other at the same OD levels and with the same frequency. In short, each is exactly reciprocating the other's gravitational force, until the distance between them is greater than $2^{255}+32\Phi$, which is to say greater than either one's ontological depth.

In contrast to that, let's suppose Cal is a multi-A-combo with an ontological depth of $2^{128} + 2^{127}$ and thus moving at half the speed of light. The instant Cal's center is more than $(2^{128} + 2^{127})\Phi$ from Hal's center, not a one of Cal's logically concentric forms extends to Hal's center; and yet, of the forms logically concentric in Hal, $(2^{255}+32) - (2^{128} + 2^{127})$ of them (a huge number indeed) still extend to Cal's center as well as far beyond it.

Clearly, then, when it comes to gravity, there are two very different situations possible. <u>On the one hand</u>, we have two multi-combos fully or nearly equal to one another in ontological depth, and their distance from one another in Φ's is less than the ontological depth of either one of them. <u>On the other hand</u>, we have two multi-combos the 1st of which has an ontological depth vastly inferior to that of the 2nd and their distance from one another in Φ's exceeds that of the ontological depth of the 1st but not of the 2nd.

What might be the significance of that difference? For now, I can do no more than suspect it's of importance to the internal structure of the atom where all the particles, except for the electron,[1] are very few Φ's apart, but all have an ontological depth of at least c. 2^{121}. As for the reason they are so close together, remember what Esoptrics says about categorical forms, other than OD1, operating in, or below, that MT zone consisting of the OD's 2^{128} to $2^{129}-1$: They *attract* rather than *repel* one another as they do when operating above that MT zone. That, of course, is the result of Esoptrics' contention that the way they operate at and below MT is the opposite (***i.e.:*** mirrored image = reverse) of the way they operate above the MT. It's a situation, I suspect, which implies that the inverse square law may not apply to the atomic nucleus.

Compare that with what Prof. Greene says on pages 398-399 of ***The Fabric Of The***

---

[1] Since all the duo-combos concentric in the electron are native to OD1, it's immaterial what its distance is from the nucleus' multi-combos. Between them and it, there's a kind of *anti*-gravity ever forbidding them to pull it in. The only reason it stays with the nucleus at all is that, being native to OD1, it ever receives its field of locomotion from some multi-combo in the nucleus until it's accelerated to $OD2^{128}$, and then flies away at light speed. If, as Esoptrics says, 8^{256} (***i.e.:*** $1.55\text{x}10^{231}$) duo-combos were created native to OD1, then, even with 2^{128} (***i.e.:*** $3.4\text{x}10^{38}$) OD1 duo-combos per electron, there can be $4.56\text{x}10^{192}$ electrons. If so many is far too many, vast must be the number of free OD1 duo-combos (***i.e.:*** neutrinos) flooding the cosmos. Science says it's so. Between Esoptrics and Science, it's an agreement indicating Esoptrics is right to equate its duo-combos with neutrinos.

Cosmos where he seems to be saying that, under some circumstances, the gravitational attraction between two extremely close tiny objects may, to a certain point, vary as their separation and thereafter remain constant. So he *seems to me* to be saying, and, at the top of page 398, adds that the validity of gravity's inverse square law has been tested experimentally down to no further than a tenth of a millimeter. So far, though, I'm still very much in the dark regarding both what he's truly saying and regarding what is truly implied by the two different situations I've described 2 paragraphs earlier.

Assuming Esoptrics' version of gravity's inverse square law is sufficiently clear, the reader can perhaps now follow me as I contrast that version with the one which apparently prevails among today's theoretical physicists. Prof. Greene's ***The Fabric Of The Universe*** shall serve as the source for the latter.

On page 419 of his book, Prof. Greene mentions the notion of "gravitational waves" which move *"within"* rather than *"through* space" [Italics are in the original – ENH.]. On page 341, he mentions the notion that "the gravitational force" is conveyed at the microscopic level by particles called "gravitons". On page 72, he tells us that, for Einstein, the "warps and ripples" constituting gravity *"travel at exactly the speed of light"* [Italics are in the original – ENH.] instead of acting instantaneously as they do in Newton's view of gravity. On pages 394 thru 400, the terminology is "gravitational force", "gravitational field", "gravitational field lines", and "gravitational pull". He then seems to take the inverse square law as it pertains to electromagnetic radiation (as in the case of light) and applies it to the gravitational inverse square law in exactly the same way, thus making that law wholly contingent upon the kind of space which has three dimensions. Gravity, as he puts it: "is thus a reflection of a geometrical property of spheres in three space dimensions." On page 396, he even gives us a little drawing (***i.e.:*** "Figure 13.4") illustrating the notion that, as the gravitational field lines from the Sun "spread out uniformly", the distance between them increases as their distance from their source increases; and so, the number of them which can strike a three dimensional object depends upon that object's distance and how much of its surface area can be struck by the Sun's gravitational field lines. Since the latter is determined by the square of the three dimensional object's distance from the *source* of the gravitational field lines, the gravitational force is reduced—not merely to the number of spread out lines touching the three dimensional object—but, rather, to "the total number of field lines divided by the sphere's area" (bottom of pg. 395). Since the latter is the square of the distance, that explains how the inverse square law applies to gravity.

That, of course, is not even remotely close to what Esoptrics says about gravity. First of all, for Esoptrics, there's no such thing as a "graviton" or a "gravity wave"; there are only the many forms logically concentric with one another in a multi-combo and distinguished from one another by their current OD (***i.e.:*** 4th dimension level). Each of the forms, thru its actuality, is the vehicle for their multi-combo's gravitational force, and the total number of them squared = the maximum gravitational force which their multi-combo can exert. That *maximum* force is one their multi-combo can exert only upon whatever has a center no more than 1Φ from the center of their multi-combo, and the intensity of that *maximum* force is in no way contingent upon whatever might be the number of dimensions in the target of their gravitational force. All that matters is the ontological depth of the bodies in question (***i.e.:*** the number of levels in their 4th dimension) and what their distance from one another in Φ's is compared to their ontological depth.

Secondly, for Esoptrics, Newton was much closer than Einstein to the truth regarding gravity's velocity. For Esoptrics, gravity is never something *traveling* in any sense whatsoever, let alone away from the center of the multi-combo producing it. Because every multi-combo's gravitational force is exerted by its logically concentric forms, it always *is present* both at the multi-combo's center and—at any given OD level—all the way out in every direction for that number of Φ's equal to the given OD level.[1]

Neither the Sun nor any other heavenly body *radiates* gravity. Each is wrapped in a four dimensional mantle of attracters (***i.e.:*** logically concentric forms) each at a discrete OD in that mantle and each having both a discrete "heart beat" frequency plus a discrete radius in Φ's. Thus, how much of its total gravitational force a first body can exert upon a second depends upon: (1) which of the attracters in the first body's mantle has a radius equal in Φ's to its distance—center to center—from the second body and (2) the frequency of the attracter's "heart beat" in K. Gravity being such, the number of dimensions in either body is perfectly irrelevant. Whatever the number of dimensions in either body, the gravitational "pull" of one upon the other diminishes as the inverse of the square of their distance in Φ's. It's merely a question of understanding the kind of mathematics (***i.e.:*** the algebraic math of the mirror) produced by God's *a priori* coordinates.

Yes, each of the forms in the four dimensional mantle has a discrete "heart beat" (***i.e.:*** discrete frequency at which it alternates between two diameters). That, though, does not mean their gravitational force *travels thru space*. For a discrete number of K, each form has a discrete diameter, and then *instantaneously* has a second one for that same number of K, and then *instantaneously* has the first diameter for that same number of K, and then *instantaneously* has the second one for that same number of K, and on and on, until a change in current OD is thrust upon it by external forces.

The difference between gravity per the scientists and Esoptrics can be illustrated even further by turning to what Prof. Greene tells us about what he calls the "Higgs ocean" (pg. 262), the "Higgs field" (pg. 257), and "a non zero Higgs field vacuum expectation value" (pg. 257). On page 262, "the Higgs ocean fills all of space". On 270, "empty space" . . . "may always be filled with an ocean of Higgs field." On 492, a Higgs ocean "permeates all of space". For Esoptrics, there is no such thing as "all of space"; there is only an astronomical number of u-spaces envelopes, each of which is the actuality of a form as presented to the one or more carrying generators looking to it for their u-spaces. Thus, for Esoptrics, every u-space in the Universe is not merely *filled* with the actuality of a form; rather, it <u>is</u> the actuality of some form as presented to some one or more generators. So called "empty space" means, for Esoptrics, u-spaces envelopes provided by forms whose current OD's are far above 2^{128} and whose centers are logically concentric at or near the center of some distant heavenly body such as a planet, the Sun, a black hole at the center of a galaxy, or the like. Esoptrics, then, could tolerate the idea of referring to empty space as a four dimensional Higgs field in which anywhere from 1 to 2^{256}-1 levels of that 4th dimension are devoid of the presence of the actuality of any form but that of Omega.

On page 262, Prof. Green says no particle can escape the influence of the Higgs ocean. Esoptrics' says no *generator* can escape the influence of a form. After all, no generator can exist save as what's performing a microstate of excitation, and no generator can perform a microstate of excitation save where some form provides it with a u-spaces envelope even as another form (***i.e.:*** the generator's piggyback one) tells it what ought to be the current OD of the form from which to seek its u-spaces envelope.

Again on page 262, Prof. Green informs us: "the Higgs field resists only accelerated motion." There would be no "friction" between the Higgs field and any "particle moving through outer space at constant speed". That's precisely the relationship between a carry-

[1] Friends of Einstein's vs. Newton's take on gravity say the former can, but the latter not, explain *exactly* why—in orbiting the Sun—Mercury's closest point to it (***i.e.:*** its perihelion) shifts forward with each orbit to the extent it does. For Esoptrics, the real explanation lies in 2 factors unknown to both: (1) which multi-h-combo in the Sun (central or no?) supplies the form moving Mercury's carrying generator around the Sun, and (2) how *that* combo's carrying generator is moving *it*.

ing generator and the many layered, 4 dimensional mantle of forms thru which it's moving. For, a generator moving at constant velocity would be using only the u-spaces envelope provided by a particular one of that mantle's logically concentric forms. As long as it's thus *moving by means of* that form rather than trying *to transit beyond* it, it's moving at one level of the 4th dimension rather than trying to move up or down in it; and so, there's no resistance or friction. Oppositely, should the generator try to accelerate, its own piggyback form must increase its current OD, and the generator itself must seek its u-spaces from those offered by a form at an OD and level of the 4th dimension higher than the one it was previously using. That may not be easy to do, since that requires finding a u-space which is—in the higher OD form's u-spaces envelope—logically contiguous to the u-space in the u-spaces envelope it had been using. The result of that must be resistance. Naturally, the multiple of the acceleration shall determine how many forms, levels of the 4th dimension, and u-spaces envelopes the accelerating generator must transit in order to find a form and u-spaces envelope at the higher OD demanded by its multiple of acceleration. Because the number of forms, levels of the 4th dimension, and u-spaces envelopes to be transited thus rises, the level of resistance and friction shall rise in proportion to the multiple of the acceleration. As Prof. Green says of the Higgs field, deceleration, too, would produce the same resistance. Esoptrics, then, could tolerate the idea of referring to the forms logically concentric in a multi-combo as a four dimensional Higgs field which offers no resistance to any particle save one trying either to accelerate upward from a lower to a higher—or to decelerate from a higher to a lower—level of that 4th dimension.

On page 263, Prof. Greene states "most physicists" believe no fundamental particle would have any mass at all save for the Higgs ocean. On page 373, last paragraph, mass is said dependent upon what "drag force" is experienced wading thru the Higgs ocean. That's similar to Esoptrics' contention that every multi-combo's mass = the square of the number of generators concentric in it; but, many can be concentric only as a result of the many levels of the 4th dimension provided by the forms. Esoptrics, then, could tolerate the idea of referring to the forms logically concentric in a multi-combo as the four dimensional Higgs field without which no multi-combo can have the mass of many generators.

There is one point on which Esoptrics would never agree with what Prof. Green says about the Higgs field. On page 254 of his book, he's detailing Prof. Alan Guth's theory regarding the initial burst of inflation which occurred in the Big Bang. There, we are told that, "for the briefest of instants," the Higgs field was on a "plateau, in the high-energy, negative-pressure state". That initial burst of inflation, then, was the result of "the outward push supplied by the Higgs field's repulsive gravity". Higgs gave the Big Bang its bang.

As I have explained—quite clearly, I dare say—there was, for Esoptrics, no such thing as significant gravity until there were accelerations causing the production of multi-combos containing a vast multitude of logically concentric forms and their generators. As I set forth in chapter 10, that could not have occurred prior to 10^{-18} sec. into creation. For Esoptrics, then, that initial bit of inflation was not, and could not have been, the result of any kind of gravity whether negative or positive. It thus could not have been the result of anything but the way God went about creating duo-combos in the Universe's first 9 epochs. Therefore, to those self-proclaimed paragons of superior intellectual integrity and utterly unbiased thinking who—despite what they *say* they are—abide no cosmological theory unless it excludes God from the picture, say: God gave the Big Bang its bang, and chapter 10 tells you exactly how.

Chapter Twenty-Four:

THE GRAND UNIFICATION OF 6 FORCES:

Yes, I am quite well aware physicists commonly talk in the terms of *four* rather than *six* forces—namely: gravitation, electromagnetism, the strong nuclear force, and the weak nuclear force. As I understand Esoptrics, though, anti-gravitation must be added in as a fifth force, and—if one is to fully understand the weak nuclear force and electromagnetism—one must first treat magnetism as a separate and, therefore, *sixth* force, and, additionally, one must explain the weak nuclear force in a way of which modern science has no knowledge whatsoever.[1]

I'm reasonably certain I've already made it sufficiently clear what gravity and anti-gravity are in Esoptrics; and so, I shall merely synopsize them so: Gravity is—in every multi-combo's logically concentric forms not native to OD1—the attractive force exerted by its generators thru its forms upon all other duo- and multi-combos, save where the multi-combos acting upon one another are multi-***g***-combos operating in the same generic zone from the standpoint of their leading form's current OD, or where the relationship between 2 multi-***g***-combos is that of the provider vs. the recipient of a u-spaces envelope.

If that's understood, it's rather easy to describe what the strong nuclear force is for Esoptrics: It's the reverse of Esoptrics' anti-gravity. As I've stated more than once, anti-gravity prevails between categorical forms operating *above* the MT zone. Where they—other than the electron—operate *below* the MT zone, they operate in the reverse manner and, therefore, attract one another in a way which brings them as close to one another as possible. In saying that, Esoptrics seems to agree quite well with what Wikipedia says in its article ***Fundamental Interaction***. There, in its sub-section ***Strong Interaction***, it says the strong nuclear force "holds only inside the atomic nucleus" and must be "strong enough to squeeze the protons into a volume that is 10^{-15} of that of the entire atom."

For Esoptrics, the weak nuclear force, I suspect, is what occurs when the leading form of a multi-combo in the atom either increases or decreases its current OD, which is to say moves up or down in the 4th dimension and, thereby, increases or decreases the number of 4th dimension levels within its multi-combo. An increase causes it to *ingest* single or multiple duo-combos and to make them logically concentric with itself at its newly acquired 4th dimension levels. A decrease causes it to *emit* single or multiple duo-combos in the form of neutrinos or anti-neutrinos, or photons, or the like as it forfeits some of its 4th dimension levels. In some cases, the decrease may cause both the emission and extreme acceleration of what then becomes for a while the leading form of a multi-combo with an ontological depth of perhaps c. 2^{146} (***i.e.:*** c. 2^{128}x2^{18}) and thus having a mass c. 34.4 to 68.7

---

[1] Wikipedia, in the sub-section ***Beyond The Standard Model*** of its article ***Fundamental Interaction***, says: "Some theories beyond the standard model include a hypothetical fifth force."

billion times that of a multi-A-combo.[1] Esoptrics' take on the weak nuclear force thus seems to concur with Wikipedia's statement (sub-section ***Weak Interaction*** of its article ***Fundamental Interaction***) that the weak nuclear force causes beta decay.

For Esoptrics, magnetism and the magnetic force are the result of the way rational (and, in some cases, irrational) forms rotate in the primary and secondary planes. As I have explained elsewhere, these rotations take place at velocities so high,[2] they affect this our world of the tertiary plane in much the same way a water-skier affects the surface of the water over which he's racing—namely: Tracks appear briefly in that surface. In our world, it means magnetic lines of force appear.

Those rotations in the primary and secondary plane have two other important results. On one hand, they impart a wave motion to whatever the weak nuclear force is emitting beyond the atom, and thus does Esoptrics explain how the magnetic force works with the weak nuclear force to produce the wave characteristics of electromagnetic radiation. On the other hand, as I explained in chapter 21, those rotations determine the electrical charges this or that multi-combo shall have.

In its article ***Electromagnetic Radiation***, Wikipedia tells us that such radiation: "has both electric and magnetic components, which stand in a fixed ratio of intensity to each other, and which oscillate in phase perpendicular to each other and perpendicular to the direction of energy and propagation." Those statements accord rather well with Esoptrics' contentions: (1) that rotations in each of the three planes are logically perpendicular to those in the other two, and (2) that there is a definite ratio between them.

For Esoptrics, then: If by magnetic force one means the way forms rotate in the primary and secondary planes, then the magnetic force is the cause of both the wave characteristics of what the weak nuclear force emits (or ingests) and whatever electrical charges the various multi-combos have. It is thus also the ultimate cause of whatever results those electrical charges produce. In the end, then, Esoptrics perhaps has only five *fundamental* forces: gravity, anti-gravity, the strong nuclear force, the weak nuclear force, and the magnetic force (Say tripartite rotational force, if you prefer.) which is the more basic source for electrical charges, wave characteristics, magnetism, and magnetic lines of force.

Regardless of how skimpy and perhaps not completely accurate the above may be, I think it rather strongly established that whatever forces there may be, each and every one of them ultimately stems one way or another from the same one source—namely: the forms and either: (1) in the way they behave, whether that be the way they rotate in the three planes or the way they move up and down in current OD or the rate at which they change from act to act, or (2) in whether or not their native OD is categorical, or (3) in what zone the categorical ones are acting.

✡✞✡✞✡✞✡✞✡✞✡✞✡✞✡✞✡✞✡✞✡✞✡✞✡✞✡✞✡✞

[1] On page 425 of his book, Prof. Greene tells us that, on Oct. 15, 1991, the Fly's Eye cosmic ray detector measured a particle with an energy equal to 30 billion proton masses.

[2] Remember: These high velocities do not violate what Einstein says about light speed. For, it is the u-spaces envelopes and their u-spaces which are moving as such speeds. The generators using those u-spaces are by no means changing from u-space to u-space more rapidly than $1\Phi/2^{128}$K.

Chapter Twenty-Five:

SHAPE OF THE WHOLE VS. THE OCCUPIED UNIVERSE:

On page 243 of his book ***The Fabric Of The Cosmos***, Prof. Greene (Ignoring the above distinction?!) *seems to me* to be saying we have no "final answer" regarding the shape of the Universe. I say *seems* because he uses the phrase "shape of the cosmic fabric" rather than "shape of the Universe". In the paragraph running from page 249 to 250 of his book, he *seems to me* to be saying the leading entrant regarding the Universe's shape is the one promoting a "flat, infinitely large spatial shape". Here again, though, instead of using the phrase "shape of the universe", he uses "structure of spacetime." On page 290, he says "the fabric of space" may curve like a ball's surface or like a saddle or—like the top of a table—not curve at all. Is he, thereby, admitting the Universe might be spherical in shape? That's not likely since, on page 406, he remarks that, though Prof. Richard Tolman (1881-1948) of the California Institute of Technology advocated "a spherical universe" in the 1930's, it's "been ruled out by observations." On page 434, he adds precise measurements by COBE[1] and WMAP[2] give strong support to the proposal that "space is *flat.*" [Italics in the original – ENH]. I dare imagine, of course, he doesn't mean it's flat in the usual sense of "2 dimensional"; rather, he means its "horizontal" measurement is much greater than its "vertical" one. In other words, it's not all that deep or thick *relative* to how wide it is.

As I've said repeatedly: Esoptrics holds neither the Universe (say "space as a whole", if you prefer) nor any of its visible occupants has any kind of *inherent* shape whatsoever. Whatever shape they have, it's the *collective effect* of the logical way the u-spaces—and, thru them, the generators—are outside of one another. All 3 dimensionality is the *real* and *observable effect* of the fact that basic realities (***i.e.:*** forms and generators) are distinct from, separate from, and outside of, one another solely as a result of having a discrete set of internal characteristics dictated by the antecedent influence of God's *a priori* coordinates—coordinates which not only set them apart but, rather, determine where each is logically in relationship to each and every one of all the others. That's much the same as when our minds determine where the number 1 is logically in relationship to each and every one of all other numbers no matter how many of them there are.

---

[1] COBE = Cosmic Background Explorer a/k/a Explorer 66 = a satellite sponsored by NASA, built at the Goddard Space Flight Center, and launched into Earth orbit from Vandenberg Air Force Base on Nov. 18, 1989, with the goal of finding evidence to support the Big Bang theory.

[2] WMAP = Wilkinson Microwave Anisotropy (***i.e.:*** having properties dependent upon direction as opposed to what has identical properties in all directions) Probe a/k/a Explorer 80 = a spacecraft sponsored by NASA with the Goddard Space Flight Center and Princeton University, and launched June 30, 2001, from the Cape Canaveral Air Force Station. In its article ***Wilkinson Microwave Anisotropy Probe***, Wikipedia states: "As of October 2010, the WMAP spacecraft is derelict in a heliocentric graveyard orbit after 9 years of operations."

Some, of course, will insist nothing can have a unique internality until, first of all, it has a unique locality in time and space (Say "spacetime", if you prefer.). Esoptrics insists there can be no unique places in time and space—or even any such thing as time, space, or spacetime—until first there are basic realities each with a unique internality which periodically changes slightly without ever forfeiting its uniqueness. Which comes first: the chicken (***i.e.:*** spacetime) or the egg (***i.e.:*** unique internality)? Say time or space or spacetime, and you necessarily haul in the notion of the continuum, infinite divisibility, and the absurd math such entails. It's a mental nightmare thundering with unbearable loudness the madness of putting the "chicken" before the "egg". For all of that, hardly any of even history's most brilliant thinkers have ever heard or hear the noise.[1]

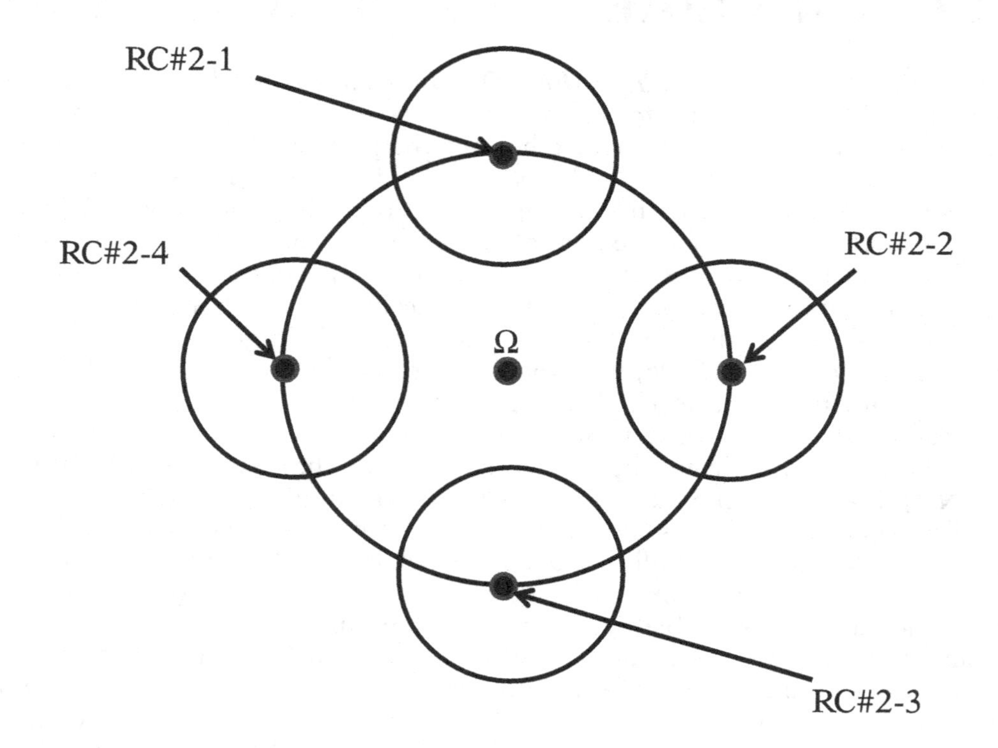

DRAWING #25-1: SHAPE OF THE UNIVERSE WITH A FULLY DEVELOPED REVERSE CATEGORY #2

What, then, according to Esoptrics, is the *collectively effected* shape of the whole Universe vs. its occupied area? The former is far more so spherical than anything else, and drawing #25-1 above is an attempt to illustrate the sense of that.

---

[1] On his page 471, Prof. Greene repeats the popular opinion that, for Immanuel Kant (1724-1804), it's "downright impossible" to think about and describe the Universe, if you "do away with space and time". Esoptrics utterly refutes that Kantian line. It fully explains how—if you do away with God's *a priori* coordinates—you can't think about and describe anything cosmic save in a way which, however useful, draws in so pathetic, infantile and wrong a view of time and space, the result is mathematical buffoonery and buffoons too dishonest to see the buffoonery for the utterly untenable nonsense it most brazenly is. But, with God in the mix, one *can* "do away with space and time" *as grossly misinterpreted* by Kant and almost all of humanity and—as Esoptrics *manifestly* does—still think about and describe the Universe quite well.

In the above drawing, were I able to draw spheres instead of circles, the larger circle would be a sphere sporting 8 spheres instead of 4—each of which, as the arrows indicate, represents one of the multi-g-combos anchoring the 2nd reverse category.[1] As with each of the 4, so would it be with each of the 8: At maximum distance from Omega, one half extends from the larger sphere's outer perimeter toward its center, while the other half extends outward beyond the larger sphere's outer perimeter. Why?!

Each of the dots marked by an arrow represents the center of its sphere and, therefore, the center of the generator carrying its sphere. Since each of those generators has no source of u-spaces other than Omega; and since each generator is at the outermost of the u-spaces provided by Omega as a u-spaces envelope—the center of each of the 8 smaller spheres is as far from the center of Omega as it can ever go throughout the course of the 9th epoch's duration of 2^{385}K (***i.e.:*** c. 18 trillion Earth years by our way of reckoning).

As a sphere in imagination, the larger circle represents Omega which—as repeated *ad nauseam*—is the one form created by God native to OD2^{256}. For probably at least the first trillion years by our reckoning, no multi-g-combo's generator traveled far enough away from Omega's center to enable its leading form's u-spaces envelope to extend beyond Omega's u-spaces envelope; and so, for probably at least the first trillion years, the *effective* shape of the Universe was Omega's *effectively* spherical shape. More precisely, the effective shape was the *collectively* effected spherical shape of Omega's u-spaces envelope.[2]

That being so, drawing #25-1 depicts what would, to some extent, be the effective shape of the Universe after the generators of each of the 8 RC#2 multi-g-combos has reached the outermost of the u-spaces provided by Omega. I say "to some extent" because it does not reflect a very important possibility. How so?!

In the drawing, the diameter of each of the smaller circles is roughly half that of the larger circle. If that be taken as to scale, it declares that, in each of the 8 RC#2 multi-g-combos, the leading form's current OD is at or near 2^{255}—an OD which would give each of them a radius of c. $2^{255}\Phi$ as opposed to Omega's radius of $2^{256}\Phi$. Needless to say, nothing *requires* such a circumstance. On the contrary, each of them could have a current OD of anywhere from 2^{255} to 2^{256}–1 and thus a radius anywhere from 2^{255} to 2^{256}–1Φ. If each one's leading form were currently at OD2^{256}–1, the actuality of each one's leading form would extend from Omega's center to 9 trillion light years beyond Omega's perimeter (***i.e.:*** the diameter of each would equal Omega's). The result would again be a Universe whose effective shape is almost perfectly spherical. Oppositely, if each one's leading form had a unique current OD, the Universe's effective shape would be that of a balloon with 8 different size bumps on its surface. Can anyone sanely expect me to produce a drawing illustrating that?

Such, then, is the story of the Universe's shape as it is affected by the RC#2 multi-g-combos? What, then, of the RC#3, RC#4, RC#5, RC#6, RC#7, and RC#8 multi-g-combos? I'm not about to tackle the task of producing drawings pertaining to every one of the remaining 7. Even if I were masochistic enough to want to undertake such a project, the size limits of this book's pages would not allow it. I must, then, beg my reader to be content with two more drawings addressing how the RC#3 and RC#4 multi-g-combos affect the

---

[1] Remember: In multi-g-combos anchoring the 2nd reverse category (***i.e.:*** RC#2 multi-g-combos), the ontological depth is at least 2^{255} but no more than 2^{256}–1, and each one's leading form is at least at OD2^{255}+1 and may be anywhere from OD2^{255} to OD2^{256}–1.

[2] That implies that, for perhaps a trillion years, a vast mantle of "empty space" surrounded that area of the Universe in which duo- and multi-combos were active. Such space, though, would be empty only in the sense that the u-spaces comprising it were devoid of every OD and 4th dimension level other than 2^{256}, which is to say the OD of their provider, Omega. In other words, Omega's actuality, and many of the u-spaces Omega's actuality <u>*is*</u>, are the only realities present there.

Universe's shape. As the reader shall see in drawing #25-3 on page 207, even the attempt to depict the latter cannot be carried out in anything close to full detail.

Drawing #25-2 below is an attempt to give the reader some idea of what the Universe's effective shape would be, if the RC#3 multi-g-combos reached the outermost limits of the RC#2 multi-g-combos after the latter have reached the outermost limits of Omega.

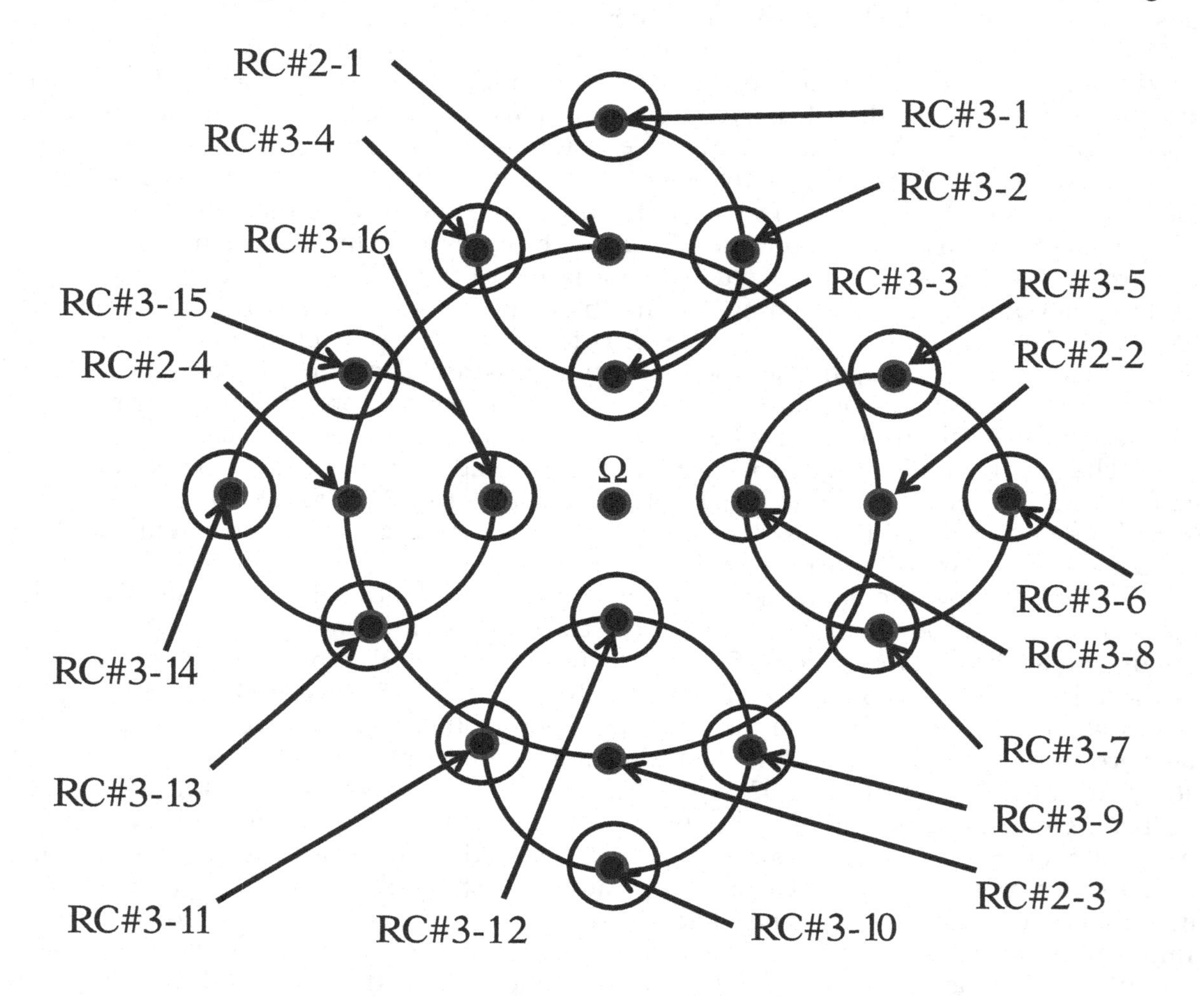

DRAWING #25-2: SHAPE OF THE UNIVERSE WITH FULLY DEVELOPED REVERSE CATEGORIES #2 & #3

Here again, I'm using circles to portray globes. Here again, then, what's drawn as 4 circles must be imagined as double that many spheres. Here again, the drawing does not and cannot reflect the following: (1) Though Omega's dia. is $2^{257}\Phi$ throughout the 9th epoch, the dia. of each of the 8 RC#2 multi-g-combos may—as u-spaces envelopes—vary from 2^{256} to 2^{257}–2Φ at OD2^{255} and 2^{256}–1 respectively; and (2) the dia. of each of the at least 64 RC#3 multi-g-combos may—as u-spaces envelopes—vary from 2^{255} to 2^{256}–2Φ at OD2^{254} and 2^{255}–1 respectively.

In drawing #25-3, next page, I'm making no attempt to present a *complete* picture of what the Universe's effected shape is with fully developed reverse categories of the 2nd, 3rd, and 4th orders. I dare suppose every reader will agree with me on that strategy upon seeing that drawing #25-3 is so complicated as to be almost unintelligible.

Because there's no room for it, I have not, in #25-3 below, tried to depict every RC#3 multi-g-combo as having RC#4 multi-g-combos in potency to it. Only 4 of the 12 are drawn as doing such. For the same reason, I've not tried to show each RC#2 multi-combo as having more than 3 RC#3 multi-g-combos in potency to it. As drawing #25-2 does with regard to the diameters of the RC#2 and #3 multi-g-combos, so does drawing #25-3 do with regard to the diameters of the RC#4 multi-g-combos: It does not and cannot reflect the fact that—with 2 *generic* zones in RC#4—the diameters of the latter may vary: (1) anywhere from 2^{253} to $2^{254}-2\Phi$ as long as they remain in the 4th reverse category's *253rd generic* zone, or (2) anywhere from 2^{254} to $2^{255}-2\Phi$ as long as they remain in the 4th reverse category's *254th generic* zone. Remember the *diameter* in Φ's = 2 x current OD.

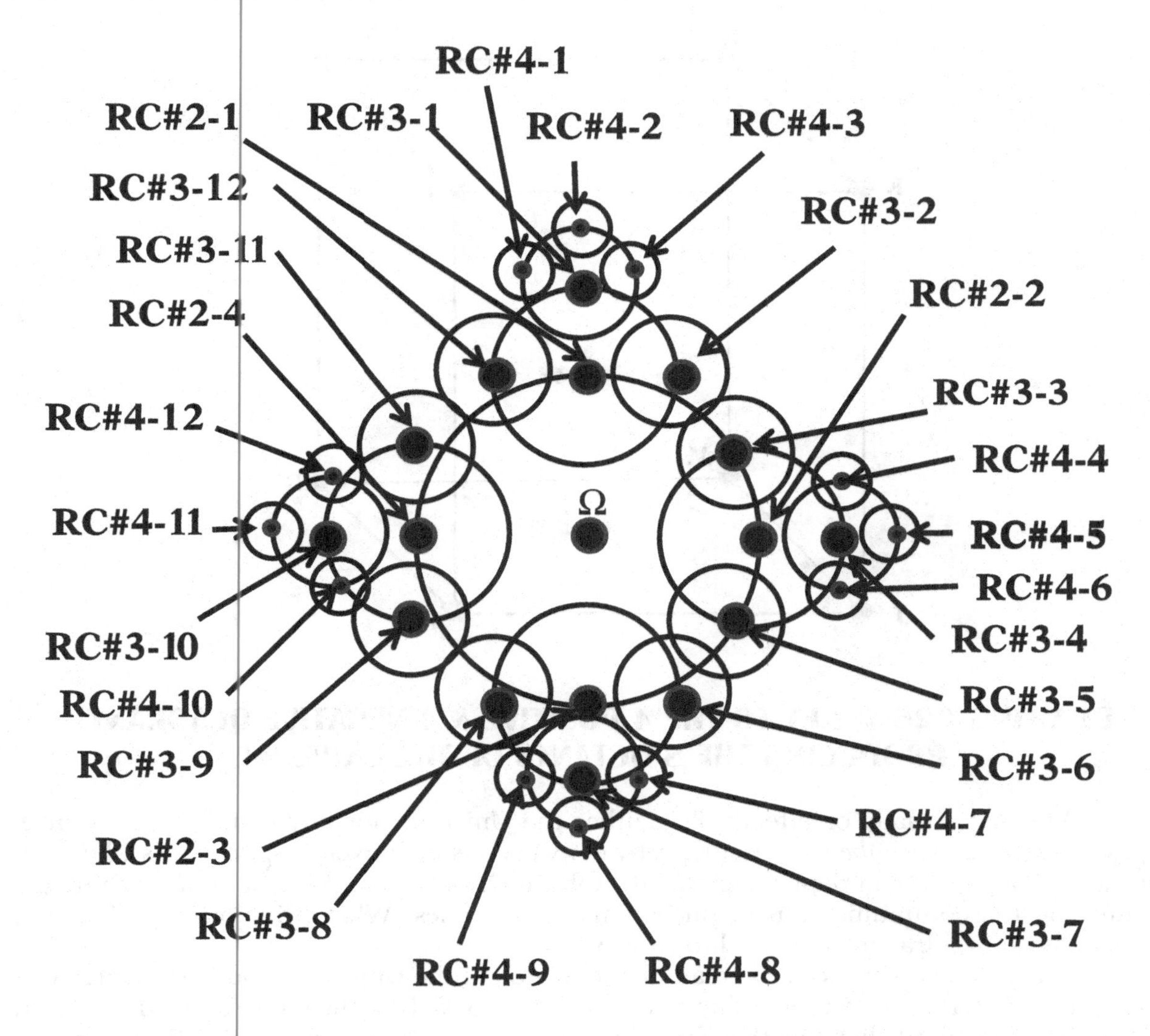

DRAWING #25-3: SHAPE OF THE UNIVERSE WITH FULLY DEVELOPED REVERSE CATEGORIES #2, #3 & #4

How, then, does one reconcile Esoptrics' effectively spherical Universe with Science's "flat" one? The simplest and perhaps most glaring answer is that Esoptrics speaks of what the *whole* Universe effectively is *from the standpoint of the effective shape of its u-spaces envelopes*. That leaves wide open the possibility that the duo- and multi-combos operate solely within an area whose horizontal measurement is much greater than its vertical one and thus leaving the vertical axis of its *occupied* territory a tiny fraction of what is available. But, if that's so and one can observe only what has duo- and multi-combos active in it, then the shape of the *observable* Universe cannot be the same as the shape of the *whole* Universe. Why might the Universe's observably occupied area take on a "flattened" shape rather than the mostly spherical shape of the whole? Let's answer using this:

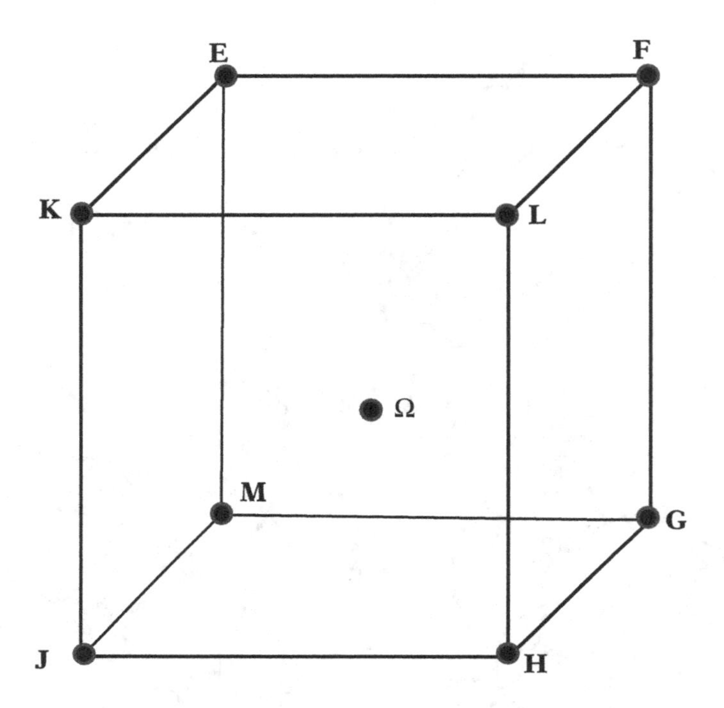

DRAWING #25-4: KEY TO THE 4 POSITIVE & 4 NEGATIVE QUADRANTS PRODUCING THE 8 OCTANTS OF THE UNIVERSE

Whether actually or effectively, neither the Universe nor any of its u-spaces envelopes is a *square* or a *cube* (I do not thereby deny each is effectively a *sphere*.). As I have often said, though, we can depict each as metaphorically such for the sake of a teaching aid facilitating the communication of one or more principles. What principle or principles, then, is the above drawing designed to convey?

When God commenced to create the 9th epoch of the Universe, God first created one form native to $OD2^{256}$ and then 8 duo-combos native to each of the OD's of 2^{255} thru $2^{256}–1$. I suggest the result of that was this: (1) One of the 8 sets raced toward point E as a second but reverse set raced toward point H thus producing one stream and its opposite in a flat plane relative to one another; (2) a third one of the 8 sets raced toward point F as a fourth but reverse set raced toward point J thus producing a second stream and its opposite in a second flat plane relative to one another; (3) a fifth one of the 8 sets raced toward point L as a sixth but reverse set raced toward point M thus producing a third stream and its opposite in a third flat plane relative to one another; and (4) a seventh one of the 8 sets raced toward point K as an eighth but reverse set raced toward point G thus producing a fourth stream and its opposite in a fourth flat plane relative to one another. The result was thus 4 sets each of a positive and a negative quadrant of the Universe each racing away from its opposite in opposite directions in a flat plane relative to one another. Notice the distinction there: Each stream is in a flat plane relative to its opposite (***i.e.:*** 180^{0} apart); but, as they do that, each is using a *u-spaces envelope* which is quite the opposite of *flat*.

If one has followed that so far, one can perhaps readily see it's throwing 2 rather glaring issues at us: (1) Is there any kind of communication whatsoever between either the two reverse halves of a single stream or between any two of the four sets of 2 streams each? (2) If each stream started out in a flat plane, which is more likely: (a) that each shall produce a spherical octant in which the duo- and multi-combos are active throughout that octant's spherical area, or (b) that each shall produce a spherical octant in which the duo- and multi-combos are active solely throughout an area of that octant whose vertical depth relative to its horizontal expanse is flat?

If the answers are as I suspect they should be, then we live in a Universe in line with, and racing away from, a mirrored image Universe. Meanwhile, the midpoint between us and our reverse Universe is being intersected by three other Universe-reverse-Universe planes each at a 45^{0} angle to our Universe-reverse-Universe plane. Thus, though the shape of the Universe *as u-spaces envelopes* is effectively spherical, its shape as an "inhabited" Universe is effectively that of an 8-pointed star each arm of which is relatively flat and isolated from and oblivious of, the other 7.

Again, if the answers are as I suspect they should be, then it becomes rather evident: (1) why the *observable* Universe's effective shape is relatively flat rather than spherical, (2) why the observable Universe's shape is at best the shape of an eighth of the whole, and (3) why the shape of those 8 arms—whether individually or collectively—is radically different from the shape of the Universe *as one or more u-spaces envelopes*. In short, if the answers are as I suspect they should be, Prof. Greene and company are far from the truth regarding the effective shape of the *whole* Universe and are such simply because they know nothing about the distinction between the whole vs. the occupied Universe.

Have I, think you, said enough on this issue? Forgive me if I conclude I have and, therefore, bring this chapter to a close.

Is it not reasonable to say that motion is not without the mind, since if the succession of ideas in the mind become swifter, the motion, it is acknowledged, shall appear slower without any alteration in any external object?
——**GEORGE BERKELEY:** ***Principles Of Human Knowledge***, Section 14 as given on page 415, lower right-hand column, of vol. 35 of ***Great Books Of The Western World*** as published by Encyclopædia Britannica, Inc.; Chicago, 1952 (That's right, Mr. Berkely. But, did you also wonder if maybe there are circumstances in which "ideas in the mind" *can in fact* become swifter, indeed become as much as 10^{92} swifter? Probably, you did not, because that would require realizing that consciousness is a *series* of infrequently changing acts rather than a single, continuously changing one.).

For anything that moves round about in a circle, in less times than our ideas are wont to succeed one another in our minds, is not perceived to move; but seems to be a perfect entire circle of that matter or colour, and not a part of a circle in motion.
——**JOHN LOCKE:** ***Concerning Human Understanding***, Book II: Chap. XIV, Sec. 8, as given on page 156, lower right-hand column, of Vol. 35 of ***Great Books Of The Western World*** as published by Encyclopædia Britannica, Inc.; Chicago, 1952 (So have many often observed, Mr. Locke. But, why did neither you nor they ever question if maybe our ideas could "succeed one another in our minds" at speeds far, far greater than the one with which we are familiar?)

Physicists spend a large part of their lives in a state of confusion. It's an occupational hazard.
——**BRIAN GREENE:** ***The Fabric Of The Cosmos***, as found on page 470 of the 2005 paperback edition published by Vintage Books, a division of Random House, Inc., New York. 2005. ©2004 Brian Greene (Such must ever be true of all who leave God out of their struggle to figure out what the fabric of the cosmos *really* is.).

Chapter Twenty-Six:

CONSCIOUSNESS' STROBOSCOPIC EFFECT:

As with so much else in the universe, every act of consciousness—in so far as it is consciousness of *sensation*—is a well-defined act on the part of a form performing a macro-state of excitation. Each act has a definite duration in K and—as a u-spaces envelope—a "size" relative to all other acts by all other forms. Consciousness is thus a series of what some might call "quanta". Which form, then, performs our acts of consciousness (***i.e.:*** *sensory* consciousness throughout this chapter), and what is its "size" and duration in K?

Esoptrics says consciousness is a series of individual macrostates of excitation on the part of a form *native* to some irrational, major but generic OD (***ex. gr.:*** OD2^{127}) and *currently* at an OD somewhere between 2^{155} and 2^{156}. For what follows, it's not necessary to know either *exactly* where it is or *why* it's there. Let's call this form Phil.

Because Phil is native to an OD below 2^{128}, Phil is a major form and, as a form, a u-spaces envelope. As a u-spaces envelope, Phil is *effectively* a sphere whose diameter in Φ = his current OD doubled (***vid.:*** formula #9-1 on pg. 68). His dia. is, therefore, somewhere between 2^{156} (***i.e.:*** 9.1343852×10^{46}) & 2^{157} (***i.e.:*** $1.82687704 \times 10^{47}$)Φ, and his volume in u-spaces would, as the cube of that, = somewhere between 2^{468} (***i.e.:*** 7.621456×10^{140}) and 2^{471} (***i.e.:*** $6.097165137 \times 10^{141}$) u-spaces a/k/a Φ's.

Since, according to Esoptrics 1 cm. = $1.361128468 \times 10^{46}$Φ, Phil's diameter translates into somewhere between 6.7108913 cm. at OD2^{155} and 13.4217826 cm. at OD2^{156}. That = 2.642 to 5.2842 in. in dia. where 2.54 cm. = 1 in.. Within that range is the human brain's size which Wikipedia's article ***Human Brain*** gives as 1,130 cubic cm. (***i.e.:*** the cube of 10.4158 cm.) in women and 1,260 cubic cm. (***i.e.:*** the cube of 10.8008 cm.) in men.

Since Phil is an irrational form, his every cycle would have a *minimum* duration of $2[(2^{155})^2] = 2(2^{310}) = 2^{311} = 4.1718496 \times 10^{93}$K and a *maximum* duration of $2[(2^{156})^2] = 2(2^{312}) = 2^{313} = 1.66873987 \times 10^{94}$K. He would have that minimum if he were exactly at 2^{155} and that maximum if he were exactly at 2^{156}.

For Esoptrics, 1 sec. = $1.388543802 \times 10^{95}$K. Dividing that by 4.1718496×10^{93} (***i.e.:*** 2^{311}) and then by $1.66873987 \times 10^{94}$ (***i.e.:*** 2^{313}), we calculate Phil, per sec., performs between 33.28 and 8.32 *cycles* at 2^{155} and 2^{156} respectively. Since each cycle has 2 phases, Phil, in one sec., performs between 66.56 *acts* at 2^{155} and 16.64 *acts* at 2^{156}, and each *act* lasts between 1/66.56 sec. (***i.e.:*** .015024 sec.) at 2^{155} and 1/16.64 sec. (***i.e.:*** .060096 sec.) at 2^{156}. Obeying the logic of a *mirror*, half of those acts are of one kind, and the other half of an opposite kind. Half are awareness of something and half awareness of nothing (***i.e.:*** no *awareness* at all). Consciousness *as awareness of something* is thus between 8.32 & 33.28 acts per second.

That's interesting to note for a very important reason: If I remember correctly, it's been well established by observation that we humans cannot detect the flashing nature of light which flashes more than 20 times a second. For example, an old note of mine says that, on page 72 of vol. 23 of the 1967 edition of ***Encyclopedia Britannica***, we're told that, when extremely bright flashes of light are played on the eye, the eye is totally unable to detect the flashes once the rate reaches approximately 20 per second. If that's correct then the form we're describing here would correlate nicely with that. For, we are saying there are somewhere between 8.32 and 33.28 acts of actual awareness per second, and the rate of "approximately 20 per second" falls within that range. If you rightly understand how static each act of actual consciousness is, it follows that, if awareness occurs *less* than 20 times a second, then it won't be able to detect the flashing mode of a light flashing *more* than 20 times a second. I'll go into that more thoroughly on the next page.

Thus does Esoptrics' appeal to a form somewhere between 2^{155} and 2^{156} correlate nicely with what scientific observation says about the size of the human brain and the human inability to detect the flashing nature of light once it flashes 20 times a second or more. But, that correlation comes about only if Esoptrics' formulas for calculating size and absolute rate of change are correct when giving the human form: (1) a size between 2^{156} and $2^{157}\Phi$, and (2) a change rate of between once every $4.1718496x10^{93}$K and once every $1.66873986x\ 10^{94}$K. The correctness of those formulas is based upon these two hypotheses: (1) The minimum size = $7.34683969x10^{-47}$ cm., and (2) the minimum time length = $7.201789375\ x\ 10^{-96}$ sec., thus making the maximum rate of change $1.388543802x10^{95}$ times per second. Consequently: (1) In fitting the size of the form at 2^{155} to 2^{156} to the size of the brain, Esoptrics strongly upholds its contention regarding the minimum unit of space, and (2) in explaining the flashing light syndrome by means of formula #8-9 on page 61, Esoptrics thus strongly upholds its contention regarding the minimum unit of time and maximum rate of change. That it could correlate two ways like that is no small matter.

So then, we have two time frames: one involving 66.56 acts (***i.e.:*** phases) per second and another involving 16.64 acts per second. Let us now calculate how far light travels in 1/66.56 vs. 1/16.64 seconds.

If we take light speed as 29,979,245,800 cm./sec. and divide by 16.64, the result is 1,801,637,367.79 cm. per act at 2^{156}. Divide by 66.56, and the result is 450,409,341.95 cm. per act at 2^{155}. Since we receive light from opposite directions side to side, those 2 results must be doubled so: 1,801,637,367.79 x 2 = 3,603,274,735.58 cm., and 2 x 450,409,341.95 = 900,818,683.9 cm..

So then, if Phil is exactly at 2^{155}, he's performing 66.56 acts (***i.e.:*** phases) per second and, per act, receiving the *bi-directional* influx of light from an area 900,818,683.9 cm. in diameter. This light he must pack into a sphere only 6.71 cm. in diameter. Obviously, we're going to have to do a lot of miniaturizing. Dividing 900,818,683.9 cm. by 6.71, we find the number of times such light must be reduced—namely: 134,250,176.438 times. What's that multiple's square root? My calculator says it's 11,586.6378. Reduce 6.71 cm. by that square root. The result will be .00 057 9 cm.. That, I dare *suggest*, is about the size of the smallest object consciousness can see per act. Next, reduce .00 057 9 cm. by that same 11,586.6378. The result will be .00 000 004 998 cm., and that puts us at the "size" of the hydrogen atom. That amounts to saying this: Take the atomic diameter (***i.e.:*** the mirror threshold's diameter a/k/a 2^{128} steps from 0 actuality), multiply it by 11,586.6378, and that gives you the *minimum* diameter of what consciousness can see per act; next multiply the atomic diameter by $11{,}586.6378^2$, and that gives you the *maximum* diameter of what consciousness can see per act; next multiply the atomic diameter by $11{,}586.6378^4$, and that gives you the bi-directional distance light travels per act of consciousness.

Does that progression sound familiar? It should, because, in Esoptrics, the progression N, N^2, N^4, etc., is the one found among the categorical forms, and the categorical form native to $OD2^{128}$ rules the complete atom.

Suppose Phil is at 2^{156}. He would not be *exactly* there, since he would have to be at least 1 step short of that. Still, working with 2^{156} rather than something such as $2^{156}–1$ is more than adequate for our purposes. At 2^{156}, Phil's performing 16.64 *acts* per second and, per act, receiving the *bi-directional* influx of light from an area 3,603,274,735.58 cm. in diameter. This light he must pack into a sphere 13.42 cm. in diameter. Here, too, we're going to have to do a lot of miniaturizing. Dividing 3,603,274,735.58 by 13.42, we find the number of times such light must be reduced, and that number is 268,500,352.87. What's that multiple's square root? My calculator says 16,385.98 (2^{14} = 16,384). Reduce 13.42 cm. by that square root. The result will be .00 081 9 cm.. That, I dare *suggest* is about the size of the smallest object consciousness can see per act. Now, reduce .00 081 9 cm. by that same 16,385.98. The result will be .00 000 004 998 cm.. Once again we're at the "size" of the hydrogen atom. Once again, that amounts to saying this: Take the atomic diameter, multiply it by 16,385.98, and that gives you the *minimum* diameter of what consciousness can see per act; next multiply the atomic diameter by $16{,}385.98^2$, and that gives you the *maximum* diameter of what consciousness can see per act; next multiply the atomic diameter by $16{,}385.98^4$, and that gives you the bi-directional distance light travels per act of consciousness. Here again, it's the familiar progression of N, N^2, N^4.

Thus does Esoptrics take its basic formulas for size and duration in chapter 9 and correlate them with: (1) the size of the human brain, (2) the maximum and minimum diameters of what consciousness can see per act, (3) the fact consciousness cannot detect the *flashing* nature of light flashing more than 20 times per second, (4) the notion that the minimum unit of time is $7.201789375\text{x}10^{-96}$ seconds, (5) the notion that the minimum unit of space is $7.34683969\text{x}10^{-47}$ cm., (6) the diameter of the hydrogen atom, and (7) the speed of light. Who else has done as much?

Just how static is Phil's every act of awareness? To answer, let's first compare the duration of Phil's two act cycle (***viz:*** $2[(2^{155})^2] = 2^{311} = 4.1718496 \text{ x } 10^{93}\text{K}$) with the duration of the two act cycle of a form native to OD1, still at it, and capable of consciousness. Its every cycles' duration is only 2K; and so, for every cycle by Phil at $OD2^{155}$, an OD1 form performs c. 10^{93} cycles. What Phil, thru his positive phases, *experiences* as a mere 20th of a sec., the OD1 form *experiences* as lasting what we'd call c. 10^{93} seconds = c. 10^{83} millennia of $3.1557\text{x}10^{10}$ sec. each. That's how astronomically static each of Phil's phases is.

But how can Phil at $OD2^{155}$ experience a duration that *astronomically* lengthy and experience it as being quite brief? It's because, as long as Phil at $OD2^{155}$ is performing the same one primary act (***i.e.:*** macrostate of excitation), he undergoes no internal change whatsoever and, consequently, is aware of no kind of change whatsoever. With no awareness of any kind of change whatsoever, he has no clue whatsoever to the duration of his primary act; and so, his act seems, to him, wholly without duration. As you know, if you sleep soundly, 8 hours go by in a flash. Here's a repeat of how one scientist expresses it:

> The existence of time thus relies on the *absence* of a particular symmetry: things in the universe must *change* from moment to moment for us even to define a notion of *moment to moment* that bears any resemblance to our intuitive conception. If there were perfect symmetry between how things are now and how they were then, if the change from moment to moment were of no more consequence than the change from rotating a cue ball, time as we normally conceive it wouldn't exist. That's not to say the spacetime expanse, schematically illustrated in Figure 5.1, wouldn't exist; it could. But since everything would be completely uniform along the time axis, there'd be no sense in which the universe evolves or changes. Time would be an abstract feature of this reality's arena—the fourth dimension of the spacetime continuum—but otherwise, it would be unrecognizable.
> ——**BRIAN GREENE:** ***The Fabric Of The Cosmos***, page 470 of the 2005

paperback edition by Vintage Books, a division of Random House, Inc., New York. ©2004 by Brian Greene.

That's exactly what Esoptrics is saying if one makes a few minor changes in wording and then says: The sense images given in an act of sensory consciousness must *change* from one instant of K to its next for us even to define a notion of successive instances of K which bear any resemblance to our intuitive conception of time. If there were perfect equivalence between how our sense images appear in each of several successive instances of K, then time as we normally conceive of it wouldn't exist for us relative to any act of consciousness in which such a condition prevails. That's not to say the passage of u-time would not exist; it would. But, as long as every iota of what's sensed in a given act of consciousness remains unchanged throughout a given number of sequential instances of K, there would be no sense in which that given act of consciousness evolves or changes; and so, throughout the course of such an internally changeless act of consciousness, the u-time duration of such an act would be unrecognizable *no matter how great that duration might be.*[1]

How, then, does Phil at $OD2^{155}$ experience change? Like all forms, Phil at $OD2^{155}$ performs 2 kinds of primary acts per cycle: an expanded one followed by a contracted one. In his expanded act, Phil at $OD2^{155}$ is aware of what's frozen and unchanging in his form's u-spaces; whereas, in his contracted act, he is aware of nothing; and so, as far as he is *aware*, he performs only one act per cycle. While in his contracted act, an immense series of generators each act an astronomical number of times and, thereby, deposit units of force (tension) in each of the $(2^{156})^3$ (***i.e.:*** $2^{468} = 7.62145 \times 10^{140}$) u-spaces Phil at $OD2^{155}$ will experience in his next expanded act. In sensing these units of tension—the internal order of which is set by that of the u-spaces, Herr Kant—Phil at $OD2^{155}$, feels one or more of the different kinds of sensation. For *consciousness*, matter is strictly the units of tension generated by the generators. And what a divide there thus is between intra-mental sense images and those extra-mental realities so radically much other than units of tension!

Performing that next expanded act, Phil, at $OD2^{155}$, is not observing what *is happening* but, rather, the *residue* of what *happened* in a past which, from the OD1 form's point of view, happened over the course of an astronomical number of our millennia.[2] That residue is spread out over u-spaces only *logically* outside of one another; but, all Phil at $OD2^{155}$ can observe is that they are *outside of one another* and, unable to *observe* the *manner* in which they are separate (and, for some reason, unable or unwilling to *figure out* how they are separate), merely assumes they are *physically* outside of one another in a kind of space whose internality rivals God's. Still, despite what Immanuel Kant says (***vid.:*** footnote #1 on pg. 204), it is by no means impossible to "do away with" humanity's *primitive* and *pseudo-divine* concepts of time and space and still be able to think and talk in the terms of things outside of one another. Esoptrics is *conclusive proof* of that.

---

[1] From what Prof. Greene is telling us above, Phil could be a rotating form, but would still be unable to recognize the duration of any given one of his acts as long as, throughout the given act, the content has "symmetry". That's a way of saying Phil would be unable to determine the duration of any given one of his acts in which the internal characteristics are isotropic, which is to say remain the same along all axes and, therefore, the same no matter what the *orientation* involved. As he basically expresses it: No matter which way you turn a cue ball, its internal characteristics don't vary one iota; and so, until you do more to it than *rotate* it, there is no recognition of time's duration.

[2] Does that prove God a liar? The one fact *most manifestly* and *persistently* given us is: Since the present endures not, we experience nothing save memory of what happened *in the past*—whether a micro-sec. ago or a micro-sec. before that, etc.. How dishonest, then, one must be to fail to see *very early* in life that *consciousness* is of what was and never of what is except as an *inference*!

Still, how is there awareness of change? The brain, thru memory, somehow allows Phil at $OD2^{155}$ to compare the content of his subsequent primary acts with the content of his previous ones. How might that be possible? I suspect it has in part to do with the difference in the volumes of the contracted vs. the expanded act. Since the volume of the latter vs. that of the former is possibly $(2^{156} + 2^{127})^3$ vs. $(2^{156})^3$, it's possible that, in the latter, Phil and every such form experience the content of both their latest and some of their prior acts. Or maybe it's like this: Half of Phil's c. 10^{140} u-spaces present him with the extra-mental world's impressions, while the other half presents him with his intra-mental world's impressions—some of which come from a kind of form not native to our world. What do I imply by that appended remark?

> This tricky issue—to what extent is our personal identity tied to our physical being?—has been debated for years in a variety of guises without being answered to everyone's satisfaction. While I believe identity all resides in the physical, others disagree, and no one can claim to have the definitive answer.
> ——**BRIAN GREENE: *The Fabric Of The Cosmos*,** page 441 of the 2005 paperback edition by Vintage Books, a division of Random House, Inc., New York. ©2004 Brian Greene.

Certain it is that I view Esoptrics as clearly insisting that not all of any human being's identity resides in the "physical"—even if that includes duo-combos. The human soul governing Phil's piggyback form at $OD2^{155}$ is not itself a piggyback one.

How does Phil at $OD2^{155}$ live in and experience a world radically different from the world an OD1 form lives in and experiences? In other words, how can the latter be, not merely a part of the former but, rather, a *subservient* part of it? The answer is first this: Phil at $OD2^{155}$ lives in, and experiences, a world populated by other forms each having a current OD somewhat close to Phil's. They and Phil function basically like strobes causing the rapidly rotating "fan blades" to appear stationary and thus to seem continuously in only one out of the astronomical number of positions they actually occupy in the course of each one of the acts of consciousness Phil and his kind execute at or near $OD2^{155}$. As a result, Phil and his kind thus experience, and agree upon the content of, the same slow-motion world produced by the stroboscopic effect they—and all forms at or near their current OD—have upon the chaotic world of microscopic things.

> Somewhere between the tiny world of individual atoms and subatomic particles and the familiar world of people and their equipment, the rules change because the sizes change.[1] The motivation for asserting this division is clear: a tiny particle, according to quantum mechanics, can be located in a fuzzy mixture of here and there, yet we don't see such behavior in the big, everyday world. But exactly where is the border? And, of vital importance, how do the two sets of rules interface when the big world of the everyday confronts the minuscule world of the atomic, as in the case of a measurement?
> ——**BRIAN GREENE: op. cit.:** page 203.

Where exactly _is_ the border? As I said earlier: Esoptrics says it's at the MT's OD of 2^{128} which, for us, is at the "size" of $2^{129}\Phi$ and thus 5×10^{-8} cm.. And how do the rules *below*

---

[1] Around 50 years ago, I said to my older brother that time and space are different in the world of the tiniest realities. He, with a degree in physics, scoffed at my assertion which at least basically said: "the rules change because the sizes change." I've been saying that since 1947 at 11 years old.

the MT interface with the rules *above* the MT? When irrational but generic forms are accelerated from below to above the MT in OD in order to form the molecules, they—as the result of their greatly diminished rate of change from act to act—impose their stroboscopic effect upon the MT and below and, thereby, cause the "fuzzy mixture of here and there" to appear as a very staid instance of mostly here for long periods of time. It's ultimately merely a matter of learning all the effects of that 4th dimension called ontological distance.

Let's now close this chapter with a little analogy designed to summarize the nature of consciousness. To that end, imagine one of the old style cameras using film instead of a memory chip. Oddly, it has two different lenses each with its own shutter. One lens and its shutter are on the face of the camera and pointing their front at the outside world and their rear at the film; the other and its shutter are inside the camera and pointing their front at the film's rear and their rear at an observer called Phil. As long as the outside lens' shutter is open, the inside lens' shutter is closed, and vice versa. As long as the outside lens' shutter is open, light is entering the camera, exposing the film, and creating a time exposure. After the outside lens' shutter has been doing that for c. 10^{93}K, it closes for c. 10^{93}K, and the inside lens now opens to allow Phil, for c. 10^{93}K, to observe on the film the time exposure which remains utterly frozen and changeless throughout the c. 10^{93}K Phil spends observing it. For Phil, though, the duration of the time spent observing the time exposure is "unrecognizable" due to the fact that—for him to determine the duration of that time—what's on the exposed film "must *change* from moment to moment for" him "even to define a notion of *moment to moment* that bears any resemblance to" his "intuitive conception" (Greene pg. 470) of a time segment. Thus, though the duration of each of Phil's acts of consciousness—each coupled with an act of unconsciousness—is an astronomically lengthy couplet of c. 10^{93}K, the duration of each as far as Phil can tell is *zero* seconds. As a result, as far as Phil can *observe*, there is no *sequence* of acts; there is only one, continuous, infinitely divisible act taking place in a single, continuous, infinitely divisible stream of time and change. To *figure out* what consciousness and time *really* are, Phil must be an introspective philosopher using the reflexive half of bi-directional empirical observation rather than a scientist limited to mono-directional empirical observation.

Imagine another such camera with this difference: In each cycle, the outside shutter opens for only 1K before it closes and allows the inside shutter to open for 1K. Its observer is Matt. Suppose that, for Matt, Phil, and all intelligent observers, every experience of c. 20 successive cycles of consciousness/unconsciousness is called a second. In that case, wherever Phil says a second of time has transpired, Matt will disagree vigorously and insist trillions of trillions of trillions of trillions of trillions of trillions of trillions of billions of seconds have passed. Time at $OD2^{155}$ is nothing like what it is at OD1, which is to say what we call time is nothing like what it is for the souls of the dead not with God in infinity.

Hell and Purgatory are not in the center of the Earth in the sense of "center of the *planet*"; rather, they are in Earth's center in the sense of "center of every one of the *tiniest particles* of earth (Notice the change from an upper to a lower case "e", so that "Earth" means "planet" but "earth" means each piece of the material constituting Earth.). For those able to think in the terms of 4 dimensional u-spaces, that means a very low level of OD, if not OD1. All who died a second ago by our reckoning may have already—in the blaze of deaths all-revealing light—been facing an astronomically vast plenitude of truth for many, many trillions and trillions of eons. Oh, and if truth can hurt, what think you of so great a burst of it?! It shall be sheer terror, says Isaiah, to understand the message (28:19).

Chapter Twenty-Seven:

THE ULTIMATE MYSTERY:

With such progress, physicists have been motivated to see how much further inflationary cosmology[1] can go. Can it for example, resolve the ultimate mystery, encapsulated in Leibniz's question of why there is a universe at all.
——**BRIAN GREENE:** ***The Fabric Of The Cosmos***, page 310 of the 2005 paperback edition by Vintage Books, a division of Random House, Inc., New York. ©2004 by Brian Greene.

Yes, for some scientists, such as Prof. Greene, it's not merely a mystery; it's the *ultimate* one. For some, such as myself, it's not a mystery at all. Why is that?

Prof. Greene himself gives us the key to the answer when, on page 456 of his book, he asserts that free will's role in "fundamental physical law" is still "unresolved." Its role may be "unresolved" for some scientists such as Prof. Greene; but, for some, such as myself, it's totally resolved leaving us fearlessly convinced of this: The reason "why there is a universe at all" is that God—above all else whatsoever—loves free will as God's *Infinitely* Informed Intellect defines free will.

I, for one, am unshakably sure that The *Infinitely* Informed's take on free will is incalculably superior to that of any *finite* individual and any number of such. Guided by Catholic Doctrine as I understand it, God defines free will as the *presence* of *self*-determination rather than the *absence* of *all predetermination*. But, guided by Catholic Doctrine as I understand it, I am aware that *infinitely* informed God's view of every individual's *self* is infinitely superior to every living individual's knowledge of his or her *self*. Why is that?

This side of the grave, one knows nothing of one's self save one's current self. But, what is that? It's merely what memory says has happened to one's self in the past. Since that changes from microsecond to microsecond, one's knowledge of one's *current* self is knowledge of an astronomically insignificant fraction of one's *total* self, if "total self" means the simultaneous presence of all the past, present, and future instants of one's life.

Quite the opposite is true of God's knowledge of each one of us whether living or dead. For, God—without beginning or end—unceasingly and omnisciently knows the *total* self of every individual whether living, dead, actual, or only possible. That's because, for God, every moment of time—whether past, present, future, actual or merely possible—is *equally present* to God. The self any given individual knows *here and now* is no more ob-

---

[1] The term "cosmology" is commonly defined as dealing with the evolution of the Universe *as well as* its structure. "Cosmogony" (Greek cosmos = world + gonos = creation) is commonly defined as dealing strictly with the *origin* and *development* of the Universe, galaxies, solar system, and the Earth-Moon system. That being so, it seems to me Prof. Greene repeatedly uses "cosmology" where it would be more to the point to use "cosmogony"—a term I've yet to find in his book. Can it be he's unaware of the difference between "cosmology" and "cosmogony"? That's hard to believe.

servable by, and present to, God than is the self one was at conception or the self one shall be $10^{googleplex}$ eons from now. Review St. Thomas' circle bit on pages 136 & 137.

If you wish, take your *total* self as known by God and call it your "karma". That done, ask yourself: In the view of an *infinitely* informed observer, which is more perfectly a case of free will and self-determination: OPTION#1: every choice at any and every instant in one's life is determined by one's karma (***i.e.:*** karma-determined will), or OPTION #2: every choice at any and every given instant in one's life is determined by one's self at that given instant? On the face of it, unless, like God, one is *infinitely* informed (or at least as astronomically well-informed as are the dead bathed in death's all-revealing light), you cannot possibly say *for sure* which case is *more perfectly* a case of free will and self-determination. Perhaps more to the point, you cannot possibly say *for sure* there shall never be a point in your life at which you will acknowledge the superiority of option #1. And what if The *Infinitely* Informed observes that every one, in death, comes to a future instant in which one's total self—whether with joy or ferocious anger over it—will observe with absolute certitude that option #1 (***i.e.:*** karma determined will) is, by very, very far, more perfectly a case of free will and self-determination than is option #2.

Some will be tempted to say: "You're saying that—always and everywhere—every one's every choice is predetermined by a version of one's self which cannot possibly choose otherwise than as it does." I'm saying no such thing. Everyone's karma has the *power* to choose other than it does. All it lacks is a *reason* to choose other than it does.

But why does it lack that reason? Is it because it has no power to *come up* with a reason to choose differently? By no means! It's because it has no reason to come up with such a reason and too many rapturously beloved reasons to reject coming up with such a reason. Every total self is an instance of absolutely and outstandingly unique individuality, and to be such is to be what's so fantastically desirable, one has far too much reason to reject every reason to be anything else. Since only God knows one's total self, only God knows what choice at what point in life preserves what one's total self is.

The above is not a line of thinking I either *can* sell or should even *try* to sell to others, unless—whether by Divine Revelation or personal observation—I can know for a fact that their karma calls for them to be sold on the above line of thinking. It is for their total self, not me, to determine what I can sell to them. And how I rejoice for it to be so!

Does the notion of a current choice determined by the future sound outrageous? And yet, it's basically a principle which Prof. Greene himself sets forth in his book. He starts off describing a gentleman who owns a particular painting he's reluctant to sell; but, after talking to a Mr. Smithers, he changes his mind and sells the painting to Mr. Smithers. Commenting on this change of mind and choice, he writes:

> Even though you remember that before your visitor's arrival you had been thinking that you didn't know what to do, from your current vantage point it seems as though you really did. It is not quite that future events have affected the past, but your enjoyable meeting with Mr. Smithers and your subsequent declaration of your willingness to sell have illuminated the past in a way that makes definite particular things that seemed undecided at the time. It is as though the meeting and your declaration helped you to accept a decision that was already made, one that was waiting to be ushered forth into the light of day. The future has helped you tell a more complete story of what was going on in the past.
>
> ——**BRIAN GREENE:** ***The Fabric Of The Cosmos***, page 187 of the 2005 paperback edition by Vintage Books, a division of Random House, Inc., New York. ©2004 Brian Greene.

For now, one may—though with no possibility of justified certitude—conclude currently determined choice is alone rightly called free will. In the post mortem future, death's all-revealing light shall present one with an astronomically vast plenitude of light. That future vantage point's vast knowledge shall so illuminate the past as to make certain what, in the past, seemed uncertain about free will. Thus shall that light make sure that a decision regarding the true and most desirable nature of free will had already been made and was merely waiting to emerge into the light, and that decision was: Karma-determined free will is, by very, very far, the superior kind. The future shall thus help you to arrive at the complete story of what was going on the past as a result of God's infinitely informed knowledge of what all—whether joyfully or angrily—will eventually admit is, by very, very far, the true and most desirable nature of free will.

Do you see the two radically different scenarios that suggests?

On the one hand, if, prior to death, one *truly* chose to know the truth of free will, then, from death's vantage point, one will be absolutely certain that—though one may not have been aware of it at the time—one really did, in life, choose to define free will the way God defines it, and any failure to do so was merely an honest mistake born of *excusable* ignorance. Where that's so, death's all-revealing light shall not embarrass one on this point.

On the other hand, if, prior to death, one's choice to know the truth of free will was a *farce*, then, from death's vantage point one will be absolutely certain that—though one may not have been aware of it at the time—one really did, in life, choose to reject the superior definition of free will, and the failure to accept that truth was due to a perverse opposition to truth—an opposition blind to its perversity only thru <u>in</u>excusable ignorance. Where that's so, death's all-revealing light shall embarrass one exceedingly much on this point.

How shall one then react to such humiliation? It depends upon how intensely perverse and protracted was one's opposition to truth.

If the opposition was not greatly intense or protracted, there may be only shame and a humility enabling one to accept the humiliation, to acknowledge one's error, to embrace God's definition of free will, and then, in the temporary torments of Purgatory, to seek a way to erase the shame.

If the opposition was greatly intense and protracted, there will be only hatred for the light which dares to prove one's supposed love of truth an exceedingly perverse and protracted sham. So intense shall that hatred be, one shall choose nothing a decillionth (***i.e.:*** 10^{-33}) as much as one shall choose a place able to guarantee there shall *never* be the slightest reduction in the ferocity with which one hates death's all-revealing, humiliating light and hates and foul-mouths The Author of it.

Why, then, is there the Universe at all? Being love as perfectly loving as The *Infinitely* Informed can devise, God—*ex necessitate infallibilitatis* and not *ex necesssitate coactionis*—loves above all else free will and self-determination as defined by God's Infinitely Informed Intellect. As a result, God is the *principal* cause of why the Universe is at all. Then comes the *proximate* cause behind every detail which ever was, is, or will be found in the Universe. By God's choice, that proximate cause is—with God's assistance—what the occupants' *total* selves choose those details should be. God's Will is the reason why the Universe is at all; but, <u>by</u> God's Will, the wills of our *total* selves are the reason why all and every one of the Universe's details are what they are. In short, Prof. Greene, by God's Will and with God's help, we—not the quantum laws or Newton or Maxwell or Einstein—govern the Universe (***vid.:*** his pg. 186).

2013 ESOPTRICS UPDATE

Assessing cutting-edge proposals for deep physical laws may well require the ferocious might of particle accelerators able to re-create violent conditions unseen since moments after the big bang. But for me, there would be nothing more poetic, no outcome more graceful, no unification more complete, than for us to confirm our theories of the ultra-small—our theories about the ultramicroscopic make up of space, time, and matter—by turning our most powerful telescopes skyward and gazing silently at the stars.
——**BRIAN GREENE: *The Fabric Of The Cosmos***, closing words on page 493 of the 2005 paperback edition published by Vintage Books, a division of Random House, Inc., New York. ©2004 Brian Greene (Though I've already harped on this issue too much, I will still here harp on it some more. The issue is how to go about discovering "the ultramicroscopic makeup" of time, space, particles, locomotion, and change. Scientists—knowing only <u>*mono*</u>-directional empirical observation—advocate spending billions and billions of dollars upon gigantic particle accelerators, telescopes, microscopes, space probes, and large numbers of scientists paid to look at the world around us thru such highly expensive monster machines. But, if Esoptrics proves correct, what has happened here? Inspired by Catholic Doctrine's praise of Philosophy, the contemplative life [***i.e.:*** introspection and thus <u>*bi*</u>-directional empirical observation] and prayer for the help of God's grace, a grossly amateurish, would-be philosopher-theologian—with no formal education beyond high school and one semester of music at Loyola University of the South and without the slightest knowledge of higher math [thus making him effectively a mere intellectual *child* compared to Prof. Greene & co.]—turns utterly away from all the monster machines and probes of the *extra*-mental world, focuses solely on his own self's innermost depths in an effort to see if it's possible to learn something about whatever's performing his acts of consciousness, and—*low and behold!* Though myriads of scientists peering outward with the aid of multi-billion dollar monster machines have yet to produce the slightest trace of a description of the "ultramicroscopic makeup" of time, etc., he produces a highly detailed and mathematically precise one able to infer impressive answers to a multitude of cosmological puzzles, to some of which, Science gives no answer. Oh, Catholic Doctrine! How right you were to sing the praises of Philosophy over Science, and wrong the self-proclaimed "intellectually superior" scientists to denounce you as the enemy of humanity's intellectual advancement! Let's see now if they—they who struggled so persistently and diligently to vilify the Catholic Church with cries of: "Arch enemy of reason and humanity's intellectual advancement!"—can cope with the disgrace of the sudden, astonishing, and complete reversal of their fortunes by a Catholic "child", if Esoptrics is on the mark. ***Job*** 22:19; ***Psalms*** 2:4, 59:8, & 112:10; ***Rev.*** 2:16 & 19:15.).

Chapter Twenty-Eight:

ESOPTRICS VS. GOTTFRIED WILHELM LEIBNIZ:

Among the few who have listened briefly to Esoptrics, some quickly dismissed it saying: "Oh, Leibniz already said all of that 300 years ago." I probably should not be wasting either my own or my reader's time replying to such a ridiculous comparison. Still, I will do so on the outside chance it might be helpful to some. Most readers, then, would probably do well to skip this chapter.

To be sure, Leibniz (1646-1716) vaguely prefigured a tiny bit of what Esoptrics says; but, to equate his few indefinite details with Esoptrics' multitude of mathematically precise ones amounts to equating the mass of an anthill with that of Mount Everest. It amounts to equating the finger paintings of a 3 year old with the masterpieces of Michelangelo on the grounds that each applied paint to a surface. So massive is the difference between the mastery of detail in the works of the latter vs. that of the former, to equate the one with the other is as outrageously ludicrous as can be. Consider now a few examples of just how massive is the difference between Esoptrics & Gottfried Wilhelm Leibniz.

> . . . . Leibniz ponders the nature of change of place by interrogating three concepts of motion: discontinuous motion (or motion by leaps), continuous motion, and continuous motion with interspersed rests. Motion occurs by leaps when a moving body is transferred immediately from place a to place b without passing through all the intermediary places a', a", etc. This type of motion proves incoherent. Since scale makes no difference to the logical possibility of an event, it is no different for a body to be teleported across a tiny interval than for us to suddenly find ourselves in Rome, attending John Paul II's funeral. Such motion defies all our experience, and introduces more confusion than it dispels. So without further ado, Leibniz dismisses motion by leaps, and turns his attention to continuous motion. Continuous motion requires a state of change, that is, an intermediary place a' between any two places a and b where the body simultaneously occupies (at least part of) both places. However, this means that the state of change entails that two contradictory states of affairs obtain at the same time. Replace a' with 'dying,' a with 'alive' and b with 'dead' in the above sentence; clearly "if there was a moment in common to the living and the nonliving, then the same person would be simultaneously living and not living, which I acknowledge to be absurd" (LoC 147). Hence, the state of change is nonsensical and continuous change impossible.
>
> Finally, Leibniz explores the possibility of motions interrupted by rests. Unfortunately, this hypothesis is not a serious alternative to the other two options, for the motion between rests must be either discontinuous or

continuous, and both of these have proven indefensible. Leibniz's analysis leads us to an impasse, for we have lost both leaps and continuous motion. He overcomes this difficulty by appealing to durationless time and extensionless place; that is, rather than conceiving of these measures as units however small, he defines them as pure boundaries or limits, like the endpoint of a line. This allows him to introduce a different concept of leap, one Levey calls' locus proximus': the transfer of a body from one point to another, where those points are at no distance from one another. The body does not leap over any intervening space because the boundaries of the two spaces touch one another; hence local motion occurs by infinitesimal 'leaps' from one contiguous space to another. Change, then, is an aggregate of two distinct states with no distance between them. These pseudo-leaps, however, imply that a moving body has no continuous existence, and so we must appeal to something else to sustain the existence of phenomenal things:

> I do not think we can explain this better than by saying that the body *E* is somehow extinguished and annihilated at *B*, and is actually created anew and resuscitated at *D*, which you may call by the new but very beautiful name *transcreation*. Moreover, although there is indeed a sort of a leap from one sphere *B* into the other *D*, it is not the kind of leap that we refuted above, since the two spheres are not distant ...But no cause can be conceived for why a thing that has ceased to exist in one state should begin to exist in another...except a kind of permanent substance that has both destroyed the first state and produced the new one, since the succeeding state does not necessarily follow from the proceeding [Sic! "proceeding" for "preceding"? - ENH] one.
> ——**GOTTFRIED WILHELM LEIBNIZ** (1646-1716)**:** *The Labyrinth of the Continuum: Writings on the Continuum Problem,* 1672-1686, ed., trans. Richard T. W. Arthur. (Yale UP, 2001).

. . . . . This passage suggests that the conception of monads is not far from Leibniz's mind, even as he goes on to affirm that transcreation is the work of God.
——**PATRICK RIESTERER:** ***Lost In The Labyrinth: Spinoza, Leibniz And The Continuum***, pages 87 & 88 of: "A Thesis Submitted to the School of Graduate Studies In Partial Fulfillment of the Requirements For the Degree Master of Arts McMaster University, paper 5706. ©2006 Patrick Riesterer. I found this at http://digitalcommons.mcmaster.ca/opendissertations on the Internet February 23, 2013.

Yes, the above, at first, bears striking resemblance to what Esoptrics says; but, look closer, and notice the vast differences. Locomotion is not "motions interrupted by rests"; it's instantaneous changes from one temporarily changeless *state of excitation to another*, and states of *excitation* can never rightly be called "rests". Esoptrics makes no appeal to "extensionless place", but to states of excitation which are *physically* extensionless but, nevertheless, *logically* extended 6 different ways as u-spaces and generators and 7 different ways as forms. It's not a case of "points" "at no distance from one another" but, rather, a case of states of excitation at no *physical* distance from one another but, nevertheless, at definite *logical* distances from one another. It's not a case of: "local motion occurs by infinitesimal 'leaps' from one contiguous <u>space</u> to another"; it's a case of changes, on the generators' part,[1] from one temporarily changeless unbalanced *state of excitation* to one of the 80 (***i.e.:*** 26 at the same OD + 27 at the next highest OD + 27 at the next lowest OD) *logically contiguous to it*. It's not a case of: "Change, then, is an aggregate of two distinct states with no distance between them"; it's a case of: change on the part of the generators is an aggregate of distinct states of excitation with no *physical* distance between them but, with a mathematically well-defined *logical* distance.

It's not a case of: "These pseudo-leaps, however, imply that a moving body has no continuous existence"; it's a case of these changes imply only that, in <u>loco</u>motion, generators undergo only *accidental* and never *substantial* changes,[2] and, therefore, each is forever the same generator it has always been from creation. When an external influence affects a generator's rate of change among microstates, it's no more a case of going in and out of existence than is a Newtonian, externally forced change in state of rest or direction of locomotion. Going from one state to a *completely different* one might imply "no continuous existence"; but, even as low as $OD2^{128}$, the difference between any 2 contiguous microstates is *astronomically* insignificant—namely: c. 10^{-38} per 2, 4, or 6 kinds of excitation.

Though Esoptrics manifestly gives out *6-way coordinated* and *mathematically precise* descriptions of *exactly* what is the internality of each of locomotion's states and of how any 2 of them—whether at the same level of the 4th dimension or across 2 adjacent levels—are contiguous, neither Leibniz nor any of his commentators can give out even the slightest inkling of either how to describe the internality of any of locomotion's successive states or of how to describe what makes some of them contiguous without being contiguous in space. And where in him or them is there even so little as the slightest trace of anything prefiguring the difference between contiguous in a *three*-dimensional vs. a *four*-dimensional space? That devoid of knowledge of *both* the internality of locomotion's successive states *and* the differences between them, how is it possible to say anything even remotely intelligent about whether or not stop and go locomotion implies "no continuous existence"? No, transcreation has no part in Esoptrics, and to equate Leibniz's take on stop and go locomotion with Esoptrics' is a level of intellectual blindness rarely, if ever, exceeded.

Perhaps the most telling difference between Leibniz and Esoptrics is in his statement that "the succeeding state does not necessarily follow from the preceding one." What else might one expect to hear from one who knows nothing whatsoever about either the internality of locomotion's various states or about the differences in them making non-spatial contiguity possible? In stark contrast, because of how much it knows about locomotion's various states, Esoptrics gives us—in extensive and mathematically precise detail—exactly why the succeeding state does indeed *necessarily* follow the preceding one. It's a contrast loudly thundering that, relative to Esoptrics, Leibniz knew nothing about the internality of the Universe's ultimates, and that's why he could not even begin to understand how existence could be continuous even as time, space, locomotion, and every manner of change are intermittent. How could it have been otherwise, since, prior to Esoptrics, no human being ever had the slightest knowledge regarding God's divine *a priori* coordinates? And without that knowledge, there is no such thing as knowing the real, fundamental "ultramicroscopic makeup" of existence, time, space, matter, locomotion, and change.

---

[1] There is, of course, in Leibniz's works not the slightest trace of Esoptrics' distinction between the generators (These can be called "sextads" in contrast to his monads.) and the forms (These can be called "hebdomads" in contrast to his monads). There is also not the slightest trace in Leibniz' works of Esoptrics' duo- (These can be called "triscadecads" in contrast to his monads.) vs. multi-combos.

[2] Substantial changes occur only: (1) in individual *multi*-combos when they undergo a massive disintegration into their component *duo*-combos, and (2) in a *complex* of multi-combos when it undergoes a massive disintegration into its component multi-combos.

> Leibniz introduces the numerical analogy to explain the relationship among attributes and the essence they express. If six is the essence of God, there are several ways one can combine the attributes to express that essence.
> ——**PATRICK RIESTERER:** op. cit. page 83.

No, his "six is the essence of God" is nothing like Esoptrics' 6 divine coordinates whereby God imparts 6 kinds of potency to the 2^{256} kinds of forms and, thru them, to the generators. And where is there any mention of the 7th kind of coordinates whereby God—in the case of the forms—imparts to each of those 6 potencies 2^{256} levels of intensity? Then, too, where is there the slightest trace of him showing how the "several ways one can combine the attributes to express that essence" produce u-spaces each logically outside of, and related to, all the others in a logical sequence so mathematically precise it can be nailed in place by means of triads of couplets? Does he somewhere explain how these ultimate units of space can be logically *four* rather than *three* dimensional, and why they can be found at c. 10^{-47} cm.? Does he say anything about the 3 planes of rotation or about OD, or about how OD is the 4th dimension, or about gravity, anti-gravity, mass, duo-combos vs. the 6 kinds of multi-combos, dark matter & energy, the Universe's first 10^{-18} sec., microscopic chaos vs. macroscopic order, the generators' rectilinear motion vs. the u-spaces' curvilinear motion, the applicability of light speed to the latter vs. former, and on and on and on?

> 'Mirroring' and 'mirror' (*miroir, speculum, Spiegel*) are profusely employed metaphorically by Leibniz both in texts dealing with the nature and functions of language and other semiotic systems and in texts where he explains the complex system of inter-monadic relations and those between the created and the divine monad. . . . .
>
> The mirror of choice, for Leibniz, is the one he characterizes as being 'alive' and 'active' – as against the usual idea that a mirror is a 'passive' device, faithfully reproducing a reality external to it in which it does not interfere. . . . .
>
> . . . .
>
> The metaphor of the mirror is indeed employed by Leibniz in order to conceptualize the core-relation of a metaphysics that seeks to combine plurality with unity, autonomy and interdependence, dynamism and completeness. The autonomy, unity, and dynamic character of each monad is often characterized by him by means of expressions such as 'miroir *vivant*' and 'miroir *actif* et *indivisible*', which underline the inner principle of activity whereby the unitary monad unfolds itself in time, thereby 'mirroring' dynamically, rather than statically, a universe of equally dynamic monads. The unity of the universe, on the other hand, as perceived/mirrored by each monad, is comparable to that of a *cabinet de glaces*. Albeit they are 'active' and 'living' (properties usual mirrors do not possess), the monads retain a fundamental property of mirrors, namely, their distance from the source mirrored. . . . .
>
> . . . . .
>
> . . . . . As already pointed out, monads mirror each other and through their inter-mirroring mirror the whole universe.
> ——**CHRISTINA MARRAS:** ***On The Metaphorical Network Of Leibniz's Metaphysics***, a Ph.D. dissertation, Tel Aviv University, 2003, pages XI-XII & XIV and found Feb. 18, 2013, at www.tau.ac.il/huanities/philos/.

Here again, a first glance may see a resemblance to Esoptrics. Esoptrics, though, uses the mirror metaphor 3 ways untouched above: (1) to illustrate how God produces 6 of the 7 divine *a priori* coordinates, (2) to account for the progressions involving the number 2, and (3) to account for some major differences between the behavior of forms native to OD's 1 thru 2^{128} vs. that of forms native to the higher OD's. That the 2 behaviors mirror each other somewhat does not come even remotely close to saying Esoptrics' ultimates mirror each other and, thereby, the whole universe. We are, however, in total aggrement when it comes to the nature of the mirror with which Esoptrics deals: It is indeed a "miroir vivant". For, a self-mirroring mirroring activity is indeed "vivant".

Yes, on a few points, Leibniz's metaphysics resembles Esoptrics; but, in mathematically precise details, Esoptrics goes so far beyond Leibniz, that to dismiss Esoptrics as what was already said 300 years ago by Leibniz is consummate buffoonery.

✡✞✡✞✡✞✡✞✡✞✡✞✡✞✡✞✡✞✡✞✡✞✡✞

And hence we see the reason why it is pretty late before most children get ideas of the operations of their own minds; and some have not any very clear or perfect ideas of the greatest part of them all their lives. Because, though they pass there continually, yet, like floating visions, they make not deep impressions enough to leave in their mind clear, distinct, lasting ideas, till the understanding turns inward upon itself, reflects on its own operations, and makes them the objects of its own contemplation. Children when they come first into it, are surrounded with a world of new things, which, by a constant solicitation of their senses, draw the mind constantly to them; forward to take notice of new, and apt to be delighted with the variety of changing objects. Thus the first years are usually employed and diverted in looking abroad. Men's business in them is to acquaint themselves with what is to be found without; and so growing up in a constant attention to outward sensations, seldom make any considerable reflection on what passes within them, till they come to be of riper years; and some scarce ever at all.
——**JOHN LOCKE:** ***Concerning Human Understanding***, Book II: Chap. I: Sec. 8, as found on pages 122 and 123 of vol. 35 of ***Great Books Of The Western World*** as published by Encyclopædia Britannica, Inc.; Chicago, 1952 (Well, Mr. Locke, would you agree Esoptrics involves "considerable reflection on what passes within", or would you not? Indeed, might you be at least tempted to say that never before did anyone ever give anything near to that much "considerable reflection on what passes within them"?)

Instead of a traditional bibliography, the back matter includes an "Index of Quotes" that includes author names, quote locations, and the context of the quotes within this text.
——**ANNA CALL:** ***Clarion Review*** (I've done that for what I consider a desirable convenience. How so? Going to the many quotes included in the book at the page mentioned presents one with a full description of where the quote occurs originally on which page of which work by which author and when it was published by which publisher. In effect, the bibliography is spread out among the in-text quotes themselves. As one is reading thru the book for the first time, each quote thus relieves the reader of any need to seek a full description of the quote's origin by breaking off and turning to some other part of the book before returning to the original location. The use of footnotes instead of endnotes reflects the same courtesy. How often I've read books in which I wished the author had done the same for me! Prof. Greene's book aggravated the heck out of me on this point. But, to each his own!)

The *corpus caeleste*, 'the heavenly body' *par excellence* with him, is the tenth outermost crystalline sphere, which by its diurnal motion from east to west controls the motion of all inferior material things, and is called the *primum mobile*. St. Thomas argues that this outermost sphere itself is moved by some intelligence, either by a soul animating it, or by an angel, or immediately by God. Through this *primum mobile*, St. Thomas thinks, God governs the motions and fixes the qualities of the whole material universe.
——**FR. JOSEPH RICKABY, S. J.** (1845-1932): ***Of God And His Creatures***, footnote#546, page 365 of the PDF file at www.catholicprimer.org. ©public domain (For Esoptrics, the "heavenly body *par excellence*", and "*primum mobile*" is the tenth outermost categorical form which, as a u-spaces envelope, is effectively a sphere crystalline in the sense of transparent. By its rotations in the primary and secondary planes, it controls the quantum world motions of all inferior forms as its own motions in the primary and secondary planes are controlled by a heavenly sextet moved in turn by God as, thru it, God governs the motions and fixes the qualities of all the Universe's lower ontological distances. So great are the similarities between St. Thomas' "tenth outermost crystalline sphere" and Esoptrics' "tenth categorical form", one must gasp in awe: "Wow, St. Thomas! What an astonishingly prescient leap for one whose attempts at Cosmology were mostly child's play!").

Epilogue:

On October 22, 2012, my brother Gordon sent me an e-mail saying:

Edward

Is it accurate to say that Esoptrics in part identifies those elements of consciousness that allow it to be repeatedly fooled by "physical reality"?

Gordon.[1]

Here, in part, follows my reply begun Tuesday, October 23, 2012, and ended the same day:

No, it is not accurate. What's accurate is this: Esoptrics in part identifies those elements of consciousness which allow us to live at, and enjoy living at, the level of ontological distance (OD for short) at which we do live. What does that mean?

We enjoy watching movies and television. We are able to do so because, due to the way our eyes function, our brains present us with whole, smoothly moving pictures which, though present on our retinas, are by no means present on either the movie theatre's or the television's screens. Yes, there are whole pictures on the theatre's screen; but, they are a series of still frames by no means streaming the continuous way our eyes say they do. On the TV screen, the difference is even far more dramatic, since, at any given instant, nothing is present on the screen but a single, tiny pixel of light.

How, then, comes to pass the dramatic transformation to whole, smoothly moving pictures? It's basically a kind of stroboscopic effect produced as the human retina retains flashes of light longer than the screens do. Twirl a flashlight fast enough, and your brain shall present you with a "solid" ring of light, despite the fact no such "solid" ring is present in the extra-mental world.

---

[1] **NOTE OF AUGUST 15, 2013:** On page 199 of his book, ***The Fabric Of The Cosmos***, Prof. Greene speaks of the "truth of quantum reality" as "guarded by nature's sleights of hand." No, Professor! Don't blame it on nature. *Nature* is not what makes it difficult to discover the truth about the tiniest levels of time and space; ***we*** are. What makes it difficult to find that truth is that—just as being a human invariably includes having the kind of body parts we all have—it also invariably includes being—metaphorically speaking—a nocturnal creature which, as such, must avoid the light lest it cease to be the self it is and, out of self-love, relishes being. In other words, Professor, for each and every one of us, the chief source of one's difficulty is not in something *outside* of one own self; it's something *inside* of it, and the worse part of it is: That something inside of each of us is there because we *want* it to be there. Then the worst part of it is: Out of *irrational* pride, we refuse to admit we want it there and, instead, ever struggle to throw the blame on anything and everything except the one source where it really belongs—namely: one's own self. Turn away, O humanity, from your irrational pride's paranoia, and, instead, have enough humility and *rational* pride to confess you are as blind as you are by your own *total* self's most willing choice.

Is that a case of consciousness allowing itself to be fooled by physical reality, or is it a case of allowing itself to be entertained while also being helped to understand how the human eye works? At the risk of seeming to resort to a pun, I would say it's in the eye of the beholder. Those dead set on seeing themselves as the victims of a malevolent world (and thus a malevolent "god"?) will choose to call it a case of being fooled; whereas, those of the opposite bent will choose to call it a case of being entertained and informed.[1]

Esoptrics is not limited to identifying those elements of consciousness which allow it to enjoy ***its*** level of ontological distance only. It also identifies those elements of macroscopic reality by means of which ***every*** macroscopic reality at or near our OD level is able to persist at or near that OD level. Those elements, it explains, function, and allow us to enjoy our fellow macroscopic realities, the same way our eyes function and allow us to enjoy movies and TV—namely: the stroboscopic effect.

Thru its stroboscopic effect, the human form capable of consciousness is to no extent fooled by "physical reality" into acquiring a false picture of the animals, vegetables, and minerals which it observes. The picture we have of them is to no extent false. For, they, too, have forms which, by their astronomically huge infrequency of change, give each of them, as a kind of time exposure, exactly the size, weight, shape, texture, behavior patterns, etc., which our naked senses observe them having. Thus does Esoptrics explain rather well what Science cannot currently explain to any degree whatsoever.

It is, my dear, dear brother, one thing to talk about those elements of consciousness which allow it to be fooled regarding physical reality and quite another to talk about those elements which allow it to be fooled regarding the *causality* behind this-or-that aspect of physical reality. It is yet a third thing to talk about those elements of consciousness which merely make it *difficult* for consciousness to discover that causality and, thereby, make it *difficult* for consciousness to avoid mistakes (as opposed to being *fooled*) regarding that causality. That latter is what Esoptrics addresses.

Since time immemorial, even many with poor ability to think knew one has no truly significant knowledge regarding physical reality until one knows well the *entire chain of causes* behind every aspect of that physical reality. The issue was never do we *observe what* we observe; but rather, do we correctly *infer* what *causes* it to be so (One of the big problems among both philosophers and scientists is a marked inability to tell what is truly nothing more than what's observed vs. what's strictly an inference or a mix of both, and instead of having anything to do with being fooled by physical reality, that has to do with a ridiculous lack of intellectual acumen.). At least many of them also knew that our sense images make it easy to err regarding that chain. The oar which appears bent in the water was ever one of their favorite evidences in that direction. As a result, many of them were to no extent fooled into thinking the quest for that entire chain was anything more than a hard row to hoe which has not yet gone very far at all; and so, they never hesitated to call their explanations anything other than *theories*.

---

[1] **NOTE OF AUGUST 17, 2013:** The "victim complex" a/k/a "martyr complex" is a crucial part of the rationalization many Atheist's use to ward off fear of death. In their delusions of grandeur, they see their inability to master fully the laws of the Universe solely the result of some kind of obstacle or obstacles in the world around them. In a manner of speaking—as was supposedly the case with Beethoven—they see even the inkwells as plotting against them. Since, in their paranoid schizophrenia, any lack of total knowledge of truth is the result of something other than themselves, they are able to rest sure they are paragons of impeccable intellectual integrity. From that, it follows—for them, that is—that even if, *per impossibile*, Christianity's God (Whom, for their egos' sake, Atheist's view as a vindictive, cruel, infantile, despotic "omnipotent ogre in the skies".) really exists, they need not fear death. For, genuine truth, they dogmatize, must prove them more righteous and godly than Christianity's sick God. How little they know of what Catholic Doctrine *really* says about God!

All well-meaning people should try to contribute as much as possible to improving such mutual understanding. It is in this spirit that I should like to ask my Russian colleagues and any other reader to accept the following answer to their letter. It is the reply of a man who anxiously tries to find a feasible solution without having the illusion that he himself knows "the truth" or "the right path" to follow. If in the following I shall express my views somewhat dogmatically, I do it only for the sake of clarity and simplicity.
——**ALBERT EINSTEIN:** page 140 of the hardbound issue of ***Ideas And Opinions*** by Bonanza Books, New York, 1954. ©1954 Crown Publishers Inc.

I am no different. I freely acknowledge that Esoptrics is merely a *theory* which my mind cannot possibly carry beyond the theoretical. Subsequent generations may do so; but, as mentally limited as I am, I cannot (Years ago, tests, I was told, put my I. Q. at 139 = gifted but no genius. Einstein's is usually estimated at 160.).

Why, then, do I even bother to continue working with Esoptrics? If nothing else, it is, by very far, the most enjoyable hobby and pastime I have ever encountered. And seeing how intense some can be in endless debates over football, basketball, hockey, soccer, movies, chess, etc., how is it somehow reprehensible for me to be equally passionate about Esoptrics? Even if it is admittedly nothing more than a highly speculative stab in the dark regarding the complete chain of causes behind time, space, matter, locomotion and whatever else goes into the mix we call the Universe, isn't it better to "fight" over it than over which movie or actor or director deserves an Oscar?[1]

I repeat then: As it ever has been with many, there never have been, in my consciousness, any elements able to fool me even once—let alone repeatedly—into thinking I know for sure what is the complete chain of causes behind physical reality. On the contrary, as it ever has been with many, there have only been, in my consciousness, elements which make it exceedingly difficult for me either to discover the supremely important complete chain of causes or to avoid mistakes in the course of that endeavor. Those elements alone are the ones with which Esoptrics deals.

Still, I will admit that there are, in some, elements of consciousness which allow it to be repeatedly fooled by physical reality. Those elements, though, are the subject matter of Depth Psychology and Moral Theology—not Esoptrics. At the risk of prolonging this beyond a reasonable length, let me say a few words in explanation of why I say that.

I was no more than 7 and probably only 6 years old when I came to the conclusion that space cannot possibly be as it is most commonly assumed to be. If even a child of 6 could see that, it cannot possibly be that there is, in the very nature of consciousness, an element or elements which allow it to be repeatedly fooled by physical reality. Or can it be, I was—out of all the billions of humans who have ever crossed this planet—the only one able to see that? What then???!!! Am I perhaps, by very far, the most perceptive genius the world has ever known? Don't make me laugh. In point of fact, throughout the pages of history, there have been many thinkers who cried out most loudly against what they de-

[1] **NOTE OF AUG. 17, 2013:** Some will chide: "Shame on you! One interested in truth will pursue and write about only what the evidences declare proven true and will not waste time on what one admits is pure speculation." In my view, whenever I write about the moral and dogmatic tenets of Catholic Doctrine, I'm writing about what the evidences declare proven true—the tenets themselves, that is, not what's plainly my personal take on them. That's why, in my view, only my writings in the field of Catholic Theology are *serious*. But, "all work and no play makes Jack a dull boy", and that harms the brain's ability to help the intellect supplant darkness with light. Therefore, in the midst of writing Theology, it's imperative to spend some time having fun writing Theoretical Physics.

nounced as the manifestly self-contradictory gibberish and absolute madness of the kind of time and space commonly touted by the majority of supposedly critical thinkers. In the final analysis, Relativity is mostly a revolt against Newton's acceptance of the commonplace notions of space and time (That's why, to this day, even some prominent scientists indignantly denounce Einstein as "that blithering idiot".).

My point here, then, is this: There are obviously among us some who ***want*** to be repeatedly fooled by physical reality, and they—*they alone!*—are the ones who truly are repeatedly fooled by physical reality when they so obstinately and dogmatically insist there can be no correct view of time and space save the ancient, commonplace one (Did you know there still are some highly educated individuals who insist Geocentrism is correct and Heliocentrism wrong?).[1] And what is the element in them which brings about so great an intellectual sin against the light?

When theorizing in the fields of Depth Psychology and Moral Theology, I say this: In a world replete with perfect knowledge of the whole chain of causality behind every bit of physical reality, every human being would be so intelligent and powerful, there can be no such thing as some who are masters and some who are servants; there can be no such thing as some who win and some who lose in this-or-that kind of competition; there can be no football or basketball or hockey or soccer champions, etc.; and, above all, there can be no verbal contests in which one wins out over the other; there can only be draws. In the instincts of every nocturnal creature, there is an element which forbids it, above all else, to seek the light, lest it destroy itself. By the same token, in the instincts of every lover of competition, there is an element which forbids it, above all else, to seek more knowledge of the chain of causes than is currently required to defeat the opposition, lest one destroy the environment which makes it possible for some to beat the other guy (The truly ultimate lover of competition is the one most interested, by very far, in beating that astronomically vast intelligence called Satan [***Ephesians*** 6:12], and the truly ultimate *insane* lover of competition is the one most interested, by very far, in beating God.). I'll say no more on this element of consciousness.

It's enough for now, dear brother. Give my love to Annamarie, and, until another time:

God bless us all, everyone.
EDWARD.

Though it shall involve a bit of repetition, let me elaborate upon what's behind what I said to my brother in the last paragraph of my e-mail to him. Yes, I know it's what moves many—mainly delusional Atheists—to call me a muckraker; but, here it is anyhow:

When theorizing in the field of Dogmatic Theology, I interpret what Catholic Doctrine (***i.e.:*** God's infallible word) says about Original Sin to be telling me this: We do not inherit from our distant progenitors merely the DNA which causes us to have bodies similar to their own; we also inherit the attitude which rendered them intelligent beings with a very deep seated "*aversio a Deo*" as Catholic Doctrine puts it in Latin. In English, it means a turning away from God or, loosely translated, an aversion for God's kind of Love. For me, "God's kind of Love" means God is above every kind of coercion, which is to say God never

---

[1] **NOTE OF AUGUST 15, 2013:** On page 42 of their book, ***The Grand Design***, Profs. Stephen Hawking & Leonard Mlodinow basically tell us that both the heliocentric and the geocentric theories explain what we observe in the heavens and that the only factor making the former preferable to the latter is "the equations of motion are much simpler" when the Sun is referenced as at rest. It amounts to saying that—despite what Galileo so arrogantly and belligerently thundered—the heliocentric theory has not been *proven*—not even today, let alone in Galileo's time.

targets others for any kind of attempt to *force* God's Will upon them. I emphasize "*targets*" because God is *Infinitely* Informed Truth, and *Infinitely* Informed Truth automatically and, *without trying,* totally annihilates any and every bit of thinking and argumentation contrary to It and does so the instant the promoters of such thinking and argumentation expose themselves to an astronomically vast plenitude of It, let alone expose themselves to an *infinitely* vast plenitude of It. That, though, is by no means a case of God *targeting* others for an attempt to force God's Will upon them. It's merely what Moral Theology calls "an unavoidable side effect" and "collateral damage". For example, the Sun does not *target* anyone for sunburn; it merely continues to be the same, one Sun for all, and any difference in its effects upon others lies solely in the ways they prepare themselves for the Sun's rays.

What, then, is meant by a turning away from That Divine Kind of Love? That should be rather obvious: It means preferring a kind of love which at least sometimes—if not always or, at least, much of the time—imposes its will on others (***i.e.:*** gets its own way) by targeting them for some kind of force. That includes defeating others in boxing, football, basketball, chess, debates, the pursuit of Olympic and Nobel medals, good looks, sexual prowess, and on and on and on. As a result of Original Sin, then, we are addicted to competing and winning. The extent to which we need to compete and win may vary greatly from individual to individual; but, still, it's an addiction so fundamental to our nature as the heirs to Original Sin's perpetrators, every one of us—*from conception, mind you!*—desperately needs to fulfill it to some extent. That's why, if one cannot fulfill it to *any* extent, the result must often be some form of mental and/or emotional derangement leading to one or more forms of behavior destructive to one's self or others or both. One of the more common results is a compulsive need to kill as many others as one can in order to get revenge for being born.

Needless to say, the one condition, which would be lethal to every one of our kind's ability to feed his or her addiction to competing and winning, would be an environment in which knowledge of self and one's environment is so total, no one can ever outdo anyone else in any kind of competition whatsoever. A deep seated and instinctive awareness of that fact is what keeps us all from uncovering any more knowledge than what is needed to unseat the current champion and to take his or her place. That, then, is the real reason why we ourselves—and not "nature's sleights of hand"—are the chief reason why we persist living in a world of incredibly crippling intellectual darkness.

Some, of course, will snarl: "What kind of sadistic God has forced us into such a situation?" The answer, of course, is: The God Who—loving above all else the most perfect kind of free will and self-determination as defined by The Infinitely Informed—has gifted us the world God infallibly knows the *total* self of each and every one of us chooses with a willingness as perfect as can be. Such a God is far from being sadistic and is to no extent *forcing* us into the situation in which we find ourselves. Every iota of every instant of every individual's situation is exactly as that individual's *total* self chooses it to be.

Still, some will growl: "Cursed the fiendish God Who makes my *total* self's choices the determining ones." To that, I reply: "In death's all-revealing light, an astronomically vast plenitude of truth shall, without trying, leave you absolutely certain that—in making your *total* self's choices the determining ones—God has done what is, by far, the most perfectly loving thing of all according to The *Infinitely* Informed. It shall then be up to you either to *love* that incontrovertible truth forever or—because it's manifestly incontrovertible—to *hate* it forever with as much willingness and ferocity as your astronomically enhanced self shall then be able to muster."

Why might our total selves want our nightmarish world? Oh, how awesome is the achievement of producing a Universe forever after devoid of every trace of pain, suffering, want, deprivation, and the like! It's an achievement so awesome, The Blessed Trinity should not reserve it to Themselves alone; rather, They should share it with Their creatures. How do that? That's so easy to figure out, I'll not bother to elaborate beyond saying "evolution". Oh, how I, for one, long to be a part of *such* a work! And, if any deny they want any part of it, say to them: With its astronomically vast plenitude of truth, death's all-revealing light shall not fail to make you absolutely certain: Oh, what a fantastically loving thing God has done in gifting us a part in the most awesome achievement of them all! That, Prof. Greene, is not only "why there is a universe at all", but, also, why it's *exactly* as it is.

Others may snicker: "If Esoptrics truly is a correct description of the fundamental realities, then you have unearthed the knowledge you say we are *all* addicted to avoiding. What, then, are you? Are you one who has put aside our inherited *'aversio a Deo'* and has climbed far above us up into the heights of moral superiority?"

To that, I reply: As competition for "top banana" pushes the level of knowledge higher and higher, there must inevitably come a day when the only way to become the new "top banana" is to push the level of knowledge to the forbidden level. The one willing to do that in pursuit of "top banana" is the very opposite of morally superior or free from the great addiction; rather, he is one in whom the addiction is so "over the top", he must win in a way which insures there can never again be one able to supplant him as the "top banana". The only way for him to do that is to be cruel enough to knowingly uncover the level of knowledge sure to destroy forever the one environment in which further competition is possible. If, then, Esoptrics is the destroyer I *hope* it is, I am simply the most merciless and fiercely competitive competitor the world ever has known and ever can know.

Seeing what a monster I perhaps am in my innermost depths, I am driven to ask: "Why would a *loving* God create the likes of me and help me attain my maximally cruel goal?" Some corner of me answers: "So pronounced now is humanity's reliance upon abortion, euthanasia, the targeting of non-combatants in warfare, homosexuality, homosexual rights, same sex marriage, fornication, adultery, birth control, divorce, and Consequentialism, it has, thereby, at long last and effectively raised to God so intense a prayer for God to send it an ultimate destroyer, that God—*out of infinitely perfect love for free will*—has sent humanity what its love of hell's ways has so passionately demanded. At your request, then, O humanity, here I am to destroy you in hell with nothing more than the words from my mouth. How frantically, then, do I pray: Be true of me, O God, this word of Yours:

> From his mouth issues forth a sharp sword with which to smite the nations.
> ——***Revelations*** 9:15."

No, I'm much the opposite of one who's climbed to the heights of moral superiority.

> This saying is reliable and worthy of all acceptance: that Christ Jesus came into the world to save sinners of whom I am the foremost. But, on account of this have I found mercy: that in me, the foremost, Jesus Christ might show forth all-surpassing patience unto the edification of those who are going to believe in him all the way to eternal life.
> ——**ST. PAUL: *1 Timothy*** 1:15-16 (My translation – ENH).

Jesus Christ was The Incarnation of Infinitely Benevolent Mercy. The reverse of Him on that most characteristic of His traits (***i.e.:*** His anti-Christ on that point), I—called into existence by the cries of a billion babies butchered—am the incarnation of an astronomically immense Tsunami of indescribably callous, ferocious, and vindictive mercilessness. No, Prof. Oppenheimer, not of you but of me did the Bhagavad Gita say: "I am become death, the destroyer of worlds." For, every sufficiently brilliant burst of truth's light—

even without trying—has, by its very nature, far more power to destroy sinners eternally than have all the nuclear bombs in the Universe.

But, how have I managed to sustain the stress of trying to be such an *ultimate* and *last* competitor of them all? Consider this: On page 340 of his ***The Fabric Of The Cosmos***, Prof. Greene talks about Leonard Susskind (b. 1940) of Stanford University. He tells us that, in 1970, when Prof. Susskind's paper on string theory was rejected by the journal to which he sent it, his reaction, in his own words, was: "I was stunned, I was knocked off my chair, I was depressed, so I went home and got drunk." I, unlike Prof. Susskind, cannot be stunned or depressed or flustered to any extent by any amount of rejection by any number of my fellow intellectual cretins. 57 years of indifference, rejection and gutter level insults have never at any time taken one iota of the wind out of my sails. If anything, all they ever did was to increase fiercely that wind's volume and velocity. My fellow humans can no more dispirit me than a gnat can build a hydrogen bomb. Why is that?

My familiarity with Catholic Doctrine is so formidable, I seriously doubt there ever was or can be a purely human person more familiar with it than I am. I am so familiar with Catholic Doctrine, I am fearlessly convinced of this: Even though my efforts to *live* 100% by Catholic Doctrine have failed miserably in one manifest regard, my *efforts* to *know* all of Catholic Doctrine, to *understand* all of Catholic Doctrine, and to *live* by all of Catholic Doctrine have been so heroically intense and persistent I am sure to receive—at least some day if my efforts continue as they have been—the approval of Catholic Doctrine and, therefore, of St. Peter's successors to his role as Bishop of Rome. That being the case, I am fearlessly convinced that I shall one day receive—as long as my efforts remain as they have been—*God's* approval. For: Oh, how readily the Church—and, therefore, Incarnate God out of commitment to the Power of the Keys—grant an A for *effort*! As long as I am sure to receive *God's* approval, what do I care how much approval, disapproval, or indifference I have evoked or shall evoke from a pack of walking septic tanks? As St. Paul (Well, *he* perhaps *may* have known more about Catholic Doctrine than I do. LOL!) expressed it:

> I can do all things in Him Who strengthens me [I added the capital letters – ENH].
> ——***Phillipians*** 4:13.

It's enough for the moment. It is now 2:32 PM, CDT, on the feast of the Assumption of the Blessed Virgin Mary, mother of God, as I finish this first complete draft of this my 21st self-published book. Thank you, dear lady, my heavenly mother.

✡ ✞ ✡ ✞ ✡ ✞ ✡ ✞ ✡ ✞ ✡ ✞ ✡ ✞ ✡ ✞ ✡ ✞ ✡ ✞ ✡ ✞

> And can the liberties of a nation be thought secure when we have removed their only firm basis, a conviction in the minds of the people that these liberties are of the gift of God? That they are not to be violated but with His wrath? Indeed I tremble for my country when I reflect that God is just; that his justice cannot sleep forever;
> ——**THOMAS JEFFERSON: *Notes On The State Of Virginia***, query 18.

It is possible, without falsehood, to deem and proclaim oneself the most despicable of men as regards the hidden faults which we acknowledge in ourselves, and the hidden gifts of God which others have. Hence, Augustine says: "Bethink you that some persons are in some hidden way better than you, although outwardly you are better than they."
——**ST. THOMAS AQUINAS** (1225-1274 AD)**:** ***Summa Theologica***, 2a/2ae, Q161: A6, ad 1.

There is, then, in the hidden depths of God's judgments, a particular reason why every mouth even of the righteous should be shut in its own praise, and opened only for the praise of God. But what this particular reason is, who can search, who investigate, who know? So "unsearchable are His judgments, and His ways beyond finding out! For, who has known the mind of the Lord? Or who has been his advisor? Or who has first given to Him, and it shall be paid back to him? For, of Him, and thru Him, and to Him, are all things—to whom be glory forever. Amen."
——**ST. AUGUSTINE OF HIPPO** (354-430 AD)**:** ***On The Spirit And The Letter*:** closing words in Vol. 5, pg. 114 of ***The Ante-Nicene Fathers*** as published by Wm. B. Eerdmans Publishing Co., Grand Rapids, Michigan, 1974. The last part in quotation marks is a quote from ***Romans*** 11:33-36.

In particle physics we may have to accept an arbitrary, complicated, not very orderly set of facts, without seeing behind them the harmony in terms of which they might be understood. It is the special faith and dedication of our profession that we will not lightly concede such a defeat.
——**J. ROBERT OPPENHEIMER:** ***The Mystery Of Matter***, on page 68 of ***Adventures Of The Mind*** from ***The Saturday Evening Post*** published by Alfred A. Knopf, New York, 1960. ©1959 by The Curtis Publishing Company (Yes, Dr. Oppenheimer, the scientist may contemplate possibly throwing in the towel, but the introspective philosopher inspired by Catholic Doctrine: ***NEVER***.).

APPENDIX A:

CHART OF CRUCIAL VALUES

UNIT	SIGN	DESCRIPTION
Alphakronon	K	$7.201789375 \times 10^{-96}$ seconds.
Alphatopon	Φ	$7.34683969 \times 10^{-47}$ cm. = 2^{-129} times the atomic dia..
Atomic Dia.	none	Diameter of the hydrogen atom with the electron in its outer-most orbit = 5×10^{-8} = 1/20,000,000 = .00 000 005 cm.
Centimeter	cm.	$1.361129468 \times 10^{46}$ Φ
Mile	mi.	$2.190524725871311 \times 10^{51}$ Φ = 160,934.399999 cm.
Parsec	ps.	3.258 light years. = 30,856,775,813,057,300 meters = 19,173,511,580,000 miles.
Dia. Of The Form Of The Universe	D_u	1.7014118×10^{31} cm. = 1.79844×10^{13} light years = 2^{128} times the atomic diameter = 2^{257} Φ
Duration Of The Ninth Epoch	none	2^{385} K = 5.67532×10^{20} sec. = 1.79844×10^{13} Earth years.
Earth Day Of 360°	none	23 hrs., 56', 4.09" = 86,164.09 sec.
Earth Day Of 360°	none	1.19636×10^{100} K.
Light Speed In Cm./Second	none	29,979,245,800 cm./sec. ± 40,000 cm./sec.
Light Speed In K/Φ	none	2^{128} K/Φ
Light Speed In Φ/Sec.	none	$4.080563489 \times 10^{56}$ Φ/sec.
Light Year	Lt. yr.	9.46051×10^{17} cm. = 5.878477×10^{12} mi. per 31,556,925.9747 sec.
Second	sec.	$1.388543802 \times 10^{95}$ K

2013 ESOPTRICS UPDATE

UNIT	SIGN	DESCRIPTION
Tropical Year	none	31,556,925.9747 sec.
Tropical Year	none	4.38158×10^{102} K
Astronomical Unit	AU	149,597,870,691 meters = 92,955,807.267 miles.
Earth	none	Mean radius = 6,371 kilometers = 3,959 miles.

✡✞✡✞✡✞✡✞✡✞✡✞✡✞✡✞✡✞✡✞✡✞✡✞✡

As advertising always convinces the sponsor even more than the public, the scientists have become sold, and remain sold, on the idea that they have the key to the Absolute, and that nothing will do for Mr. Average Citizen but to stuff himself full of electrons. . . . . But although in theory physicists realize that their conclusions are ... not certainly true, this ... does not really sink into their consciousness. Nearly all the time ... they ... act as if Science were indisputably True, and what's more, as if only science were true.... Any information obtained otherwise than by the scientific method, although it may be true, the scientists will call "unscientific," using this word as a smear word, by bringing in the connotation from its original [Greek] meaning, to imply that the information is false, or at any rate slightly phony.
——**ANTHONY SANDEN:** (1906-1993)**:** ***Science Is A Sacred Cow*** (1950), pages 26 & 176-177. (He was a British chemist best known for his above, sarcastic attack on Science. Shame on me, if I give way to the bad habit he here describes. On this issue, recall Einstein's words and mine on page XXVIII.)

APPENDIX B:

POWERS OF 2

2^{0} = 1
2^{1} = 2
2^{2} = 4
2^{3} = 8
2^{4} = 16
2^{5} = 32
2^{6} = 64
2^{7} = 128
2^{8} = 256
2^{9} = 512
2^{10} = 1,024
2^{11} = 2,048
2^{12} = 4,096
2^{13} = 8,192
2^{14} = 16,384
2^{15} = 32,768
2^{16} = 65,536
2^{17} = 131,072
2^{18} = 262,144
2^{19} = 524,288
2^{20} = 1,048,576
2^{21} = 2,097,152
2^{22} = 4,194,304
2^{23} = 8,388,608
2^{24} = 16,777,216
2^{25} = 33,554,432
2^{26} = 67,108,864
2^{27} = 134,217,728
2^{28} = 268,435,456
2^{29} = 536,870,912
2^{30} = 1,073,741,824
2^{31} = 2,147,483,648
2^{32} = 4,294,967,296
2^{33} = 8,589,934,592
2^{34} = 17,179,869,184
2^{35} = 34,359,738,368
2^{36} = 68,719,476,736
2^{37} = 137,438,953,472
2^{38} = 274,877,906,944
2^{39} = 549,755,813,888
2^{40} = 1,099,511,627,776
2^{41} = 2,199,023,255,552
2^{42} = 4,398,046,511,104
2^{43} = 8,796,093,022,208

2013 ESOPTRICS UPDATE

$2^{44} =$ 17,592,186,044,416
$2^{45} =$ 35,184,372,088,832
$2^{46} =$ 70,368,744,177,664
$2^{47} =$ 140,737,488,355,328
$2^{48} =$ 281,474,976,710,656
$2^{49} =$ 562,949,953,421,312
$2^{50} =$ 1,125,899,906,842,624
$2^{51} =$ 2,251,799,813,685,248
$2^{52} =$ 4,503,599,627,370,496
$2^{53} =$ 9,007,199,254,740,992
$2^{54} =$ 18,014,398,509,481,984
$2^{55} =$ 36,028,797,018,963,968
$2^{56} =$ 72,057,594,037,927,936
$2^{57} =$ 144,115,188,075,855,872
$2^{58} =$ 288,230,376,151,711,744
$2^{59} =$ 576,460,752,303,423,488
$2^{60} =$ 1,152,921,504,606,846,976
$2^{61} =$ 2,305,843,009,213,693,952
$2^{62} =$ 4,611,686,018,427,387,904
$2^{63} =$ 9,223,372,036,854,775,808
$2^{64} =$ 18,446,744,073,709,551,616 = 1.8447×10^{19}
$2^{65} =$ 36,893,488,147,419,103,232
$2^{66} =$ 73,786,976,294,838,201,464
$2^{67} =$ 147,573,952,589,676,412,928
$2^{68} =$ 295,147,905,179,352,825,856
$2^{69} =$ 590,295,810,358,705,651,712
$2^{70} =$ 1,180,591,620,717,411,303,424
$2^{71} =$ 2,361,183,241,434,822,606,848
$2^{72} =$ 4,722,366,482,869,645,213,696
$2^{73} =$ 9,444,732,965,739,290,427,392
$2^{74} =$ 18,889,465,931,478,580,854,784
$2^{75} =$ 37,778,931,862,957,161,709,568
$2^{76} =$ 75,557,863,725,914,323,419,136
$2^{77} =$ 151,115,727,451,828,646,838,272
$2^{78} =$ 302,231,454,903,657,293,676,544
$2^{79} =$ 604,462,909,807,314,587,353,088
$2^{80} =$ 1,208,925,819,614,629,174,706,176
$2^{81} =$ 2,417,851,639,229,258,349,412,352
$2^{82} =$ 4,835,703,278,458,516,698,824,704
$2^{83} =$ 9,671,406,556,917,033,397,649,408
$2^{84} =$ 19,342,813,113,834,066,795,298,816
$2^{85} =$ 38,685,626,227,668,133,590,597,632
$2^{86} =$ 77,371,252,455,336,267,181,195,264
$2^{87} =$ 154,742,504,910,672,534,362,390,528
$2^{88} =$ 309,485,009,821,345,068,724,781,056
$2^{89} =$ 618,970,019,642,690,137,449,562,112
$2^{90} =$ 1,237,940,039,285,380,274,899,124,224
$2^{91} =$ 2,475,880,078,570,760,549,798,248,448
$2^{92} =$ 4,951,760,157,141,521,099,596,496,896

APPENDIX B

2^{93} = 9,903,520,314,283,042,199,192,993,792
2^{94} = 19,807,040,628,566,084,398,385,987,584
2^{95} = 39,614,081,257,132,168,796,771,975,168
2^{96} = 79,228,162,514,264,337,593,543,950,336 = 7.922816×10^{28}
2^{97} = 158,456,325,028,528,675,187,087,900,672
2^{98} = 316,912,650,057,057,350,374,175,801,344
2^{99} = 633,825,300,114,114,700,748,351,602,688
2^{100} = 1,267,650,600,228,229,401,496,703,205,376
2^{101} = 2,535,301,200,456,458,802,993,406,410,752
2^{102} = 5,070,602,400,912,917,605,986,812,821,504
2^{103} = 10,141,204,801,825,835,211,973,625,643,008
2^{104} = 20,282,409,603,651,670,423,947,251,286,016
2^{105} = 40,564,819,207,303,340,847,894,502,572,032
2^{106} = 81,129,638,414,606,681,695,789,005,144,064
2^{107} = 162,259,276,829,213,363,391,578,010,288,128
2^{108} = 324,518,553,658,426,726,783,156,020,576,256
2^{109} = 649,037,107,316,853,453,566,312,041,152,512
2^{110} = 1,298,074,214,633,706,907,132,624,082,305,024
2^{111} = 2,596,148,429,267,413,814,265,248,164,610,048
2^{112} = 5,192,296,858,534,827,628,530,496,329,220,096 = 5.192297×10^{33}
2^{113} = 10,384,593,717,069,655,257,060,992,658,440,192
2^{114} = 20,769,187,434,139,310,514,121,985,316,880,384
2^{115} = 41,538,374,868,278,621,028,243,970,633,760,768
2^{116} = 83,076,749,736,557,242,056,487,941,267,521,536
2^{117} = 166,153,499,473,114,484,112,975,882,535,043,072
2^{118} = 332,306,998,946,228,968,225,951,765,070,086,144
2^{119} = 664,613,997,892,457,936,451,903,530,140,172,288
2^{120} = 1,329,227,995,784,915,872,903,807,060,280,344,576 = 1.329228×10^{36}
2^{121} = 2,658,455,991,569,831,745,807,614,120,560,689,152
2^{122} = 5,316,911,983,139,663,491,615,228,241,121,378,304
2^{123} = 10,633,823,966,279,326,983,230,456,482,242,756,608
2^{124} = 21,267,647,932,558,653,966,460,912,964,485,513,216 = 2.126765×10^{37}
2^{125} = 42,535,295,865,117,307,932,921,825,928,971,026,432
2^{126} = 85,070,591,730,234,615,865,843,651,857,942,052,864 = 8.507059×10^{37}
2^{127} = 170,141,183,460,469,231,731,687,303,715,884,105,728 = 1.701412×10^{38}
2^{128} = 340,282,366,920,938,463,463,374,607,431,768,211,456
2^{129} = 680,564,733,841,876,926,926,749,214,863,536,422,912
2^{130} = 1,361,129,467,683,753,853,853,498,429,727,072,845,824
2^{131} = 2,722,258,935,367,507,707,706,996,859,454,145,691,648
2^{132} = 5,444,517,870,735,015,415,413,993,718,908,291,383,296
2^{133} = 10,889,035,741,470,030,830,827,987,437,816,582,766,592
2^{134} = 21,778,071,482,940,061,661,655,974,875,633,165,533,184
2^{135} = 43,556,142,965,880,123,323,311,949,751,266,331,066,368
2^{136} = 87,112,285,931,760,246,646,623,899,502,532,662,132,736
2^{137} = 174,224,571,863,520,493,293,247,799,005,065,324,265,472
2^{138} = 348,449,143,727,040,986,586,495,598,010,130,648,530,944
2^{139} = 696,898,287,454,081,973,172,991,196,020,261,297,061,888 = 6.968983×10^{41}
2^{140} = 1,393,796,574,908,163,946,345,982,392,040,522,594,123,776
2^{141} = 2,787,593,149,816,327,892,691,964,784,081,045,188,247,552

2^{142} = 5,575,186,299,632,655,785,383,929,568,162,090,376,495,104
2^{143} = 11,150,372,599,265,311,570,767,859,136,324,180,752,990,208 = 1.1×10^{43}
2^{144} = 22,300,745,198,530,623,141,535,718,272,648,361,505,980,416
2^{145} = 44,601,490,397,061,246,283,071,436,545,296,723,011,960,832
2^{146} = 89,202,980,794,122,492,566,142,873,090,593,446,023,921,664
2^{147} = 178,405,961,588,244,985,132,285,746,181,186,892,047,843,328 = 1.784×10^{44}
2^{148} = 356,811,923,176,489,970,264,571,492,362,373,784,095,686,656
2^{149} = 713,623,846,352,979,940,529,142,984,724,747,568,191,373,312
2^{150} = 1,427,247,692,705,959,881,058,285,969,449,495,136,382,746,624
2^{151} = 2,854,495,385,411,919,762,116,571,938,898,990,272,765,493,248
2^{152} = 5,708,990,770,823,839,524,233,143,877,797,980,545,530,986,496
2^{153} = 11,417,981,541,647,679,048,466,287,755,595,961,091,061,972,992 = 1.14×10^{46}
2^{154} = 22,835,963,083,295,358,096,932,575,511,191,922,182,123,945,984
2^{155} = 45,671,926,166,590,716,193,865,151,022,383,844,364,247,891,968
2^{156} = 91,343,852,333,181,432,387,730,302,044,767,688,728,495,783,936
2^{157} = 182,687,704,666,362,864,775,460,604,089,535,377,456,991,567,872
2^{158} = 365,375,409,332,725,729,550,921,208,179,070,754,913,983,135,744
2^{159} = 730,750,818,665,451,459,101,842,416,358,141,509,827,966,271,488
2^{160} = 1,461,501,637,330,902,918,203,684,832,716,283,019,655,932,542,976
= 1.4615×10^{48}
2^{161} = 2,923,003,274,661,805,836,407,369,665,432,566,039,311,865,085,952
2^{162} = 5,846,006,549,323,611,672,814,739,330,865,132,078,623,730,171,904
2^{163} = 11,692,013,098,647,223,345,629,478,661,730,264,157,247,460,343,808
2^{164} = 23,384,026,197,294,446,691,258,957,323,460,528,314,494,920,687,616
2^{165} = 46,768,052,394,588,893,382,517,914,646,921,056,628,989,841,375,232
2^{166} = 93,536,104,789,177,786,765,035,829,293,842,113,257,979,682,750,464
2^{167} = 187,072,209,578,355,573,530,071,658,587,684,226,515,959,365,500,928
2^{168} = 374,144,419,156,711,147,060,143,317,175,368,453,031,918,731,001,856
2^{169} = 748,288,838,313,422,294,120,286,634,350,736,906,063,837,462,003,712
2^{170} = 1,496,577,676,626,844,588,240,573,268,701,473,812,127,674,924,007,424
= 1.496578×10^{51}
2^{171} = 2,993,155,353,253,689,176,481,146,537,402,947,624,255,349,848,014,848
2^{172} = 5,986,310,706,507,378,352,962,293,074,805,895,248,510,699, 696,029,696
2^{173} = 11,972,621,413,014,756,705,924,586,149,611,790,497,021,399,392,059,392
2^{174} = 23,945,242,826,029,513,411,849,172,299,223,580,994,042,798,784,118,784
2^{175} = 47,890,485,652,059,026,823,698,344,598,447,161,988,085,597,568,237,568
2^{176} = 95,780,971,304,118,053,647,396,689,196,894,323,976,171,195,136,475,136
2^{177} = 191,561,942,608,236,107,294,793,378,393,788,647,952,342,390,272,950,272
= $1.9156194261 \times 10^{53}$
2^{178} = 383,123,885,216,472,214,589,586,756,787,577,295,904,684,780,545,900,544
2^{179} = 766,247,770,432,944,429,179,173,513,575,154,591,809,369,561,091,801,088
2^{180} = 1,532,495,540,865,888,858,358,347,027,150,309,183,618,739,122,183,602,176
= $1.53249554 \times 10^{54}$
2^{181} = 3,064,991,081,731,777,716,716,694,054,300,618,367,237,478,244,367,204,352
2^{182} = 6,129,982,163,463,555,433,433,388,108,601,236,734,474,956,488,734,408,704
2^{183} = 12,259,964,326,927,110,866,866,776,217,202,473,468,949,912,977,468,817,408
2^{184} = 24,519,928,653,854,221,733,733,552,434,404,946,937,899,825,954,937,634,816
2^{185} = 49,039,857,307,708,443,467,467,104,868,809,893,875,799,651,909,875,269,632
2^{186} = 98,079,714,615,416,886,934,934,209,737,619,787,751,599,303,819,750,539,264

APPENDIX B

2^{187} = 196,159,429,230,833,773,869,868,419,475,239,575,503,198,607,639,501,078,528

2^{188} = 392,318,858,461,667,547,739,736,838,950,479,151,006,397,215,279,002,157, 056 = 3.923189×10^{56}

2^{189} = 784,637,716,923,335,095,479,473,677,900,958,302,012,794,430,558,004,314, 112

2^{190} = 1,569,275,433,846,670,190,958,947,355,801,916,604,025,588,861,116,008,628 ,224

2^{191} = 3,138,550,867,693,340,381,917,894,711,603,833,208,051,177,722,232,017,256,448

2^{192} = 6,277,101,735,386,680,763,835,789,423,207,666,416,102,355,444,464,034,512,896 = $6.27710736 \times 10^{57}$

2^{193} = 12,554,203,470,773,361,527,671,578,846,415,332,832,204,710,888,928,069,025,792 = 1.25542×10^{58}

2^{194} = 25,108,406,941,546,723,055,343,157,692,830,665,664,409,421,777,856,138,051,584

2^{195} = 50,216,813,883,093,446,110,686,315,385,661,331,328,818,843,555,712,276,103,168

2^{196} = 100,433,627,766,186,892,221,372,630,771,322,662,657,637,687,111,424,552,206,336 = 1.0043363×10^{59}

2^{197} = 200,867,255,532,373,784,442,745,261,542,645,325,315,275,374,222,849,104,412,672

2^{198} = 401,734,511,064,747,568,885,490,523,085,290,650,630,550,748,445,698,208,825,344

2^{199} = 803,469,022,129,495,137,770,981,046,170,581,301,261,101,496,891,396,417,650,688

2^{200} = 1,606,938,044,258,990,275,541,962,092,341,162,602,522,202,993,782,792,835,301,3
76 = 1.606938×10^{60}

2^{201} = 3,213,876,088,517,980,551,083,924,184,682,325,205,044,405,987,565,585,670,602,7
52

2^{202} = 6,427,752,177,035,961,102,167,848,369,364,650,410,088,811,975,131,171,341,205,5
04

2^{203} = 12,855,504,354,071,922,204,335,696,738,729,300,820177,623,950,262,342,682,411,0
08 = 1.28555×10^{61}

2^{204} = 25,711,008,708,143,844,408,671,393,477,458,601,640,355,247,900,524,685,364,822,
016

2^{205} = 51,422,017,416,287,688,817,342,786,954,917,203,280,710,495,801,049,370,729,644,
032

2^{206} = 102,844,034,832,575,377,634,685,573,909,834,406,561,420,991,602,098,741,459,288
,064 = 1.02844×10^{62}

2^{207} = 205,688,069,665,150,755,269,371,147,819,668,813,122,841,983,204,197,482,918,576
,128

2^{208} = 411,376,139,330,301,510,538,742,295,639,337,626,245,683,966,408,394,965,837,152
,256

2^{209} = 822,752,278,660,603,021,077,484,591,278,675,252,491,367,932,816,789,931,674,304
,512

2^{210} = 1,645,504,557,321,206,042,154,969,182,557,350,504,982,735,865,633,579,863,348,6
09,024 = 1.645505×10^{63}

2^{211} = 3,291,009,114,642,412,084,309,938,365,114,701,009,965,471,731,267,159,726,697,2
18,048

2^{212} = 6,582,018,229,284,824,168,619,876,730,229,402,019,930,943,462,534,319,453,394,4
36,096

2^{213} = 13,164,036,458,569,648,337,239,753,460,458,804,039,861,886,925,068,638,906,788,
872,192 = 1.316404×10^{64}

2^{214} = 26,328,072,917,139,296,674,479,506,920,917,608,079,723,773,850,137,277,813,577,
744,384

2^{215} = 52,656,145,834,278,593,348,959,013,841,835,216,159,447,547,700,274,555,627,155,
488,768

2^{216} = 105,312,291,668,557,186,697,918,027,683,670,432,318,895,095,400,549,111,254,310,977,536 = 1.053123 × 10^{65}

2^{217} = 210,624,583,337,114,373,395,836,055,367,340,864,637,790,190,801,098,222,508,621,955,072

2^{218} = 421,249,166,674,228,746,791,672,110,734,681,729,275,580,381,602,196,445,017,243,910,144

2^{219} = 842,498,333,348,457,493,583,344,221,469,363,458,551,160,763,204,392,890,034,487,820,288

2^{220} = 1,684,996,666,696,914,987,166,688,442,938,726,917,102,321,526,408,785,780,068,975,640,576 = 1.684997 × 10^{66}

2^{221} = 3,369,993,333,393,829,974,333,376,885,877,453,834,204,643,052,817,571,560,137,951,281,152

2^{222} = 6,739,986,666,787,659,948,666,753,771,754,907,668,409,286,105,635,143,120,275,902,562,304

2^{223} = 13,479,973,333,575,319,897,333,507,543,509,815,336,818,572,211,270,286,240,551,805,124,608 = 1.3479973 × 10^{67}

2^{224} = 26,959,946,667,150,639,794,667,015,087,019,630,673,637,144,422,540,572,481,103,610,249,216

2^{225} = 53,919,893,334,301,279,589,334,030,174,039,261,347,274,288,845,081,144,962,207,220,498,432

2^{226} = 107,839,786,668,602,559,178,668,060,348,078,522,694,548,577,690,162,289,924,414,440,996,864 = 1.0783979 × 10^{68}

2^{227} = 215,679,573,337,205,118,357,336,120,696,157,045,389,097,155,380,324,579,848,828,881,993,728

2^{228} = 431,359,146,674,410,236,714,672,241,392,314,090,778,194,310,760,649,159,697,657,763,987,456

2^{229} = 862,718,293,348,820,473,429,344,482,784,628,181,556,388,621,521,298,319,395,315,527,974,912

2^{230} = 1,725,436,586,697,640,946,858,688,965,569,256,363,112,777,243,042,596,638,790,631,055,949,824 = 1.7254359 × 10^{69}

2^{231} = 3,450,873,173,395,281,893,717,377,931,138,512,726,225,554,486,085,193,277,581,262,111,899,648

2^{232} = 6,901,746,346,790,563,787,434,755,862,277,025,452,451,108,972,170,386,555,162,524,223,799,296

2^{233} = 13,803,492,693,581,127,574,869,511,724,554,050,904,902,217,944,340,773,110,325,048,447,598,592 = 1.3803493 × 10^{70}

2^{234} = 27,606,985,387,162,255,149,739,023,449,108,101,809,804,435,888,681,546,220,650,096,895,197,184

2^{235} = 55,213,970,774,324,510,299,478,046,898,216,203,619,608,871,777,363,092,441,300,193,790,394,368

2^{236} = 110,427,941,548,649,020,598,956,093,796,432,407,239,217,743,554,726,184,882,600,387,580,788,736 = 1.1042794 × 10^{71}

2^{237} = 220,855,883,097,298,041,197,912,187,592,864,814,478,435,487,109,452,369,765,200,775,161,577,472

2^{238} = 441,711,766,194,596,082,395,824,375,185,729,628,956,870,974,218,904,739,530,401,550,323,154,944

2^{239} = 883,423,532,389,192,164,791,648,750,371,459,257,913,741,948,437,809,479,060,803,100,646,309,888

2^{240} = 1,766,847,064,778,384,329,583,297,500,742,918,515,827,483,896,875,618,958,121,6

06,201,292,619,776 = 1.7668471 × 10^{72}

2^{241} = 3,533,694,129,556,768,659,166,595,001,485,837,031,654,967,793,751,237,916,243,212,402,585,239,552

2^{242} = 7,067,388,259,113,537,318,333,190,002,971,674,063,309,935,587,502,475,832,486,424,805,170,479,104

2^{243} = 14,134,776,518,227,074,636,666,380,005,943,348,126,619,871,175,004,951,664,972,849,610,340,958,208 = 1.4134777 × 10^{73}

2^{244} = 28,269,553,036,454,149,273,332,760,011,886,696,253,239,742,350,009,903,329,945,699,220,681,916,416

2^{245} = 56,539,106,072,908,298,546,665,520,023,773,392,506,479,484,700,019,806,659,891,398,441,363,832,832

2^{246} = 113,078,212,145,816,597,093,331,040,047,546,785,012,958,969,400,039,613,319,782,796,882,727,665,664 = 1.1307821 × 10^{74}

2^{247} = 226,156,424,291,633,194,186,662,080,095,093,570,025,917,938,800,079,226,639,565,593,765,455,331,328

2^{248} = 452,312,848,583,266,388,373,324,160,190,187,140,051,835,877,600,158,453,279,131,187,530,910,662,656

2^{249} = 904,625,697,166,532,776,746,648,320,380,374,280,103,671,755,200,316,906,558,262,375,061,821,325,312

2^{250} = 1,809,251,394,333,065,553,493,296,640,760,748,560,207,343,510,400,633,813,116,524,750,123,642,650,624 = 1.8092514 × 10^{75}

2^{251} = 3,618,502,788,666,131,106,986,593,281,521,497,120,414,687,020,801,267,626,233,049,500,247,285,301,248

2^{252} = 7,237,005,577,332,262,213,973,186,563,042,994,240,829,374,041,602,535,252,466,099,000,494,570,602,496

2^{253} = 14,474,011,154,664,524,427,946,373,126,085,988,481,658,748,083,205,070,504,932,198,000,989,141,204,992 = 1.4474011 × 10^{76}

2^{254} = 28,948,022,309,329,048,855,892,746,252,171,976,963,317,496,166,410,141,009,864,396,001,978,282,409,984

2^{255} = 57,896,044,618,658,097,711,785,492,504,343,953,926,634,992,332,820,282,019,728,792,003,956,564,819,968

2^{256} = 115,792,089,237,316,195,423,570,985,008,687,907,853,269,984,665,640,564,039,457,584,007,913,129,639,936 = 1.1579209 × 10^{77}

2^{257} = 231,584,178,474,632,390,847,141,970,017,375,815,706,539,969,331,281,128,078,915,168,015,826,259,279,872 = 2.3158418 × 10^{77}.

2^{264} = 29,642,774,844,752,946,028,434,172,162,224,104,410,437,116,074,403,984,394,101,141,506,025,761,187,823,616 = 2.9642775 × 10^{79}

2^{265} = 59,285,549,689,505,892,056,868,344,324,448,208,820,874,232,148,807,968,788,202,283,012,051,522,375,647,232

2^{266} = 118,571,099,379,011,784,113,736,688,648,896,417,641,748,464,297,615,937,576,404,566,024,103,044,751,294,464 = 1.1845711 × 10^{80}

2^{274} = 30,354,201,441,027,016,733,116,592,294,117,482,916,287,606,860,189,680,019,559,568,902,170,379,456,331,382,784 = 3.0354201 × 10^{82}

2^{275} = 60,708,402,882,054,033,466,233,184,588,234,965,832,575,213,720,379,360,039,119,137,804,340,758,912,662,765,568

2^{276} = 121,416,805,764,108,066,932,466,369,176,469,931,665,150,427,440,758,720,078,238,275,608,681,517,825,325,531,136 = 1.21416806 × 10^{83}

2^{284} = 31,082,702,275,611,665,134,711,390,509,176,302,506,278,509,424,834,232,340,028,998,555,822,468,563,283,335,970,816 = 3.1082702 × 10^{85}

2^{285} = 62,165,404,551,223,330,269,422,781,018,352,605,012,557,018,849,668,464,680,057,997,111,644,937,126,566,671,941,632

2^{286} = 124,330,809,102,446,660,538,845,562,036,705,210,025,114,037,699,336,929,360,115,994,223,289,874,253,133,343,883,264 = 1.2833081×10^{86}

2^{294} = 31,828,687,130,226,345,097,944,463,881,396,533,766,429,193,651,030,253,916,189,694,521,162,207,808,802,136,034,115,584 = 3.1828687×10^{88}

2^{295} = 63,657,374,260,452,690,195,888,927,762,793,067,532,858,387,302,060,507,832,379,389,042,324,415,617,604,272,068,231,168

2^{296} = 127,314,748,520,905,380,391,777,855,525,586,135,065,716,774,604,121,015,664,758,778,084,648,831,235,208,544,136,462,336 = 1.2731475×10^{89}

2^{304} = 32,592,575,621,351,777,380,295,131,014,550,050,576,823,494,298,654,980,010,178,247,189,670,100,796,213,387,298,934,358,016 = 3.8321796×10^{91}

2^{305} = 65,185,151,242,703,554,760,590,262,029,100,101,153,646,988,597,309,960,020,356,494,379,340,201,592,426,774,597,868,716,032

2^{306} = 130,370,302,485,407,109,521,180,524,058,200,202,307,293,977,194,619,920,040,712,988,758,680,403,184,853,549,195,737,432,064 = 1.3037030×10^{92}

2^{314} = 33,374,797,436,264,220,037,422,214,158,899,251,790,667,258,161,822,699,530,422,525,122,222,183,215,322,508,594,108,782,608,384 = 3.3374797×10^{94}

2^{315} = 66,749,594,872,528,440,074,844,428,317,798,503,581,334,516,323,645,399,060,845,050,244,444,366,430,645,017,188,217,565,216,768

2^{316} = 133,499,189,745,056,880,149,688,856,635,597,007,162,669,032,647,290,798,121,690,100,488,888,732,861,290,034,376,435,130,433,536 = 1.3349919×10^{95}

2^{324} = 34,175,792,574,734,561,318,320,347,298,712,833,833,643,272,357,706,444,319,152,665,725,155,515,612,490,248,800,367,393,390,985,216 = 3.4175793×10^{97}

2^{325} = 68,351,585,149,469,122,636,640,694,597,425,667,667,286,544,715,412,888,638,305,331,450,311,031,224,980,497,600,734,786,781,970,432

2^{326} = 136,703,170,298,938,245,273,281,389,194,851,335,334,573,089,430,825,777,276,610,662,900,622,062,449,960,995,201,469,573,563,940,864 = 1.3670317×10^{98}

2^{334} = 34,996,011,596,528,190,789,960,035,633,881,941,845,650,710,894,291,398,982,812,329,702,559,247,987,190,014,771,576,210,832,368,861,184 = $3.4996012 \times 10^{100}$

2^{335} = 69,992,023,193,056,381,579,920,071,267,763,883,691,301,421,788,582,797,965,624,659,405,118,495,974,380,029,543,152,421,664,737,722,368

2^{336} = 139,984,046,386,112,763,159,840,142,535,527,767,382,602,843,577,165,595,931,249,318,810,236,991,948,760,059,086,304,843,329,475,444,736 = $1.3998405 \times 10^{101}$

2^{344} = 35,835,915,874,844,867,368,919,076,489,095,108,449,946,327,955,754,392,558,399,825,615,420,669,938,882,575,126,094,039,892,345,713,852,416 = $3.5835916 \times 10^{103}$

2^{345} = 71,671,831,749,689,734,737,838,152,978,190,216,899,892,655,911,508,785,116,799,651,230,841,339,877,765,150,252,188,079,784,691,427,704,832

2^{346} = 143,343,663,499,379,469,475,676,305,956,380,433,799,785,311,823,017,570,233,599,302,461,682,679,755,530,300,504,376,159,569,382,855,409,664 = $1.4334366 \times 10^{104}$.

2^{354} = 36,695,977,855,841,144,185,773,134,324,833,391,052,745,039,826,692,497,979,801,421,430,190,766,017,415,756,929,120,296,849,762,010,984,873,984 = $3.6695978 \times 10^{106}$

2^{355} = 73,391,955,711,682,288,371,546,268,649,666,782,105,490,079,653,384,995,959,602,842,860,381,532,034,831,513,858,240,593,699,524,021,969,747,968

2^{356} = 146,783,911,423,364,576,743,092,537,299,333,564,210,980,159,306,769,991,919,205,685,720,763,064,069,663,027,716,481,187,399,048,043,939,495,936 = $1.4678391 \times 10^{107}$

APPENDIX B

2^{364} = 37,576,681,324,381,331,646,231,689,548,629,392,438,010,920,782,533,117,931,316, 655,544,515,344,401,833,735,095,419,183,974,156,299,248,510,959,616 = 3.7576813 $\times 10^{109}$

2^{365} = 75,153,362,648,762,663,292,463,379,097,258,784,876,021,841,565,066,235,862,633, 311,089,030,688,803,667,470,190,838,367,948,312,598,497,021,919,232

2^{366} = 150,306,725,297,525,326,584,926,758,194,517,569,752,043,683,130,132,471,725,266 ,622,178,061,377,607,334,940,381,676,735,896,625,196,994,043,838,464 = 1.5030673 $\times 10^{110}$

2^{374} = 38,478,521,676,166,483,605,741,250,097,796,497,856,523,182,881,313,912,761,668, 255,277,583,712,667,477,744,737,709,244,389,536,050,430,475,222,646,784 = 3.8478522 $\times 10^{112}$

2^{375} = 76,957,043,352,332,967,211,482,500,195,592,995,713,046,365,762,627,825,523,336, 510,555,167,425,334,955,489,475,418,488,779,072,100,860,950,445,293,568

2^{376} = 153,914,086,704,665,934,422,965,000,391,185,991,426,092,731,525,255,651,046,673 ,021,110,334,850,669,910,978,950,836,977,558,144,201,721,900,890,587,136 = 1.5391409 $\times 10^{113}$

2^{384} = 39,402,006,196,394,479,212,279,040,100,143,613,805,079,739,270,465,446,667,948, 293,404,245,721,771,497,210,611,414,266,254,884,915,640,806,627,990,306,816 = 3.9402006 $\times 10^{115}$.

2^{385} = 78,804,012,392,788,958,424,558,080,200,287,227,610,159,478,540,930,893,335,896, 586,808,491,443,542,994,421,222,828,532,509,769,831,281,613,255,980,613,632 = 7.8804012 $\times 10^{115}$

2^{512} = 13,407,807,929,942,597,099,574,024,998,205,846,127,479,365,820,592,393,377,723, 561,443,721,764,030,073,546,976,801,874,298,166,903,427,690,031,858,186,486,050 ,853,753,882,811,946,569,946,433,649,006,084,096 = $1.3407808 \mathrm{x} 10^{154}$

2^{771} = 12,420,144,738,405,671,481,191,835,907,700,020,442,055,088,136,933,572,889,112, 416,304,208,407,621,491,015,090,647,027,270,629,171,823,603,901,845,577,048,585 ,649,372,640,352,918,515,131,554,298,200,329,449,113,635,639,808,166,799,244,40 2,122,285,052,787,558,602,103,993,549,731,750,007,142,774,830,528,462,848 = $1.2420145 \mathrm{x} 10^{232}$

Lord Kelvin was so satisfied with this triumph of science that he declared himself to be as certain of the existence of the ether as a man can be about anything.... "When you can measure what you are speaking about, and express it in numbers, you know something about it...." Thus did Lord Kelvin lay down the law. And though quite wrong, this time he has the support of official modern Science. It is NOT true that when you can measure what you are speaking about, you know something about it. The fact that you can measure something doesn't even prove that that something exists.... Take the ether, for example: didn't they measure the ratio of its elasticity to its density?
——**ANTHONY SANDEN** (1906-1993)**:** ***Science Is A Sacred Cow*** (1950), pages 69-70 & 85. (He was a British chemist best known for his above, sarcastic attack on Science.).

The supposed astronomical proofs of the theory [of relativity], as cited and claimed by Einstein, do not exist. He is a confusionist. The Einstein theory is a fallacy. The theory that ether does not exist, and that gravity is not a force but a property of space can only be described as a crazy vagary, a disgrace to our age.
——**CHARLES LANE POOR** (1866-1951) as quoted by Elizabeth Dilling in ***A "Who's Who" and Handbook of Radicalism for Patriots*** (1934), page 49. (He was a professor of Celestial Mechanics at Columbia University 1903-1944.).

If the 'Principle of Relativity' in an extreme sense establishes itself, it seems as if even Time would become discontinuous and be supplied in atoms, as money is doled out in pence or centimes instead of continuously;—in which case our customary existence will turn out to be no more really continuous than the events on a kinematograph screen;—while that great agent of continuity, the Ether of Space, will be relegated to the museum of historical curiosities.
——**SIR OLIVER JOSEPH LODGE** (1851-1940)**:** ***Continuity***. His 1913 ***Presidential Address to the British Association***, pages 40-41 (He was a British physicist and professor of Physics & Mathematics at University College, Liverpool, 1881-1900. Together, these 3 quotes illustrate a crucial point about knowing and testing thru *indirect* means: Even where it's done by highly respected scientists, the bit of concluding what's inferred by empirical observation is still risky and leads all too often to mistakes which, for decades, their devotees insist are as certain as anything can be. The rise and fall of the Ether is powerful testimony to that. I found these quotes, on the internet at todayinsci.com/QuotationsCategories/E_Cat/Ether-Quotations.htm April 3, 2014).

APPENDIX C:

POWERS OF 6

2	36
3	216
4	1,296
5	7,776
6	46,656
7	279,936
8	1,679,616
9	10,077,696
10	60,466,176
11	362,797,056
12	2,176,782,336
13	13,060,694,016
14	78,364,164,096
15	470,184,984,576
16	2,821,109,907,456 = 2.82111×10^{12}.
17	16,926,659,444,736
18	101,559,956,668,416
19	609,359,740,010.496
20	3,656,158,440,062,976
21	21,936,950,640,377,856
22	131,621,703,842,267,136
23	789,730,223,053,602,816
24	4,738,381,338,321,616,896
25	28,430,288,029,929,701,376
26	170,581,728,179,578,208,256
27	1,023,490,369,077,469,249,536
28	6,140,942,214,464,815,497,216
29	36,845,653,286,788,892,983,296
30	221,073,919,720,733,357,899,776
31	1,326,443,518,324,400,147,398,656
32	7,958,661,109,946,400,884,391,936 = $7.95866111 \times 10^{24}$
33	47,751,966,659,678,405,306,351,616
34	286,511,799,958,070,431,838,109,696
35	1,719,070,799,748,422,591,028,658,176
36	10,314,424,798,490,535,546,171,949,056
37	61,886,548,790,943,213,277 031,694,336
38	371,319,292,745,659,279,662,190,166,016
39	2,227,915,756,473,955,677,973,140,996,096
40	13,367,494,538,843,734,067,838,845,976,576
41	80,204,967,233,062,404,407,033,075,859,456
42	481,229,803,398,374,426,442,198,455,156,736
43	2,887,378,820,390,246,558,653,190,730,940,416
44	17,324,272,922,341,479,351,919,144,385,642,496

45	103,945,637,534,048,876,111,514,866,313,854,976
46	623,673,825,204,293,256,669,089,197,883,129,856
47	3,742,042,951,225,759,540,014,535,187,298,779,136
48	22,452,257,707,354,557,240,087,211,123,792,674,816
49	134,713,546,244,127,343,440,523,266,742,756,048,896
50	808, 281,277,464,764,060,643,139,600,456,536,293,376
51	4,849,687,664,788,584,363,858,837,602,739,217,760,256
52	29,098,125,988,731,506,183,153,025,616,435,306,561,536
53	174,588,755,932,389,037,098,918,153,698,611,839,369,216
54	1,047,532,535,594,334,222,593,508,922,191,671,036,215,296
55	6,285,195,213,566,005,335,561,053,533,150,026,217,291,776
56	37,711,171,281,396,032,013,366,321,198,900,157,303,750,656
57	226,267,027,688,376,192,080,197,927,193,400,943,822,503,936
58	1,357,602,166,130,257,152,481,187,563,160,405,662,935,023,616
59	8,145,612,996,781,542,914,887,125,378,962,433,977,610,141,696
60	48,873,677,980,689,257,489,322,752,273,774,603,865,660,850,176
61	293,242,067,884,135,544,935,936,513,642,647,623,193,965,101,056
62	1,759,452,407,304,813,269,615,619,081,855,885,739,163,790,606,336
63	10,556,714,443,828,879,617,693,714,491,135,314,434,982,743,638,016
64	63,340,286,662,973,277,706,162,286,946,811,886,609,896,461,828,096
64	$63.34028666 \times 10^{48} = 6.334028666 \times 10^{49} = (7.95866111 \times 10^{24})^2$
128	$4{,}011.992 \times 10^{96} = 4.011992 \times 10^{99} = (63.34029 \times 10^{48})^2$
256	$16.09608 \times 10^{198} = 1.61 \times 10^{199} = (4.011992 \times 10^{99})^2$

APPENDIX D:

EFFECTIVE ACTUALITY DIAMETERS AT SELECT ONTOLOGICAL DISTANCES

(2xOD = dia. in Φ & 50% of each dia. = radius in Φ converted to cm. & mi..)

DIAMETER OF ITS ACTUALITY DURING FIRST HALF OF A FORM'S CYCLE

OD	**CENTIMETERS & Φ**	**MILES**
2^{128}	0.00 00 00 05 = $2^{129}\Phi$ (& = OD2^{129} radius)	
2^{129}	0.00 00 00 10[1] = $2^{130}\Phi$(& = OD2^{130} radius)	
2^{130}	0.00 00 00 20 = $2^{131}\Phi$(& = OD2^{131} radius)	
2^{131}	0.00 00 00 40 = $2^{132}\Phi$(& = OD2^{132} radius)	
2^{132}	0.00 00 00 80 = $2^{133}\Phi$(& = OD2^{133} radius)	
2^{133}	0.00 00 01 60	
2^{134}	0.00 00 03 20	
2^{135}	0.00 00 06 40	
2^{136}	0.00 00 12 80	
2^{137}	0.00 00 25 60	
2^{138}	0.00 00 51 20= $2^{139}\Phi$(& = OD2^{139} radius)	
2^{139}	0.00 01 02 40	
2^{140}	0.00 02 04 80	
2^{141}	0.00 04 09 60	
2^{142}	0.00 08 19 20	
2^{143}	0.00 16 38 40	
2^{144}	0.00 32 76 80	
2^{145}	0.00 65 53 60	
2^{146}	0.01 31 07 20= $2^{147}\Phi$(& = OD2^{147} radius)	
2^{147}	0.02 62 14 40	
2^{148}	0.05 24 28 80	
2^{149}	0.10 48 57 60	
2^{150}	0.20 97 15 20= $2^{151}\Phi$(& = OD2^{151} radius)	
2^{151}	0.41 94 30 40	
2^{152}	0.83 88 60 80	
2^{153}	1.67 77 21 60	
2^{154}	3.35 54 43 20	
2^{155}	6.71 08 86 40 = $2^{156}\Phi$(& = OD2^{156} radius)	
2^{156}	13.42 17 72 80	
2^{157}	26.84 35 45 60	
2^{158}	53.68 70 91 20	
2^{159}	107.37 41 82 40 = $2^{160}\Phi$	
2^{160}	214.74 83 64 80 ÷ 160,934.72 =	0.00 13 34 38 18 213
2^{161}	429.49 67 29 60 ÷ 160,934.72 =	0.00 26 68 76 36 43

[1]From this line onward, the figure in this column is merely the previous line multiplied by 2.

2^{162}	858.99 34 59 20 ÷ 160,934.72 =	0.00 53 37 52 72 85
2^{163}	1,717.98 69 18 40 ÷ 160,934.72 =	0.01 06 75 05 45 71
2^{164}	3,435.97 38 36 80 ÷ 160,934.72 =	0.02 13 50 10 91 42
2^{165}	6,871.94 76 73 60 ÷ 160,934.72 =	0.04 27 00 21 82 84
2^{166}	13,743.89 53 47 20 ÷ 160,934.72 =	0.08 54 00 43 65 70
2^{167}	27,487.79 06 94 40 ÷ 160,934.72 =	0.17 08 00 87 31 39
2^{168}	54,975.58 13 88 80 ÷ 160,934.72 =	0.34 16 01 74 62 78
2^{169}	109,951.16 27 77 60 ÷ 160,934.72 =	0.68 32 03 49 25 56
2^{170}	219,902.32 55 55 20 ÷ 160,934.72 =	1.36 64 06 98 51 13
2^{171}	439,804.65 11 10 40 ÷ 160,934.72 =	2.73 28 13 97 02 26
2^{172}	879,609.30 22 20 80 ÷ 160,934.72 =	5.46 56 27 94 04 52
2^{173}	1,759,218.60 44 41 60 ÷ 160,934.72 =	10.93 12 55 88 09 03
2^{174}	3,518,437.20 88 83 20 ÷ 160,934.72 =	21.86 25 11 76 18 07
2^{175}	7,036,874.41 77 66 40 ÷ 160,934.72 =	43.72 50 23 52 36 15
2^{176}	14,073,748.83 55 32 80	87.45 00 47 04 72 30[1]
2^{177}	28,147,497.67 10 65 60	174.90 00 94 09 44 60
2^{178}	56,294,995.34 21 31 20	349.80 01 88 18 89 20
2^{179}	112,589,990.68 42 62 40= $2^{180}\Phi$	699.60 03 76 37 78 40
2^{180}	225,179,981.36 85 24 80	1,399.20 07 52 75 56 80
2^{181}	450,359,962.73 70 49 60	2,798.40 15 05 51 13 60
2^{182}	900,719,925.47 40 99 20	5,596.80 30 11 02 27 20
2^{183}	1,801,439,850.94 81 98 40	11,193.60 60 22 04 54 40
2^{184}	3,602,879,701.89 63 96 80	22,387.21 20 44 09 08 80
2^{185}	7,205,759,403.79 27 93 60	44,774.42 40 88 18 17 60
2^{186}	14,411,518,807.58 55 87 20	89,548.84 81 76 36 35 20
2^{187}	28,823,037,615.17 11 74 40	179,097.69 63 52 72 70 40
2^{188}	57,646,075,230.34 23 48 80 = $2^{189}\Phi$	358,195.39 27 05 45 40 80
2^{189}	115,292,150,460.68 46 97 60	716,390.78 54 10 90 81 60
2^{190}	230,584,300,921.36 93 95 20	1,432,781.57 08 21 81 63 20
2^{191}	461,168,601,842.73 87 90 40	2,865,563.14 16 43 63 26 40
2^{192}	922,337,203,685.47 75 80 80	5,731,126.28 32 87 26 52 80
2^{193}	1,844,674,407,370.95 51 61 60	11,462,252.56 65 74 53 05 60
2^{194}	3,689,348,814,741.91 03 23 20	22,924,505.13 31 49 06 11 20
2^{195}	7,378,697,629,483.82 06 46 40	45,849,010.26 62 98 12 22 40
2^{196}	14,757,395,258,967.64 12 92 80	91,698,020.53 25 96 24 44 80
2^{197}	29,514,790,517,935.28 25 85 60	183,396,041.06 51 92 48 89 60
2^{198}	59,029,581,035,870.56 51 71 20	366,792,082.13 03 84 97 79 20
2^{199}	118,059,162,071,741.13 03 42 40	733,584,164.26 07 69 95 58 40
2^{200}	236,118,324,143,482.26 06 84 80 = $2^{201}\Phi$	1,467,168,328.52 15 39 91 16 80
2^{201}	472,236,648,286,964.52 13 69 60	2,934,336,657.04 30 79 82 33 60
2^{202}	944,473,296,573,929.04 27 39 20	5,868,673,314.08 61 59 64 67 20
2^{203}	1,888,946,593,147,858.08 54 78 40	11,737,346,628.17 23 19 29 34 40
2^{204}	3,777,893,186,295,716.17 09 56 80	23,474,693,256.34 46 38 58 68 80
2^{205}	7,555,786,372,591,432.34 19 13 60	46,949,386,512.68 92 77 17 37 60
2^{206}	15,111,572,745,182,864.68 38 27 20	93,898,773,025.37 85 54 34 75 20
2^{207}	30,223,145,490,365,729.36 76 54 40	187,797,546,050.75 71 08 69 50 40
2^{208}	60,446,290,980,731,458.73 53 08 80	375,595,092,101.51 42 17 39 00 80
2^{209}	120,892,581,961,462,917.47 06 17 60	751,190,184,203.02 84 34 78 01 60

[1]From this line onward, the figure in this column is merely the previous line multiplied by 2.

2^{210}	241,785,163,922,925,834.94 12 35 20	1,502,380,368,406.05 68 69 56 03 20
2^{211}	483,570,327,845,851,669.88 24 70 40	3,004,760,736,812.11 37 39 12 06 40
2^{212}	967,140,655,691,703,339.76 49 40 80	6,009,521,473,624.22 74 78 24 12 80 = 1.0222885 ly.[1]
2^{213}	1,934,281,311,383,406,679.52 98 81 60	12,019,042,947,248.45 49 56 48 25 60 = 2.044577 ly.
2^{214}	3,868,562,622,766,813,359.05 97 63 20	24,038,085,894,496.90 99 12 96 51 20 = 4.089154 ly.
2^{215}	7,737,125,245,533,626,718.11 95 26 40	48,076,171,788,993.81 98 25 93 02 40 = 8.178308 ly.
2^{216}	15,474,250,491,067,253,436.23 90 52 80	96,152,343,577,987.63 96 51 86 04 80 = 16.356616 ly.
2^{217}	30,948,500,982,134,506,872.47 81 05 60	192,304,687,155,975.27 93 03 72 09 60 = 32.713232 ly.
2^{218}	61,897,001,964,269,013,744.95 62 11 20	384,609,374,311,950.55 86 07 44 19 20 = 65.426 464 ly.
2^{219}	123,794,003,928,538,027,489.91 24 22 40	769,218,748,623,901.11 72 14 88 38 40 = 130.852928 ly.
2^{220}	247,588,007,857,076,054,979.82 48 44 80	1,538,437,497,247,802.23 44 29 76 76 80 = 261.705856 ly.
2^{221}	495,176,015,714,152,109,959.64 96 89 60	3,076,874,994,495,604.46 88 59 53 53 60 = 523.411712 ly.
2^{222}	990,352,031,428,304,219,919.29 93 79 20	6,153,749,988,991,208.93 77 19 07 07 20= 1,046.823424 ly.
2^{223}	1,980,704,062,856,608,439,838.59 87 58 40 = $2^{224}\Phi$ (& = OD2^{224} radius)	12,307,499,977,982,417.87 54 38 14 14 40 = 2,093.646848 ly.
2^{224}	3,961,408,125,713,216,879,677.19 75 16 80	24,614,999,955,964,835.75 08 76 28 28 80 = 4,187.293696 ly.
2^{225}	7,922,816,251,426,433,759,354.39 50 33 60	49,229,999,911,929,671.50 17 52 56 57 60 = 8,374.587392 ly.
2^{226}	15,845,632,502,852,867,518,708.79 00 67 20	98,459,999,823,859,343.00 35 05 13 15 20= 16,749.174784 ly.
2^{227}	31,691,265,005,705,735,037,417.58 01 34 40 = $2^{228}\Phi$ (& = OD2^{228} radius)	196,919,999,647,718,686.00 70 10 26 30 40 = 33,498.349568 ly.
2^{228}	63,382,530,011,411,470,074,835.16 02 68 80	393,839,999,295,437,372.01 40 20 52 60 80 = 66,996.699136 ly.
2^{229}	126,765,060,022,822,940,149,670.32 05 37 60	787,679,998,590,874,744.02 80 41 05 21 60 = 133,993.398272 ly.
2^{230}	253,530,120,045,645,880,299,340.64 10 75 20	1,575,359,997,181,749,488.05 60 82 10 43 20= 267,986.796544 ly.
2^{231}	507,060,240,091,291,760,598,681.28 21 50 40	3,150,719,994,363,498,976.11 21 64 20 86 40 = 535,973.593088 ly.
2^{232}	1,014,120,480,182,583,521,197,362.56 43 00 80 = $2^{233}\Phi$ (& = OD2^{233} radius)	6,301,439,988,726,997,952.22 43 28 41 72 80= 1,071,947.186176 ly.

[1]Ly = the distance light travels in a tropical year of 31,556,925.9747 seconds, if the speed of light per second is 29,979,300,000 cm., which is to say 946,054,550,873,323,710 cm. per year = 9,460,545, 508,733.23710 km. per yr.. If 9,460,545,508,733.2 be divided by 1.6093472, then light travels 5,878,498,753,241.8 miles in a tropical year. If 6,009,521,473,624.2 miles be divided by 5,878,498, 753,241.8, then the diameter of the actuality at 2^{212} is 1.022288466 light years.

2^{233}	2,028,240,960,365,167,042,394,725.12 86 01 60	12,602,879,977,453,995,904.44 86 56 83 45 60 = 2,143,894.372352 ly.
2^{234}	4,056,481,920,730,334,084,789,450.25 72 03 20	25,205,759,954,907,991,808.89 73 13 66 91 20 = 4,287,788.744704 ly.
2^{235}	8,112,963,841,460,668,169,578,900.51 44 06 40	50,411,519,909,815,983,617.79 46 27 33 82 40 = 8,575,577.489408 ly.
2^{236}	16,225,927,682,921,336,339,157,801.02 88 12 80	100,823,039,819,631,967,235.58 92 54 67 64 80 = 17,151,154.978816 ly.
2^{237}	32,451,855,365,842,672,678,315,602.05 76 25 60 = $2^{238}\Phi$ (& = 2^{238} radius)	201,646,079,639,263,934,471.17 85 09 35 29 60 = 34,302,309.957632 ly.
2^{238}	64,903,710,731,685,345,356,631,204.11 52 51 20	403,292,159,278,527,868,942.35 70 18 70 59 20 = 68,604,619.915264 ly.
2^{239}	129,807,421,463,370,690,713,262,408.23 05 02 40	806,584,318,557,055,737,884.71 40 37 41 18 40 = 137,209,239.830528 ly.
2^{240}	259,614,842,926,741,381,426,524,816.46 10 04 80	1,613,168,637,114,111,475,769.42 80 74 82 36 80 = 274,418,479.661056 ly.
2^{241}	519,229,685,853,482,762,853,049,632.92 20 09 60	3,226,337,274,228,222,951,538.85 61 49 64 73 60 = 548,836,959.322112 ly.
2^{242}	1,038,459,371,706,965,525,706,099,265.8 4 40 19 20	6,452,674,548,456,445,903,077.71 22 99 29 47 20 = 1,097,673,918.644224 ly.
2^{243}	2,076,918,743,413,931,051,412,198,531.6 8 80 38 40 = $2^{244}\Phi$(& = 2^{244} radius)	12,905,349,096,912,891,806,155.42 45 98 58 94 40 = 2,195,347,837.288448 ly.
2^{244}	4,153,837,486,827,862,102,824,397,063.3 7 60 76 80	25,810,698,193,825,783,612,310.84 91 97 17 88 80 = 4,390,695,674.576896 ly.
2^{245}	8,307,674,973,655,724,205,648,794,126.7 5 21 53 60	51,621,396,387,651,567,224,621.69 83 94 35 77 60 = 8,781,391,349.153792 ly.
2^{246}	16,615,349,947,311,448,411,297,588,253. 50 43 07 20	103,242,792,775,303,134,449,243.39 67 88 71 55 20 = 17,562,782,698.307584 ly.
2^{247}	33,230,699,894,622,896,822,595,176,507. 00 86 14 40	206,485,585,550,606,268,898,486.79 35 77 43 10 40 = 35,125,565,396.615168 ly.
2^{248}	66,461,399,789,245,793,645,190,353,014. 01 72 28 80	412,971,171,101,212,537,796,973.58 71 54 86 20 80 = 70,251,130,793.23 03 36 ly.
2^{249}	132,922,799,578,491,587,290,380,706,028 .03 44 57 60	825,942,342,202,425,075,593,947.17 43 09 72 41 60 = 140,502,261,586.460672 ly.
2^{250}	265,845,599,156,983,174,580,761,412,056 .06 89 15 20 = $2^{251}\Phi$(& = 2^{251} radius)	1,651,884,684,404,850,151,187,894.34 86 19 44 83 20 = 281,004,523,172.921344 ly.
2^{251}	531,691,198,313,966,349,161,522,824,112 .13 78 30 40	3,303,769,368,809,700,302,375,788.69 72 38 89 66 40 = 562,009,046,345.842688 ly.
2^{252}	1,063,382,396,627,932,698,323,045,648,2 24.27 56 60 80	6,607,538,737,619,400,604,751,577.39 44 77 79 32 80

		= 1,124,018,092,691.685376 ly.
2^{253}	2,126,764,793,255,865,396,646,091,296,4 48.55 13 21 60	13,215,077,475,238,801,209,503,154.78 89 55 58 65 60 = 2,248,036,185,383.370752 ly.
2^{254}	4,253,529,586,511,730,793,292,182,592,8 97.10 26 43 20	26,430,154,950,477,602,419,006,309.57 79 11 17 31 20 = 4,496,072,370,766.741504 ly.
2^{255}	8,507,059,173,023,461,586,584,365,185,7 94.20 52 86 40 = $2^{256}\Phi$ (& = 2^{256} radius)	52,860,309,900,955,204,838,012,619.15 58 22 34 62 40 = 8,992,144,741,533.483008 ly.
2^{256}	17,014,118,346,046,923,173,168,730,371, 588.41 05 72 80 = 1.7014118×10^{31}	105,720,619,801,910,409,676,025,238.31 16 44 69 24 80 = 1.0572063×10^{26} = 17,984,289,483,066.966016 ly.

✡✞✡✞✡✞✡✞✡✞✡✞✡✞✡✞✡✞✡✞✡✞✡

. . . . forms are a mere fiction of the human mind, unless you will call the laws of action by that name.
——**FRANCIS BACON** (1561-1626)**:** ***Novum Organum***, First Book: Paragraph 51, upper right-hand column of page 11 in Vol. 30 of ***Great Books Of The Western World*** as published by Encyclopædia Britannica, Inc.; Chicago, 1952 (Ah, Sir Francis! With Esoptrics, I dare say, your assessment of forms is far from the truth.).

INDEX OF TOPICS

(UM = ultramicroscopic makeup of time, space, matter, locomotion, & change)

A.

ABORTION: effectively a supremely intense prayer begging God to withdraw The Hand of Divine Mercy & to allow the ultimate monster, me, to come into the world: 232.

ABSOLUTE CERTITUDE: I claim no such: XXVI, 175, 229.

ABSTRACTABLE, PRIME: see: PRIME ABSTRACTABLE.

AVERSIO A DEO: = a turning away from God's kind of coercion free love: 230-231; I'm not free of it: 231-232.

ACCELERATION: caused mainly by collision (q. vd.): 39; in forms = increasing the current OD = moving up in the 4th dimension thereby decreasing the frequency with which u-time affects it: 130; in generators = increase the frequency of change from one u-space to a logically contiguous one thereby increasing the frequency with which u-time affects it: 130; reduces the apparent size of the u-spaces: 130, 143, 143 ftn. #1; whether in a form or a generator, affects the ratio of u-times per i-time of the accelerating form or generator: 130.

ACCELERATION, THE GREAT: 111-120; as a motion thru the 4th dimension, the whole process occurs in only 10^{-19} sec.: 111-112; what it effects above the MT it effects below the MT in a reverse manner: 112; what it effects below the MT re-described July 2, 2013: 180-183.

ACCORDIAN UNIVERSE: also see: UNIVERSE: Esoptrics' view of anti-gravity rules it out along with unending expansion: 196-197.

ACT, PRIMARY: see: STATE OF EXCITATION, OVERALL.

ACTION AT A DISTANCE: a major puzzle for scientists but not Esoptrics: 145; for Einstein, was "spooky" & "ludicrous": 145, 145 ftn. #1; the result of a form's actuality being wholly in every one of the u-spaces in the envelope being used by 2 or more entangled generators: 146-150 .

ACTUALITY: = in forms & generators, their current overall state of excitation: 35; = sum total of a form's or generator's current internal characteristics characterized by smoothness & continuousness: 11, 17; = what acts: 10; contracted vs. expanded phases of: 72-78, 98, 103; diameters of in a pyramid of molecular forms: 160; dia. of in the Milky Way: 164; drag inducing differential between it & potentiality in generators produces units of tension: 18; effective radius of generally = form's current OD: 68, 130; in every form's primary act is 6-way: 27; in God is 100%: 21; of a form = a u-spaces envelope: V, 35, 46, 54, 65, 67, 68, 130, 145, 199; of a form is every "empty space": 36, 205 ftn. #2; of a form is wholly in each of the u-spaces it provides as a u-spaces envelope: 149; of Omega: 148; of the forms vs. the Higgs field: 199-200; ratio of to potentiality in the infinite: 18-19; source of gravity: 198; vs. potentiality: 17, 18-19; way a form's rotates corrects Prof. Feynman: 153; what's experienced of it by The Infinite is Its Real Internality Itself: 15; what's experienced of it by the finite is a projected "ghostly" mirrored image of its real internality: 11-12, 14-15.

AIRPLANE: light traveling from tail to nose: 142-143.

ALGEBRA: also see COSMOLOGY: it not Geometry is Esoptrics' mathematical logic: XXI, 50-53; its calculations make separate some u-spaces shown identical by Geometry's pictures: 40.

ALPHAKRONON: = 1 u-time c. 10^{-96} sec. = K: 7, 31; # per unit of rectilinear motion varies inversely as generator's velocity: 8, detailed calculation of: 95;

no fractional ones: 1, 95.
ALPHATOPON: = 1 u-space = c. 10^{-47} cm. = 1Φ = 2^{-129} x atom's dia.: V, 2, 7, 49; detailed calculation of: 95; no fractional ones: 2 ftn. #2, 92 ftn. #1, 93 ftn. #1, 95, 188.
ANGELS: 9 choirs of: 29; St. Thomas Aquinas not so wrong saying they move the planets: 76, ftn. #1; vs. piggyback form = form not vs. form yes logically concentric with a generator: 2 ftn. #4, 3 ftn. #1, 31.
ANIMATION, STOP ACTION: locomotion as: 47.
ANTI-GRAVITY: also see: GRAVITY: also see: INVERSE SQUARE LAW: = force with which categorical forms in the same generic zone above the MT zone repel one another & any categorical form supplying them with u-spaces: 90, 113, 195, 201; double barreled: 114, 119, 195-196; is always exerted by OD1 categorical forms against all other categorical ones: 90, 197 ftn. #1; reverse of is the strong nuclear force (q. vd.): 201.
APHELION VS. PERIHELION: defined 107.
A PRIORI: coordinate vs. proposition: XXV ftn. #1; proposition = true independently of all experiences save of language: XXV ftn. #1; St. Thomas Aquinas on: XXV-XXVI; Wikipedia on: XXV-XXVI.
A PRIORI VS. A POSTERIORI: Wikipedia says distinction not cut & dry: XXV-XXVI.
AQUINAS, ST. THOMAS: also see ARISTOTELIAN: his Hylemorphism vs. Esoptrics' Hylegenemorphism: 2 ftn. #4; his vs. Esoptrics' explanation of how forms have potency & act: 31 ftn. #1; his vs. Esoptrics' view of angels: 2 ftn. #4, 3 ftn. #1; not so wrong saying angels move the planets: 76 ftn. #1; on *a priori* vs. *a posteriori*: XXV-XXVI; on Aristotle's vs. Plato's definition of "motion": 11 ftn. #1; startling similarity between his 10th crystalline sphere & Esoptrics' 10th categorical form: 226; would say the finite encounterer sees of himself an essence (***i.e.:*** internality) not the same as the essence of his existence (***i.e.:*** the actuality of his act of mirroring himself), whereas, The Infinite Encounterer sees of Himself an essence identical to the essence of his existence: 13-15.
ARGUE VOCIFEROUSLY: that the UM is the realm of Philosophy & Theology but not Physics: 174.
ARISTOTLE: definition of substance: 9 ftn. #1; his Hylemorphism vs. Esoptrics' Hylegenemorphism: 2 ftn. #4; his vs. Plato's use of "motion": 11 ft. #1; initial error later multiplies 1,000 times: 131 ftn. #1; on Science's superior methodology: 30; Spinoza's bad take on his view of extent to which substances are independent: 10; unaware of the distinction between parasitic vs. non-parasitic substances: 12 ftn. #1.
ARISTOTELIAN-THOMISTIC NOTION: every change must be initiated by God as The First Unmoved Mover: 49 ftn. #2, 66.
ATHEISTS: also see: LeMAITRE: also see: MENDEL: also see: THEISTS: no *deus ex machina* more fantastic then theirs: XXII, 155 ftn. #1; Theists are all idiots according to them thereby indicating how far they themselves are from being as intellectually superior as they fancy they are: 138, 138 ftn. #1; victim complex = their way to rationalize away their fear of death: 228 ftn. #2; view Christianity's God as a cruel omnipotent ogre in the skies: 228 ftn. #1; wrong to denounce the Catholic Church as arch enemy of reason & humanity's intellectual advancement: 220.
ATOMS: also see HYDROGEN ATOM: reverse effect below MT of the great acceleration above MT: 112-113.
AU: = astronomical unit = avg. Earth-Sun distance = 92,955,807.267 mi.: 107.

B.

BACON, FRANCIS: a, for me, seemingly confused opinion regarding the contemplative life as focus on the mind vs. as the contemplation of nature: 106; shades of Aristotle as quoted by me on my page 30: 106.
BALL: bouncing points of on an airplane: 142.
BARYCENTER: heavenly body's main multi-combo but one not at that body's center: 71; is not merely an imaginary point: 102-103; relative to Earth & Moon: 103-105.
BEETHOVEN: example of paranoia: 228 ftn. #1.
BENCHMARKS, HEAVENLY: see: HEAVENLY BENCHMARKS.
BHAGAVAD GITA: 232.
BIG BANG: father of is Belgian priest Fr. Georges LeMaitre: 138 ftn. #1; for Science, what an incredibly complex Calculus says but, for Esoptrics, what simple Algebra says: 87 ftn. #1; is not creation event itself & arose in a preexisting Universe says Prof. Greene: V, 88; only Esoptrics gives a detailed description of what preceded it: V, 88; starting with time zero, Esoptrics tells in detail how God caused what to bang, why & how in 8 mini-bangs & then the 9th big bang: V; 79-88.
BLACK HOLES: in Esoptrics = multi-combos in which anywhere from 2^{192} to $2^{256}-1$ duo combos are logically concentric: 71, 112, 118, 140.
BLESSED VIRGIN: 176.
BOHM, DAVID: refutes Quantum Theory's insistence on probabilities only: 154-155; similarity of his thinking to Esoptrics': 156.
BOHR, NIELS: also see: SCHRÖDINGER: collapse of the stage 2 formulation of quantum theory is because experimenters & their equipment are different from elementary particles says Greene: 162; only Esoptrics explains what that difference is & the exact location of the border at which that difference begins: 162.
BONDING, INTIMATE: see: INTIMATE BONDING.
BORDER: at which the everyday big world confronts the tiny atomic world is described by Esoptrics only: 215.
BOSONS: = force particles: 42; as forms: XXI, 2, 43, 49; integer spin: 43; vs. God's ability to be self-created: XXII.
BREATHE: in empty space: XXIII.
BROTHER, OLDER: 50 years ago, scoffed at my assertion that the rules change because the sizes change: 215 ftn. #1.
BUFFOONERY: consummate = to dismiss Esoptrics as merely what Leibniz said: 225; mathematical = result of trying to describe time & space before one knows about God's 7 *a priori* coordinates: 204 ftn. #1.

C.

CAMERA: double shuttered one as figurative explanation of how consciousness works at our OD vs. at OD1: 216.
CARRYING GENERATOR: see: GENERATOR.
CATEGORICAL OD'S: list: 29.
CATEGORICAL FORMS: see: FORMS, CATEGORICAL.
CATEGORIES: list of 8 original vs. 8 reverse: 111; reverse 8 produced in 10^{-19} sec. when MT forms accelerate far enough up in the 4th dimension to become multi-g-combos: 111-120.
CATEGORIES, REVERSE also see: MULTI-g-COMBOS: detailed descriptions of the 8: 111-120; I'm not certain of the number of multi-g-combos per: 114 ftn. #1; presence of multi-h-combos in, using the Milky Way, Sun & planets as examples: 118-119; the 64 types of multi-g-combos involved: 116; when their u-spaces envelopes extend beyond Omega's: 115 ftn. #1.
CATEGORIES, PRESERVATION LAWS: 113; basis of anti-gravity in Esoptrics: 90, 113-114.
CATHOLIC DOCTRINE: right to sing the praises of Philosophy over Science: 220; wrongly denounced by some scientists as arch enemy of reason &

humanity's intellectual advancement: 220.
CAUSALITY: also see: PHYSICAL REALITY'S ABILITY TO FOOL US: even the ancients admitted there's no significant knowledge of any physical reality without knowledge of the complete chain of causes behind it: 228; irrational pride alone allows us to be deceived about it: 229-231; with nothing able to fool me into thinking I know for sure anything's whole chain of causes, I've never denied Esoptrics is anything more than theoretical inferences: XXVIII, 229.
CAUSE: principle one of the Universe vs. proximate cause of its every detail = God vs. one's total self with God's support: 219.
CHANGE: accidental vs. substantial: 45 ftn. #1, 223 ftn. #2; continuous existence of forms & generators not affected by their ways of changing: 45, 223; fastest rate of is at OD1 & = 1 change per K: 135.
CHANNEL OF CREATION: thru categorical & generic forms only: 81.
CHICKEN VS. EGG: also see: LOCALITY: either locality in spacetime comes before there can be separate identities or separate identities come before there can be locality in spacetime. Esoptrics explains how the latter is possible, while none can explain how the former can be possible without begetting the nightmare of the continuum & the absurd math of infinite divisibility: 204.
CHILD, INTELLECTUAL: me as: 220.
CHILD'S PLAY: in St. Thomas: 226.
CLOWN ACT: also see: JOKE: openly confessing ignorance of the UM, scientists pretend to tell us "space itself" is stretching & the idea of such is too beautiful to be wrong: 126.
COBE: also see: WMAP: 203 ftn. #1.
COLLISION: most common kind =2 generators trying to activate the same u-space simultaneously: 39, 48; why none between 2 multi-combos rotating in opposite directions around the same multi-combo at the same OD: 168 ftn. #1.
CONSCIOUSNESS: also see: CAMERA: also see: GORDON: also see: IRRATIONAL PRIDE: also see: PHYSICAL REALITY'S ABILITY TO FOOL US: also see: STROBOSCOPIC EFFECT: = act by an irrational generic major form performing a macrostate of excitation c. $OD2^{155}$ to $OD2^{156}$: 211; acts of correlated with light speed, size of the atom, size of the Φ, length of K, & flashing lights: 210-213; as the focus of Esoptrics vs. Depth Psychology & Moral Theology: 229; duration of its acts compared to acts at OD1: 213; is most manifestly of what is past only: 214, 214 ftn. #2; its explanation of = Esoptrics' main achievement: 3; number of cycles & phases per sec.: 211; of change explained: 214, 215; phases alternate between aware of nothing vs. aware of a frozen time exposure created in the prior phase: 211; Prof. Greene's bit on time indicates the form could rotate & still be unaware of time: 214 ftn. #1; seems a continuous flow of change thru time only because u-time registers not with it as long as its form is in a changeless state of excitation: 133-134, 213-214; sensory kind is limit of forms: 130; the necessary triousious structure of its every act: 10; thru its form's stroboscopic effect it allows us to live with & enjoy myriads of humans, animals, vegetables & minerals for whom i-time is equal to or close to our own: 215.
CONTINUUM: also see: GEOCENTRISM: can't be confirmed by observation: XIX 49 ftn. #2; demands no limit to the smallness of the smallest segment of time, space or locomotion: XIX, 49 ftn. #2; Esoptrics ends the claim no theory can work without it: XIX; infinite divisibility of begets absurd math say Locke & Hume: II, VIII, XIX; most contested issue of all for 2,500 years:

XIX; most still cling to it: XIX; supporters appeal to what sense images imply: XIX.
COORDINATES: God's *a priori* ones produce 7 kinds of logical sequence: 1; God's *a priori* vs. humans' *a posteriori* ones: XIX-XXI, 40; Philosophy's introspection is only path to knowledge of God's: XXI-XXII, 40; with no knowledge of *a priori* ones, most cling to *a posteriori* ones: XX.
CORRELATIONS: a few interesting ones: 69-70.
COSMOLOGY: also see: THEORETICAL PHYSICS: = knowledge of the origin & structure of the Universe: 1; as algebraic vs. geometrical: XXIII, XXIV, 22; is purely algebraic in Esoptrics: XXI, 40.
COSMOLOGY, PATH TO KNOWLEDGE OF: see: CAUSALITY: also see: PHILOSOPHY VS. SCIENCE.
COSMOLOGY VS. COSMOGONY: Prof. Greene seems ignorant of the distinction: 217 ftn. #1.
CO-TENANTS: more accurate description of subjects in a triousious object: 9.
CREATION also see: BIG BANG: also see: EPOCHS: also see: INFLATION: from time zero to first 10^{-18} sec.: 79-87, 121-123; my change from 6 to 8 forms per act of per channel: 79, 112; similarity between Esoptrics' & Science's time lines: 87 ftn. #1; takes place only thru forms native to a generic or categorical OD: 81; total number of duo-combos created: 84-86.
CRIPPLED: by a predilection for whatever Cosmology excludes God from the picture: 156; by too little knowledge of God, 40, 53, 54.
CRUTCH, MENTAL: space as it is for minds crippled by too little knowledge of God: 40, 53, 54.

D.

DARK MATTER: = multi-combos emitting duo-combos at a level of the 4th dimension too far above the highest OD to which any of the leading forms in our locality are current: 163-165; not particles other than any of the kinds we know: 165; second party denying dark matter: 110.
DARK ENERGY: = duo- &/or multi-combos traveling at a level of the 4th dimension far above the highest OD to which any of the leading forms in our locality are current: 163-165; not a kind of energy other than any of the kinds we know: 165.
DECIMAL PLACES: I leave it to the reader to decide which ones are significant: 95.
DIAGONAL LOCOMOTION: 47-48; used by some to attribute physical extension to u-spaces: 50-51.
DEUS EX MACHINA: 40.
DIMENSION, FOURTH: see: FOURTH DIMENSION.
DIMENSIONS: also see: THREE-DIMENSIONALITY: also see SEPARATION, also see: PROXIMITY: manmade expressions of atom's dia. vs. Esoptrics' real expression: 53; physical vs. logical at UM: XX, 204.
DION: also see: SUB-SEXTONS: any form native either to OD2 or OD4 & every particle in which the leading form is a Dion: 167; two of the 3 kinds of sub-sextons: 167; what Esoptrics classes with Science's leptons: 167.
DIVERSITY: see: MULTIPLICITY.
DIVINE INTERVENTION: only hope for Esoptrics: 88.
DIVISIBILITY: see: INDIVISIBLES.
DIVISIBILITY, INFINITE: see: INFINITE DIVISIBILITY.
DOGMATIC: Einstein explains if he speaks so it's only for the sake of clarity & simplicity without actually claiming he knows the right path: 229.
DOUBLE SLIT EXPERIMENTS: described by H&M: 151-152; indicate "paths that go across the universe and back" = Esoptrics' on high speed rotations in the primary & secondary planes: 152-153.
DRAWINGS: Esoptrics vs. Geometry in

using squares, cubes, circles, & spheres: 51-53.
DREXEL UNIVERSITY MATH FORUM: Bill Gaede quotes its ***Dr. Math Introduces Geometry***: 41.
DUO-COMBO: an ultimate *occupant* of the Universe in which a form & a generator are logically concentric: V, 31, 42; as generator = inertial mass in law of motion & as form = gravitational mass in law of gravitation & source of action at a distance & cause of radiation, electrical charges & magnetic lines of force: 42; as electrical corpuscles: 45; as neutrinos & anti-neutrinos: 44; as photons: 44; eliminates Newton's need to view particles, light, & electrical corpuscles as 3 different kinds of points: 43-44; similarity with supersymmetry's implication force & matter are 2 facets of the same thing: 42; total # in Universe charted: 84-87; waves vs. particles easily explained by the duo-combo's 2 sides: 45.
DURATION: can only be defined relatively: 135.
DWARF GALAXIES: see: GALAXIES, DWARF.

E.

EARTH: Esoptrics on the daily rotation of: 95-105; has more than 1 multi-h-combo: 103; inner core's rotation: 97; vast majority of its orbital speed of 66,616 MPH belongs to the curvilinear velocity of the u-spaces envelope being used by its generator: 76, 77, 141; vast majority of its rotation's speed of 1,040 MPH is the curvilinear velocity of the u-spaces being used by the carrying generators in its atoms' multi-A-combos: 100.
EDDINGTON, SIR ARTHUR: coined the term "arrow of time" says Prof. Greene: 139.
EINSTEIN, ALBERT: also see: MERCURY: occurrences of his name: XVIII, XXIII, XXIV, 1, 4, 6, 8, 26, 40, 41, 42, 44, 45, 71, 76, 78, 87, 121, 127, 128, 129, 130, 135, 142, 145, 145 ftn. #2, 146, 154, 166, 198, 199 ftn. #1, 202 ftn. #1, 219, 229, 230; if speaks dogmatically, it's only for clarity's sake, since he claims not to know the truth: 229; says phenomena uniquely determine Theoretical Physics: 40; IQ=160: 229; says Physics deals only with sense experience: 40; scorned as "that blithering idiot" by some scientists: 229; trying to describe nature apart from the space-time continuum = trying to breathe in empty space: XXIII; what "material point" might have meant to him: 41-42; wrong about gravity's velocity: 198-199.
EINSTEIN VS. ESOPTRICS: on space's connection with matter: 4; on space curving near matter vs. thickening in the 4th dimension to become matter: 4; Science treating directly with sense experiences vs. Esoptrics treating directly with introspection's understanding of God's 7 *a priori* coordinates: 40.
ELECTRON: see: MONONS: also see: MULTI-e-COMBOS: mass of: 183; multi-combo in which every one of its logically concentric forms is native to OD1: 45; multi-e-combo in Esoptrics' terminology: 91; positron is not truly an anti-electron says Esoptrics: 172; produced when an OD1 form accelerates to remain in potency to the OD2 form which has been accelerated: 113, 185; races away from the atom if accelerated to $OD2^{128}$: 91; repels all other categorical forms: 45, 64, 90, 91, 197 ftn. #1; shells & sub-shells of in Esoptrics: 183-188; should alternately dive partly toward & race away from the nucleus as the u-spaces envelope providing it with u-spaces expands & contracts: 188; taus & muons do not decay into electrons says Esoptrics: 172.
ELECTRON SHELLS & SUB-SHELLS: Esoptrics detailed description of: 183-188; Wikipedia's erroneous explanation how far electrons travel from the

nucleus vs. Esoptrics': 188.
ELECTRICAL CHARGES: see: LEADING FORMS.
EMPIRICAL FACT: nothing below 10^{-47} cm. or 10^{-18} sec. is such: 4.
EMPIRICAL OBSERVATION: also see INFERENCE: note on: XXXII.
ENCOUNTERABLENESS: = sum total of some subject's internal characteristics: 11 ftn. #2.
ENCOUNTEREDS, DIRECT: = the sense images: 9; not predicable of what's encountering them: 9.
ENCOUNTERED, REFLEX: = what encounterer knows of himself on the rebound: 9-10.
ENCOUNTERER: = what's directly aware of its sense images but unable to be directly aware of itself: 9; not predicable of its encountereds: 9; thesis seeking synthesis thru the anti-thesis of himself: 15; unaware of activity unless he is some kind of activity: 10.
ENCOUNTERER, FINITE: mirroring himself with finite intensity, he sees a kind of ghostly mask projected by his face: 13; the result is mirrored images which are not equally real: 13; what he sees of himself is a set of internal traits radically different from those of his mirroring activity itself & so his self's essence is not the same as the essence of his act of existence: 15.
ENCOUNTERER, INFINITE: a single act with 3 functions: 14; mirroring Himself with infinite intensity, He sees His actuality exactly as It is: 13; the act of encountering is encountering Himself: 13; the result is mirrored images which are equally real: 14; what He sees of Himself is a set of internal traits identical to those of His mirroring activity Itself & so His Self's essence is the same as the essence of His act of existence: 15.
ENTANGLEMENT also see: ACTION AT A DISTANCE: result of 2 generators having their home position in the same atom despite the astronomical number of positions they may have as a result of high speed rotations in the primary & secondary planes by the u-spaces they're using: 149-150.
ENTROPY: for me, a Rube Goldberg contrivance: 144.
ENVELOPES: see: U-SPACES.
EPOCHS: also see: CREATION: creation of the 9: 79-87; durations in K & ontological depths of the 9: 82; durations in sec. of the 9: 83.
EPR PARADOX: initials for Einstein, Podolsky, Rosen: 145 ftn. #2.
EPR EFFECT: Sir John Polkinghorne's misnomer for the EPR paradox: 145 ftn. #2.
EQUATIONS: "said none predicting a 4th dimension" ≠ "said nothing at all about a 4th dimension": 190 ftn. #2.
ESOPTRICS: also see: COSMOLOGY: also see: GORDON = logic of the mirror cuz deals with reverse images & multiples of 2: V, 1, 12; 2 issues it ends: God exists & Phil. bests Science in UM's description: XXI-XXII; first theory to end the need for the continuum: XIX; in its description of MT, gives the only explanation of the exact border separating the lower sizes which obey QT's rules from the upper sizes which obey Newton's & Einsteins rules: 162; is an algebraic logic: V, XXI, 40; only detailed description of the pre-inflation era preceding the Big Bang: V, 120; obviates Newtonian need for more than one kind of point for particles, another for light & another for electrical corpuscles: 44-45; only detailed description of UM: V, 54; overthrows no equations fundamental to Relativity or Quantum Mechanics: XXVI; primarily Philosophy & Theology: V, 1; quantity vs. quality of its math: 87 ftn. #1; replaces space with u-spaces envelopes: V; supreme principle of: 15, 17, 168; vs. Geometry in using squares, cubes, circles & spheres: 51-53.
ESOPTRICS AS COSMOLOGY: describes basics of time, space, etc.: 1.
ESOPTRICS AS PHILOSOPHY: starts

with introspection: 1, 22, 40.
ESOPTRICS AS THEOLOGY: lists God's *a priori* coordinates & how, thru them, God creates the ultimates & their traits: 1, 40, 54.
ESOPTRICS, NAME OF: from Greek for mirror: V, 1.
ESOPTRICS, REJECTION OF: 49 ftn. #1, 88, 120, 126.
ESOPTRICS, START OF: 1957: 1.
ESOPTRICS VS. SCIENCE: also see: PHILOSOPHY VS. SCIENCE: theoretical system dealing directly only with introspection's contemplative understanding of God's 7 *apriori* coordinates vs. one dealing directly only with sense experiences: 9, 40.
ESSENCE: = internal characteristics: XX, XXXI, 11 ftn. #2; precedes & begets locality: XXVI ftn. #1.
ETHER: example of empirical observation leading to bogus inferences: 246
EVOLUTION: God's way of sharing with us the glory of an awesome achievement: 232.
EXCITATION, STATES OF: see: STATES OF EXCITATION.
EXISTENCE: essence of in God = God's essence: 13-15; kind of dependence of in a parasitic vs. a non-parasitic subject: 12, 12 ftn. #1; is continuous in every form & generator: 45 ftn. #1; vs. in Leibniz: 223.
EXPERIENCE: also see: PERCEPTION: total of perception's cognitions says Webster's: XXV ftn. #1.
EXPERIENCE, IMMEDIATE: = what's observed face to face = self & its states only: 9.
EXPERIENCE, SENSE: also see: RISKY BUSINESS: irrational pride & not the nature of consciousness is the element of consciousness allowing sensation to deceive us regarding its causality: 229-231; is of the units of tension generated by generators due to the drag inducing differential between their actuality & potentiality: 2, 18, 36, 37, 130; lies not about 3-dimensional shapes: 42, 228; merely can't present the ultramicroscopic why of those shapes: 42; with it only does Physics deal says Einstein: 40.
EXTENSION: also see: DIMENSIONS: also see: THREE DIMENSIONALITY: in what ultimates are *individually* vs. *collectively*: V, XX, 2, 3, 37, 39, 41, 42, 44, 51, 203; physical vs. logical: V, XXII, XXV ftn. #1, 1, 2, 36, 37, 41, 46, 53-53, 147, 148, 190 ftn. #1, 222.

F.

FAIR USE, PRINCIPLE OF: 26.
FAME: too often lethal to creative thinking & writing: 88.
FERMIONS: = matter particles: 42; as generators: XXI, XXII, 2, 43, 49; half integer spin: 43.
FEYNMAN, PROF. RICHARD: also see: H&M: way off the mark from not knowing the difference between rectilinear vs. curvilinear paths: 153; why the paths can seem simultaneous: 153.
FIRST PRINCIPLES: only Philosophy's introspective methodology can discover those which are truly such: 38.
FORCE: Science's 4 = gravitation, electromagnetism, strong nuclear force & the weak nuclear force: 201; some hypothesize a 5th says Wikipedia: 201 ftn. #1; strong nuclear = result of how categorical forms attract one another below the MT zone says Esoptrics: 201; strong vs. weak nuclear: 178 ftn. 1; weak nuclear = result of leading forms accelerating up or decelerating down in the 4th dimension says Esoptrics: 201; Wikipedia's description of electromagnetic radiation accords well with Esoptrics' bit on how forms rotate in the 3 planes perpendicular to one another & in definite ratios to one another: 202.
FORM: also see: FORMS; also see: PATTERNS OF DIVINE ROTATION: also see: STATE OF EXCITATION, OVERALL: = a macro-ultimate in potency to God: 21, 27, 29, 31-32, 51, 55; = a piggyback form, except in the case of an-

gels: 2 ftn. #4, 3 ftn. #1, 31; 4 kinds of behavior = accelerate, decelerate, pulsate, rotate: 70; actuality of is a u-spaces envelope: V, 35, 44, 46, 47, 53, 61, 64, 65, 67, 68, 71, 78, 80, 113, 114, 115, 118, 124, 130, 141, 187, 188, 196, 196 ftn. #1, 199, 200, 201, 205, 206, 211, 226; an ultimate *constituent* of the Universe always logically concentric with a generator in a duo-combo: V, 3; as a u-spaces envelope, its Φ radius = its current OD: 68; Esoptrics version of Science's boson: 2, 43; its actuality *is* space: 199; perform macrostates of excitation from c. 10^{-47} cm. to c. 10^{31} cm. in dia. as u-spaces envelopes: 2, 68; piggyback can't be a human soul: 130 ftn. #1; related to God 6 ways gives each 3 sets of 2 mutually opposed kinds of component excitation & 6 logical dimensions: 20-21, 31, 55; rotations of in the 3 planes bespeak native vs. current OD: 56-65, 59; similarity of their whole states of excitation to bosons integer spin: 43; the symmetry of: 38; thru their generators' rectilinear motions, forms at the same OD are also logically distinct: 55, 55 ftn. #1; vs. a Higgs field: 199-200; why forms generate no units of tension: 38.

FORMS, CATEGORICAL: = ones native to one of the 10 categorical OD's 2^0, 2^1, 2^2, 2^4, 2^8, 2^{16}, 2^{32}, 2^{64}, 2^{128}, 2^{256}: 29; unless native to OD1, no categorical one is logically concentric in a multi-combo except as the leading form: 45.

FORMS, IRRATIONAL: = those native to an OD, such as 2, 5, 6, not having a rational square root: 60; rate at which they pulse is determined by the square of their current OD: 61, 194.

FORMS, MAJOR: = those native to an OD of 2^{128} or lower: 60, 67; can be accelerated to a current OD as high as 2^{128} times their native OD: 64; deceleration limits are still a puzzle for me: 64.

FORMS, MINOR: = those native to an OD greater than 2^{128}: 60; acceleration & deceleration limits are still a puzzle for me: 64.

FORMS, MOLECULAR: = major generic but irrational forms accelerated to somewhere between the MT zone's top at OD 2^{129} & the 8th reverse category's floor at OD2^{192}: 158, 162; example: 158-162; explain how the chaotic microscopic realm can produce the orderly macroscopic realm: 158-162.

FORMS, MT: see: MT FORMS.

FORMS PER OD: switch from 6 to 8 per act of channel creation: 79, 79 ftn. #1.

FORMS, PULSATION OF: also see: ORBITS: as a u-spaces envelope, has one dia. in the 1st phase of each cycle & a different dia. in the 2nd phase: 61; dia. shift among major forms reverses that among minor forms resulting in envelopes anywhere from almost circular to highly elliptical: 68-78; effect = expanding & contracting envelope causing generators using it to move in & out rhythmically relative to the envelope's center: 70; effect on bodies orbiting the Sun: 68-78 examples of the effect using Earth & the comet Hyakutake: 71-78; u-spaces from the expanded phase remain somehow in the contracted phase: 73; vacillation on my part regarding it: 78 ftn. #1.

FORMS, RATIONAL: = those native to an OD, such as 4, 9, 16, having a rational square root: 60: only accelerating rational major forms produce multi-combos: 64, 67; produce also the 4th dimension of multi-combos & action at a distance: 67; rate at which they pulse is determined by their current times their native OD: 60, 194.

FOURTH DIMENSION: = 2^{256} levels of intensity possible to each of 6 ways to be excited by God in the Universe's 9th epoch: V, 27, 28; depth of increases to 2^{512} levels in the Universe's 10th epoch: 27-28; gross ignorance of among scientists: 143; how allows absorption and radiation of light: 105; in each multi-combo, # of levels extending beyond its center diminishes one level per Φ from that center: 67; movement

thru is one level per shift & at a velocity of 1 level per K: 99-100; movement thru is still contained with the c. 10^{-47} cm. of a single u-space: 130; no observable distance between its levels: XXXIV, 67; note on: XXXIV; resists movement up or down thru its various levels but not movement thru a single level: 200; switching from one multi-combo's 4th to another's requires finding a logical juncture: 101-102; ultimate shock to physics = evidence for more than the familiar 3 says Prof. Greene: 190 ftn. #2; unlike with u-spaces, the kind of logical separation between its levels has no 3D effect: XXXIV, 67; vs. String Theory's one dimensional strings: 94, 105.

FRACTIONS: also see: DECIMAL PLACES: none for alphatopons: 2 ftn. #2, 92 ftn. #1, 95, 188; none for alphakronons: 1, 95; none for OD's: 188.

FREE USE, PRINCIPLE OF: hopefully, those I quote will honor it: 26.

FREE WILL: also see: SELF: also see: TIME, ARROW OF: absence of all predetermination vs. presence of self-determination: 139, 217; karma determined vs. current self determined: 218; power vs. reason to choose: 139, 218-219; Prof. Greene's example of a future event's effect on a past choice: 218; some object there's too many people for each to have his or her own way & others that injustice becomes meaningless: 140; some rail against God as sadistic to support karma determined choice & will continue to do so even in the face of a post-mortem, astronomically vast plenitude of truth proving conclusively God has done the most loving thing Infinitely Informed Love can do: 231.

G.

GAEDE, BILL: citing the Drexel University Math Forum's definition of "point" as infinitesimal, rants against "Einstein's Idiots": 41; defines infinitesimal as zero dimensions = 0D: 41.

GALAXIES: result of the great acceleration & the 64 types of multi-g-combos it produced: 111-120; using how its law of the preservation of the categories defines anti-gravity, Esoptrics gives a simpler answer than Science to why the galaxies race away from one another while their component heavenly bodies do not: 119.

GALAXIES, DWARF: Esoptrics agrees with Science that most galaxies are dwarf types: 119.

GAPS: how every multi-combo fills whatever ones it has: 72 ftn. #1; may be the gap $OD2^{128}$ is never filled: 90.

GENERATOR: also see: STATE OF EXCITATION, OVERALL = micro-ultimate: 29, 31; each an ultimate *constituent* of the Universe always logically concentric with a form in a duo-combo: V, 3; Esoptrics' version of Science's fermions: 1, 43; generate 1, 2, or 3 kinds of units of tension: 2, 37, 130; in potency to a form not its piggyback one: 32, 33-34, 51; perform microstates of excitation all c. 10^{-47} cm. for sensation dependent minds: 2; similarity of their half states of excitation to fermions half-integer spin: 43; similarity to, & difference from, String Theory's strings: 94; to which form one is in potency is determined by its piggyback form's current OD: 32, 91-92; why they generate units of tension: 38.

GENERATORS, VELOCITY OF: rule governing all generators not traveling at light speed: 92-93.

GENERIC OD'S: powers of 2: 29.

GENIUS: to imagine me the most perceptive one ever is merely a cause to laugh: 229; yes, tests show my I. Q. is high but still far from the highest: 229.

GEOCENTRISM: also see: RISKY BUSINESS: as with the notion of the continuum, its supporters appeal to what sense images imply: XIX; H&M say Heliocentrism is preferable only because the equations of motion are

much simpler if the Sun is referenced as at rest: 230 ftn. #1.

GEOMETRY: also see: LENGTH: also see: STRAIGHT LINE: a change it can't depict: 40; in Esoptrics is a metaphorical teaching aid only: XXI, XXIV, 16, 21, 22, 50, 51-53, 79 ftn. #2, 113 ftn. #2, 124, 190, 191; relying on it & senses vs. on Algebra & God's input to determine a ball's bounce points: 142; relying on it & senses vs. on Algebra & God's input to determine distance light is traveling: 143; though its pictures show some u-spaces not separate, Algebra's calculations show they are: 40; vs. Esoptrics in depicting u-spaces envelopes as squares, cubes, circles & cubes: 51-53, 79 ftn. #2, 208; vs. Physics is Geometry for Relativity say H&M: XXIV.

GEOMETRY SPEAK: 44, 49 ftn. #1, 56, 57, 62, 80, 86, 99, 105.

GILSON, ETIENNE: on St. Thomas' explanation of how forms have potency & act: 31 ftn. #1.

GNAT: & a hydrogen bomb: 233

GOD: also see: COORDINATES: also see: FREE WILL: also see: UNIVERSE, CAUSE OF: as principle vs. proximate cause of the Universe: 219; every moment of time is equally present to: 136-137; excluded if ultimates are inherently extended: XX ftn. 1; existence of is necessary if Esoptrics is valid: XXII; gave the Big Bang its bang: 200; has not forced us into this world against our *total* self's will: 231; implicated in a lie: 49 ftn. #2, 133 ftn. #1, 214 ftn. #2; infinite version of the self-mirroring mirror's mirroring activity: V, 13-14; "in potency to" for forms = get from God their 6 kinds of excitation & the 2^{256} levels of intensity possible to each of the 6: XXIII ftn. #1; minds crippled by too little knowledge of: 40, 53, 54; never coerces but, as truth, destroys every lover of a lie without *targeting* them for such: 230-231; no need to ask who created God since God is a regression *ad Infinitum*: 155 ftn. #1; produced 8 mini-bangs prior to Big Bang: V, 79-87; source of logical dimensions, extension & separation thru 7 *apriori* coordinates: V, XIX-XX, XXI, XXII, XXIII, XXIII ftn. #1, 2, 40, 44, 46, 47, 54, 94, 131, 131 ftn. #1, 132, 134, 141, 142, 143, 165, 190, 190 ftn. #1, 199, 203, 204, ftn. #1, 223, 225; The First Mirror: 14; when created the Universe: 137 ftn. #1; why created the Universe: 217, 219.

GOOGLEPLEX: XIX.

GORDON HAAS: also see: IRRATIONAL PRIDE: e-mail asking if Esoptrics identifies those elements of consciousness which allow it to be fooled by physical reality: 227; e-mail from me replying Esoptrics identifies those elements allowing us to live at our level of OD & to enjoy our fellow inhabitants: 227; e-mail saying I'm subordinating reality to consciousness & my correction saying I'm subordinating our ability to understand reality to our ability to understand consciousness: XXII; e-mail saying I've described the mindset required to start understanding Esoptrics: XXII; letting itself be fooled vs. entertained & enlightened by physical reality: those dead set on playing victim choose the former, their opposites, the latter: 228; "unless you is a quark": 83; what allows consciousness to be fooled by reality vs. fooled about its causality vs. what merely makes it difficult to discover the causality: 228.

GRAND UNIFICATION: also see: FORCE: for Science it's of 4 forces, but 5 or even 6 for Esoptrics: 201; whatever the kind or number of forces, the unifying factor is the forms either as to the 4 ways they behave or the OD to which they are native or the zone in which they are acting: 202.

GRAVITY: also see: ANTI-GRAVITY: also see: HIGGS FIELD: also see: INVERSE SQUARE LAW: = a force generated by each multi-combo's concentric generators, equal in intensity to

the square of the number of those generators, & made possible & transmitted by the forms concentric in the multi-combo: 38, 189-190; Esoptrics' explanation of how it diminishes: 189-200; Esoptrics vs. Greene & Einstein on how gravity "travels": 198; H&M wrongly claim the inverse square law of depends upon # of space's dimensions: 189; Newton was correct on gravity's instantaneous velocity: 198.

GRAVITY'S 2 KINDS: between multi-combos having the same or nearly the same ontological depth vs. between a multi-combo & ones having a greatly smaller ontological depth: 197.

GREENE, BRIAN: IN PASSING: also see: IRRATIONAL PRIDE: also see: MOVIES: Big Bang is not creation event itself & occurred in a pre-existent Universe: V, 120; Big Bang theory says not what banged or how, why, or if: 88; Fly's Eye cosmic ray detector measured a particle = to 30 billion protons: 201 ftn. #2; gravitational force twixt 2 very close tiny objects may vary to a certain point: 197-198; Higgs field gave big bang its bang: 88, 200; Higgs field is the source of every particle's mass: 200; Higgs ocean fills all of space: 199; Higgs ocean resists only accelerated motion: 199-200; key to String Theory's success is switch from dots of no size to strings with spatial extent: 190 ftn. #1; muses over whether Universe moves thru space: 196 ftn. #1; no final answer on the Universe's shape, but leading entrant says flat, infinitely large, spatial one & a spherical shape has been ruled out by observation: 203; no particle escapes Higgs field's influence: 199; observable Universe is a tiny fraction of the whole: 165 ftn. #1; on Joe Polchinski's regions which trap String Theory's strings: 196: ftn. #2; on Kant's impossible to describe the Universe if you do away with time & space: 204 ftn. #1; prior to String Theory no theory said anything at all about 4th dimension: 190 ftn. #2; Science has no idea re: UM: V, 54, 126; Science has no insight into Universe's start: 120; Science knows not what space *really* is: 65; since c. 7 billion years old, Universe's expansion is accelerating: 114; space's regions flee each other at greater than light speed: 125 ftn. #1; space swirls: 65 ftn. #1; speed of light limit doesn't apply to space itself: 65 ftn. #1; too tiny for Science, UM is for Phil. & Theol. say some: V; Universe shall be a vast, empty, lonely place in 100 billion years if inflationary cosmologists are right: 196 ftn. #3; validity of gravity's inverse square law has been tested no further than a 10th of a millimeter: 198; wants to experience the Universe at every possible level & not just those we can sense: 65; ultimate shock to physics = evidence for a 4th dimension: 190 ftn. #2; wrong explanation of how gravity travels: 198.

H.

H&M: = Hawking & Mlodinow: 124: beg the question of how microscopic chaos begets macroscopic order: 157-158; Esoptrics rejects their take on gravity in 4 dimensions: 189; not so much interested in whatever thinking explains the observed world as in whatever thinking does it without God: 156; speed of light limit doesn't apply to space itself: 65 ftn. #1, 121; saying God created the Universe deflects the question to who created God: 155, 155 ftn. #1; the only reason to choose Heliocentrism over Geocentrism is the equations of motion are much simpler if the Sun is referenced as at rest: 230 ftn. #1; use *"simultaneously"* in one quote but "rather simultaneously" in another re: Feynman's explanation of the double slit experiments: 153 ftn. #1.

HAAS, GORDON: see: GORDON.

HAPPY: if #9 is so: 54.

HARARI, HAIM: also see: WIKIPEDIA: Esoptrics sextons & sub-sextons compared to his quarks & leptons: 171-174; his puzzle over nature's charges of 1, 1/3 & 2/3 is solved by Esoptrics theme of the 3 axes & the way sextons vs. sub-sextons rotate in them: 173; the connection he puzzles over between charge & color is explained by Esoptrics: 173.

HEAVENLY BENCHMARKS also see: PATTERNS OF DIVINE ROTATION: enable sextets to spin: 55-57; vs. heavenly sextets: 21, 55-57.

HEAVENLY SEXTETS: distinct from one another cuz each is forever associated with a particular form: 21, 29, 55; each a created finite reflection of God: 55; heavenly benchmarks enable them to spin: 55-57, 65; vs. heavenly benchmarks: 21, 55-57, 65.

HELL: also see FREE WILL: at the center of Earth vs. earth: 216; one sec. here may = many trillions of trillions of eons there: 216; place of refuge for those hating God & God's truth for exposing their lie for the lie it is: 219.

HIGGS FIELD: vs. God as what gave the Big Bang its bang: 200; vs. the actuality of the forms as an influence no particle escapes: 199; vs. the actuality of the forms as what fills all of space: 199; vs. the actuality of the forms as what gives each fundamental particle its mass: 200; vs. the actuality of the forms as the 4th dimension which resists only accelerated motion: 199-200.

HUBBLE, EDWIN: showed cursory gazes at the heavens can mislead says Greene: 166.

HYAKUTAKE COMET: Wikipedia on: 72 ftn. 2.

HYDROGEN ATOM: also see: ELECTRON SHELLS: its structure is the issue ever puzzling me the most: 175; I've always been convinced its leading form is native to 2^{128} & its carrying generator the carrier of the whole atom: 175; my recent change re: where the atomic categories occur: 175-177, 180-183; not sure it's 255 or 27 quarks: 177, 179-180; three ways a quark may circle the quark supplying it with a u-spaces envelope: 178-180; using a telescope's image to explain how the great acceleration produced the 8 accelerated atomic categories: 181 ftn. #1.

HYDROGEN BOMB: & a gnat: 233.

HYDROGEN NUCLEUS: very compact as categorical forms other than OD1 attract each other below MT zone: 178-179.

HYLEGENEMORPHISM: Esoptrics' reply to Hylemorphism: 2 ftn. #4.

HYLEMORPHISM: name for the Aristotelian-Thomistic notion every substance is a combination of matter & form: 2 ftn. #4.

I.

IDIOTS: Bill Gaede on Einstein's: 41; "blithering" one is some scientists description of Einstein: 230; what Atheists insist all Theists are thereby indicating how far they themselves are from being as intellectually superior as they fancy they are: 138, 138 ftn. #1.

ILLUSION: prime abstractable is not such: 11.

IMMEDIATE: see: EXPERIENCE, IMMEDIATE.

INDIVISIBLES: also see: INFINITESIMALS: also see PLANCK SACLE: below Planck scale, divisibility may be meaningless says Prof. Greene: 54; wholly so vs. physically but not logically so: XXV ftn. #1; vs. irreducible: 42.

INERTIAL SYSTEMS: rate at which the generators are changing from one microstate of excitation to another is different in each system: 143.

INFERENCE: also see: RISKY BUSINESS: often used by scientists to support erroneous conclusions: 4, 246.

INFINITE DIVISIBILITY: 6 years old when I saw infinite this-that-&-the other is brazenly self-contradictory

nonsense: 49 ftn. #2; have always been those who said the same: 229-230; such makes it obvious some *want* to be fooled by physical reality: 230; why want to be fooled= love of an environment making competition possible: 230-231.
INFINITESIMALS: also see: INDIVISIBLES: logically vs. physically so: XXII, XXV ftn. #1, 1, 7, 8, 41-42, 51; most assume all such are also indivisibles: XXV ftn. #1.
INFLATION: also see: CREATION: also see: PRE-INFLATION ERA: also see: "SPACE ITSELF": 2 kinds in Esoptrics: 121-126; 1st kind = newly created generators forcing previous ones to move away from the centers of creation's 256 sequential channel levels at 2^{128} x light speed: 121-123; 1st kind's 256 episodes lasted a total of 2^{256}K = c. $8.3x10^{-19}$ sec.: 123; 2nd kind = multi-g-combos as u-spaces envelopes fleeing each other: 124-126; H&M say inflation was not completely uniform: 121; H&M say physicists not sure exactly how inflation happened: 121; like H&M, Esoptrics says inflation was not completely uniform: 122.
INHERE: see: SUBJECT, PARASITIC.
IN POTENCY TO: = seeking its states of excitation from = for a form, God &, for a generator, a form not its own piggyback one: XXIII ftn. #1, 21, 27, 31,32, 35, 38, 71, 80.
INTERGALACTIC SPACE TRAVEL: key may somehow lie in the high-speed rotations of the forms in the primary & secondary planes: 65.
INTIMATE BONDING: only level of dependence not necessary to any ultimate finite reality: 10, 12 ftn. #1.
INTROSPECTION: also see: PHILOSOPHY: = introverted examination of what's maximally intra-mental: 9, 54; = Philosophy's methodology: V, XIX, 9, 54; only path to knowledge of God's *a priori* coordinates: XXI-XXII.
INVERSE SQUARE LAW FOR GRAVITY: also see: GRAVITY: as defined by Esoptrics: 189; as wrongly explained by Prof. Greene, Wikipedia, & H&M: 189; Esoptrics denies it's affected by # of dimensions of space: 189; the detailed account of that denial: 189-196; the law applies to what happens in both phases of a single cycle: 195; the law may not apply to the atomic nucleus: 197; the law may not apply to 2 very close tiny objects says Greene: 197-198; validity of has been tested no further down than 10th of a millimeter says Greene: 198.
IN VS. OUT OF PHASE: u-spaces in a form's expanded vs. contracted phases: 73.
IRRATIONAL PRIDE: also see: GORDON: also see: MONSTER: also see: OBSERVATION VS. INFERENCE: denies we want an intellectually dark world & throws the blame on everything but ourselves: 227, 227 ftn. #1; Prof. Greene blames nature's sleights of hand with concealing the truth of quantum reality: 227 ftn. #1; some, for its sake, see themselves as victims if they infer the chain of causes erroneously while their opposites, as Einstein does, never deny their inferences are merely theories: 227-229.
I-TIME: its undetectable vs. calculable duration: 133-134; time as experienced by consciousness: 133-134; result of the rate at which forms at any OD above OD1 change from one macrostate of excitation to another: 135; ratio of u-time to for Alphon & Bogon: 143, 143 ftn. 1; time as it applies to generators & forms as long as they are performing a temporarily changeless state of excitation: 130.

J.

JOKE: also see: CLOWN ACT: confessing no knowledge of what happened at Universe's start, scientists still try to explain gravity, anti-gravity, & fleeing galaxies: 120.

K.

K = one unit of u-time: 1; = 1 akphakronon = c. 10^{-96} sec.: 7; duration relative to one sec. varies as the inverse of the square of the observer's current OD divided by the number of cycles his consciousness' form performs per sec.: 95-96, 95 ftn. #2.

KALUZA, THEODOR: 190 ftn. #2.

KANT, IMMANUEL: also see MINDS: Esoptrics imitates his reversal of spectator's role in observation: 9; he rightly calls space, as commonly conceived, a mental crutch: 40; on sensations: XXXII; Prof. Greene on Kant's bit it's impossible to describe the Universe if you do away with time & space: 204 ftn. #1, 214; without his kind of time & space, buffoons cannot, but, with God's kind, Esoptrics can describe it: 204 ftn. #1, 214.

KARMA: total self (q. vd.): 218.

KELVIN, LORD WILLIAM THOMPSON: the ether is maximally certain: 246.

L.

LADDER: a metaphor for the 4th dimension: 68, 102, 163, 165; as such, multi-combos offer other combos' generators 3 options: 68; switching from one to a taller one requires finding a logical juncture: 101-102.

L-COMBOS: also see: W-COMBOS: = light combos = every multi-a- (q. vid.), multi-b- (q. vd.), & multi-e-combo (q. vd.) = every multi-combo in an atom but not a multi-A-combo: 91; rule governing the OD of the form from which their carrying generator shall seek its u-spaces: 91-92; rule governing the velocity of their carrying generators: 92-93.

LEADING FORMS: also see: HARARI, HAIM: also see: SEXTONS: after years, I decided multi-combos can have 2 such native to the same OD, currently at the same OD but rotating in opposite directions: 168; the number jumps to 6 by distinguishing between 3 kinds of forms distinguished from one another by which of the 3 axes is the axis of the tertiary plane & each of the 3 either a rotating or a counter-rotating sub-kind: 169; then the number jumps to 18 if each of the 3 has 6 instead of 2 modes: 169-170; the twelve charged & 3 neutral types possible to each of the 3 kinds & the source of their fractional charges: 171; with the total types thus raised to 36, the number of forms created to the categorical OD's other that 2^{256} must be multiplied by 36: 173-174.

LEAP: also see: PRESCIENT: only figuratively applicable to Esoptrics description of rectilinear motion: 3, 8, 32, 39, 47, 49, 80, 147, 168 ftn. #1.

LEAPFROGGING: same as LEAP: 47, 113.

LEIBNIZ, GOTTFRIED: Esoptrics dismissed by some as merely a repeat of what he said 300 years ago: 221; mentioned (as Leibnitz) by Einstein re: acceleration problem & space: 129; no similarity of his monads to Esoptrics generators (sextads) or forms (hebdomads) or duo-combos (triscadecads) or multi-combos: 223 ftn. #1; no trace in him of Esoptrics' mathematically precise definition of the 26 states logically contiguous in the same level of the 4th dimension to a given state + the 27 logically contiguous at each of the 2 logically contiguous levels of the 4th dimension: 223; no trace in him of what states necessarily follow from a preceding one & why: 223; no triads of couplets defining each state nor anything about 3 planes of rotation or gravity, anti-gravity, etc.: 224; six is God's essence is nothing like Esoptrics' 6 divine coordinates: 224; use of the mirror metaphor in him vs. in Esoptrics: 225.

LeMAITRE, FR. GEORGES: father of the Big Bang theory &, according to Atheists, was an idiot because a Theist: 138 ftn. #1.

LENGTH: also see: PLANCK LENGTH: also see: STRAIGHT LINE: in Esoptrics = # of logically sequential u-spaces vs. Geometry's # of manmade units of measurement: 53.

LEPTONS: as OD1, 2 & 4 forms: 63; as sub-sextons: 167: cluster of duo-combos: 170; Haim Harari on: 171-173; vs. sub-sextons: 167, 170-173, 177-178; Wikipedia on: 171.

LIGHT: see PHOTONS.

LIGHT, SIN AGAINST: 230.

LIGHT SPEED: = 1Φ per 2^{128}K in Esoptrics: 7, 91, 92, 64, 65; detailed calculation of what 1Φ/2^{128}K = in cm. per sec.: 95-96; effect had on it by the generic zones: 93; re-cast as $2^{-257}\Omega_D$ per $2^{128}\forall$: 143; the limit applies to the generators' rectilinear & not the u-spaces' curvilinear motion: 65, 65 ftn. #1, 121, 128, 202 ftn. #1; why the same in every inertial system: 143.

LOCALITY: according to Prof. Greene, scientists' name for that "feature of the universe" making space "the medium that separates & distinguishes one object from another": 146; essence precedes & begets it: XXVI ftn. #1; recent experiments verify Universe has no locality: 146; sensation addicts' take on it is sadly mistaken: 156; such have I been trumpeting since grammar school & only Esoptrics explains how, without locality, there can still be locations at various distances from one another: XXVI, 146-147; that's possible because each u-spaces envelope is the actuality of a form whose actuality is wholly & equally present in each of its envelope's u-spaces: 148-149.

LOCKE, JOHN: can't say what space & extension are: 46; his view of motion without space totally refuted by Esoptrics: 54.

LOCOMOTION, CONTINUOUS: also see: INFINITE DIVISIBILITY: 6 when I first saw it's blatantly untenable nonsense: 49 ftn. #2; can't be confirmed by observation that there is no limit to the smallness of the smallest segment of it: 49 ftn. #2; not an empirical fact: 49 ftn. #2; ultimates so inherently dynamic they can change on their own at a particular rate vs. the Aristotelian-Thomistic notion all change requires initiation by The First Unmoved Mover: 49 ftn. #2.

LOCOMOTION, CURVILINEAR: = kind of motion unique to the u-spaces & produced solely by the forms as each causes its u-spaces envelope to rotate around its center: 47; light speed limit does not apply to it: 65, 65 ftn. #1, 121, 128, 149-150; vs. circular rectilinear locomotion: 127-128.

LOCOMOTION, RECTILINEAR: also see: DIAGONAL: also see: GENERATORS, VELOCITY OF: also see: MOTION: also see: ZIG-ZAG: = kind of motion unique to generators: 47, 55; Esoptrics' 1st law of vs. Newton's = change of *rate* vs. change of *state*: 3-4, 49; figurative graphic illustration of how it occurs: 47-48; God's stop action animation presentation: 47; ingest, digest, excrete a u-space: 47; my change of mind about diagonal paths: 47; occurs as generators pause between durationless instants of change each c. 10^{-47} cm.: V, XIX, 3, 8, 47, 48, 55; speed limit is light speed: 65, 65 ftn. #1, 121, 128; speed limit is light speed after 1st 10^{-18} sec. from time zero: 80, 121, 122, 123, 128.

LOGICAL JUNCTURE: point at which a u-space provided by one of the forms in one multi-combo of one ontological depth is logically contiguous to a u-space provided by a form in a different multi-combo with a different ontological depth & thereby offering to a generator a different range of 4th dimension levels at which to traverse the Universe: 101-102.

LOGICAL SEQUENCE: at same 4th dimension level vs. across adjacent levels: 39; established by God's 7 *a priori* coordinates & ontological quantum (q. vd.): 38-39.

LOYOLA UNIVERSITY OF THE SOUTH: 220.

M.

MACH, ERNST: Esoptrics' answer to his question on how "spinning" has any meaning in an empty Universe: 65; referenced by Einstein re: acceleration problem: 129.
MARUPITER: fictional planet invoked to explain effect of an explosively eliminated multi-h-combo: 109.
MARY: see: BLESSED VIRGIN: also see: VISITATION, FEAST OF.
MASS: Esoptrics predicted 50 years ago neutrinos & anti-neutrinos would eventually be found to have it: 188; generators not forms have any, but it's the forms which enable the generators to have any by serving as the only means by which generators can become 4 dimensionally concentric: 38, 200; inertial vs. gravitational: 41-42; of every multi-combo = the square of its ontological depth: 4, 189, 200; of a multi-A-combo vs. the electron: 183; of a multi-combo vs. a single duo-combo: 183; most massive of multi-combos has an ontological depth of 2^{256} = c. 2.7×10^{17} times the Sun's: 4.
MATH: Esoptrics' kind is far simpler than that of the scientists though it may be daunting in quantity: 87 ftn. #1.
MATTER: in Esoptrics = units of tension generated by the generators & felt by us as sense images: 44.
MATTER, DARK: see: DARK MATTER.
MEMORY: all one ever observes is its testimony to what was a micro-sec. ago & a micro-sec. before that, etc.: 214.
MENDEL, FR. GREGOR: father of Genetics &, according to Atheists, was an idiot because a Theist: 138 ftn. #1.
MERCURY: on Einstein's explanation of the precession of it orbit: 199 ftn. #1.
METHODOLOGY: see: PHILOSOPHY VS. SCIENCE.
MICHELANGELO: if Esoptrics = Leibniz, his art = a child's finger painting as both applied paint to a surface: 221.
MICROSCOPES: role in Science's rise: 30.
MILKY WAY: 101, 102, 118, 119, 121, 140, 149, 164; velocity of: 141.
MINDS: also see: KANT: crippled by too little knowledge of God: 40, 53, 54; my fifth-rate one: XVIII, 56, 78 ftn. #1, 87 ftn. #1, 229; sensation addicted: XXII; sensation dependent: 2, 16; with too little knowledge of God, they have only primitive, pseudo-divine concepts of time & space as proven conclusively by Esoptrics: 214.
MIRROR: also see: REALITY, ULTIMATE: its 4 progressions involving the number 2: 24-25; microscopic world inside the mirror = OD's 1 thru 2^{128}: 67; macroscopic world outside the mirror = OD's $2^{128}+1$ thru 2^{256}: 67; predisposed to multiples of 2: V, 23; still, some insist I have not the slightest reason to use multiples of 2: 25 ftn. #1; why multiples of 2: 1, 23.
MOLECULES: = m-clusters: 91; product of a major generic but irrational form accelerating above MT c. 2^{32} times or less: 70 ftn. #1; slanted & parked axes produce a pyramid whose number of layers = the leading form's # of generic zones above MT: 158-161.
MOMENTS: 2 kinds = durationless moments of time & timeless moments of duration: 133.
MONON: also see: SUB-SEXTON: every form native to OD1 or multi-combo in which every form is such: 167; one of the 3 kinds of sub-sextons: 167; what Esoptrics classes with Science's electrons: 167.
MONSTER: also see: REJECTION: my description of myself for those accusing me of thinking I'm above the competitive spirit: 231-232; why might a God of Love aid one as cruel as I am: 232.
MOON: orbit of: 102-103; rotation of & entanglement with Earth: 102 ftn. #1.
MOTION: also see: AQUINAS: also see LOCOMOTION: as what moves = actuality is what acts: 10-11, 11 ftn. #1;

as shifts up & down in the 4th dimension: 44; Esoptrics' first law of in the primary plane: 58.
MOTIVATION: behind my pursuit of Theoretical Physics is recreation: 229, 229 ftn. #1.
MOUNT EVEREST: = an anthill if Esoptrics = Leibniz: 221.
MOVIES: as illustration of the stroboscopic effect: 227; from the 1936 ***Things To Come***: *"All* the universe or *nothing!"*: 65, 88; motto uttered the year I was born, 1936: 65.
MT = mirror threshold = $OD2^{128}$: 67; being a mirror, OD's run from 2^{-128} MT to the reverse 2^{128} MT: 67; MT generic zone = OD's 2^{128} thru $2^{129}-1$: 90; serves as the basis for Esoptrics' distinction between gravity vs. anti-gravity: 90.
MT COMBOS: = multi-A-combos (q. vd.): 91.
MT FORMS: = those native to $OD2^{128}$: 67.
MUCKRAKER: 49 ftn. #1, 120, 230, 285.
MULTI-A-COMBO: atom's main multi-combo = MT combo & in a class by itself = provider of whole atom's carrying generator: 91; not certain it has a form currently at OD1: 112 175 ftn. #2; rule governing the velocity of its carrying generator: 92-93; unique rule governing the OD of the form from which its carrying generator will try to receive its u-spaces: 92.
MULTI-a-COMBO: 1 of 3 kinds of L-combos (q. vd.) = any multi-combo in the atom's nucleus having a leading form native to some categorical OD other than 2^0 & 2^{128}: 91.
MULTI-b-COMBO: 1 of 3 kinds of L-combos (q. vd.) = any multi-combo in the atom's nucleus having a leading form native to a major generic OD: 91.
MULTI-COMBO: also see: LEADING FORMS: = 2 or more duo-combos logically concentric with one another but each at a discrete OD: V, 4, 32, 39, 43, 44, 49; 6 kinds defined: 90-91; as a ladder, offers other combos 3 options: 68; at vs. not at an orbiting body's center: 71; has gaps & how it fills them: 72 ftn. #1; metaphorically depicted as an onion in which the Φ radius of each layer = the current OD of the form producing it: 67-68; most massive one: 4; produced by a rational major form rising to an OD above the OD at which it was created: V, 3, 64, 67; weighty vs. light ones: 91; which of its generators carries it: 49.
MULTI-COMBOS, VELOCITY OF: rule governing the velocity of every kind's carrying generator: 92-93.
MULTI-e-COMBO: also see: ELECTRON: 1 of 3 kinds of L-combos (q. vd.) = electron = multi-combo in which only forms native to OD1 are logically concentric: 91.
MULTI-g-COMBO: 1 of 2 kinds of W-combos (q. vd.) = a galaxy class multi-combo having a leading form native to $OD2^{128}$: 90; not certain it has a form currently at OD1 though certain it can't be native to OD1: 112, 192 ftn. #1; the 64 types found in the reverse categories: 116.
MULTI-h-COMBO: also see: MARUPITER: 1 of 2 kinds of W-combos (q. vd.) = every multi-combo holding together a heavenly body: 90; may be more than one in a heavenly body: 103, 108; presence of in a reverse category: 118; truncated vs eliminated explosively: 109.
MULTIPLICITY: also see: PRINCIPLE, SUPREME: result of Esoptrics' supreme principle applied to its thus synthetic self: 15-16.
MYSTERY, ULTIMATE: see: ULTIMATE MYSTERY.

N.

NASA: on Earth's orbital velocity: 77; on its web site, Johns Hopkins Univ. affirms debris in Pluto's vicinity: 109.
NEWTON, SIR ISAAC: his 1st law of motion is the foundation of atheistic Science: 8; his 1st law of motion vs. Esoptrics': 3-4, 49; his law only an inference: 4, 49 ftn. #2; more correct

than Einstein on gravity's velocity: 198; not provable by observation that his law holds at 10^{-47} cm.: 4, 49; occurrences of his name: XXI, 3, 4, 8, 41, 42, 43, 49, 49 ftn. #2, 71, 76, 78, 128, 129, 130, 151, 152, 157, 198, 199 ftn. #1, 219, 223, 230.
NOBEL PRIZE: in 1957 to Tsung Dao Lee & Chen Ning Yang for overturn of law of parity just after I did something perhaps similar: 13 ftn. #1.
NOISE: loud one few ever hear: 204.
NUMBERS, LARGE: those who enabled me to use them: 71 ftn. #1, 98 ftn. #2.

O.

OAR: appearing bent in water was used even by the ancients as an example of how risky it is to rely on sense experience: 228.
OBJECT, POLYOUSIOUS: an undivided whole enclosing 2 or more inseparable subjects: 12, 17, 18.
OBJECT, TRIOUSIOUS: also see: SIAMESE; also see: SUBJECT: = an undivided whole enclosing 3 inseparable subjects: 9; necessary structure of every act of consciousness & every ultimate reality: 10, 16.
OBJECT, THE FIRST TRIOUSIOUS: 3 frames of reference in: 19-20; 6 points of reference in: 20.
OBSERVATION VS. INFERENCE: also see: CAUSALITY: also see: IRRATIONAL PRIDE: do we observe what we observe vs. do we rightly infer what we infer: 228; Einstein basically admits his ideas are the latter & not the former: 229; many can't tell what's one or the other or a mix of the two: 228; Newton's 1st law of motion is the latter & not the former: 4, 49 ftn. #2; with nothing able to fool me into thinking I know for sure anything's whole chain of causes, I've never denied Esoptrics is anything more than theoretical inferences: XXVIII, 229.
OBSERVED: nothing below 10^{-15} cm. or 10^{-18} sec. can be such by us: 54.
OD: see: ONTOLOGICAL DISTANCE.
OMEGA: = the one form native to $OD2^{256}$: 111 & others too numerous to list; never made concentric with any other form: 64, 112; not a *multi*-combo: 115; primary u-spaces envelope: 121, 122.
ONION: metaphorical depiction of a multi-combo: 67-68.
ONTOLOGICAL DEPTH: of an epoch of the Universe = # of OD's available in that epoch: 28, 82; vs. of a multi-combo: 112, 191 ftn. #1.
ONTOLOGICAL DISTANCE: = the 4th dimension of space & the 2^{256} levels of intensity possible to each of the 6 ways to be excited by God: V, 4, 27; native vs. current: 29 ftn. #2, 32 ftn. #1; specific, generic & categorical: 29.
ONTOLOGICAL QUANTUM: = inverse of the square of a form's current OD & determines which u-spaces are contiguous in logical sequence in that form's u-spaces envelope: 38-39.
ORBITS: also see: EARTH: also see: FORMS, PULSATION OF: at times is mainly the result of the curvilinear motion of the u-spaces envelope being used by the orbiting body's main carrying generator: 71, 75; St. Thomas Aquinas not so wrong to attribute those of the planets to angels: 76 ftn. #1.
ORIGINAL SIN: Catholic Doctrine on says, in my view, we inherit an *aversion a Deo*: 230.
OSCAR: better to debate Theoretical Physics than who merits one: 229.

P.

PARANOIA: 227 ftn. #1; Beethoven as example of: 228 ftn. #1.
PARASITIC VS. NON-PARASITIC: see: SUBJECT.
PATTERNS VS. ANTI-PATTERNS OF DIVINE ROTATION: effect of how heavenly benchmarks (q. vd.) affect the heavenly sextets (q. vd.) which, in turn, affect the forms: 56-75; number of pauses per cycle = number of shifts

per cycle: 60; number of phases per cycle = 2: 60, 61; rotations in the tertiary plane indicate a form's current OD: 57-58; rotations in the primary & secondary planes indicate a form's native OD: 57-59; rotations in the primary & secondary planes of the categorical forms native to OD's 1, 2 & 4 are indistinct & thereby set those 3 apart from the other 6 & lay the basis for Esoptrics' version of Science's leptons vs. quarks: 62-64; shifts are purely logical: 57; shifts in the primary plane all = 1 per 1K thus being always & everywhere the fastest rate of change possible: 58-59.

PAUL, ST.: may have known more about Catholic Doctrine than I do: 233.

PERCEPTION: also see: EXPERIENCE: whatever is apprehended whether by senses or mind says Webster's: XXV ftn. #1.

PERIHELION VS. APHELION: defined 107.

PHENOMENA: world of uniquely determines Theoretical Physics says Einstein: 40.

PHILOSOPHY: also see: INTROSPECTION: attacks on usefulness of in cosmology: 4-7; methodology is introspection: V, XIX, 9, 54, 220; only path to most basic first principles: XXII.

PHILOSOPHY VS. SCIENCE: in pursuit of Cosmology = individual, inspired by Catholic Doctrine & asking God's help, probing his own innermost depths vs. millions of scientists probing the extra-mental world with myriads of multi-billion dollar machines: V, XIX, 9, 30, 54, 220.

PHOTONS: absorption & radiation of: 44, 105, 188; in Esoptrics: 44-45, 170, 201.

PHYSICAL REALITY'S ABILITY TO FOOL US: also see: GENIUS: also see: MONSTER: only element of consciousness enabling it to fool us about its causality is a pride too irrational to admit we want to remain ignorant of its causality: 230-231; never is that element the nature of consciousness itself as evidenced by what I saw as a child of 6: 229; that element of consciousness is the subject matter of Depth Psychology & Moral Theology & not Esoptrics: 230; that element never fooled some into imagining our knowledge of the causality was anything other than speculation: 228-229; the elements making it difficult to discover the causality are what Esoptrics addresses: 229.

PIGGYBACK FORM: see: FORM.

PLANCK LENGTH: 10^{-33} cm. & 10^{18} smaller than what Science can observe & 10^{14} times Esoptrics 10^{-47} cm.: 54; shrinking past it, space may so transform as to make division meaningless says Prof. Greene: 54.

PLANCK TIME: 10^{-43} sec.: 54.

PLANES: primary, secondary & tertiary: 56-65; tertiary plane is the plane of the world we experience: 64; with high speed rotations in the primary & secondary planes, Esoptrics explains electrical charges, magnetic lines of force at right angles to the tertiary plane, & the chaos Quantum Theory detects in the sub-atomic world: 64.

PLANE, PRIMARY: first law of motion in: 58-59.

PLUTO: after 75 years, is no longer classed a planet, but: 108; how further from the Sun than a much larger planet: 108; NASA web site confirms debris in its vicinity: 109.

PODOLSKY, BORIS: one of Einstein's colleagues in the issue of the EPR paradox: 145 ftn. #2.

POINT: Bill Gaede on: 41; defined as infinitesimal = zero dimensions: 41; zero dimensions for some vs. zero physical but not zero logical dimensions for Esoptrics: 41-42.

POINT, MATERIAL: forms & generators are not such: 44; I'm not sure what it meant to Einstein: 42; maybe meant irreducible ultimate constituent: 42; problem for classical mechanics says Einstein: 41.

POLCHINSKI, JOE: Prof. Greene says Joe

& co-workers "showed" the endpoints of strings would be trapped "within certain regions" but, unlike Esoptrics with its u-spaces envelopes, does not explain what those certain regions are: 196 ftn. #2.

POLYOUSIOUS OBJECT: see: OBJECT POLYOUSIOUS.

POSITRON: see ELECTRON.

POTENCIES: also see: IN POTENCY TO: = 6 kinds of component states of excitation caused in the forms by infinity & in the generators by the forms: 20, 21, 33, 38; named: 27.

POTENTIALITY: is a subject & not merely an abstraction: 18; not equally as real as is a non-inhering subject: 18; ratio of to actuality in the infinite: 18-19; vs. actuality: 17.

PRE-INFLATION ERA: only Esoptrics gives a description of it: V, 120; preceded Big Bang: V, 120; Prof. Greene says Science has no idea what it was like: V, 120.

PRESCIENT LEAP: St. Thomas' 10th crystalline sphere: 226.

PRIDE, IRRATIONAL: see: IRRATIONAL PRIDE.

PRIMARY ACT: also see STATE OF EXCITATION, OVERALLL: = a form's or generator's current overall state of excitation: 13, 27, 29, 29 ftn. #2, 31-38, 36 ftn. #1, 43, 47, 53, 55 ftn. #1, 213-215.

PRIME ABSTRACTABLE: also see: ENCOUNTERER, FINITE: = continuousness seeming to underlie motion thru time & space: 9; Encounterer's own continuousness projected outward: 9; not an illusion: 11; parasitic subject produced by an encounterer mirroring himself with less than infinite efficiency: 11-15; third kind of reality not predicable of the other 2 cotenants of a triousius whole: 10.

PRINCIPLE, SUPREME: Esoptrics' is *a priori* & synthetic, defines the logical essence of God, & is the ultimate ground of whatever is possible: 16.

PRISON: OD1 as: 143 ftn. #1, 216.

PROXIMITY: in Esoptrics, as logical in nature among u-spaces, is established by Algebra's space-free calculations & not by Geometry's spatial pictures: 39, 40; within the same level of the 4th dimension vs. across 2 adjacent levels: 39.

PURGATORY: at the center of Earth vs. earth: 216; one sec. here may = many trillions of trillions of eons there: 216; place of refuge for those reacting to God's truth with a hate free shame: 219.

PYRAMID: see: MOLECULES.

Q.

QT: = Quantum Theory.

QUANTUM LEAPS: also see: LEAPS: erroneous terminology only figuratively applied to the rectilinear motion of the generators: 8, 47, 49, 49 ftn.2, 80.

QUANTUM, ONTOLOGICAL: see: ONTOLOGICAL QUANTUM.

QUANTUM MEASUREMENT PROBLEM: see: STAGE 2 WAVE FUNCTION COLLAPSE.

QUANTUM THEORY: insists there's only probabilities in the sub-atomic world: 153-154; David Bohm refutes QT's claim says Polkinghorne: 154-155; Esoptrics insists QT's claim reflects its proponents ignorance of important details: 154; QT's claim reflects its proponents ignorance of the difference between the rectilinear motion of the generators vs. the curvilinear motion of the u-spaces: 156; QT sees in the sub-atomic world the same dizzying gyrations Esoptrics sees but without Esoptrics' knowledge of what space really is: 157; the chaos it detects in the sub-atomic world is the result of the high speed rotations of the forms, as u-spaces envelopes, in the primary & secondary planes: 64-65, 128, 152-155; the chaos it detects raises the question of how such microscopic chaos produces macroscopic order: 157-158; why obstinate

clinging to QT's dogmatism = whatever thinking excludes God: 155-156.
QUANTUM THEORY'S PROBLEM: = re: how the cloudy & fitful tiny can give rise to the clear & reliable large, Esoptrics answers the stroboscopic effect of the molecular forms: 158-161, 215-216; only Esoptrics, in its MT, gives the exact border separating the lower sizes which obey QT's rules from the upper sizes which obey Newton's & Einstein's rules: 162, 215-216.
QUARKS: as $OD2^4$, 2^8, 2^{16}, 2^{32}, 2^{64} & 2^{128} forms: 63; as sextons: 171-173, 177-178, 180, 183; cluster of duo-combos: 170; Haim Harari on 171-173.
QUOTATION: the *parole* of literary men everywhere says Samuel Johnson: 26.

R.

RADIATION, ELECTRO-MAGNETIC: one or more duo-combos expelled as a multi-combo's leading form decelerates from its current OD: 32.
RATIONALIZATION: see: ATHEISTS.
RC: = reverse category: 111; list of the 8: 111; result of MT forms accelerating upward in the 4th dimension: 111-118.
REALITY, A BASIC: = what can't be broken up in fact either into smaller pieces or into more basic realities: 10; independent only in the sense it, unlike accidents & parasitic subjects, need not be intimately bonded to another to avoid annihilation: 10; structure of each is triousious: 10; ultimate constituent vs. occupant: 31; ultimately either a form, a generator, or units of tension generated by a generator: 44.
REALITY, THE ULTIMATE: 3-way structure of a mirroring activity mirroring itself: V, 1, 11-12; finite vs. God's infinite version: V, 11-14;
REJECTION: also see: ESOPTRICS, REJECTION OF: why my relationship with Catholic Doctrine insures no amount of it can ever dispirit me to any extent: 232-233.
RELATIVITY: makes Physics Geometry: XXIV; its light speed limit applies to the rectilinear velocity of generators & not the curvilinear velocity of u-spaces: 65, 65 ftn. #1.
RISKY BUSINESS: going by what sense images imply: XIX, 166, 174, 246.
ROSEN, NATHAN: one of Einstein's 2 colleagues in the issue of the EPR paradox: 145 ftn. #2.
RUBE GOLDBERG CONTRIVANCE: my opinion of entropy: 144.

S.

SCHRÖDINGER, ERWIN: his 1926 formula collapses in stage 2: 161.
SCIENCE: Aristotle on: 30; can't observe what's smaller than c. 10^{-24} cm.: 54, 94, 174 ftn. #1; methodology is "extrospective" observation: XXII; methodology is extroverted examination of what's deemed extra-mental: 9, 30, 54, 220; surge of = result of which: change in attitude vs. advance in instruments of observation & means of communication: 30.
SCIENCE DAILY.COM: difference twixt Earth's inner & outer rotations: 97.
SCIENCE VS. PHILOSOPHY: see: PHILOSOPHY VS. SCIENCE.
SELF: also see FREE WILL: also see: SOUL, HUMAN: current vs. total: 217; complete vs. partial: 139; only The Infinitely Informed knows the complete: 140, 217-218; total self as karma: 218.
SENSATION: also see: EXPERIENCE, SENSE: also see: RISKY BUSINESS: minds addicted to it: XXII; minds dependent on it: 2.
SEPARATION: also see: CHICKEN VS. EGG: also see: DIMENSIONS: also see: LOCALITY: ultimates separate from one another solely because each has a unique God-given internality vs. a unique locality: XXVI, 36, 40, 204.
SEXTETS, HEAVENLY: see: HEAVENLY SEXTETS.
SEXTONS: also see: LEADING FORMS: every form native to one of the 6 OD's

of 2^4, 2^8, 2^{16}, 2^{32}, 2^{64}, 2^{128} & any particle in which the leading form is such: 167; I class these with Science's quarks: 167; rotating vs. counter-rotating Aleph, Beth, & Daleth types & 6 modes of each: 169-170; the 6 each become 36 types: 168-171.
SIAMESE: triplets: 9, 10; twins: 9.
SIMULTANEITY: though Science cannot determine it, Esoptrics can: 132-133.
SOUL, HUMAN: also see: SELF: at OD1: 143; not a piggyback form: 130 ftn. #1; reflection capable, it can define its targets in the terms of the acts of will it used to focus on them: 130 ftn. #1; whole of human identity lies in the physical says Prof. Greene but not Esoptrics: 215.
SPACE: also see: CONTINUUM: also see: KANT: also see: LOCALITY: also see: U-SPACES: as commonly conceived is a *deus ex machina*: 40; as commonly conceived is a mental crutch for minds crippled by too little knowledge of God: 40, 53, 54; as commonly conceived is a primitive, pseudo-divine concept proven such by Esoptrics: 214; as commonly conceived was rejected by me at the age of 6: 49 ftn. #2, 229; John Locke can't say what it or extension are: 46; most insist it's inherently extended: XX; only Esoptrics explains how it can have parts & what extension is: 46; replaced in Esoptrics by $2x10^{231}$ u-spaces envelopes: V, 35; Science admittedly doesn't know what space *really* is: 65; smallest observable: 54, 174; swirls says Prof. Greene: 65 ftn. #1; very much a reality in a way unknown to Science: 35.
SPACE, EMPTY: where Omega's u-spaces are the only reality present: 205 ftn. #2.
"SPACE ITSELF", EXPANSION OF: also see: INFLATION: Greene's terminology: 126; H&M's terminology: 121; no such thing in Esoptrics & replaced by multi-g-combos as space envelopes expanding away from one another due to the categorical preservation law's anti-gravity: 124-126.
SPECIFIC OD'S: 29.
SPEED OF LIGHT: see: LIGHT SPEED.
SPHERE: also see: TENTH CRYSTALLINE: u-spaces collectively make their envelopes effectively such: 42, 52, 68, 205, 208, 209, 211, 226.
SPINNING: see: MACH, ERNST.
SPINOZA: also see: REALITY, BASIC: also see: INTIMATE BONDING: misunderstood Aristotle's take on extent of every substance's degree of independence: 10.
STAGE 2 WAVE FUNCTION COLLAPSE: also see: BOHR: result of the stroboscopic effect of the experimenter's supra-MT molecular form: 161-162.
STATE OF EXCITATION, OVERALL: = primary act: 27, 29, 29 ftn. #2, 31, 32, 43; in every form & generator, is composed of the 6 component states of excitation: 21, 27; in every form's overall state, each of the 6 component states is activated equally: 27, 31; in every generator's overall state, no component state & its opposite are simultaneously activated: 32. Metaphorical illustration of a form's balanced vs. a generator's unbalanced primary act: 32-34; number of discrete primary acts possible to a form vs. a generator: 33-34.
STATES OF EXCITATION, COMPONENT: also see: U-SPACES: also see: PROXIMITY: for forms, 6 macrostates from c. 10^{-47} cm. to c. 10^{31} cm. in dia. for sensation dependent minds: 2; for generators, 6 microstates all c. 10^{-47} cm. in dia. for sensation dependent minds: 2; none possible without God's input: 8; not rightly called states of rest: 222; six basic kinds found in every overall state named: 27; the triads of couplets describing each: 36-40, 47-48.
STROBOSCOPIC EFFECT: compared to motion picture & TV screens: 227; Esoptrics' answer to the quantum measurement problem (q. vd.): 161-162; Esoptrics' explanation of the dif-

ference between microscopic chaos & macroscopic order: 158; its uniformity in effect is what produces the apparent uniformity of the environment: 144; molecular forms &: 158-162; of consciousness so slows the microscopic world's chaos as to allow us to live in & enjoy a world of things whose effect upon the microscopic world is the same or close to the same as ours: 215, 227-228.
STROBE LIGHTS: how they work: 160-161, 215.
STRAIGHT LINE: also see: LENGTH: in a flat plane = any series of shifts in which each shift involves changes solely in the same one of the 3 couplets used to define a u-space: 53.
STRING THEORY VS. ESOPTRICS: each string is a single "tiny filament of energy" vs. generators, as a tiny "filament of energy", may be as many as 2^{256} within the confines of a single u-space but logically outside of one another because each is at a discrete level of the 4th dimension: 94; strings have physical size & shape individually vs. generators have only logical size & shape collectively: 94; strings produce particles depending upon how they vibrate vs. generators produce particles, heavenly bodies & galaxies depending upon how many are logically concentric in the same u-space, the native OD of each one's leading form, & how that leading form rotates & pulsates: 94.
SUBJECT: = what's not merely a characteristic or part of another: 9 ftn. #1;
SUBJECT, PARASITIC: = a substance annihilated if not intimately bonded to a non-parasitic substance: 12 ftn. #1.
SUBJECT, NON-PARASITIC: = a substance not needing to be intimately bonded to another to avoid annihilation: 12 ftn. #1.
SUB-SEXTONS: classification including the monons & the 2 kinds of dions: 167; Esoptrics equates them with Science's leptons: 167; like leptons, all have a charge of 1: 167, 170-171; no distinct rotations in the primary & secondary planes: 167; vs. sextons: 167, 170-171.
SUBSTANCE: = what's not merely a characteristic or part of another but needs to be intimately bonded to another if parasitic: 9 ftn. #1, 12 ftn. #1.
SUMMA THEOLOGICA: XXV, 25, 234.
SUN: XIX, XXI, 4, 72-78, 72 ftn. #1, 95, 100-104, 107, 108, 108 ftn. #1. 109, 119, 140, 141, 149, 152, 164, 198, 199, 199 ftn. #1, 230 ftn. #1; 231; distance from Milky Way's center: 119; mass of: 4, 71.
SURPRISE: 137.
SUSKIND, LEONARD: also see REJECTION: rejected, got drunk: 232.
SWINBURNE CENTER: dwarf ones are the Universe's most abundant type of galaxy & Esoptrics agrees: 119.

T.

TELESCOPES: role in Science's rise: 30.
TELEVISION: as illustration of the stroboscopic effect: 227.
TENTH CRYSTALLINE SPHERE: St. Thomas' concept of compared to Esoptrics' concept of Omega: 226.
THEISTS: calling them all idiots, Atheists (q. vd.) bespeak just how superior is their own vaunted level of intellectual excellence: 138 ftn. #1.
THEORETICAL PHYSICS: also see: RISKY BUSINESS: mostly inferences implied by observed facts &, as such, that's what Esoptrics is: XXVI; my reason for pursuing is the kind of fun which helps me think about Catholic Doctrine: 229, 229 ftn. #1.
THOMAS AQUINAS: see: AQUINAS, ST. THOMAS.
THREE-DIMENSIONALITY: math as example of logical kind: XXI ftn. #1; not an illusion: XX, 42; collective & real effect of u-spaces outside of one another solely in logical sequence as a result of each having a unique internality given it by 6 of God's 7 *a priori*

coordinates: V, XX, 2, 3, 39, 42, 44, 51, 203.
TIME: also see: MOMENTS: also see: I-TIME, also see U-TIME: = the measure of change without which there is no time: 133; doesn't apply to any state of excitation as long as the state is changeless: 133; durationless moments of vs. timeless moments of duration: 133; every instant of is equally present to God: 136-137; how it affects forms is more important to us than how it affects generators: 130; slows for an accelerating form but speeds up for the same's generator: 130-131; smallest detected: 54; though Science can't say what time is or give a universally binding measure, Esoptrics can: 132, 135; when God created the Universe: 137 ftn. #1; with too little knowledge of God, most have only primitive, pseudo-divine concepts of time & space as proven conclusively by Esoptrics: 214.
TIME, ABSOLUTE: is forever 2^{385} K per cycle of the Universe's 9th epoch & 1K per change at OD1 as the result of the rate at which forms at OD1 always change from one macrostate of excitation to another: 135.
TIME, ARROW OF: one of modern Physics' deepest mysteries says Prof. Greene: 139; Physics can't explain it, but Esoptrics can: 139-140; term taken from Sir Arthur Eddington says Prof. Greene: 139; the directionality is determined by God's infinitely informed love of free will as defined by God's infinitely informed intellect: 139-140.
TIME, EVERY INSTANT OF: equally present to God vs. atheistic scientists: 136-137, 138; Theists thought of it long before the Atheists did: 138; Theists' vs. Atheists' view of human capacity to observe such: 138-139.
TIME, RELATIVE: to an observer is defined as the inverse of the square of his form's current OD divided by the number of acts of consciousness in what he calls a second & detailed calculation of what that is for us: 134-135; result of the rate at which forms at any OD above OD1 change from one macrostate of excitation to another: 135.
TIME, UNIFORMITY OF: what Prof. Greene sees as inferring such is actually the result of the uniform effect of the stroboscopic function of the molecular forms: 143.
TOLMAN, RICHARD: advocated a spherical Universe in the 1930's: 203.
TOP BANANA: 232.
TOWER OF BABEL: tower of pseudo-scientific babble: 155.
TRANSCREATION: Leibniz's term: 222; no place in Esoptrics: 223.
TREE: sound of one falling: 2 ftn. #4.
TRIADS OF COUPLETS: 36-40, 47-48.
TRIOUSIOUS OBJECT: see: OBJECT, TRIOUSIOUS.

U.

ULTIMATE MYSTERY: also see: FREE WILL: also see: TIME, ARROW OF: according to Prof. Greene = why there is a Universe at all: 217; no mystery at all for Esoptrics which answers: God's love of free will as defined by God's Infinitely Informed Intellect: 217, 219.
ULTIMATE REALITY: see: REALITY, ULTIMATE: also see: UNIVERSE.
ULTRAMICROSCOPIC MAKEUP: = UM for short= billion billion times too small for science: V, 54, 94, 174.
UM: = ULTRAMICROSCOPIC MAKEUP OF TIME & SPACE: in Esoptrics, Philosophy & Theology have done a far better job searching for it than Science has: 174, 220, 223.
UNIFICATION: see: GRAND UNIFICATION.
UNIVERSE: also see: ACCORDIAN UNIVERSE: also see: CAUSE: also see: INFLATION: also see: IRRATIONAL PRIDE: also see: MOVIES: also see: ULTIMATE MYSTERY: alpha areas of: 3; at Omega's level, has no place in space or ability to move thru space

until the 10th epoch: 196 ftn. #1; duration = 2^{385}K: 1, 28; gamma areas of: 3; history of involves timeless moments of duration intertwined with durationless moments of time: 133; its ultimate realities are either forms or generators: 2; nine epochs of & OD limits of each: 28; ontological depths (q. vd.) of its 9 epochs: 28; time spent going from time zero to start of 9th epoch = c. 10^{-38} sec.: 28; traits of its 10th epoch: 28-29; ultimate *constituent* = form or generator but ultimate *occupant* = duo-combo: 31, 43; whereabouts of the 9th epoch in the 10th: 29 ftn. #1; why say 9th is current epoch: 29.

UNIVERSE, CAUSE OF: neither quantum laws nor Newton nor Maxwell nor Einstein: 219; principal cause of why it is at all is God's love of free will &, by God's choice, the proximate cause of each & every detail of the Universe is the occupants' free will as defined by God: 219.

UNIVERSE'S ORIGIN: beyond Science's cope according to Hawking but not beyond Esoptrics': 87.

UNIVERSE, SHAPE OF: also see: TOLMAN: also see WMAP: is the collective effect of the logical way the u-spaces & envelopes are outside of one another: 203; no final answer says Greene, but leading entrant says flat, infinitely large spatial one & observation rules out spherical: 203; spherical from the standpoint of its u-spaces envelopes but relatively flat from the standpoint of its utilized u-spaces: 208-209; utilized u-spaces may give the inhabited Universe the shape of an 8-pointed star: 209.

UNIVERSE, ZONES OF: categorical vs. generic: 89; OD's included in each: 89.

UNKNOWABLE: Kirkus misuse of: XXIX.

"U-SPACE": short for "microstate of excitation: 7, 8.

U-SPACES: also see: STATES OF EXCITATION: also see: PROXIMITY: = actuality of a form: V, 33, 35-36; at 2^{-129} x atom's dia. = c. 10^{-47} cm.: V, 2, 35; collectively add up to 3-dimensional shapes in real effect: V, XX, 2, 41-42, 79 ftn. #2; individually are unextended physically but extended logically: V, XX, 51; no diagonal measurement: 50-53, 190; no fractional ones: 2 ftn. #2; not really units of space but rather microstates of excitation: 8, 42; ones shown identical by Geometry's pictures are separate for Algebra's calculations: 40; *potential* micro-states of excitation until activated by a generator: 35-36; Relativity's light speed limit does not apply to them: 65, 65 ftn. #1, 121, 128, 149-150; senses lie not about what they are collectively: 42; which are logically contiguous to which is determined by the logical sequence dictated by a triad of couplets in conjunction with the ontological quantum (q. vd.): 36-40, 38-39, 47-48.

U-SPACES ENVELOPES: # of u-spaces per envelope = cube of 2 x providing form's current OD: V, 35, 67; effectively & really 3-dimensional: 42, 52, 52 ftn. #1, 205, 208, 209, 211, 226; fields of locomotion: 113, 114, 115, 191; inflation is the result of them fleeing each other rather than space itself stretching: 121, 124, 125, 126; Science knows nothing of them: 157; what depicting them as squares, cubes, circles & spheres means for Esoptrics vs. Geometry: 51-53.

U-TIME: also see: I-TIME: also see: TIME, ABSOLUTE: detailed calculation of for those at our current OD: 134-135; example of how u-time's impact on a form or generator is relative to the latter's rate of change: 2 ftn. #1; smallest units of at c. 10^{-96} sec.: V, 1; no fractional units of: 1; no impact on forms or generators as long as their states of excitation are changeless: 1-2, 133; ratio of to i-time: 130; ratio of to i-time in Alphon & Bogon: 143, 143 ftn. #1; result of the rate at which forms at OD1 always change from one macrostate of excitation to another: 135.

V.

VELOCITY, CURVILINEAR: applies only to u-spaces: 47; light speed limit doesn't apply: 65, 65 ftn. #1.
VELOCITY, RECTILINEAR: Earth's around the Sun is mostly the velocity of the u-spaces its carrying generator is using: 141; pertains only to generators rate of changing from u-space to u-space: 65, 65 ftn. #1; rule governing that of every multi-combo's generator not at light speed = velocity is relative to # of steps into the current generic zone: 92-93; velocities of Earth, Sun, the Milky Way, etc., have nothing to do with the velocity of a generator's rectilinear motion despite what H&M say: 141.
VELOCITY THRU 4TH DIMENSION: movement up & down thru it may be at as much as 2^{128} times light speed: 99-100.
VISITATION, FEAST OF: 176.

W.

WALKING SEPTIC TANKS: 233.
WAVES VS. PARTICLES: in Esoptrics, no difficulty in accounting for the difference: 45.
W-COMBOS: also see: L-COMBOS: = weighty multi-combos = every multi-h- (q. vd.) or multi-g-combo (q. vd.) = multi-combos far more massive than those in the atom: 91; rule governing the OD of the form from which its carrying generator shall seek to receive its u-spaces: 91-92; rule governing the velocity of their carrying generators: 92-93..
WIKIPEDIA: calls Bill Gaede industrial spy: 41; Esoptrics denies its statement that muons & taus decay into electrons: 172; human brain's size: 211; human eye can't detect flashes greater than 20 per sec.: 211-212; on *a priori* in St. Thomas Aquinas: XXV-XXVI; on electromagnetic radiation: 202; on electron shells & sub-shells: 183-184; on the comet Hyakutake: 72 ftn. #2; says distinction between *a priori* & *a posteriori* is not cut & dry: XXVI; smallest time interval measured: 54; some theories include a fifth force: 201 ftn. #1; strong nuclear force holds only inside the atomic nucleus: 201; Sun 24,900 ± 1,000 light years from Milky Way's center: 119; verboten says Anna Call's review: XXVIII; weak nuclear force causes beta decay: 201-202.
WILL, FREE: see: FREE WILL.
WMAP: also see: COBE: 203 FTN. #2..
WOLK-STANLEY, JESSICA: 41.

Z.

ZIG-ZAG LOCOMOTION: 48.
ZONES: categorical vs. generic: 89; OD's included in each: 89; planetary orbits as evidence of generic zones: 107-110.
ZOOM-IN: a view those in infinity with God can enjoy with regard to every point in time but which Prof. Greene, like all without God, declares "a fictitious vantage point": 139.

"Of Dr. Goldsmith he said, 'No man was more foolish when he had not a pen in his hand or more wise when he had."
—**JAMES BOSWELL** (1740-1795)**:** ***Life Of Samuel Johnson***, as found on page 451 middle left-hand column of Vol. 44 of ***Great Books Of The Western World*** as published by Encyclopædia Britannica, Inc.; Chicago, 1952 (There's a good reason to be solely a writer and never a talker.).

JOHNSON: Goldsmith had no settled notions upon any subject; so he talked always at random. It seemed to be his intention to blurt out whatever was in his mind, and see what would become of it. He was angry too, when catched in an absurdity; but it did not prevent him from falling into another the next minute . . . [Skip!] . . . Goldsmith, however, was a man, who, whatever he wrote, did it better than any other man could do.
—**JAMES BOSWELL** (1740-1795)**:** ***Life Of Samuel Johnson***, as found on page 380 lower left & upper right columns of Vol. 44 of ***Great Books Of The Western World*** as published by Encyclopædia Britannica, Inc.; Chicago, 1952 (That sounds a lot like me, if I try to converse with several others at once. I simply do not have the kind of intelligence which allows one, as they say, to think on one's feet.).

In conversation you never get a system.
—**JAMES BOSWELL** (1740-1795)**:** ***Life Of Samuel Johnson***, *April 16, 1775*, as found on page 257 top left column of Vol. 44 of ***Great Books Of The Western World*** as published by Encyclopædia Britannica, Inc.; Chicago, 1952 (There's another good reason to be solely a writer and never a talker).

In the *Olla Podrida*, a collection of Essays published at Oxford, there is an admirable paper upon the character of Johnson, written by the Reverend Dr. Horne, the last excellent Bishop of Norwich. The following passage is eminently happy: "To reject wisdom, because the person of him who communicates it is uncouth, and his manners are inelegant;—what is it, but to throw away a pineapple, and assign for a reason the roughness of its coat?"
—**JAMES BOSWELL** (1740-1795)**:** ***Life Of Samuel Johnson***, as found in footnote 2 at the bottom of the right-hand column of page 585 in Vol. 44 of ***Great Books Of The Western World*** as published by Encyclopædia Britannica, Inc.; Chicago, 1952 (Among those with alphabet soups behind their names, it is rare indeed to find one who will agree with that principle—at least not when it comes to me. Yes, of course, they, in their defense, will snicker: "Nothing from him is wisdom.").

INDEX OF QUOTES

ADLER, MORTIMER J.: Phil. builds no bridges: 5.

ALZOG, REV. DR. JOHN: according to Hugh of St. Victor, contemplation is the path to perfect Science: 22.

AQUILA, RALPH: review: XXXIII.

AQUINAS, ST. THOMAS: all things reflected in God the mirror: 25; every moment of time is equally present to God: 136; right to deem oneself the most despicable of men: 234.

ARISTOTLE: initial error later multiplies 1,000 times: 131 ftn. #1; scientific observation vs. abstruse argumentation: 30.

ATIYAH, SIR MICHAEL: as teaching aids, Geometry's pictures are important even to algebraists: 22.

AUGUSTINE OF HIPPO, ST.: a secret reason to praise only God: 234.

BACON, FRANCIS: adoration of the mind leads some to withdraw too much from the observations of experience: 106; forms are a fiction of the mind: 254; in Scriptures, God's favor went to Abel as image of the contemplative life: 106.

BERKELEY, GEORGE: if mind's ideas moved quicker, external objects would appear slower: 210.

BITTLE, FR. CELESTINE N.: God's essence = ultimate ground of intrinsic possibility: 25; on modern Thomism's *praemotio physica*: 66.

BOILEAU, DR. DAVID: Esoptrics review: II.

BOSWELL, JAMES: quotes Samuel Johnson's praise of quotation: 30; quotes Bishop Horne's remark regarding Johnson's manner of speech: 282; quotes Samuel Johnson's remark regarding conversation: 282; quotes Samuel Johnson's remarks regarding Oliver Goldsmith: 282.

CALL, ANNA: ***Clarion Review*** parts: II, XXVIII, 26, 66, 88, 110, 226.

DONCEEL, J. F.: the more knowledge comes from within, the more perfect it is: 7.

DULLES, AVERY ET AL: metaphysics as quest for real explanations = a fool's errand: 5.

EINSTEIN, ALBERT: because it eliminates the continuum, an algebraic description of nature seems for now an attempt to breathe in empty space: XXIII; need to repeat frequently: XVIII; Newton's mighty work can't be superseded: 8; problem of the material point for classical mechanics: 41; recourse to speaking dogmatically is only for the sake of clarity & simplicity: XXVIII, 229; that pure speculation *hasn't* doesn't prove it *can't* gain knowledge of reality: 6; to what acceleration is relative is a problem for Newton's concept of space: 129; was an illusion to believe mere reflection could find everything knowable: 5-6; what happens when Newton's light as material points is absorbed: 43-44, 105; world of phenomena determines theoretical physics: 40.

EINSTEIN, ALBERT & INFELD, LEOPOLD: ideas come before formulas: 26.

FRANKLIN, BENJAMIN: author's greatest pleasure = being quoted by learned authors: 30.

FREEDMAN, DAVID H.: what a joke on us if one of the biggest clues was in the very fabric of thought itself: 7.

GILSON, ETIENNE: Philosophy merely expresses an attitude: 5.

GREENE, BRIAN: admits Science still struggling to understand space & time: 131; all human identity resides in the physical: 215; Big Bang theory says nothing about time zero itself or what banged or why or if: 88; dark matter & energy totally different from our ordinary kind & much more abundant: 165; Einstein's static universe error overturned by Hubble in 1929: 166; every instant of time is always there: 135-136; fabric of space ex-

panding is an idea almost too beautiful to be wrong: 126; fictitious zoom in view of all time we will never have: 138; for David Bohm, not particles *or* waves, but particles & waves: 156; future event's effect on past choice: 218; light speed barrier applies only to material objects: 128; on extremely small scales, time & space may morph into more fundamental concepts making questions of further division as meaningless as asking if #9 is happy: 54; on simultaneity: 132; mystery of time's arrow: 139; on action at a distance: 146; Niels Bohr's (q. vid.) explanation of the quantum measurement problem: 162; on entropy: 144; scientists spend much time in confusion: 210; stage 2 formulation of quantum theory remains mysterious: 161-162; String Theory replaces the notion of indivisible dots with tiny filaments shaped like a string: 93-94; String Theory's strings are "a billion billion times" too small to be seen, leading some to insist that's the realm of Philosophy & Theology & not Physics: 174; time = the measure of change without which time as normally conceived wouldn't exist: 133, 213; ultimate mystery = Leibniz's question of why there is a universe at all: 217; uniformity of our environment infers time's uniformity: 144; what really inhibits shrinking smaller than the Planck scale: 54; what the pursuit of deep physical laws requires: 220; where is the border at which the everyday big world confronts the minuscule atomic world: 215.

HAWKING, STEPHEN: admits he can't say what time really is: 131; all known particles are fermions or bosons: 43; ancients thought pure thought could work out all the Universe' laws but, today, Philosophy's sole task is the analysis of language: 6; his thinking called nonsense: 78; no measure of time all will agree on: 132; scientific theories are first put forward for aesthetic or metaphysical reasons: 26; ultramicroscopic limit: 150; Universe's origin beyond Science's scope: 87.

HAWKING, STEPHEN & MLODINOW, LEONARD: 60 MPH on Earth is 18 MPS to observer in the Sun: 140; bouncing ball on jet shows disagreement on distance between 2 events: 142; description of the double slit experiments: 151-152; every version of the Universe is at once: 137; General Relativity makes physics Geometry: XXIV; gravity's inverse square law would hold one way in 3 dimensions & another in 4: 189; inflation = Universe expanded by a factor of 10^{30} in 10^{-35} sec.: 121; light from tail to nose of jet shows disagreement on distance light has travelled: 142; light speed limit doesn't apply to "space itself": 121; no complete quantum theory of gravity: 121; no need for God in explaining how universe arose: 155-156; on Feynman's "simultaneously takes every possible path": 153; physicists not sure exactly how inflation happened: 121; quantum physics claims no amount of information allows us to predict with certainty: 153-154; quantum physics claim vs. ignorance of some pertinent details: 154; quantum physics microscopic chaos begets Newton's macroscopic order because the former's effect is inversely proportional to the latter's size: 157-158; supersymmetry implies force & matter are 2 facets of the same thing: 42.

HEATH, H. L.: since earliest times, it's been widely known relying on what sense imagery infers is risky business: 166.

HOLMAN, TOM: his e-mail to me: II.

HUME, DAVID: the absurdity of infinite divisibility XXVII.

ISAIAH: sheer terror: 216.

JEFFERSON, THOMAS: liberties secure only if people convinced they are God's gift: 233.

JOB, BOOK OF: 22:19; referenced only: 220.

JOHNSON, SAMUEL: See: BOSWELL.
KAPLAN, ABRAHAM: saying it could thru armchair speculations answer questions about things, Philosophy incurred a bad name: 5.
KIRKUS INDIE: ***Kirkus Review***: parts of: II, VII, XXIX, XXX.
LEIBNIZ, GOTTFRIED WILHELM: as quoted by Paul Riesterer regarding transcreation. 222.
LEMONICK, MICHAEL: a gutter level slur for MOND: 110.
LOCKE, JOHN: absurdity of infinite divisibility: VIII; can't say what space & extension are: 46; motion can neither be nor be conceived without space: 54; senses are so distracting, they greatly inhibit the acquiring of ideas regarding the mind's operations: 225; what moves more quickly than our minds' ideas do is not observed to move: 210.
LODGE, OLIVER JOSEPH: on the ether: 246.
MARITAIN, JACQUES: metaphysics useless to Science of nature: 4.
MARRAS, CHRISTINA: on Leibniz's use of the mirror metaphor: 224.
MOSES, DR. GREGORY J.: Esoptrics review: II.
MASSEY, RAYMOND: "All the universe or nothing.": 88.
OPPENHEIMER, J. ROBERT: we may have to accept an arbitrary set of facts: 234.
OWENS, JOSEPH: St. Thomas' avoidance of new terms: XVIII.
PAUL, ST.: ***I Timothy*** 1:15-16: foremost sinner: 232. ***Phillipians*** 4:13: can do all in Him: 233.
POLKINGHORNE, SIR JOHN: action at a distance & EPR "effect": 145; quantum probabilities may be due to physicists' not knowing all the details of what's happening: 154; quantum probabilities refuted by David Bohm in 1954: 154-155; the discrete has replaced the continuous: XXIV; the problem is how the cloudy & fitful tiny gives rise to the clear & reliable large: 158.
POOR, CHARLES LANE: on the ether: 246.
PSALMS, BOOK OF: 2:4, 59:8, 112:10, referenced only: 220.
REVELATIONS, BOOK OF: 2:16, 19:15, referenced only: 220; 9:15: 232.
RICKABY, JOSEPH, S. J.: on St. Thomas Aquinas' 10th crystalline sphere: 226.
RIESTERER, PATRICK: on Leibniz's attempts to explain locomotion: 221-222; on Leibniz's six is God's essence: 224.
SANDEN, ANTHONY: Lord Kelvin & the ether: 246; scientists' duplicity: 226.
TERESA OF AVILA, ST.: I submit all to the Church: XVIII.
WIKIPEDIA: *a priori* vs. *a posteriori* in St. Thomas Aquinas: XXV.

✡✝✡✝✡✝✡✝✡✝✡✝✡✝✡✝✡✝✡✝✡✝✡✝

His In Omnibus Rebus,
Judicum Praevaleat Ecclesiae Catholicae
Quod Est, Judicio Ecclesiae Catholicae,
Judicium Cathedrae Petri.
Amen!

Ad Majorem Christi Suaeque Ecclesiae Gloriam,
Consummatum Est In Die Festo
Resurrectionis Domini Nostri, Jesu Christi,
Die 20 Aprilis Anno Domini 2014.
Amen!

ABOUT THE AUTHOR:

I was born on April 13, 1936, in New Orleans, Louisiana, at the Hotel Dieu hospital to which my father drove my mother from our home in Haaswood, Louisiana. From early on, perhaps the chief driving force in my life has been a fierce hunger to stuff my head with as much knowledge as possible. Necessarily, that was a longing which entailed a desperate struggle to find—without injuring my physical health—as much leisure time as possible to devote to reading and writing. It was a struggle so desperate, it has always ruled in the life of a recluse and ruled out every possibility of either marrying or in any other way sharing my daily life with another human being.

I graduated from Jesuit high school, in New Orleans, in 1953. A single fruitless semester studying music at Loyola University of the South in New Orleans was followed by almost two years of floundering in a sea of confusion, and I then joined the U. S. Air Force on Dec. 7, 1955. In less than a year's time, that precipitated Atheism (It lasted for 6 years.) and an intense urge to commit suicide. It was an urge I managed to control only by devoting every spare moment to reading books in the many fields of Science and the Humanities in search of the ability to be absolutely certain there is no afterlife.

Honorably discharged in April of 1960, I underwent another two and a half years of floundering so severe, I came extremely close to a mental breakdown. In desperation, I gave away everything I owned and, for thirteen years, beginning November 9, 1962, I took mainly to the life of a wandering hermit. In search of as much time and energy as possible for reading and inner reflection upon self, God, and the nature and purpose of reality, I crisscrossed the United States on foot four times. At first, I lived off of whatever food and clothing I could beg; but, after learning how to live on a dollar a day or less, I turned mainly to working at various monasteries[1] in the winter in exchange for the two to three hundred dollars required to feed and to clothe myself during the next spring, summer, and fall of walking. The monasteries also provided access to libraries in which I could read, and extract notes from, the great writings of the Catholic Church. In the course of that 13 year odyssey, there was a four year period during which I refused to speak to anyone (except on very rare occasions) and communicated only by means of written notes.

In August of 1975, my father lost his mind, and my siblings insisted I was the only one in the family with the time and ability to tend to our father in his hour of need and to manage the rather large number of assets our father had amassed. Thus, after thirteen years, my *preferred* lifestyle came to an end. Dire poverty then gave way to economic independence, and total seclusion gave way to what little privacy can be enjoyed by bachelors who prefer to avoid partying and to stay home and—as much as possible—to bury themselves in as much reading and writing as the world around them will allow. My 21 self-published books in print as of December, 2013, are evidence of how much reading and writing I managed to do despite the many demands placed on me by my family.

After our father's death in 1981, I took care of our mother until her death in 1996. I now continue managing my family's assets until such time as I can sell all of them off at a price acceptable and fair to my siblings.

In my 21 self-published books, this self-educated author seeks to share with others the avenues of thought down which my mind was lead by *thirteen* years of *heroically* intense inner concentration followed by (to date Apr. 2014) *thirty-eight* years of *moderately*

---

[1] Never can I adequately thank the Discalced Carmelite fathers and brothers of Oakville, CA, the Trappist fathers and brothers near Ava, MO, and Huntsville, UT, and the Capuchin Franciscan fathers of Broken Arrow, OK, and Oak Ridge, NJ.

intense inner concentration. Those avenues pertain to a wide range of philosophical, theological, and psychological topics and, last but not least, to the unique cosmological theory set forth in this and another 2 of my self-published books. Those other 2 are: ***Introspective Cosmology II*** and ***Esoptrics: The Logic Of The Mirror*** sub-titled: ***The Divine Algebraic Logic Used By God To Create And To Maintain The Universe***. Five earlier books on the theory are unpublished and shall remain that way, since they have no value save as a record of how the theory progressed. Of course, I suppose the case could be made for saying the same of the two books just mentioned.

Until another time, dear reader (recklessly assuming there shall ever be one), God bless us all—everyone.

EDWARD N. HAAS
39193 HAAS ROAD – HAASWOOD (my rural location)
PEARL RIVER, LA 70452-3383 (post office's urban location)
E-MAIL: htoknow@yahoo.com

Haaswood is a small, unincorporated, historical area basically named after my paternal grandfather who settled here no later than 1913 and perhaps a year or two prior to that. Haaswood, LA, can be Internet searched as one of the federal government's quadrangle maps is named after it. If one Internet searches my name with the middle initial and the last name spelled with two A's and one S (***i.e.:*** Haas), instead of two S's and one A (***i.e.:*** Hass), dozens of pages about myself and my writings arise. I'm perhaps one of the world's more widely publicized utter unknowns. LOL.